STATISTICS IN MEDICINE
Statist. Med. 2008; **27**:1798–1800
Published online in Wiley InterScience (www.interscience.wiley.com).

BOOK REVIEWS
Editor: Petra Macaskill

1. *Alex Dmitrienko, Christy Chuang-Stein and Ralph D'Agostino (eds)*, Pharmaceutical Statistics using SAS®: A Practical Guide.
2. *Sophia Rabe-Hesketh and Brian S. Everitt*, A Handbook of Statistical Analyses using Stata (2nd edn).

1. PHARMACEUTICAL STATISTICS USING SAS®: A PRACTICAL GUIDE. Alex Dmitrienko, Christy Chuang-Stein and Ralph D'Agostino (eds), SAS® Press, Cary, NC, 2007. No. of pages: ix + 443. Price: $ 69.95. ISBN: 978-1-59047-886-8

This book provides a well-organized overview of important statistical topics applied in pharmaceutical drug development. Most impressive is the wide range of coverage—from drug discovery to late-stage clinical studies. The book is well written by a set of authors well known in the pharmaceutical industry. There is a good balance between explanation of the statistical theory and solutions to examples of practical problems. The statistical methodology is covered in-depth and is well explained but goes far beyond an introduction of how to use SAS in pharmaceutical drug development, which might be suggested by the title of the book. The solutions are up-to-date and well illustrated by down-to-earth examples using SAS code and SAS output.

The first introductory chapter is followed by 13 chapters that can more or less be grouped into five main topics that deal with a particular area.

Drug discovery (Chapters 2–4): The focus in Chapter 2 is on two popular methods for classification: boosting and partial least squares (PLS) for discrimination. Background information, theory and implementation of an example of permeability of compounds are provided mainly based on self-written SAS macros and PROC PLS.

Chapter 3 discusses techniques used in early stages to optimize the model building process, i.e. selecting training and test sets, selecting variables and determining valid prediction using an example of solubility data using PROC GAM and SAS Enterprise Miner.

Chapter 4 describes some advanced statistical methodology to understand, estimate and interpret various criteria for assessing the validity of an analytical method. Both linear and nonlinear models, and graphs are illustrated for a sandwich ELISA assay.

Animal toxicology (Chapter 5): This chapter discusses key statistical aspects in the design and analysis of toxicology studies. It starts with describing randomization schemes for the commonly used parallel design and Latin square. Then it presents the statistical power evaluation in the example of a two-factor ANOVA model of a QT interval in a large animal toxicology study by simulation. The statistical methods are illustrated by the analysis of treatment-related effects on body weight using repeated measures analysis based on PROC MIXED.

Nonparametric methods (Chapter 6) and optimal design of experiments (Chapter 7): Chapter 6 starts with the two-sample setting (Wilcoxon rank sum and Fligner–Policello test). It also discusses the one-way layout where the general alternative is a change in location (Kruskal–Wallis), as well as pair-wise comparisons and comparisons to a control that can be conducted with specific SAS macros. The last section introduces a power determination in a 'nonparametric' sense.

Chapter 7 provides an understandable introduction to the theory of optimal designs. It discusses examples arising in various pharmaceutical applications: a binary dose–response model, a four-parameter logistic model as used for ELISA, the beta regression model to deal with ordinal data, a bivariate probit model for correlated binary responses and pharmacokinetic models with and without cost constraints.

Clinical pharmacology studies (Chapter 8): Pharmacokinetic terminology is introduced and examples concentrated on area under the curve and

maximum concentration to test bioequivalence and dose linearity are analyzed using PROC MIXED.

Clinical studies (*Chapters 9–14*): Chapter 9 discusses allocation in randomized clinical trials. It is easy to read and is mostly devoted to permuted block designs and also discusses covariate-adaptive allocation procedures.

Sample size calculations are discussed in a rather unconventional way (Chapter 10). Starting with classical error rates it then introduces and explains the crucial type I error rate. Rather than providing an overview, two examples are covered in depth.

Chapter 11 reviews popular statistical methods used in dose-ranging clinical trials. Contrast-based approaches to assess dose-related trends, linear and sigmoid models to estimate the dose–response shape and close testing to determine the minimum effective dose are well illustrated using examples of a cross-over and parallel group design.

Chapter 12 extensively discusses how to handle incomplete longitudinal data. It presents the formal framework of missing data (e.g. MCAR and MAR) and promotes the use of direct-likelihood methods against the use of a popular method like Last Observation Carried Forward. Clinical examples are illustrated using generalized estimating equations (GEE) implemented in PROC GLIMMIX.

Basic approaches for assessing the reliability and validity of rating scales are reviewed in Chapter 13. Inter-rater reliability (e.g. kappa statistic) and internal consistency (Cronbach's α), convergent, divergent and discriminant validity as well as responsiveness are discussed.

Decision analyses in different parts of drug development with emphasis on clinical development are illustrated in Chapter 14 using e.g. PROC DTREE. Examples range from the design of a single trial (sample size and sequential design) and choosing the optimal dose to project prioritization.

The book is excellent for statisticians working in the pharmaceutical industry because of the real-life examples and SAS code. The complete SAS code and data sets used in this book are available on the book's companion web site at http://support.sas.com/publishing/bbu/companion_site/60622.html. In particular, it might be useful for the isolated statistician working for a company without senior colleagues.

The book serves as a helpful resource for anybody engaged in statistics used in the pharmaceutical industry due to its good coverage of the theory and many valuable references at the end of each chapter.

EGBERT BIESHEUVEL
Biometrics
NV Organon
The Netherlands

(DOI: 10.1002/sim.3076)

---

2. A HANDBOOK OF STATISTICAL ANALYSES USING STATA (2nd edn). Sophia Rabe-Hesketh and Brian S. Everitt, Chapman & Hall/CRC, Boca Raton, FL, 2007. No. of pages: ix+342. Price: $49.46. ISBN 1-58488-756-7

This book describes how to use Stata to perform a variety of statistical analyses, taking real-life data examples and going through their analyses one by one in each chapter. It starts with a concise introductory chapter on Stata and its basic commands that allow readers to start using the software. The introductory chapter also refers to extensive Stata resources that are useful for both current and new Stata users. Following the introductory chapter, the book provides 14 independent chapters that can be read in any order. Each chapter illustrates the use of a particular type of statistical methodology with Stata by following the authors' actual analysis of a scientific data set step by step. The essence of this book is in this step-by-step illustration of Stata-based analyses. A very nice feature is the availability of all data sets, including those used for exercises, from www.stata.com/texts/stas4 so that readers can actually practice the Stata-based analyses described in the book. The 14 chapters cover a variety of statistical methods including: simple tests (*t*-test, Chi-square test); ANOVA; multiple linear and logistic regression, and generalized linear models; longitudinal data analysis; random effects models; generalized estimating equations; basic epidemiological analyses; survival analysis; finite mixture models; principal component analysis; and cluster analysis.

All of the 14 chapters are organized under the same structure. Each chapter starts with a concise explanation of its illustrative data, followed by a brief methodological description of the statistical methodology used in the analysis of the data. The authors' Stata-based analysis of the data is

Praise from the Experts

"*Pharmaceutical Statistics Using SAS* contains applications of cutting-edge statistical techniques using cutting-edge software tools provided by SAS. The theory is presented in down-to-earth ways, with copious examples, for simple understanding. For pharmaceutical statisticians, connections with appropriate guidance documents are made; the connections between the document and the data analysis techniques make 'standard practice' easy to implement. In addition, the included references make it easy to find these guidance documents that are often obscure.

"Specialized procedures, such as easy calculation of the power of nonparametric and survival analysis tests, are made transparent, and this should be a delight to the statistician working in the pharmaceutical industry, who typically spends long hours on such calculations. However, non-pharmaceutical statisticians and scientists will also appreciate the treatment of problems that are more generally common, such as how to handle dropouts and missing values, assessing reliability and validity of psychometric scales, and decision theory in experimental design. I heartily recommend this book to all."

> **Peter H. Westfall**
> **Professor of Statistics**
> **Texas Tech University**

"The book is well written by people well known in the pharmaceutical industry. The selected topics are comprehensive and relevant. Explanations of the statistical theory are concise, and the solutions are up-to-date. It would be particularly useful for isolated statisticians who work for companies without senior colleagues."

> **Frank Shen**
> **Executive Director**
> **Global Biometric Sciences**
> **Bristol-Myers Squibb Co.**

"This book covers an impressive range of topics in clinical and non-clinical statistics. Adding the fact that all the datasets and SAS code discussed in the book are available on the SAS Web site, this book will be a very useful resource for statisticians in the pharmaceutical industry."

> **Professor Byron Jones**
> **Senior Director**
> **Pfizer Global Research and**
> **Development, UK**

"The first thing that catches one's attention about this very interesting book is its breadth of coverage of statistical methods applied to pharmaceutical drug development. Starting with drug discovery, moving through pre-clinical and non-clinical applications, and concluding with many relevant topics in clinical development, the book provides a comprehensive reference to practitioners involved in, or just interested to learn about, any stage of drug development.

"There is a good balance between well-established and novel material, making the book attractive to both newcomers to the field and experienced pharmaceutical statisticians. The inclusion of examples from real studies, with SAS code implementing the corresponding methods, in every chapter but the introduction, is particularly useful to those interested in applying the methods in practice, and who certainly will be the majority of the readers. Overall, an excellent addition to the SAS Press collection."

José Pinheiro
Director of Biostatistics
Novartis Pharmaceuticals

"This is a very well-written, state-of-the-art book that covers a wide range of statistical issues through all phases of drug development. It represents a well-organized and thorough exploration of many of the important aspects of statistics as used in the pharmaceutical industry. The book is packed with useful examples and worked exercises using SAS. The underlying statistical methodology that justifies the methods used is clearly presented.

"The authors are clearly expert and have done an excellent job of linking the various statistical applications to research problems in the pharmaceutical industry. Many areas are covered including model building, nonparametric methods, pharmacokinetic analysis, sample size estimation, dose-ranging studies, and decision analysis. This book should serve as an excellent resource for statisticians and scientists engaged in pharmaceutical research or anyone who wishes to learn about the role of the statistician in the pharmaceutical industry."

Barry R. Davis
Professor of Biomathematics
University of Texas

Pharmaceutical Statistics Using SAS®

A Practical Guide

Edited by
Alex Dmitrienko
Christy Chuang-Stein
Ralph D'Agostino

THE
POWER
TO KNOW

The correct bibliographic citation for this manual is as follows: Dmitrienko, Alex, Christy Chuang-Stein, and Ralph D'Agostino. 2007. *Pharmaceutical Statistics Using SAS®: A Practical Guide*. Cary, NC: SAS Institute Inc.

Pharmaceutical Statistics Using SAS®: A Practical Guide

Contents

Preface

Introduction

The past decades have witnessed significant developments of biostatistical methodology applied to all areas of pharmaceutical drug development. These applications range from drug discovery to animal studies to Phase III trials. The use of statistical methods helps optimize a variety of drug development processes and ultimately helps ensure that new chemical entities are pure and stable, new therapies are safe and efficacious.

For the most part, new developments in biostatistical theory and applications are scattered across numerous research papers or books on specialized topics and rarely appear under the same cover. The objective of this book is to offer a broad coverage of biostatistical methodology used in drug development and practical problems facing today's drug developers.

Each chapter of the book features a discussion of methodological issues, traditional and recently developed approaches to data analysis, practical advice from subject matter experts, and review of relevant regulatory guidelines. The book is aimed at practitioners and therefore does not place much emphasis on technical details. It shows how to implement the algorithms presented in the various chapters by using built-in SAS procedures or custom SAS macros written by the authors. In order to help readers better understand the underlying concepts and facilitate application of the introduced biostatistical methods, the methods are illustrated with a large number of case studies from actual pre-clinical experiments and clinical trials. Since many of the statistical issues encountered during late-phase drug development (e.g., survival analysis and interim analyses) have been covered in other books, this book focuses on statistical methods to support research and early drug development activities.

Although the book is written primarily for biostatisticians, it will benefit a broad group of pharmaceutical researchers, including biologists, chemists, pharmacokineticists, and pharmacologists. Most chapters are self-contained and include a fair amount of high-level introductory material. This book will also serve as a useful reference for regulatory scientists as well as academic researchers and graduate students.

We hope that this book will help close the gap between modern statistical theory and existing data analysis practices in the pharmaceutical industry. By doing so, we hope the book will help advance drug discovery and development.

Outline of the Book

The book begins with a review of statistical problems arising in pharmaceutical drug development (Chapter 1, "Statistics in Drug Development"). This introductory chapter is followed by thirteen chapters that deal with statistical approaches used in drug discovery experiments, animal toxicology studies, and clinical trials. Since some of the chapters discuss methods that are relevant at multiple stages of drug development (e.g., nonparametric methods), the chapters in this book are not grouped by the development stage. Table helps the reader identify chapters that deal with a particular area in pharmaceutical statistics.

Chapter list

Chapter	Drug discovery	Animal toxicology	Clinical trials
Chapter 2, "Modern Classification Methods for Drug Discovery"	•		
Chapter 3, "Model Building Techniques in Drug Discovery"	•		
Chapter 4, "Statistical Considerations in Analytical Method Validation"	•		
Chapter 5, "Some Statistical Considerations in Nonclinical Safety Assessment"		•	
Chapter 6, "Nonparametric Methods in Pharmaceutical Statistics"	•		•
Chapter 7, "Optimal Design of Experiments in Pharmaceutical Applications"	•		•
Chapter 8, "Analysis of Human Pharmacokinetic Data"			•
Chapter 9, "Allocation in Randomized Clinical Trials"			•
Chapter 10, "Sample-Size Analysis for Traditional Hypothesis Testing: Concepts and Issues"			•
Chapter 11, "Design and Analysis of Dose-Ranging Clinical Studies"			•
Chapter 12, "Analysis of Incomplete Data"			•
Chapter 13, "Reliability and Validity: Assessing the Psychometric Properties of Rating Scales"			•
Chapter 14, "Decision Analysis in Drug Development"			•

List of Contributors

This book represents a collaborative effort with contributions from 42 scientists whose names are listed below.

Caroline Beunckens, Universiteit Hasselt
Bruno Boulanger, Eli Lilly
Carl-Fredrik Burman, AstraZeneca
John Castelloe, SAS Institute
Alan Chiang, Eli Lilly
Christy Chuang-Stein, Pfizer
Kimberly Crimin, Wyeth
Ralph D'Agostino, Boston University
Viswanath Devanaryan, Merck
Walthère Dewé, Eli Lilly
Alex Dmitrienko, Eli Lilly
Douglas Faries, Eli Lilly
Valerii Fedorov, GlaxoSmithKline
Kathleen Fritsch, Food and Drug Administration
Robert Gagnon, GlaxoSmithKline
Andy Grieve, Pfizer
Kevin Guo, Eli Lilly
Wherly Hoffman, Eli Lilly

Jason Hsu, The Ohio State University
Anastasia Ivanova, University of North Carolina
Ivy Jansen, Universiteit Hasselt
Kjell Johnson, Pfizer
Paul Juneau, Pfizer
Michael Kenward, London School of Hygiene and Tropical Medicine
Olga Kuznetsova, Merck
Cindy Lee, Eli Lilly
Sergei Leonov, GlaxoSmithKline
Geert Molenberghs, Universiteit Hasselt
Daniel Ness, Eli Lilly
Ralph O'Brien, Cleveland Clinic
Scott Patterson, GlaxoSmithKline
William Rayens, University of Kentucky
Stephen Ruberg, Eli Lilly
Stephen Senn, University of Glasgow
Brian Smith, Amgen
Wendell Smith, B2S Consulting
Herbert Thijs, Universiteit Hasselt
Kristel Van Steen, Universitiet Gent
Geert Verbeke, Katholieke Universiteit Leuven
Thomas Vidmar, Pfizer
Yuehui Wu, GlaxoSmithKline
Ilker Yalcin, Eli Lilly

Feedback

We encourage you to submit your comments, share your thoughts, and suggest topics that we can consider in other editions of this book on the Biopharmaceutical Network's web site at

`http://www.biopharmnet.com/books`

We value your opinions and look forward to hearing from you.

Acknowledgments

We would like to thank the reviewers, including Dr. Ilya Lipkovich (Eli Lilly and Company) and Professor Peter Westfall (Texas Tech University), who have provided a careful review of selected chapters and valuable comments.

Alex Dmitrienko wants to thank Drs. Walter Offen, Gregory Sides, and Gregory Enas for their advice and encouragement throughout the years.

Christy Chuang Stein wishes to thank Drs. Bruce Stein and Mohan Beltangady for their support and encouragement.

Last, but not least, we would like to thank Donna Faircloth, acquisitions editor at SAS Press, for all the work and time she has invested in this book publishing project.

Alex Dmitrienko, Global Statistical Sciences, Eli Lilly and Company, USA.

Christy Chuang-Stein, Midwest Statistics, Pfizer, USA.

Ralph D'Agostino, Department of Mathematics and Statistics, Boston University, USA.

Chapter 1

Statistics in Drug Development

Christy Chuang-Stein
Ralph D'Agostino

1.1   Introduction

In the past 50 years, the value of medicine has been clearly demonstrated by a longer life expectancy, a lower infant mortality rate, and the higher quality of life many of our senior citizens have been enjoying. Since the introduction of stomach-acid-blocking H2 antagonist drugs in the late 70's, the number of surgeries to treat ulcer has been greatly reduced. Childhood vaccination has literally wiped out diphtheria, whooping cough, measles, and polio in the U.S. Deaths from heart disease have been cut by more than half since 1950 and continue to decline. Even though we still face great challenges in combating cancer, great strides have been made in treating childhood leukemia. Early detection has led to successful treatment of some types of cancer such as breast cancer. Treatments for schizophrenia and bipolar disorder have allowed many patients to live almost normal lives. A report (2006) on the value of medicine can be found at the Pharmaceutical Research and Manufacturers of America (PhRMA) website.

The use of statistics to support discovery and testing of new medicines has grown rapidly since the Kefauver-Harris Amendments, which became effective in 1962. The Kefauver-Harris Amendments required drug sponsors to prove a product's safety and efficacy in controlled clinical trials in order to market the product. Since the Amendments, the number of statisticians working in the pharmaceutical industry has greatly increased. This increase took another jump when the manufacturing process came under close scrutiny. As we move into the 21st century, the lure and the promise of genomics and

Christy Chuang-Stein is Site Head, Midwest Statistics, Pfizer, USA. Ralph D'Agostino is Professor, Department of Mathematics and Statistics, Boston University, USA.

proteomics will further intensify scientists' reliance on statistics. The need to enhance our overall knowledge about diseases, the need to insert more points into the decision-making process, and the need to bring economics into development strategy considerations will undoubtedly present new opportunities for statisticians.

Even in the face of new opportunities, there are many well-established roles for statisticians in the pharmaceutical industry. The term "well-established" is a relative term since new roles will become more established over time. For example, trial simulation and modeling, viewed as new advancements a decade ago, have now become common practice to help design better trials across the pharmaceutical industry.

Concerned that the current medical product development path may have become increasingly challenging, inefficient, and costly, the U.S. Food and Drug Administration (FDA) issued a document in March 2004 entitled "Challenge and Opportunity on the Critical Path to New Medical Products". The document attempts to bridge the technological disconnect between discovery and the product development process. The disconnect is thought to be largely due to the fact that the pace of development work has not kept up with the rapid advances in product discovery. The document addresses three major scientific and technical dimensions in the critical path of product development. The three dimensions relate to safety assessment, demonstration of a product's medical utility (benefit or effectiveness), and the product's industrialization (scaling up). In addition to understanding the challenges, establishing the right standards and developing better toolkits for each dimension will be key to our ultimate success in overcoming the perceived stagnation in getting new drugs and biologics to the market. Statisticians, with their training in quantification and logical thinking, can play a major role in the preparation and the execution of the action plan.

The call for innovation is nothing new for the pharmaceutical industry. The industry as a whole has made great strides in its basic science research in recent years. Cutting edge techniques are being developed on a daily basis to probe into the biologic origin and genetic connection of diseases. The research on microarrays and genomics has produced more data than could be perceived just a few years ago. With the race to unlock the mysteries of many diseases and finding cures for them, statistical support needs to be broadened in dimensions and increased in depth. Time has never been more right for statisticians to work alongside with their colleagues, being discovery scientists, clinical personnel, manufacturing engineers, or regulatory colleagues. The collaboration should not only help transform data to knowledge, but also help use knowledge for better risk-based decisions.

In this chapter, we will briefly cover some traditional statistical support to show how statistics has been used in many aspects of drug development. Our coverage is by no means exhaustive. It is simply an attempt to illustrate how broad statistical applications have been. We will also highlight some areas where a statistician's contribution will be crucial in moving forward, in view of the FDA's Critical Path initiative and the pharmaceutical industry's collective effort to take advantage of the FDA's call for innovation.

1.2 Statistical Support to Non-Clinical Activities

In an eloquent viewpoint article, Dennis Lendrem (2002) discussed non-clinical statistical support. Traditionally, statistical thinking and approaches are more embraced in areas where regulators have issued guidelines. Examples are pre-clinical testing of cardiac liability, carcinogenicity, and stability testing. Recently, Good Manufacturing Practice has also become a subject of great regulatory interest. The latter captured public attention when manufacturing problems created a shortage of the flu vaccines for the 2004–2005 season. By comparison, statistical input in areas such as high-throughput screening, chemical development, formulation development, drug delivery, and assay development is being sought only when the scientists feel that statisticians could truly add value. This mentality could limit statisticians' contributions since researchers will not know how statisticians could help unless they have previously worked with statisticians or have been referred to statisticians by their grateful colleagues. For example, scientists who are used to experimenting with one factor at a time won't know the value of factorial experiments. Similarly, even though statisticians well versed in Six Sigma and Design for Six Sigma are well aware of the many applications of the Six Sigma principles, they need to actively sell the applications to potential clients.

The non-clinical support model differs from that in the clinical area because of the usually large client-to-statistician ratio. As a result, after a statistician completes a particular job, he/she often looks for opportunity to consolidate the techniques and institutionalize the tools for the client to use on a routine basis. The automation allows statisticians to focus on opportunities for new collaboration and developing new methodologies for applications.

Non-clinical statisticians often work individually with their clients. Lendrem (2002) described them as "pioneers" because of the frequent needs to venture into unknown areas of new technology. Quantifying gene expression via the microarray technology is one such example. Another is industry's (and government's alike) investment in identifying biomarkers for testing mechanism of action of new molecular or biologic entities. In both cases, the findings will have great clinical implications, but the work starts in the research laboratories and our non-clinical statisticians are the first to deal with the need to measure, to quantify, and to validate the measurements from the technical perspective.

Because of the small number of non-clinical statisticians in many pharmaceutical companies, it is useful for non-clinical statisticians to form an inter-company network to benefit mutual learning. Some of this networking has been in existence for some time. In the U.S., a CMC (Chemistry, Manufacturing, and Control) Statistical Expert Team was formed in the late 60's to focus on the chemistry and control issues related to the manufacturing of pharmaceutical products. Another example is the Pharmacogenomics Statistical Expert Team that was formed in the fall of 2003. Both teams are sanctioned by PhRMA and consist of statisticians from major pharmaceutical companies.

1.3 Statistical Support to Clinical Testing

Clinical testing is typically conducted in a staged fashion to explore the effect of pharmaceutical products on humans. The investigation starts with pharmacokinetic and pharmacodynamic studies, followed by proof-of-concept and dose-ranging studies. Some specialty studies such as drug effect on QT/QTc prolongation and drug-drug interactions studies are conducted at this early stage. The identification of common adverse reactions and early signs of efficacy are the objectives of such trials. The early testings, if satisfactory, lead to the confirmatory phase where the efficacy and safety of the product candidate are more thoroughly investigated in a more heterogeneous population.

Despite common statistical principles, different stages of clinical testing focus on different statistical skill sets. For early proof-of-concept and dose-ranging efforts, study designs could be more flexible and the goal is to learn as efficiently and effectively as possible. Adaptations, in terms of dose allocation, early termination, and study population give great flexibility to such trials. Extensive modeling that incorporates accumulated learning on a real-time basis can lead a sponsor to decision points in a more expedited fashion. Because the purpose of this phase of development is primarily to generate information to aid internal decisions, the developers are freer to use innovative approaches as long as they can successfully defend the decisions that become the basis for later development.

By comparison, statistical approaches for the confirmatory phase need to be carefully pre-planned, pre-specified, and followed in order to give credibility to the results. A pharmaceutical sponsor needs to decide *a priori* study designs, primary endpoints, primary analysis population, success criteria, handling of missing data, multiple comparisons, plus many others. ICH E9 (1998) gives a very detailed description of all aspects of trial design and analysis that a statistician should consider at this stage. When adaptation is planned, the rule needs to be clearly specified in advance. When interim analysis is anticipated, a sponsor's access to the interim results needs to be tightly controlled.

The confirmatory phase is the place where knowledge about a new molecular or biologic entity is solidified to support a target label. The knowledge, along with the approved label, becomes the basis for recommendations to prescribing physicians and the medical community. The confirmatory trials are also the place where the risk-benefit and cost-effectiveness of a new pharmaceutical product are first delineated. The greater number of subjects studied at this stage gives a sponsor a decent chance to study adverse actions that have a rate between 0.1% and 1%. This phase overlaps somewhat with the life cycle management phase where new indications are being explored and drug differentiation is being sought. If there is a post-marketing study commitment, additional studies will be initiated to fulfill the conditions for approval.

Increasingly, statisticians are participating in promotion review and educational communications to the general public. In addition, many statisticians contribute to activities related to pharmacovigilance and epidemiology.

1.4   Battling a High Phase III Failure Rate

The attrition rate of compounds in the pharmaceutical industry is extremely high. Setting aside compounds that fail the preclinical testing, it is generally recognized that less than 12% of compounds entering into the human phase testing will eventually make it to the market place. The rate is a composite figure formed as the product of the success rates of passing the Phase I testing, passing the Phase II testing, passing the Phase III testing, and passing the regulatory review. Among failures at the various stages, Phase III attrition has the greatest impact. This is so not only because of all the accumulated resources expended up to this point, but it is also because Phase III failure represents a great disappointment to the sponsor, leaving the sponsor short of a defendable marketing application.

In a recent article, Chuang-Stein (2004) conducted a root cause analysis of the Phase III failure rate that was reported to be running at the 50% level. This most recent figure is higher than the 32% rate reported in DiMasi, Hansen, and Grabowski (2003). Chuang-Stein attributed the cause to three major factors: the candidate factor, the sponsor factor, and the environmental factor. While we can't dismiss the pipeline problem, and we have admittedly very little control over the behaviors of some corporate decision-makers at the highest level, many of the causes indeed relate to how clinical development is being conducted and how decisions are made to move compounds through different phases of the development. Chuang-Stein discussed what statisticians could do to help reduce the attrition rate at the late stage. One area where the methodology is well developed and statisticians could make immediate contributions is the judicious use of adaptive designs, or at least group sequential designs, in Phase III trials. The goal of such designs is to give the Phase III trials the best chance for success or to terminate them early if the trials are not likely to meet their objectives. Implicit in such designs is the inclusion of more decision points based on interim results to allow evidence-based decisions. The need to incorporate regular decision points is not limited to Phase III testing. It should be part of every stage of the drug development continuum. These decision points serve as reality checks on the long and costly development journey.

The industry is at a crossroad, and changes are critically needed. Statisticians should take advantage of the challenges and fully engage themselves in looking for better ways to support clinical development of pharmaceutical products.

1.5   Do Statisticians Count?

In a soul-searching article, Andy Grieve (2002) asks whether statisticians count. Even though the number of statisticians working in the pharmaceutical industry has increased by 50-fold since the late 70's, Grieve felt that the influence statisticians had in their respective companies had not increased proportionally. Grieve looked at the barriers that prevented statisticians from contributing as much as they could and offered some solutions.

Particularly noteworthy is the assertion that it is the statistician, and not statistics, that is important. Statistics, as a discipline, does not influence, does not persuade, does not design studies, does not analyze data, does not interpret findings, and does not report results. Statisticians are the ones who make the discipline meaningful by doing all of the above. In other words, statisticians, through their own behavior and communication, spread the statistical principles and promote the statistical thinking. So, when we discuss the successful use of statistics in drug development, we need to bear in mind that as statisticians working in the pharmaceutical industry, we need to be the champions for such causes through our passion for the statistics profession.

1.6   Emerging Opportunities

The Critical Path initiative document describes many opportunities to improve the efficiency of product development. We will mention just a few here. On better tools to assess a compound's safety, FDA states the need for new techniques to evaluate drug liver toxicity, new methods to identify gene therapy risk, better predictors of human immune responses to foreign antigens, new methods to further enhance the safety of transplanted human tissues, and efficient protocols for qualifying biomaterials. On better tools to demonstrate the medical utility of a compound, FDA shares the agency's successful experience with biomarkers in HIV infection and duodenal ulcer healing. FDA states the need for more biomarkers and surrogate markers that can guide product development. In addition, FDA discusses the need for better animal models to combat bioterrorism, more clinically relevant endpoints, better imaging technologies, more innovative designs and analysis methods, and the need for implementing the concept of model-based drug development. The latter involves building mathematical and statistical characterization of the time course of the disease and the drug effect, using available clinical data.

The Critical Path initiative document also discusses the need for better methods to characterize, standardize, control, and manufacture medical products on the commercial scale. Since manufacturing expenses could exceed the research and development investment, there is a need for a better validation process that follows the risk-based inspection paradigm advocated by the FDA in recent months. The latter includes more attention to setting specifications and shifting from detailed data analysis to overall process quality assessment. The same philosophy suggests moving toward acceptance of a probabilistic definition, rather than a pass or fail on the manufacturing process. Most important, FDA wants to encourage the manufacturers to integrate state-of-the-art science and technology into their manufacturing processes.

Following the issuance of the Critical Path initiative document, different centers within the FDA have further identified areas for innovations and have presented opportunities to the FDA's Science Board on November 5, 2004. Since May 2004, many workshops have directed at least part of their agenda towards more efficient and effective ways to test and develop new treatments. Common to many of the discussions are the needs to apply quantitative thinking and techniques. Taking the clinical phase of product development as an example, we see that a major emphasis is to use mathematical and statistical models to help guide drug development and approval. The central idea is to pool data from multiple trials to augment our knowledge base and actively incorporate such knowledge in subsequent studies. Interestingly enough, the concept of pooling data has now been extended to pooling data of drugs that belong to the same class. The pooling of information across companies, while challenging, will undoubtedly facilitate the collective learning of the pharmaceutical industry.

The opportunities for statisticians to make substantial contributions at the strategic level are beyond what one could have imagined 20 years ago. Along with the opportunities come expectations that statisticians will help solve the puzzle faced by modern-day scientists in the pharmaceutical industry.

1.7 Summary

Statistics, as a discipline, has broadened its scope significantly over the past 20 years. Wherever there is a need for quantification, statistics has a role. The ability to think in terms of variability, to separate signals from noise, to control sources of bias and variation, and to optimize under given conditions, makes statisticians a valuable partner in the development of new pharmaceutical and biological products.

Mining historical data to add to our cumulative knowledge is a low-cost and high-yield activity. Many companies realize the value of this activity and are actively pursuing it. For example, Bristol-Myers Squibb (Pink Sheet, December 13, 2004) formed a discovery toxicology group and retrospectively analyzed approximately 100 development compounds that failed during a 12-year period. Bristol-Myers Squibb hoped to use the acquired knowledge to decide what assays and technology to implement early to reduce compound attrition. Bristol-Myers Squibb concluded that a combination of in vitro, in vivo, and in silico techniques was needed to improve productivity and reduce attrition. According to the same report in the Pink Sheet (December 13, 2004), other pharmaceutical companies have reached similar conclusions.

Data mining is also expected to help us look for better predictors for hepatotoxicity and cardiovascular toxicity such as Torsade de pointes. Data mining examples can go on and on. We can't think of any scientists who are more poised and qualified to lead this data-based learning endeavor than statisticians!

The challenge is definitely on us, statisticians!

References

Chuang-Stein, C. (2004). "Seize the opportunities." *Pharmaceutical Statistics*. 3, 157–159.

DiMasi, J., Hansen, R.W., Grabowski, H.G. (2003). "The price of innovation: new estimates of drug development costs." *Journal of Health Economics*. 22, 151–185.

"FDA 'Critical Path' May Lead To Changes In Dosing, Active-Control Trials." December 13, 2004. *Pink Sheet*. 31.

Grieve, A. (2002). "Do statisticians count? A personal view." *Pharmaceutical Statistics*. 1, 35–43.

International Conference of Harmonization of Pharmaceuticals for Human Use (ICH) E9. (1998). "Statistical Principles for Clinical Trials." Available at http://www.ich.org.

Lendrem, D. (2002). "Statistical support to non-clinical." *Pharmaceutical Statistics*. 1, 71–73.

Pharmaceutical Research and Manufacturers of America (PhRMA) Report. (2006). *Value of Medicines*. Available at http://www.phrma.org/education/.

The Critical Path to New Medical Products. (March 2004). Available at http://www.fda.gov/oc/initiatives/criticalpath/whitepaper.html.

Chapter 2

Modern Classification Methods for Drug Discovery

Kjell Johnson
William Rayens

This chapter focuses on two modern methods for classification: boosting and partial least squares for discrimination. For each of these methods, we provide detailed background information about their theory as well as instruction and guidance on their implementation in SAS. Finally, we apply these methods to a common drug discovery-type data set and explore the performance of the two methods.

2.1 Introduction

Drug discovery teams are often faced with data for which the samples have been categorized into two or more groups. For example, early in the drug discovery process, high throughput screening is used to identify compounds' activity status against a specific biological target. At a subsequent stage of discovery, screens are used to measure compounds' solubility, permeability, and toxicity status. In other areas of drug discovery, information from animal models on disease classification, survival status, and occurrence of adverse events is obtained and scrutinized. Based on these categorizations, teams must decide which compounds to pursue for further development.

In addition to the categorical response, discovery data often contain variables that describe features of the samples. For example, many computational chemistry software packages have the ability to generate structural and physical-property descriptors for any defined set of compounds. In genomics and proteomics, expression profiles can be measured on tissue samples.

Kjell Johnson is Associate Director, Nonclinical Statistics, Pfizer, USA. William Rayens is Professor, Department of Statistics, University of Kentucky, USA.

Given data that contain descriptors and a categorical response, teams desire to uncover relationships between the descriptors and response that can provide them with scientific intuition and help them predict the classification of future compounds or samples.

In drug discovery, data sets are often large and over-described (more descriptors exist than samples or highly correlated descriptors). And often, the relationships between descriptors and classification grouping are complex. That is, the different classes of samples cannot be easily separated by a line or hyperplane. Hence, methods that rely on inverting the covariance matrix of the descriptors, or methods that find the best separating hyperplane are not effective for this type of data. Traditional statistical methods such as linear discriminant analysis or logistic regression are both linear classification methods and rely on the covariance matrix of the descriptors. For the reasons mentioned previously, neither is optimal for modeling many discovery type data sets.

In this situation, one approach is to reduce the descriptor space in a way that stabilizes the covariance matrix. Principal component analysis, variable selection techniques such as genetic algorithms, or ridging approaches are popular ways to obtain a stable covariance matrix. When the matrix is stable, traditional discrimination techniques can be implemented. In each of these approaches, the dimension reduction is performed independently of the discrimination. An alternative approach, partial least squares for linear discrimination, has the added benefit of simultaneously reducing dimension while finding the optimal classification rule.

Another promising classification method is support vector machines (Vapnik, 1996). In short, support vector machines seek to find a partition through the data that maximizes the margin (the space between observations from separate classes), while minimizing classification error. For data in which classes are clearly separable into different groups, it is possible to maximize the margin while minimizing classification error, regardless of the complexity of the boundary. However, for data whose classes overlap, maximizing the margin and minimizing classification error are competing constraints. In this case, maximizing the margin produces a smoother classification boundary, while minimizing classification error produces a more flexible boundary. Although support vector machines are an effective classification tool, they can easily overfit a complex data set and can be difficult to interpret.

Recursive partitioning seeks to find individual variables from the original data that partition the samples into more pure subsets of the original data (Breiman et al., 1984). Thus, a recursive partition model is a sequence of decision rules on the original variables. Because recursive partitioning selects only one variable at each partition, it can be used to model an overdetermined data set. This method can also find a more complex classification boundary because it partitions the data into smaller and smaller hypercubes. While a recursive partition model is easy to interpret, its greedy nature can prevent it from finding the optimal classification model for the data.

More recently, methods which combine classifiers, known as ensemble techniques, have been shown to outperform many individual classification techniques. Popular ensemble methods include bagging (Breiman, 1996), ARCing (Breiman, 1998), and boosting (Freund and Schapire, 1996a), which have been shown to be particularly effective classification tools when used in conjunction with recursive partitioning.

Because of their popularity, effectiveness, and ease of implementation in SAS, we will limit the focus of this chapter to boosting and partial least squares for discrimination.

In Section 2.2, we will motivate our methods with an example of a typical drug discovery data set. Section 2.3 will explore boosting and its implementation in SAS as well as its effectiveness on the motivating example and Section 2.4 will review model building techniques. Section 2.5 will discuss the use of partial least squares for discrimination and SAS implementation issues, and apply this method to the motivating example.

To save space, some SAS code has been shortened and some output is not shown. The complete SAS code and data sets used in this book are available on the book's companion Web site at `http://support.sas.com/publishing/bbu/companion_site/60622.html`.

2.2 Motivating Example

Generally speaking, permeability is the ability of a molecule to cross a membrane. In the body, key membranes exist in the intestine and brain, and are composed of layers of molecules and proteins organized in a way to prevent harmful substances from crossing while allowing essential substances to pass through. The intestine, for instance, allows substances such as nutrients to pass from the gut into the blood stream. Another membrane, the blood-brain barrier, prevents detrimental substances from crossing the blood stream into the central nervous system. While a molecule may have the correct characteristics to be effective against a specific disease or condition, it may not have the correct characteristics to pass from the gut into the blood stream or from the blood stream into the central nervous system. Therefore, if a potentially effective compound is not permeable, then its effectiveness may be compromised.

Because a compound's permeability status is critically important to its success, pharmaceutical companies would like to identify poorly permeable compounds as early as possible in the discovery process. These compounds can then be eliminated from follow-up, or can be modified in an attempt to improve permeability while keeping their potential target effectiveness.

To measure a compound's ability to permeate a biological membrane, several in-vitro assays such as PAMPA and Caco-2 have been developed (Kansy et al., 1998). In each of these assays, cell layers are used to simulate a body-like membrane. A compound is then placed into solution and is added to one side of the membrane. After a period of time, the compound concentration is measured on the other side of the membrane. Compounds with poor permeability will have low compound concentration, while compounds with high permeability will have high compound concentration.

These screens are often effective at identifying the permeability status of compounds, but the screening process is moderately labor- and material-intensive. At a typical pharmaceutical company, resources are limited to screening only a few hundred compounds per week. To process more compounds would require more resources. Alternatively, we could attempt to build a model that would predict permeability status.

As mentioned above, biological membranes are a complex layer of molecules and proteins. To pass through a membrane, a substance must have an appropriate chemical composition. Therefore, to build a predictive model of permeability status we should include appropriate chemical measurements for each compound.

For the example in this chapter, we have collected data on 354 compounds (the PERMY data set can be found on the book's companion Web site). For each compound we have its measured permeability status ($y = 0$ is not permeable and $y = 1$ is permeable), and we have used in-house software to compute 71 molecular properties that are theoretically related to permeability (i.e., hydrogen bonding, polarity, molecular weight, etc.). For proprietary reasons, these descriptors have been blinded and are labeled as X1, X2, ..., X71.

As is common with many drug discovery data sets, the descriptor matrix for the data is overdetermined (that is, at least one variable is linearly dependent on one or more of the other variables). To check this in SAS, we can use the PRINCOMP procedure to generate the eigenvalues of the covariance matrix. Recall that a full rank covariance matrix will have no zero eigenvalues. Program 2.1 generates the eigenvalues of the covariance matrix.

Program 2.1 Permeability data set: Computation of the eigenvalues of the covariance matrix

```
proc princomp data=permy cov n=71 outstat=outpca noprint;
    var x1-x71;
data outpca;
    set outpca(where=(_type_="EIGENVAL"));
    keep x1-x71;
proc transpose data=outpca out=outpca_t(drop=_name_) prefix=eig;
    var x1-x71;
proc print data=outpca_t;
    run;
```

Output from Program 2.1

Obs	eig1
1	531907.98
2	202322.50
3	39308.42
4	22823.33
5	9187.10
6	5087.36
7	2660.49
8	1229.59
9	814.62
10	495.17
62	.000003063
63	.000000857
64	.000000209
65	.000000149
66	0
67	0
68	0
69	0
70	0
71	0

Output 2.1 lists the first 10 and last 10 eigenvalues of the covariance matrix computed from the permeability data set. Notice that the covariance matrix for the permeability data is not full rank. In fact, six eigenvalues are zero, while more than 20 are less than 0.01. This implies that the covariance matrix is not invertible and methods that rely on the inverse of the covariance matrix will have a difficult time with these data.

Another common situation in which the descriptor matrix can be overdetermined occurs when we have more descriptors than observations. Methods that rely on a full-rank covariance matrix of the descriptors, such as linear discriminant analysis, will fail with this type of data. Instead, we must either remove the redundancy in the data or employ methods, such as boosting or partial least squares for linear discrimination, that can look for predictive relationships in the presence of overdetermined data.

2.3 Boosting

Several authors have written thorough summaries of the historical development of boosting (see, e.g., Friedman et al., 2000; Schapire, 2002) For completeness of this work, we provide a brief overview of boosting and include recent findings.

2.3.1 Evolution of Boosting

The concept of boosting originated in the machine learning community where Kearns and Valiant (1989) explored weak learning algorithms (algorithms that can classify objects better than random) and strong learning algorithms (algorithms that can classify accurately). The question of the relationship between weak and strong learning algorithms was initially posed, and was addressed soon after by Schapire (1990). In his work, Schapire showed that if a concept was weakly learnable it was also strongly learnable, and he formulated the first algorithm to "boost" a weak learner into a strong learner. Freund (1995) improved upon Schapire's initial algorithm by reducing its complexity and increasing its efficiency. However, both Schapire's algorithm and Freund's algorithms were difficult to implement in practice.

To overcome these practical implementation problems, Freund and Schapire (1996a) collaborated to produce the well-known and widely applied AdaBoost algorithm described below (the notation used in this algorithm is defined in Table 2.1).

1. Let $w_{1,1} = \cdots = w_{1,n} = 1/n$.
2. For $t = 1, 2, \ldots, T$ do:
   (a) Fit f_t using the weights, $w_{t,1}, \ldots, w_{t,n}$, and compute the error, e_t.
   (b) Compute $c_t = \ln((1 - e_t)/e_t)$.
   (c) Update the observation weights:

$$w_{t+1,i} = w_{t,i} \exp(c_t I_{t,i}) / \sum_{j=1}^{n} (w_{t,j} \exp(c_t I_{t,j})), \quad i = 1, \ldots, n.$$

3. Output the final classifier:

$$\widehat{y}_i = F(x_i) = \text{sign}\left(\sum_{t=1}^{T} c_t f_t(x_i)\right).$$

Table 2.1 Notation Used in the AdaBoost Algorithm

i	Observation number, $i = 1, 2, \ldots, n$.
t	Stage number, $t = 1, 2, \ldots, T$.
x_i	A p-dimensional vector containing the quantitative variables of the ith observation.
y_i	A scalar quantity representing the class membership of the ith observation, $y_i = -1$ or 1.
f_t	The weak classifier at the tth stage.
$f_t(x_i)$	The class estimate of the ith observation using the tth stage classifier
$w_{t,i}$	The weight for the ith observation at the tth stage, $\sum_i w_{t,i} = 1$.
$I_{t,i}$	Indicator function, $I(f_t(x_i) \neq y_i)$.
e_t	The classification error at the tth stage, $\sum_i w_{t,i} I_{t,i}$.
c_t	The weight of f_t.
$\text{sign}(x)$	$= 1$ if $x \geq 0$ and $= -1$ otherwise.

In short, AdaBoost generates a sequentially weighted additive agglomeration of weak classifiers. In each step of the sequence, AdaBoost attempts to find the best classifier according to the current distribution of observation weights. If an observation is incorrectly classified with the current distribution of weights, the observation receives more weight in the next sequence, while correctly classified observations receive less weight in the next iteration. In the final agglomeration, classifiers which are accurate predictors of the original data set receive more weight, whereas weak classifiers that are poor predictors receive less weight. Thus, AdaBoost uses a sequence of simple weighted classifiers, each forced to learn a different aspect of the data, to generate a final, comprehensive classifier, which often classifies better than any individual classifier.

Friedman, Hastie, and Tibshirani (2000) dissected the AdaBoost algorithm and revealed its statistical connections to loss functions, additive modeling, and logistic regression. Specifically, AdaBoost can be thought of as a forward stagewise additive model that seeks to minimize an exponential loss function:

$$e^{-yF(x)},$$

where $F(x)$ denotes the boosted classifier (i.e., the classifier defined in Step 3 of the algorithm). Using this framework, Friedman, Hastie, and Tibshirani (2000) generalized the AdaBoost algorithm to produce a real-valued prediction (Real AdaBoost) and corresponding numerically stable version (Gentle AdaBoost). In addition, they replaced the exponential loss function with a function more commonly used in the statistical field for binary data, the binomial log-likelihood loss function, and named this method LogitBoost. Table 2.2 provides a comparison of these methods.

Generalized Boosting Algorithm of Friedman, Hastie, and Tibshirani (2000)

1. Initialize the observation weights, $w_{1,i}$, and response for model, y_i^{work}.
2. For $t = 1, 2, \ldots, T$ do:
   (a) Fit the model $y_i^{\text{work}} = f(x_i)$ using the weights, $w_{t,i}$.
   (b) Compute $c_t(x_i)$, a contribution of x_i at stage t, to the final classifier.
   (c) Update the weights, $w_{t,i}$. Update y_i^{work} (for LogitBoost only).
3. Output the final classifier:

$$\widehat{y}_i = F(x_i) = \text{sign}\left(\sum_{t=1}^{T} c_t(x_i)\right).$$

2.3.2   Weak Learning Algorithms and Effectiveness of Boosting

As mentioned in the previous section, boosting can be applied to any classification method that classifies better than random. By far, the most popular and effective weak learners are tree-based methods, also known as recursive partitioning. Several authors have provided explanations of why recursive partitioning can be boosted to produce effective classifiers (see, for example, Breiman, 1998). In short, boosting is effective for classifiers that are unstable (produce different models when given different training data from the same distribution), but when combined, get approximately the correct solution (i.e., have low bias). This breakdown is commonly known as the bias-variance trade-off (see, for example, Bauer and Kohavi, 1999). The importance of using an unstable classifier can be easily seen by examining the AdaBoost algorithms further. Specifically, if a classifier misclassifies the same observations in consecutive iterations, then AdaBoost will be prevented from finding additional structure in the data.

To explain further, let $w_{t,i}$ represent the weight of the ith observation at the tth stage of AdaBoost. Without loss of generality, suppose that the first m observations are misclassified and $\sum_{i=1}^{n} w_{t,i} = 1$. By the AdaBoost algorithm, the updated weights are

$$w_{t+1,i} = w_{t,i}e^{c_t I_{t,i}}\left(\sum_{j=1}^{n} w_{t,j}e^{c_t I_{t,j}}\right)^{-1}.$$

Table 2.2 Comparison of the Generalized Boosting Algorithms

Algorithm parameter	Algorithm	
	AdaBoost	Real AdaBoost
$w_{1,i}$	$1/n$	$1/n$
y_i	$\in \{-1, 1\}$	$\in \{-1, 1\}$
y_i^{work}	y_i	y_i
f_t	Any weak classification method that predicts a binary response: $f_t(x_i) \in \{-1, 1\}$	Any weak classification method that predicts class probability $P(y_i = 1 \mid x_i)$, $f_t(x_i) \in [0, 1]$
$c_t(x_i)$	$\alpha_t f_t(x_i)$, where $\alpha_t = \ln \dfrac{1 - e_t}{e_t}$, $e_t = \sum_{i=1}^n w_{t,i} I(f_t(x_i) \neq y_i)$	$\dfrac{1}{2} \ln \dfrac{f_t(x_i)}{1 - f_t(x_i)}$
$w_{t+1,i}$	$\dfrac{w_{t,i} \exp(\alpha_t I(f_t(x_i) \neq y_i))}{\sum_{i=1}^n w_{t,i} \exp(\alpha_t I(f_t(x_i) \neq y_i))}$	$\dfrac{w_{t,i} \exp(-y_i c_t(x_i))}{\sum_{i=1}^n w_{t,i} \exp(-y_i c_t(x_i))}$
	Gentle AdaBoost	LogitBoost
$w_{1,i}$	$1/n$	$1/4$
y_i	$\in \{-1, 1\}$	$\in \{0, 1\}$
y_i^{work}	y_i	$4y_i - 2$
f_t	Any method that predicts a continuous response: $f_t(x_i) \in R$	Any method that predicts a continuous response: $f_t(x_i) \in R$
$c_t(x_i)$	$f_t(x_i)$	$f_t(x_i)$
$w_{t+1,i}$	$\dfrac{w_{t,i} \exp(-y_i f_t(x_i))}{\sum_{i=1}^n w_{t,i} \exp(-y_i f_t(x_i))}$	$[\exp(F_t(x_i)) + \exp(-F_t(x_i))]^{-2}$, where $F_t(x_i) = (1/2) \sum_{j=1}^t f_j(x_i)$, $y_i^{\text{work}} = \left(y_i - \dfrac{1}{1 + \exp(-2F_t(x_i))} \right) \Big/ w_{t+1,i}$

Next, suppose the first m observations are again misclassified in the $(t+1)$st stage. Then,

$$e_{t+1} = \sum_{j=1}^m w_{t,j} e^{c_t} \left(\sum_{j=1}^m w_{t,j} e^{c_t} + \sum_{j=m+1}^n w_{t,j} \right)^{-1}.$$

Notice that

$$\sum_{j=1}^m w_{t,j} e^{c_t} = \frac{1 - e_t}{e_t} \sum_{j=1}^m w_{t,j} = \sum_{j=1}^m w_{t,j} \left(1 - \sum_{j=1}^m w_{t,j} \right) \left(\sum_{j=1}^m w_{t,j} \right)^{-1} = \sum_{j=m+1}^n w_{t,j}.$$

Therefore, $e_{t+1} = 0.5$ and $c_{t+1} = 0$. This implies that the weights for iteration $t + 2$ will be the same as the weights from iteration $t + 1$, and the algorithm will choose the same model. This fact will prevent the algorithm from learning any more about the relationship between the descriptor space and response classification. Hence, methods that are stable, such as linear discriminant analysis, k-nearest neighbors, and recursive partitions with many terminal nodes, will not be greatly improved by boosting. But, methods such as neural networks (Freund and Schapire, 1996b), Naive-Bayes (Bauer and Kohavi, 1999), and recursive partitions with few terminal nodes (also referred to as stumps) will be improved by boosting. For example, Optiz and Maclin (1999) compared boosting neural networks to boosting decision trees on twenty-three empirical data sets. For a majority of the data sets, boosting neural networks or decision trees produced a more accurate classifier than any individual neural network or decision tree, respectively. However, Optiz and Maclin illustrated that the performance of boosting was data dependent; for several data sets a boosted neural network performed worse than an individual neural network. Additionally, Bauer and Kohavi (1999) illustrated that boosting a Naive-Bayes classifier improves classification, but can increase the variance of the prediction.

Because of the success of boosting with recursive partitioning using few terminal nodes, in the next section we will explore the implementation of boosting using stumps, a recursive partition with one split.

2.3.3   Implementation of Boosting in SAS

There are at least two ways to implement boosting in SAS. Conveniently, boosting can be directly implemented in SAS Enterprise Miner software. However, the use of SAS Enterprise Miner requires the purchase of this module. Boosting can also be implemented through more commonly and widely available SAS software such as Base SAS, SAS/STAT, and SAS/IML. This section will focus on the development of boosting code using widely available SAS software. But, we begin this section by exploring a few details of boosting in SAS Enterprise Miner.

2.3.4   Implementation with SAS Enterprise Miner

In SAS Enterprise Miner, an ensemble classifier is quite easy to create, and can be generated with a minimum of four nodes (see Table 2.3). SAS documentation provides thorough step-by-step directions for creating various ensembles of classifiers in the Ensemble Node Help section; we refer you to this section for details on how to implement an ensemble model. Here, we highlight a few key facts about boosting in SAS Enterprise Miner.

Table 2.3 Required Nodes to Perform Boosting in SAS Enterprise Miner

Node	Options
1. Input Data Source	Select target (response) variable
2. Group Processing	Mode = Weighted resampling for boosting
3. Model	Select Tree model for a boosted recursive partition
4. Ensemble	Ensemble node = Boosting

To generate an ensemble model, the software requires the Input Data Source, Group Processing, Model, and Ensemble nodes. In the General tab under Group Processing, one must select **Weighted resampling for boosting** and must specify the number of loops (iterations) to perform. Then, in the ensemble node, **boosting** must be chosen as the setting. It is important to note that SAS Enterprise Miner performs boosting by building trees on weighted resamples of the observations rather than buildings trees based on weighted observations. This type of boosting is more commonly referred to as adaptive resampling and combining (ARCing) and was developed by Breiman (1998). In fact, the SAS Enterprise Miner implementation of boosting is only a slight modification of Breiman's Arc-x4 algorithm. While boosting and ARCing have the same general flavor, Friedman, Hastie and Tibshirani (2000) indicate that boosting via weighted trees generally performs better than ARCing. Hence, a distinction should be made between the two methods.

Unfortunately, the constructs within SAS Enterprise Miner make it extremely difficult to implement boosting with weighted trees and any of the boosting algorithms from

Table 2.2. Instead, we implement the generalized boosting algorithms directly through more commonly used SAS software.

2.3.5   Implementation of Boosting in Base SAS

For our implementation of boosting, we will use stumps—recursive partitions with one split—as the weak learner. We begin this section by focusing on the construction of code to find an optimal recursive partition in a set of data.

EXAMPLE: Recursive Partitioning

Recursive partitioning seeks to find successive partitions of the descriptor space that separate the observations into regions (or nodes) of increasingly higher purity of the response. As an illustration, consider a data set that contains 100 four-dimensional points, (x_1, x_2, y, w), where y is the class identifier and w is the weight. The data set is generated by Program 2.2. A plot of the generated data points is displayed in Figure 2.1.

Program 2.2 Simulated data in the recursive partitioning example

```
data example1;
    do i=1 to 100;
        x1 = 10*ranuni(1);
        x2 = 10*ranuni(2);
        if (x1<5 and x2<1.5) then y=0;
        if (x1<5 and x2>=1.5) then y=1;
        if (x1>=5 and x2<7.5) then y=0;
        if (x1>=5 and x2>=7.5) then y=1;
        w=1;
        output;
    end;
* Vertical axis;
axis1 minor=none label=(angle=90 "X2") order=(0 to 10 by 1) width=1;
* Horizontal axis;
axis2 minor=none label=("X1") order=(0 to 10 by 1) width=1;
symbol1 i=none value=circle color=black height=4;
symbol2 i=none value=dot color=black height=4;
proc gplot data=example1;
    plot x2*x1=y/vaxis=axis1 haxis=axis2 frame nolegend;
    run;
    quit;
```

For this example, we desire to find a single partition of the data that creates two new regions of highest possible purity. Of course, the best partition is dependent on the measure of region impurity. An intuitive measure of impurity is misclassification error, but a more accepted impurity measure is the Gini Index. Let p_j represent the proportion of observations of Class j in a node. For a two-class problem, the misclassification error of a node is defined as $\min(p_1, p_2)$, and the Gini Index is defined as $2p_1p_2$. The total measure of impurity is a weighted sum of the impurity from each node, where each node is weighted by the proportion of observations in that node, relative to the total number of observations from its parent node.

For the EXAMPLE1 data set, a few possible partitions clearly stand out (see Figure 2.1):

$$x_1 = 5, x_2 = 7.5 \text{ or } x_2 = 1.5.$$

Classification using these cut-points yields the results in Table 2.4. The corresponding measures of misclassification error and Gini Index can be found in Table 2.5.

Figure 2.1 Plot of the data in the recursive partitioning example

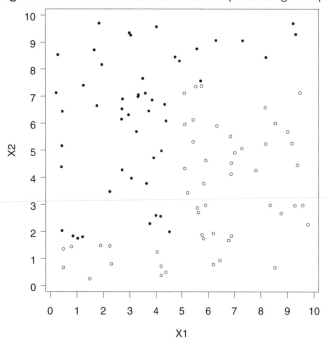

Table 2.4 Classification Results

Class	$x_1 \geq 5$	$x_1 < 5$	$x_2 < 7.5$	$x_2 \geq 7.5$	$x_2 < 1.5$	$x_2 \geq 1.5$
$y = 0$	40	11	51	0	14	37
$y = 1$	7	42	32	17	0	49

Table 2.5 Computation of the Misclassification Error and Gini Index

Partition	Misclassification error	Gini index
$x_1 \geq 5$	$\min(7/47, 40/47) = 0.15$	$2(7/47)(40/47) = 0.25$
$x_1 < 5$	$\min(11/53, 42/53) = 0.21$	$2(11/53)(42/53) = 0.33$
Total	$(0.47)(0.15) + (0.53)(0.21) = 0.18$	$(0.47)(0.25) + (0.53)(0.33) = 0.29$
$x_2 \geq 7.5$	$\min(32/83, 51/83) = 0.39$	$2(32/83)(51/83) = 0.47$
$x_2 < 7.5$	$\min(0/17, 17/17) = 0$	$2(0/17)(17/17) = 0$
Total	$(0.83)(0.39) + (0.17)(0) = 0.32$	$(0.83)(0.47) + (0.17)(0) = 0.39$
$x_2 \geq 1.5$	$\min(37/86, 49/86) = 0.43$	$2(37/86)(49/86) = 0.49$
$x_2 < 1.5$	$\min(0/14, 14/14) = 0$	$2(0/14)(14/14) = 0$
Total	$(0.14)(0) + (0.86)(0.43) = 0.37$	$(0.14)(0) + (0.86)(0.49) = 0.42$

Table 2.5 shows that the partition at $x_1 = 5$ yields the minimum value for both total misclassification error (0.18) and the Gini Index (0.29).

Unequally Weighted Observations

As defined above, misclassification error and Gini Index assume that each observation is equally weighted in determining the partition. These measures can also be computed for unequally weighted data. Define

$$p_j^* = \sum_{i=1}^{n} w_i I(y_i = j) \left(\sum_{i=1}^{n} w_i \right)^{-1}.$$

Then, for the two-class problem, the misclassification error and Gini Index are $\min(p_1^*, p_2^*)$ and $2p_1^* p_2^*$, respectively.

While three partitions for the EXAMPLE1 data set visually stand out in Figure 2.1, there are 198 possible partitions of these data. In general, the number of possible partitions for any data set is the number of dimensions multiplied by one fewer than the number of observations. And, to find the optimal partition one must search through all possible partitions. Using data steps and procedures, we can construct basic SAS code to exhaustively search for the best partition using the impurity measure of our choice, and allowing for unequally weighted data.

%Split Macro

The %Split macro searches for the optimal partition using the Gini Index (the macro is given on the book's companion Web site). The input parameters for %Split are INPUTDS and P. The INPUTDS parameter references the data set for which the optimal split will be found, and this data set must have independent descriptors named X1, X2,..., XP, a variable named W that provides each observation weight, and a response variable named Y that takes values 0 and 1 for each class. The user must also specify the P parameter, the number of independent descriptors. Upon being called, %Split produces an output data set named OUTSPLIT that contains the name of the variable and split point that minimizes the Gini Index.

Program 2.3 calls the %Split macro to find the optimal binary split in the EXAMPLE1 data set using the Gini Index.

Program 2.3 Optimal binary split in the recursive partitioning example

```
%split(inputds=example1,p=2);
proc print data=outsplit noobs label;
    format gini cutoff 6.3;
    run;
```

Output from Program 2.3

Gini index	Best variable	Best cutoff
0.293	x1	5.013

Output 2.3 output identifies $x_1 = 5.013$ as the split that minimizes the Gini Index. The Gini Index associated with the optimal binary split is 0.293.

While the %Split macro is effective for finding partitions of small data sets, its looping scheme is inefficient. Instead of employing data steps and procedures to search for the optimal split, we have implemented a search module named IML_SPLIT that is based on SAS/IML and that replaces the observation loop with a small series of efficient matrix operations.

AdaBoost Using the IML_SPLIT Module

The AdaBoost algorithm described in Section 2.3.1 is relatively easy to implement using the IML_SPLIT module as the weak learner (see the %AdaBoost macro on the book's companion Web site). Like the %Split macro, %AdaBoost has input parameters of INPUTDS and P, as well as ITER, which specifies the number of boosting iterations to perform. %AdaBoost creates a data set called BOOST that contains information about each boosting iteration. The output variables included in the BOOST data set are defined below.

- ITER is the boosting iteration number.
- GINI_VAR is the descriptor number at the current iteration that minimizes the Gini Index.
- GINI_CUT is the cut-point that minimizes the Gini Index.
- P0_LH is the probability of class 0 for observations less than GINI_CUT.
- P1_LH is the probability of class 1 for observations less than GINI_CUT.
- C_L is the class label for observations less than GINI_CUT.
- P0_RH is the probability of class 0 for observations greater than GINI_CUT.
- P1_RH is the probability of class 1 for observations greater than GINI_CUT.
- C_R is the class label for observations greater than GINI_CUT.
- ALPHA is the weight of the cut-point rule at the current boosting iteration.
- ERROR is the misclassification error of the cumulative boosting model at the current iteration.
- KAPPA is the kappa statistic of the cumulative boosting model at the current iteration.

To illustrate the %AdaBoost macro, again consider the EXAMPLE1 data set. Program 2.4 calls %AdaBoost to classify this data set using ten iterations. The program also produces plots of the misclassification error and kappa statistics for each iteration.

Program 2.4 AdaBoost algorithm in the recursive partitioning example

```
%AdaBoost(inputds=example1,p=2,iter=10);
proc print data=boost noobs label;
    format gini_cut p0_lh p1_lh p0_rh p1_rh c_r alpha error kappa 6.3;
    run;
* Misclassification error;
* Vertical axis;
axis1 minor=none label=(angle=90 "Error") order=(0 to 0.3 by 0.1) width=1;
* Horizontal axis;
axis2 minor=none label=("Iteration") order=(1 to 10 by 1) width=1;
symbol1 i=join value=none color=black width=5;
proc gplot data=boost;
    plot error*iter/vaxis=axis1 haxis=axis2 frame;
    run;
    quit;
* Kappa statistic;
* Vertical axis;
axis1 minor=none label=(angle=90 "Kappa") order=(0.5 to 1 by 0.1) width=1;
* Horizontal axis;
axis2 minor=none label=("Iteration") order=(1 to 10 by 1) width=1;
symbol1 i=join value=none color=black width=5;
proc gplot data=boost;
    plot kappa*iter/vaxis=axis1 haxis=axis2 frame;
    run;
    quit;
```

Output from Program 2.4

iter	gini_var	gini_cut	p0_LH	p1_LH	c_L	p0_RH	p1_RH	c_R	alpha	error	kappa
1	1	5.013	0.208	0.792	1	0.851	0.149	0.000	1.516	0.180	0.640
2	2	6.137	0.825	0.175	0	0.082	0.918	1.000	1.813	0.230	0.537
3	2	1.749	1.000	0.000	0	0.267	0.733	1.000	1.295	0.050	0.900
4	1	5.013	0.272	0.728	1	0.877	0.123	0.000	1.481	0.050	0.900
5	2	7.395	0.784	0.216	0	0.000	1.000	1.000	1.579	0.000	1.000
6	2	1.749	1.000	0.000	0	0.228	0.772	1.000	1.482	0.050	0.900
7	1	5.013	0.266	0.734	1	0.875	0.125	0.000	1.483	0.000	1.000
8	2	7.395	0.765	0.235	0	0.000	1.000	1.000	1.462	0.000	1.000
9	2	1.749	1.000	0.000	0	0.233	0.767	1.000	1.457	0.000	1.000
10	1	5.013	0.273	0.727	1	0.871	0.129	0.000	1.451	0.000	1.000

Output 2.4 lists the BOOST data set generated by the %AdaBoost macro. For the first iteration, x_1 minimizes the Gini index at 5.013. For observations with $x_1 < 5.013$, the probability of observing an observation in Class 0 is 0.208, while the probability of observing an observation in Class 1 is 0.792. Hence the class label for observations taking values of $x_1 < 5.013$ is 1. Similarly, for observations with $x_1 > 5.013$, the probabilities of observing an observation in Class 0 and Class 1 are 0.851 and 0.149, respectively. Therefore, the class label for observations taking values of $x_1 > 5.013$ is 0. The stage weight for the first boosting iteration is represented by ALPHA= 1.516, while the observed classification error and kappa statistics are 0.18 and 0.64, respectively. Notice that the AdaBoost algorithm learns the structure of the relationship between the descriptor space and classification vector in seven iterations (the misclassification error and kappa statistic displayed in Figure 2.2 reach a plateau by the seventh iteration), and in these iterations the algorithm quickly identifies cut-points near the constructed splits of $x_1 = 5$, $x_2 = 7.5$, and $x_2 = 1.5$.

Figure 2.2 Misclassification error (left panel) and kappa statistic (right panel) plots for the EXAMPLE1 data set using the %AdaBoost macro

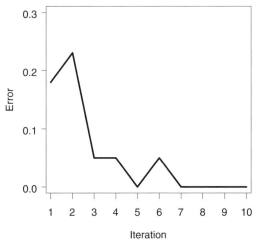 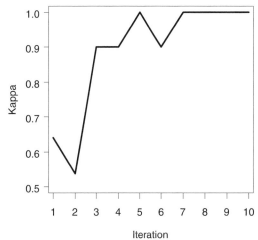

Program 2.5 models the permeability data of Section 2.2 with AdaBoost and plots the misclassification error and kappa statistics at each iteration.

Program 2.5 AdaBoost algorithm in the permeability data example

```
%AdaBoost(inputds=permy,p=71,iter=50);
proc print data=boost noobs label;
    format gini_cut p0_lh p1_lh p0_rh p1_rh c_r alpha error kappa 6.3;
    run;
* Misclassification error;
* Vertical axis;
axis1 minor=none label=(angle=90 "Error") order=(0 to 0.3 by 0.1) width=1;
* Horizontal axis;
axis2 minor=none label=("Iteration") order=(1 to 51 by 10) width=1;
symbol1 i=join value=none color=black width=5;
proc gplot data=boost;
    plot error*iter/vaxis=axis1 haxis=axis2 frame;
    run;
    quit;
* Kappa statistic;
* Vertical axis;
axis1 minor=none label=(angle=90 "Kappa") order=(0.5 to 1 by 0.1) width=1;
* Horizontal axis;
axis2 minor=none label=("Iteration") order=(1 to 51 by 10) width=1;
symbol1 i=join value=none color=black width=5;
proc gplot data=boost;
    plot kappa*iter/vaxis=axis1 haxis=axis2 frame;
    run;
    quit;
```

Output from Program 2.5

iter	gini_var	gini_cut	p0_LH	p1_LH	c_L	p0_RH	p1_RH	c_R	alpha	error	kappa
1	6	1029.7	0.294	0.706	1	0.706	0.294	0.000	0.877	0.294	0.412
2	68	82.813	0.436	0.564	1	0.805	0.195	0.000	0.431	0.294	0.412
3	33	0.992	0.129	0.871	1	0.603	0.397	0.000	0.493	0.268	0.463
4	69	45.125	0.437	0.563	1	0.940	0.060	0.000	0.335	0.274	0.452
5	37	63.938	0.356	0.644	1	0.603	0.397	0.000	0.462	0.271	0.458
6	24	0.493	0.419	0.581	1	0.704	0.296	0.000	0.443	0.249	0.503
7	43	22.930	0.448	0.552	1	0.654	0.346	0.000	0.407	0.254	0.492
8	28	0.012	0.639	0.361	0	0.403	0.597	1.000	0.392	0.246	0.508
9	23	1.369	0.496	0.504	1	0.961	0.039	0.000	0.111	0.246	0.508
10	20	1.641	0.565	0.435	0	0.000	1.000	1.000	0.320	0.249	0.503
11	23	1.369	0.449	0.551	1	0.957	0.043	0.000	0.276	0.240	0.520
12	3	1.429	0.263	0.737	1	0.569	0.431	0.000	0.320	0.251	0.497
13	69	45.125	0.452	0.548	1	0.924	0.076	0.000	0.252	0.240	0.520
14	34	5.454	0.469	0.531	1	0.672	0.328	0.000	0.297	0.226	0.548
15	31	7.552	0.529	0.471	0	0.842	0.158	0.000	0.246	0.226	0.548
16	31	7.552	0.468	0.532	1	0.806	0.194	0.000	0.234	0.220	0.559
17	32	8.937	0.572	0.428	0	0.228	0.772	1.000	0.348	0.215	0.571
18	23	1.369	0.455	0.545	1	0.966	0.034	0.000	0.244	0.206	0.588
19	20	0.836	0.615	0.385	0	0.442	0.558	1.000	0.352	0.223	0.554
20	63	321.50	0.633	0.367	0	0.447	0.553	1.000	0.357	0.189	0.621
21	68	77.250	0.488	0.512	1	0.727	0.273	0.000	0.212	0.203	0.593
22	59	46.250	0.589	0.411	0	0.159	0.841	1.000	0.400	0.201	0.599
23	23	1.369	0.463	0.537	1	0.954	0.046	0.000	0.209	0.189	0.621
24	7	736.00	0.499	0.501	1	0.768	0.232	0.000	0.126	0.192	0.616
25	42	8.750	0.363	0.637	1	0.591	0.409	0.000	0.362	0.189	0.621
26	52	1.872	0.447	0.553	1	0.654	0.346	0.000	0.309	0.192	0.616

27	61	-1.348	0.300	0.700	1	0.564	0.436	0.000	0.317	0.189	0.621
28	57	278.81	0.450	0.550	1	0.747	0.253	0.000	0.266	0.178	0.644
29	59	46.250	0.547	0.453	0	0.107	0.893	1.000	0.243	0.172	0.655
30	25	0.092	0.868	0.132	0	0.457	0.543	1.000	0.223	0.167	0.667
31	36	206.13	0.545	0.455	0	0.184	0.816	1.000	0.242	0.164	0.672
32	47	23.103	0.696	0.304	0	0.443	0.557	1.000	0.290	0.172	0.655
33	4	1.704	0.501	0.499	0	0.766	0.234	0.000	0.112	0.167	0.667
34	4	1.704	0.474	0.526	1	0.745	0.255	0.000	0.186	0.175	0.650
35	20	1.641	0.549	0.451	0	0.000	1.000	1.000	0.236	0.158	0.684
36	23	1.369	0.466	0.534	1	0.946	0.054	0.000	0.186	0.169	0.661
37	62	-0.869	0.262	0.738	1	0.548	0.452	0.000	0.254	0.150	0.701
38	6	1279.9	0.455	0.545	1	0.812	0.188	0.000	0.233	0.164	0.672
39	32	6.819	0.559	0.441	0	0.355	0.645	1.000	0.267	0.158	0.684
40	32	0.707	0.305	0.695	1	0.520	0.480	0.000	0.094	0.164	0.672
41	29	7.124	0.568	0.432	0	0.403	0.597	1.000	0.347	0.138	0.723
42	12	2.000	0.218	0.782	1	0.521	0.479	0.000	0.194	0.138	0.723
43	31	7.552	0.420	0.580	1	0.737	0.263	0.000	0.381	0.150	0.701
44	33	0.992	0.183	0.817	1	0.546	0.454	0.000	0.245	0.133	0.734
45	23	1.369	0.459	0.541	1	0.932	0.068	0.000	0.208	0.141	0.718
46	55	975.56	0.236	0.764	1	0.541	0.459	0.000	0.227	0.136	0.729
47	53	18.687	0.564	0.436	0	0.411	0.589	1.000	0.320	0.127	0.746
48	35	249.00	0.371	0.629	1	0.573	0.427	0.000	0.391	0.130	0.740
49	7	736.00	0.442	0.558	1	0.730	0.270	0.000	0.304	0.124	0.751
50	48	60.067	0.451	0.549	1	0.606	0.394	0.000	0.319	0.138	0.723

Output 2.5 provides important information for the progress of the algorithm for each iteration. Is shows that in the first 50 iterations, variable X23 is selected six times, while X20, X31, and X32 are selected three times each; boosting is focusing on variables that are important for separating the data into classes. Also, a number of variables are not selected in any iteration.

Figure 2.3 displays the error and kappa functions. Notice that after approximately 35 iterations, the error levels off at approximately 0.15. At this point, AdaBoost is learning the training data at a much slower rate.

Figure 2.3 Misclassification error (left panel) and kappa statistic (right panel) plots for the PERMY data set using the %AdaBoost macro

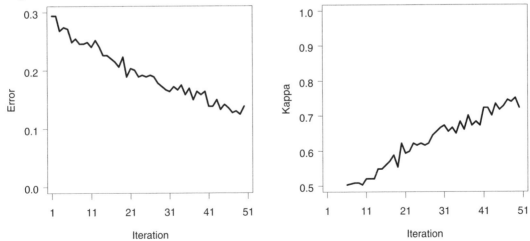

2.3.6   Implementation of Generalized Boosting Algorithms

In most model building applications, we need to evaluate the model's performance on an independent set of data. This process is referred to as validation or cross-validation. In this section we provide a more general macro, %Boost, that can be used to evaluate both training and testing data. This macro also implements the other generalized forms of boosting.

The generalized boosting algorithms described in Section 2.3.1 can be implemented using the same structure of the %AdaBoost macro by making several appropriate adjustments. To implement the Real AdaBoost algorithm, we must use predicted class probabilities in place of predicted classification, and alter the observation contributions and weights. Because the %Split macro generates predicted class probabilities, we can directly construct Real AdaBoost. However, both Gentle AdaBoost and LogitBoost require a method that predicts a continuous, real-valued response. This stipulation requires that we implement a different form of recursive partitioning. Instead of using partitioning based on a categorical response, we will implement partitioning based on a continuous response. In short, we seek to find the partition that minimizes the corrected sum-of-squares, rather than misclassification error or Gini Index, of the response across both new nodes. That is, we seek the variable, x_j, and cut-point, c, that minimize

$$\sum_{r_{1j}} (y_{1j} - \bar{y}_{1j})^2 + \sum_{r_{2j}} (y_{2j} - \bar{y}_{2j})^2,$$

where $r_{1j} = \{i : x_j < c\}$ and $r_{2j} = \{i : x_j \geq c\}$.

Like partitioning based on a categorical response, partitioning based on a continuous response requires an exhaustive search to find the optimal cut-point. To find the variable and optimal cut-point, we have created a SAS/IML module, REGSPLIT_IML. This module is included in our comprehensive macro, %Boost, that implements the AdaBoost, Real AdaBoost, Gentle AdaBoost, and LogitBoost algorithms (the macro is provided on the book's companion Web site). The input variables for %Boost are the same as %AdaBoost. In addition, the user must specify the type of boosting to perform (TYPE=1 performs AdaBoost, TYPE=2 performs Real AdaBoost, TYPE=3 performs Gentle AdaBoost, and TYPE=4 performs LogitBoost). When AdaBoost or Real AdaBoost are requested, %Boost produces an output data set with the same output variables as the AdaBoost macro. However, if Gentle AdaBoost or LogitBoost are specified, then a different output data set is created. The variables included in this data set are listed below.

- ITER is the boosting iteration number.
- REG_VAR is the variable number that minimizes the corrected sum-of-squares.
- MIN_CSS is the minimum corrected sum-of-squares.
- CUT_VAL is the optimal cut-point.
- YPRED_L is the predicted value for observations less than CUT_VAL.
- YPRED_R is the predicted value for observations greater than CUT_VAL.
- ERROR is the misclassification error of the cumulative boosting model at the current iteration.
- KAPPA is the kappa statistic of the cumulative boosting model at the current iteration.

As a diagnostic tool, the %Boost macro also creates an output data set named OUTWTS that contains observation weights at the final boosting iteration. These can be used to identify observations that are difficult to classify.

To illustrate the %Boost macro, Program 2.6 analyzes the EXAMPLE1 data set using the Gentle AdaBoost algorithm with ten iterations.

Program 2.6 Gentle AdaBoost algorithm in the recursive partitioning example

```
%boost(inputds=example1,p=2,outputds=out_ex1,outwts=out_wt1,iter=10,type=3);
proc print data=out_ex1 noobs label;
    format min_css cut_val ypred_l ypred_r error kappa 6.3;
proc print data=out_wt1;
    format weight 7.5;
    run;
```

Output from Program 2.6

```
                              OUT_EX1 data set

       iter  reg_var  min_css  cut_val  ypred_L  ypred_R  error  kappa

        1       1      0.587    5.013    0.585   -0.702   0.180  0.640
        2       2      0.528    6.137   -0.574    0.835   0.160  0.682
        3       2      0.679    1.749   -1.000    0.357   0.050  0.900
        4       1      0.600    5.013    0.603   -0.656   0.070  0.860
        5       2      0.627    7.395   -0.388    1.000   0.000  1.000
        6       2      0.702    1.749   -1.000    0.365   0.000  1.000
        7       1      0.577    5.013    0.639   -0.659   0.000  1.000
        8       2      0.669    7.395   -0.343    1.000   0.000  1.000
        9       2      0.722    1.749   -1.000    0.343   0.000  1.000
       10       1      0.566    5.013    0.651   -0.665   0.000  1.000


                              OUT_WT1 data set

                          Obs      weight

                           1      0.00019
                           2      0.01207
                           3      0.03347
                           4      0.00609
                           5      0.01820
                           6      0.00609
                           7      0.00010
                           8      0.00609
                           9      0.00295
                          10      0.00609
```

Output 2.6 provides a listing of the data set produced by the %Boost macro (OUT_EX1 data set). For the first iteration, X1 minimizes the corrected sum-of-squares (0.587) at the cut-point of 5.013. Observations that are less than this value have a predicted value of 0.585, whereas observations greater than this value have a predicted value of −0.702. For Gentle AdaBoost, the error at the first iteration is 0.18, while the kappa value is 0.64.

In addition, Output 2.6 lists the first ten observations in the OUT_WT1 data set, which includes the final weights for each observation after ten iterations. These weights are constrained to sum to one; therefore, relatively high weights represent observations that are difficult to classify, and observations with low weights are relatively easy to classify.

One can also model the permeability data using Gentle AdaBoost. But first we should split the data into training and testing sets in order to evaluate each model's performance on an independent set of data. For this chapter, the permeability data set has already been separated into training and testing sets through the SET variable. Program 2.7 creates the training and testing sets and invokes the %Boost macro to perform Gentle AdaBoost on the training data with 35 iterations.

Program 2.7 Gentle AdaBoost algorithm in the permeability data example

```
data train;
    set permy(where=(set="TRAIN"));
data test;
    set permy(where=(set="TEST"));
    run;
%boost(inputds=train,p=71,outputds=genout1,outwts=genwt1,iter=35,type=3);
proc print data=genout1 noobs label;
    format min_css ypred_l ypred_r error kappa 6.3;
    run;
* Misclassification error;
* Vertical axis;
axis1 minor=none label=(angle=90 "Error") order=(0 to 0.3 by 0.1) width=1;
* Horizontal axis;
axis2 minor=none label=("Iteration") order=(1 to 36 by 5) width=1;
symbol1 i=join value=none color=black width=5;
proc gplot data=genout1;
    plot error*iter/vaxis=axis1 haxis=axis2 frame;
    run;
    quit;
* Kappa statistic;
* Vertical axis;
axis1 minor=none label=(angle=90 "Kappa") order=(0.5 to 1 by 0.1) width=1;
* Horizontal axis;
axis2 minor=none label=("Iteration") order=(1 to 36 by 5) width=1;
symbol1 i=join value=none color=black width=5;
proc gplot data=genout1;
    plot kappa*iter/vaxis=axis1 haxis=axis2 frame;
    run;
    quit;
```

Output from Program 2.7

iter	reg_var	min_css	cut_val	ypred_L	ypred_R	error	kappa
1	6	0.838	1047.69	0.374	-0.434	0.299	0.402
2	42	0.910	8.75	0.649	-0.142	0.294	0.411
3	70	0.878	13.81	0.181	-0.734	0.238	0.523
4	33	0.933	1.17	0.706	-0.116	0.234	0.533
5	37	0.928	75.56	0.348	-0.201	0.238	0.523
6	7	0.922	745.06	0.101	-0.803	0.234	0.533
7	63	0.920	318.81	-0.325	0.241	0.196	0.607
8	43	0.943	16.74	0.333	-0.172	0.178	0.645
9	45	0.938	23.72	-0.464	0.136	0.173	0.654
10	56	0.939	452.44	0.337	-0.186	0.192	0.617
11	19	0.946	0.50	-0.349	0.157	0.154	0.692
12	71	0.944	549.46	-0.078	0.751	0.131	0.738
13	15	0.936	0.10	0.485	-0.122	0.126	0.748
14	32	0.953	8.85	-0.057	0.716	0.126	0.748
15	23	0.949	1.38	0.066	-0.924	0.112	0.776
16	1	0.950	624.00	1.000	-0.065	0.113	0.794
17	59	0.955	46.25	-0.991	-0.738	0.107	0.785
18	20	0.944	8.75	0.142	0.424	0.098	0.804
20	68	0.951	84.84	0.073	-0.704	0.107	0.785
21	34	0.955	0.71	-0.539	0.074	0.103	0.794

22	41	0.950	9.88	0.747	-0.137	0.103	0.794
23	34	0.956	0.71	-0.518	0.095	0.112	0.776
24	33	0.949	0.99	0.724	-0.137	0.107	0.785
25	13	0.957	0.02	1.000	-0.036	0.103	0.794
26	57	0.959	274.63	0.085	-0.573	0.103	0.794
27	82	0.957	1.58	0.191	-0.224	0.079	0.841

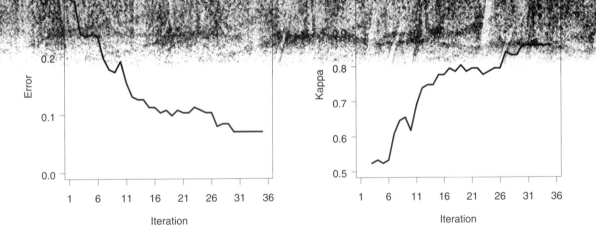

Figure 2.4 Misclassification error (left panel) and kappa statistic (right panel) on the training data for the PERMY data set using the %Boost macro (Gentle AdaBoost algorithm).

%Predict Macro

To complement the %Boost macro, we have also developed a macro, %Predict, to predict new observations using the model information generated by %Boost. This macro can be used to evaluate the performance of a model on an independent testing or validation data set. The input variables to %Predict are:

- PRED_DS is the name of the data set to be predicted. This data set must have descriptors named X1, X2, ..., XP. The response must be named Y and must assume values of 0 or 1.
- P is the number of descriptors in the input data set.
- BOOST_DS is the name of the data set with boost model information (the OUTPUTDS from the %Boost macro).
- TYPE is the type of boosting for prediction.
- ITER is the number of boosting iterations desired.

Program 2.8 calls the %Predict macro to examine the predictive properties of the classification model in the permeability data example. The program analyzes the TEST data set created in Program 2.7.

Program 2.8 Prediction in the permeability data example

```
%predict(pred_ds=test,p=71,boost_ds=genout1,outputds=gentst1,out_pred=genpred1,
    type=3,iter=35);
* Misclassification error;
* Vertical axis;
axis1 minor=none label=(angle=90 "Error") order=(0 to 0.5 by 0.1) width=1;
* Horizontal axis;
axis2 minor=none label=("Iteration") order=(1 to 36 by 5) width=1;
symbol1 i=join value=none color=black width=5;
proc gplot data=gentst1;
    plot error*iter/vaxis=axis1 haxis=axis2 frame;
    run;
    quit;
```

Figure 2.5 depicts changes in the model error across iterations. Although we saw a decrease in the misclassification error for the training set across iterations (Figure 2.4), the same is not true for the test set. Instead, the misclassification error goes up slightly across iterations. For drug discovery data, this phenomenon is not surprising—often the response in many data sets is noisy (a number of samples have been misclassified). These misclassified samples hamper the learning ability of many models, preventing them from learning the correct structure between the descriptors and the classification variable. In Section 2.4, we further explain how to use boosting to identify noisy or mislabeled observations. Removing these observations often produces a better overall predictive model.

Figure 2.5 Misclassification rates in the testing subset of the PERMY data set using the %Predict macro (Gentle AdaBoost algorithm)

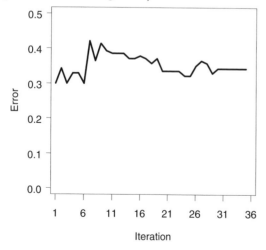

2.3.7 Properties of Boosting

Many authors have shown that boosting is generally robust to overfitting; that is, as the complexity of the boosting model increases for the training set (i.e., the number of iterations increases), the test set error does not increase (Schapire, Freund, Bartlett and Lee, 1998; Friedman, Hastie, and Tibshirani, 2000; Freund, Mansour and Schapire, 2001).

However, boosting, by its inherent construction, is vulnerable to certain types of data problems, which can make it difficult for boosting to find a model that can be generalized.

The primary known problem for boosting is noise in the response (Krieger, Long, and Wyner, 2002; Schapire, 2002; Optiz and Maclin, 1999), which can occur when observations are mislabeled or classes overlap in descriptor space. Under either of these circumstances, boosting will place increasing higher weights on those observations that are inconsistent with the majority of the data. As a check-valve, the performance of the model at each stage is evaluated against the original training set. Models that poorly classify receive less weight, while models that accurately classify receive more weight. But, the final agglomerative model has attempted to fit the noise in the response. Hence, the final model will use these rules to predict new observations, thus increasing prediction error.

For similar reasons, boosting will be negatively affected by observations that are outliers in the descriptor space. In fact, because of this problem, Freund and Schapire (2001) recommend removing known outliers before performing boosting.

Although noise can be the Achilles heel of boosting, its internal correction methods can help us identify potentially mislabeled or outlying observations. Specifically, as the algorithm iterates, boosting will place increasing larger weights on observations that are difficult to classify. Hence, tracking observation weights can be used as a diagnostic tool for identifying these problematic observations. In the following section, we will illustrate the importance of examining observation weights.

2.4   Model Building

This section discusses model building with the %Boost macro introduced in Section 2.3.

Each modification of the AdaBoost algorithm (AdaBoost, Real AdaBoost, Gentle AdaBoost, and LogitBoost) performs differently on various types of data. Because of this fact, we should evaluate the performance of each model on a training and testing set to identify the model with the best performance. For the permeability data, Program 2.9 performs each version of boosting on the training and testing sets (TRAIN and TEST data sets) created in Program 2.7. The program also computes and plots the associated misclassification errors to gauge the predictive ability of each model.

Program 2.9 Performance of the four boosting algorithms in the permeability data example

```
* AdaBoost;
%boost(inputds=train,p=71,outputds=adaout1,outwts=adawt1,iter=40,type=1);
%predict(pred_ds=test,p=71,boost_ds=adaout1,outputds=adatst1,out_pred=adapred1,
    type=1,iter=40);
* Real AdaBoost;
%boost(inputds=train,p=71,outputds=realout1,outwts=realwt1,iter=40,type=2);
%predict(pred_ds=test,p=71,boost_ds=realout1,outputds=realtst1,out_pred=realpred1,
    type=2,iter=40);
* Gentle AdaBoost;
%boost(inputds=train,p=71,outputds=genout1,outwts=genwt1,iter=40,type=3);
%predict(pred_ds=test,p=71,boost_ds=genout1,outputds=gentst1,out_pred=genprd1,
    type=3,iter=40);
* LogitBoost;
%boost(inputds=train,p=71,outputds=logout1,outwts=logwt1,iter=40,type=4);
%predict(pred_ds=test,p=71,boost_ds=logout1,outputds=logtst1,out_pred=logprd1,
    type=4,iter=40);
* Training set misclassification errors;
data adaout1;
    set adaout1; method=1;
```

```
data realout1;
    set realout1; method=2;
data genout1;
    set genout1; method=3;
data logout1;
    set logout1; method=4;
data train_error;
    set adaout1(keep=iter error method)
```

(middle portion of code obscured by scanning artifacts)

```
    set rea tst1; method
data gentst1;
    set gentst1; method=3;
data logtst1;
    set logtst1; method=4;
data test_error;
    set adatst1(keep=iter error method)
    realtst1(keep=iter error method)
    gentst1(keep=iter error method)
    logtst1(keep=iter error method);
axis1 minor=none label=(angle=90 "Error") order=(0.2 to 0.5 by 0.1) width=1;
axis2 minor=none label=("Iteration") order=(1 to 41 by 10) width=1;
symbol1 i=join value=none color=black line=1 width=3;
symbol2 i=join value=none color=black line=34 width=3;
symbol3 i=join value=none color=black line=20 width=3;
symbol4 i=join value=none color=black line=41 width=3;
proc gplot data=test_error;
    plot error*iter=method/vaxis=axis1 haxis=axis2 frame nolegend;
    run;
    quit;
```

The misclassification error curves for the four boosting algorithms are displayed in the left panel of Figure 2.6. After 40 iterations, each method has learned as much information as possible about the structure of the training set in relation to compound permeability classification (notice the flattening of each misclassification curve). These models are subsequently applied to the test set and the resulting misclassification error curves are shown in the right panel of Figure 2.6. This plot demonstrates that the predictive ability of boosting does not improve as the number of iterations increases. Notice, however, that boosting does not rapidly overfit; the test set error does not climb rapidly as iteration number increases.

Figure 2.6 Misclassification rates the training (left panel) and test (right panel) subsets of the PERMY data set (solid curve, AdaBoost; dotted curve, Real AdaBoost; dashed curve, Gentle AdaBoost; dashed-dotted curve, LogitBoost)

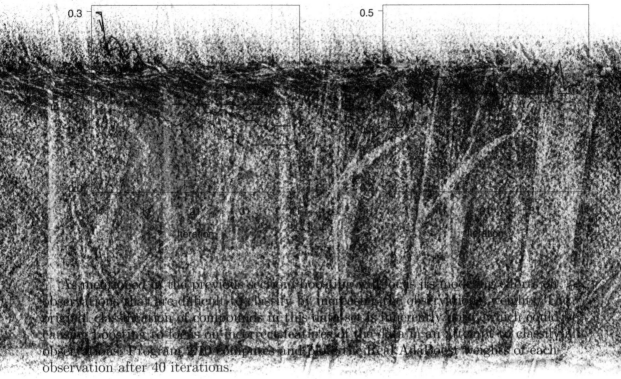

As mentioned in the previous section, boosting methods its modeling efforts on observations that are difficult to classify by increasing the observations' weights. The original classification of compounds in this data set is inherently noisy, which could be causing boosting to focus on incorrect features of the data. In an attempt to classify all observations, Program 2.10 computes and plots the Real AdaBoost weights of each observation after 40 iterations.

Program 2.10 Observation weights for the Real AdaBoost algorithm

```
%boost(inputds=train,p=71,outputds=realout1,outwts=realwt1,iter=40,type=2);
* Weights from Real AdaBoost;
proc sort data=realwt1 out=sortwts;
    by descending weight;
data sortwts;
    set sortwts;
    obsnum=_n_;
* Vertical axis;
axis1 minor=none label=(angle=90 "Weight") order=(0 to 0.03 by 0.01) width=1;
* Horizontal axis;
axis2 minor=none label=("Observations") order=(0 to 225 by 25) width=1;
symbol1 i=none value=circle color=black height=4;
proc gplot data=sortwts;
    plot weight*obsnum/vaxis=axis1 haxis=axis2 frame;
    run;
    quit;
```

Figure 2.7 displays the weights computed by Program 2.10. In this example, the Real AdaBoost algorithm focuses 50 percent of the weight on only 20 percent of the data—a telltale sign that boosting is chasing misclassified or difficult-to-classify observations.

In an attempt to find a predictive model, we have removed these observations and have re-run each boosting algorithm (Program 2.11). The misclassification error curves generated by Program 2.11 are displayed in Figure 2.8.

Figure 2.7 Observation weights for the Real AdaBoost algorithm after 40 iterations (observations are ordered by magnitude)

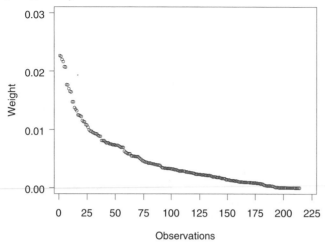

Program 2.11 Performance of the four boosting algorithms in the permeability data example (reduced data set)

```
* Real AdaBoost;
%boost(inputds=train,p=71,outputds=realout1,outwts=realwt1,iter=40,type=2);
data newtrain;
    merge train realwt1;
    if weight>0.007 then delete;
    drop weight;
* AdaBoost;
%boost(inputds=newtrain,p=71,outputds=adaout2,outwts=adawt2,iter=40,type=1);
%predict(pred_ds=test,p=71,boost_ds=adaout2,outputds=adatst2,out_pred=adapred2,
    type=1,iter=40);
* Real AdaBoost;
%boost(inputds=newtrain,p=71,outputds=realout2,outwts=realwt2,iter=40,type=2);
%predict(pred_ds=test,p=71,boost_ds=realout2,outputds=realtst2,out_pred=realpred2,
    type=2,iter=40);
* Gentle AdaBoost;
%boost(inputds=newtrain,p=71,outputds=genout2,outwts=genwt2,iter=40,type=3);
%predict(pred_ds=test,p=71,boost_ds=genout2,outputds=gentst2,out_pred=genprd2,
    type=3,iter=40);
* LogitBoost;
%boost(inputds=newtrain,p=71,outputds=logout2,outwts=logwt2,iter=40,type=4);
%predict(pred_ds=test,p=71,boost_ds=logout2,outputds=logtst2,out_pred=logprd2,
    type=4,iter=40);
* Training set misclassification errors;
data adaout2;
    set adaout2; method=1;
data realout2;
    set realout2; method=2;
data genout2;
    set genout2; method=3;
data logout2;
    set logout2; method=4;
```

```
data train_error2;
    set adaout2(keep=iter error method)
    realout2(keep=iter error method)
    genout2(keep=iter error method)
    logout2(keep=iter error method);
axis1 minor=none label=(angle=90 "Error") order=(0 to 0.3 by 0.1) width=1;
axis2 minor=none label=("Iteration") order=(1 to 41 by 10) width=1;
symbol1 i=join value=none color=black line=1 width=3;
symbol2 i=join value=none color=black line=34 width=3;
symbol3 i=join value=none color=black line=20 width=3;
symbol4 i=join value=none color=black line=41 width=3;
proc gplot data=train_error2;
    plot error*iter=method/vaxis=axis1 haxis=axis2 frame nolegend;
    run;
    quit;
* Test set misclassification errors;
data adatst2;
    set adatst2; method=1;
data realtst2;
    set realtst2; method=2;
data gentst2;
    set gentst2; method=3;
data logtst2;
    set logtst2; method=4; run;
data test_error2;
    set adatst2(keep=iter error method)
    realtst2(keep=iter error method)
    gentst2(keep=iter error method)
    logtst2(keep=iter error method);
axis1 minor=none label=(angle=90 "Error") order=(0.2 to 0.5 by 0.1) width=1;
axis2 minor=none label=("Iteration") order=(1 to 41 by 10) width=1;
symbol1 i=join value=none color=black line=1 width=3;
symbol2 i=join value=none color=black line=34 width=3;
symbol3 i=join value=none color=black line=20 width=3;
symbol4 i=join value=none color=black line=41 width=3;
proc gplot data=test_error2;
    plot error*iter=method/vaxis=axis1 haxis=axis2 frame nolegend;
    run;
    quit;
```

Figure 2.8 depicts the misclassification error curves for the four boosting algorithms in the reduced data set. The initial training error is lower than for the original training set (see Figure 2.6), and each boosting method more rapidly learns the features of the training data that are related to permeability classification. This implies that the observations that were removed from the training set were likely misclassified, but the predictive performance on the test set does not improve (see the right panel of Figure 2.8).

Undoubtedly, the test set also contains observations that are misclassified. In a final attempt to improve boosting's predictive ability, we have run Real AdaBoost on the test set and have removed the highest weighted observations. To this reduced test set, we have applied each reduced training set boosting model (Program 2.12). The misclassification error rates computed by the program are shown in Figure 2.9.

Figure 2.8 Misclassification rates in the training (left panel) and test (right panel) subsets for the reduced training set (solid curve, AdaBoost; dotted curve, Real AdaBoost; dashed curve, Gentle AdaBoost; dashed-dotted curve, LogitBoost)

```
data newtest;
    merge test realwt3;
    if weight>0.01 then delete;
    drop weight;
* AdaBoost;
%predict(pred_ds=newtest,p=71,boost_ds=adaout2,outputds=adatst3,out_pred=adapred3,
    type=1,iter=40);
* Real AdaBoost;
%predict(pred_ds=newtest,p=71,boost_ds=realout2,outputds=realtst3,out_pred=realpred3,
    type=2,iter=40);
* Gentle AdaBoost;
%predict(pred_ds=newtest,p=71,boost_ds=genout2,outputds=gentst3,out_pred=genprd3,
    type=3,iter=40);
* LogitBoost;
%predict(pred_ds=newtest,p=71,boost_ds=logout2,outputds=logtst3,out_pred=logprd3,
    type=4,iter=40);
* Test set misclassification errors;
data adatst3;
    set adatst3; method=1;
data realtst3;
    set realtst3; method=2;
data gentst3;
    set gentst3; method=3;
data logtst3;
    set logtst3; method=4;
data test_error3;
    set adatst3(keep=iter error method)
    realtst3(keep=iter error method)
    gentst3(keep=iter error method)
    logtst3(keep=iter error method);
```

```
axis1 minor=none label=(angle=90 "Error") order=(0 to 0.3 by 0.1) width=1;
axis2 minor=none label=("Iteration") order=(1 to 41 by 10) width=1;
symbol1 i=join value=none color=black line=1 width=3;
symbol2 i=join value=none color=black line=34 width=3;
symbol3 i=join value=none color=black line=20 width=3;
symbol4 i=join value=none color=black line=41 width=3;
proc gplot data=test_error3;
    plot error*iter=method/vaxis=axis1 haxis=axis2 frame nolegend;
    run;
    quit;
```

Figure 2.9 Misclassification error rates in the test subset of the reduced set (solid curve, AdaBoost; dotted curve, Real AdaBoost; dashed curve, Gentle AdaBoost; dashed-dotted curve, LogitBoost)

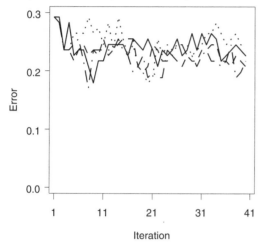

Figure 2.9 demonstrates that the misclassification error rates on this set are noticeably lower than on the original test set.

While boosting does not significantly reduce test set classification error across iterations for this example, it does allow the user to identify difficult-to-classify, or possibly misclassified observations. With these observations present, any method will have a difficult time finding the underlying relationship between predictors and response.

2.5   Partial Least Squares for Discrimination

Perhaps the most commonly used method of statistical discrimination is simple linear discriminant analysis (LDA). A fundamental problem associated with the use of LDA in practice, however, is that the associated pooled within-groups sums-of-squares and cross-products matrix has to be invertible. For data sets with collinear features or with significantly more features than observations this may not be the case. In these situations there are many different options available. By far the most common in the practice of chemometrics is to first use principal components analysis (PCA) to reduce the dimension of the data, and then to follow the PCA with a version of LDA, or to simply impose an ad hoc classification rule at the level of the PCA scores and leave it at that. While both approaches are sometimes successful in identifying class structure, the dimension reduction step was not focused on the ultimate goal of discrimination.

Of course, PCA is not the only option for collinear data. Ridging or "shrinkage" can be employed to stabilize the pertinent covariance matrices so that the classical discrimination paradigms might be implemented (Friedman, 1989; Rayens, 1990; Rayens and Greene,

1991; Greene and Rayens, 1989). Alternatively, variable selection routines based in genetic algorithms are gaining popularity (e.g., Lavine, Davidson and Rayens, 2004) and have been shown to be successful in particular on microarray data. Other popular methods include flexible discriminant analysis (Ripley, 1996) and penalized discriminant analysis (Hastie et al., 1995), which are also variations on the ridging theme.

It is well known that PCA is not a particularly reliable paradigm for discrimination since, unlike simple LDA, it is capable of identifying only gross variability and is not capable of distinguishing "among-groups" and "within-groups" variables. Indeed, a PCA approach to discrimination is typically successful only when the among-groups variability dominates the within-groups variability, as often happens with chromatography studies. This point is illustrated in a small simulation study by Barker and Rayens (2003).

It turns out that partial least squares (PLS) is much more appropriate than PCA as a dimension-reducing technique for the purposes of linear discrimination. Barker and Rayens fully established this connection between PLS and LDA, and we will refer to their use of PLS for facilitating a discriminant analysis as PLS-LDA. Some of the details of this connection will be reviewed before illustrating how PLS-LDA can easily be implemented in SAS using the PLS procedure.

To facilitate the discussion, we will use PROC PLS to perform PLS-LDA on the same permeability data set that we used to illustrate boosting.

2.5.1 Review of PLS

PLS is based on Herman Wold's original Nonlinear Iterative Partial Least Squares (NIPALS) algorithm (Wold, 1966; Wold, 1981), adapted to reduce dimensionality in the overdetermined regression problem. However, there are many different versions of the same paradigm, some of which allow the original PLS problem to be viewed as a collection of well-posed eigenstructure problems which better facilitate PLS being connected to canonical covariates analysis (CCA) and, ultimately, to LDA. This is the perspective taken on PLS in this chapter. With respect to SAS, this perspective is essentially equivalent to invoking the METHOD=SIMPLS option, which we will revisit in the discussion below.

It should be noted at the outset that under this general eigenstructure perspective it is especially obvious that different sets of constraints—and whether these constraints are imposed in both the X- and Y-spaces or only one—will lead to different PLS directions. This does not seem to be well known, or at least not often discussed in practice. With PCA the resulting "directions" or weight vectors are orthogonal and, at the same time, the associated component scores are uncorrelated. With PLS one has to choose since both uncorrelated score constraints and orthogonal weights are not simultaneously possible. Which constraint seems to be important and where you impose them depends on who you talk to. In the neurosciences, orthogonal constraints are considered essential, since the directions are often more important than the scores and having the directions be orthogonal (interpreted in practice as "independent") helps with the interpretation of the estimated "signals". Chemometricians, however, typically have no interest in an orthogonal decomposition of variable space and are much more focused on creating "new" data with uncorrelated features, which makes sense, since it is often a problem of collinearity that brings them to PLS in the first place.

For example, this latter perspective is adopted in PROC PLS where uncorrelated score constraints are implicitly assumed. We are going to simplify our presentation by largely ignoring the constraints issue. Indeed, with all the standard constraint sets the product of the sample covariance matrix and its transpose, $S_{xy}S_{yx}$, emerges at the core construct in the extraction of the PLS structure. As we will see below, it is this construct that is intimately related to Fisher's LDA. First, the definition of PLS is reviewed.

Definition of PLS

Let x and y be random p- and q-dimensional vectors with dispersion matrices Σ_x and Σ_y, respectively (the sample matrices will be denoted by S_x and S_y). Denote the covariance of x and y by Σ_{xy}. The first pair of PLS directions are defined as p- and q-dimensional vectors a and b that jointly maximize

$$\frac{\text{Cov}[(a^T x, b^T y)]^2}{(a^T a)(b^T b)}.$$

The objective of the maximization in this definition can be more clearly seen by rewriting it as

$$\frac{\text{Cov}[(a^T x, b^T y)]^2}{(a^T a)(b^T b)} = \text{Var}(a^T x)[\text{Corr}(a^T x, b^T y)]^2 \text{Var}(b^T y).$$

In this framework, it is easy to see how PLS can be thought of as "penalized" canonical correlations analysis (see Frank and Friedman, 1993). That is, the squared correlation term alone represents the objective of CCA. However, this objective is penalized by a variance term for the X-space and another for the Y-space. These variance terms represent the objective of principal components analysis. Hence, from this representation, PLS seeks to maximize correlation while simultaneously reducing dimension in the X- and Y-spaces.

PLS Theorem

The first PLS direction in the X-space, a_1, is an eigenvector of $\Sigma_{xy}\Sigma_{yx}$ corresponding to the largest eigenvalue, and the corresponding first PLS direction in the Y-space, b_1, is given by $b_1 = \Sigma_{yx}a_1$.

If orthogonality constraints are imposed on the directions in both the X- and Y-spaces, then any subsequent PLS solution, say a_{k+1}, follows analogously, with the computation of the eigenvector of $S_{xy}S_{yx}$ corresponding to the $(k+1)$st largest eigenvalue and the Y-space direction emerges as $b_{k+1} = S_{yx}a_{k+1}$. If, instead of orthogonal constraints, uncorrelated score constraints are imposed, then the $(k+1)$st X-space direction is an eigenvector corresponding to the largest eigenvalue of

$$\left(I_p - (\Sigma_x A^{(k)})\left[(\Sigma_x A^{(k)})^T \Sigma_x A^{(k)}\right]^{-1}(\Sigma_x A^{(k)})^T\right)\Sigma_{xy}\Sigma_{yx},$$

where $A^{(k)} = [a_1, a_2, \ldots, a_k]_{p \times k}$. There is a completely similar expression for the Y-space structure.

The described theorem is well known in the PLS literature and, for uncorrelated score constraints, a variant was first produced by de Jong (1993) when the SIMPLS procedure was introduced. Another proof of this Theorem under different constraint sets (and novel in that it does not use Lagrange multipliers) is provided in the technical report by Rayens (2000). For now, the point is simply that the derivation of the PLS directions will either directly or indirectly involve the eigenstructure of $S_{xy}S_{yx}$.

In the following presentation, CCA is first connected to LDA, then, using this connection between PLS and CCA, PLS is formally connected and finally, PLS to LDA.

2.5.2   Connection between CCA and LDA

When CCA is performed on a training set, X (e.g., 71 molecular properties measured on 354 compounds described in Section 2.2) and an indicator matrix, Y, representing group membership (e.g., a 1 for a permeable compound and a 0 for a non-permeable compound), the CCA directions are just Fisher's LDA directions. This well-known fact, which was first recognized by Bartlett (1938), has been reproved more elegantly by Barker and Rayens (2003) by using the following results, which are important for this presentation.

Let x_{ij} be the p-dimensional vector for the jth observation in the ith group and g be the number of groups. Denote the training set consisting of n_i observations on each of p feature variables by

$$X_{n \times p} = (x_{11}, x_{12}, \ldots, x_{1n_1}, \ldots, x_{g1}, x_{g2}, \ldots, x_{gn_g})^T.$$

Let H denote the among-groups sums-of-squares and cross-products matrix and let E be the pooled within-groups sums-of-squares and cross-products matrix,

$$H = \sum_{i=1}^{g} n_i(\bar{x}_i - \bar{x})(\bar{x}_i - \bar{x})^T, \quad E = \sum_{i=1}^{g} \sum_{j=1}^{n_i} (x_{ij} - \bar{x}_i)(x_{ij} - \bar{x}_i)^T,$$

where

$$\bar{x}_i = (1/n_i) \sum_{j=1}^{n_i} x_{ij}, \quad \bar{x} = (1/n) \sum_{i=1}^{g} \sum_{j=1}^{n_i} x_{ij}, \quad n = \sum_{i=1}^{g} n_i.$$

LDA, also known as "canonical discriminant analysis" (CDA) when expressed this way, manipulates the eigenstructure of $E^{-1}H$. The connections between CDA and a perspective on LDA that is more focused on the minimization of misclassification probabilities are well-known and will not be repeated here (see Kshiragar and Arseven, 1975).

There are two obvious ways that one can code the group membership in the matrix Y, either

$$Y = \begin{bmatrix} 1_{n_1 \times 1} & 0 & \cdots & 0 \\ 0 & 1_{n_2 \times 1} & \cdots & 0 \\ \vdots & \vdots & \ddots & \vdots \\ 0 & 0 & \cdots & 1_{n_g \times 1} \end{bmatrix} \quad \text{or} \quad Z = \begin{bmatrix} 1_{n_1 \times 1} & 0 & \cdots & 0 \\ 0 & 1_{n_2 \times 1} & \cdots & 0 \\ \vdots & \vdots & \ddots & \vdots \\ 0 & 0 & \cdots & 1_{n_{g-1} \times 1} \\ 0 & 0 & \cdots & 0 \end{bmatrix}.$$

For example, for the permeability data one could rationally choose

$$Y = \begin{bmatrix} 1_{177 \times 1} & 0_{177 \times 1} \\ 0_{177 \times 1} & 1_{177 \times 1} \end{bmatrix} \quad \text{or} \quad Z = \begin{bmatrix} 1_{177 \times 1} \\ 0_{177 \times 1} \end{bmatrix}.$$

Note that the sample "covariance" matrix S_y is $g \times g$ and rank $g - 1$, while S_z is $(g - 1) \times (g - 1)$ and rank $(g - 1)$. Regardless, the fact that both Y and Z are indicator matrices suggests that S_y and S_z will have special forms. Barker and Rayens (2003) showed the following:

$$S_z^{-1} = (n - 1)\left(\frac{1}{n_g} 1_{g-1} 1_{g-1}^T + M^{-1}\right),$$

where

$$M = \begin{bmatrix} n_1 & 0 & \cdots & 0 \\ 0 & n_2 & \cdots & 0 \\ \vdots & \vdots & \ddots & \vdots \\ 0 & 0 & \cdots & n_{g-1} \end{bmatrix}_{(g-1) \times (g-1)}, \quad S_y^c = \begin{bmatrix} S_z^{-1} & 0_{g-1} \\ 0_{g-1} & 0 \end{bmatrix}_{g \times g}.$$

With these simple expressions for S_z^{-1} and S_y^c, it is possible to relate constructs that are essential to PLS-LDA to Fisher's H matrix, as will be seen in the next section.

2.5.3   Connection between PLS and LDA

Since PLS is just a penalized version of CCA and CCA is, in turn, related to LDA (Bartlett, 1938), it is reasonable to expect that PLS might have some direct connection to LDA. Barker and Rayens (2003) gave a formal statistical explanation of this connection. In particular, recall that above the USC-PLS directions were associated with the eigenstructure of $S_{xy}S_{yx}$ or, for a classification application, $S_{xz}S_{zx}$, depending on how the membership matrix was coded. In either case, Barker and Rayens (2003) related the pertinent eigenstructure problem to H as follows:

$$S_{xy}S_{yx} = H^* = \frac{1}{n-1}\sum_{i=1}^{g} n_i^2(\bar{x}_i - \bar{x})(\bar{x}_i - \bar{x})^T,$$

$$S_{xz}S_{zx} = H^{**} = \frac{1}{n-1}\sum_{i=1}^{g-1} n_i^2(\bar{x}_i - \bar{x})(\bar{x}_i - \bar{x})^T.$$

It should be noted that when the Y block is coded with dummy variables, the presence of the Y-space penalty $(\mathrm{Var}(b^T y))$ in the definition of PLS does not seem to be all that appropriate, since Y-space variability is not meaningful. Barker and Rayens (2003) removed this Y-space penalty from the original objective function, reposed the PLS optimization problem, and were able to show that the essential eigenstructure problem that is being manipulated in this case is one that involves exactly H, and not merely H^* or H^{**}. However, in the case of two groups, as with our permeability data, it is not hard to show that

$$H^* = 2H^{**} = \frac{n-1}{2}\left(\frac{1}{n_1} + \frac{1}{n_2}\right)H$$

and, hence, that the eigenstructure problems are equivalent and all proportional to $(\bar{x}_1 - \bar{x})(\bar{x}_1 - \bar{x})^T$.

The practical upshot is simple: when one uses PLS to facilitate discrimination in the "obvious" way, with the intuition of "predicting" group membership from a training set, then the attending PLS eigenstructure problem is one that depends on (essentially) Fisher's among-groups sums-of-squares and cross-products matrix H. It is, therefore, no surprise that PLS should perform better than PCA for dimension reduction when discriminant analysis on the scores is the ultimate goal. This is simply because the dimension reduction provided by PLS is determined by among-groups variability, while the dimension reduction provided by PCA is determined by total variability. It is easy to implement PLS-LDA in SAS, and this is discussed next.

2.5.4   Logic of PLS-LDA

There are many rational ways that PLS can be used to facilitate discrimination. Consistent with how PCA has been used for this purpose, some will choose to simply plot the first two or three PLS-LDA "scores" and visually inspect the degree of separation, a process that is entirely consistent with the spirit of "territory plots" and classical CDA. The classification of an unknown may take place informally, say, by mapping it to the same two or three dimensional space occupied by the scores and then assigning it to the group that admits the closest mean score, where "closest" may be assessed in terms of Mahalanobis distance or Euclidean distance.

Alternately, the scores themselves may be viewed simply as new data that have been appropriately "prepared" for an LDA, and a full LDA might be performed on the scores, either with misclassification rates as a goal (the DISCRIM procedure) or visual separation and territory plots as a goal (the CANDISC procedure). Figure 2.10 helps illustrate the many options. In Section 2.5.5, we will discuss how to produce the PLS-LDA scores with the PLS procedure.

Figure 2.10 Schematic depicting the logic of PLS-LDA

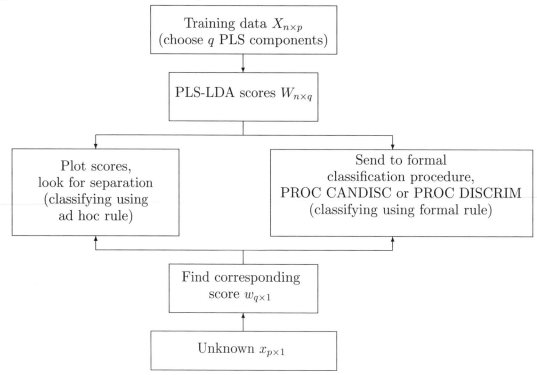

2.5.5 Application

As mentioned in the Introduction, drug discovery teams often generate data that contain both descriptor information and a categorical measurement. Often, the teams desire to search for a model that can reasonably predict the response. For this section, we apply PLS-LDA to the permeability data set (the PERMY data set can be found on the book's companion Web site) introduced in Section 2.2.

Recall that this data set contains 354 compounds that have been screened to assess permeability. While the original measure of permeability was a continuous value, scientists have categorized these compounds into groups of permeable and non-permeable. In addition, this data set contains an equal number of compounds in each group. Using in-house software, 71 molecular properties, thought to be related to compound permeability, were generated for each compound and were used as the descriptors for both modeling techniques.

This is not an uncommon drug discovery-type data set—the original continuous response is noisy and is subsequently reduced to a binary response. Hence, we fully expect some compounds to be incorrectly classified. Also, the descriptor set is computationally derived and is over-described. With these inherent problems, many techniques would have a difficult time finding a relationship between the descriptors and the response.

Implementing PLS-LDA in SAS

The purpose of this presentation is not to detail the many options that are well known in PROC DISCRIM or PROC CANDISC, or perhaps less well known in PROC PLS. Rather, this presentation is focused on simply showing the reader how to use PROC PLS in conjunction with, say, PROC DISCRIM to implement PLS-LDA as described above. To better facilitate this discussion,f we will use the permeability data, as described in the previous subsection. For this analysis, all observations are used in the training set and cross-validation is used to assess model performance.

Notice with subgroup matrices of size 177 by 71 one would not necessarily expect that any dimension reduction would be necessary and that PROC DISCRIM could be applied directly to the training set. However, many of the descriptors are physically related, so, we expect these descriptors to be correlated to some extent. In fact, each of the individual group covariance matrices, as well as the pooled covariance matrix, are 71 by 71 but only have rank 63. So, in theory, Fisher's linear discriminant analysis—whether in the form that focuses on misclassification probabilities (PROC DISCRIM) or visual among-groups separation (PROC CANDISC)—cannot be applied, owing to the singularity of these matrices. Hence, these data are good candidates for first reducing dimension and then performing a formal discriminant analysis.

There is an alternative that has to be mentioned, however. When confronted with a singular covariance matrix, PROC DISCRIM will issue an innocuous warning and then proceed to construct a classification rule anyway. It is not widely known, perhaps, but in this situation SAS will employ a so-called "quasi-inverse," which is a SAS innovation. To construct the quasi-inverse "small values" are arbitrarily added to zero variance directions, thereby forcing the corresponding covariance matrices to be nonsingular and preserving the basic rationale behind the classification paradigm. Again, the purpose of this illustration is not to compare dimension reduction followed by classification with direction classification using a quasi-inverse, but it is important that the reader be aware that SAS already has one method of dealing with collinear discriminant features.

Program 2.13 performs the PLS-LDA analysis of the permeability data using PROC PLS. The initial PROC DISCRIM is invoked solely to produce the results of using the quasi-inverse. Keep in mind that the Y variable is a binary variable in this two-group case, designed to reflect corresponding group membership. In this example the first ten PLS scores (XSCR1-XSCR10) are used for the discrimination (the LV=10 option is used to request ten PLS components). For clarity of comparison, the pooled covariance matrix was used in all cases.

Program 2.13 PLS-LDA analysis of the permeability data

```
ods listing close;
proc discrim data=permy list crosslist noprint;
    class y;
    var x1-x71;
    run;
ods listing;
proc pls data=permy method=simpls lv=10;
    model y=x1-x71;
    output out=outpls xscore=xscr;
proc print data=outpls;
    var xscr1 xscr2;
    run;
ods listing close;
proc discrim data=outpls list crosslist noprint;
    class y;
    var xscr1-xscr10;
    run;
ods listing;
```

Output from Program 2.13

```
                           The PLS Procedure

                Percent Variation Accounted for by SIMPLS Factors

        Number of
        Extracted          Model Effects          Dependent Variables
        Factors        Current       Total        Current       Total

              1        26.6959      26.6959        18.5081      18.5081
              2        16.6652      43.3611         2.9840      21.4921
              3         8.3484      51.7096         3.0567      24.5488
              4         3.5529      55.2625         3.2027      27.7515
              5         6.2873      61.5498         0.8357      28.5872
              6         4.0858      65.6356         0.8749      29.4621
              7         3.5761      69.2117         0.8217      30.2838
              8         7.0819      76.2935         0.4232      30.7070
              9         2.2274      78.5209         0.9629      31.6699
             10         4.7272      83.2481         0.4038      32.0737


              Obs        xscr1        xscr2

               1       -4.2733      -0.78608
               2       -3.7668      -5.29011
               3        2.0578       8.04445
               4       -1.0728      -0.52324
               5       -5.0157       2.37940
               6        1.7600      -0.93071
               7       -5.2317       3.22951
               8       -4.2919      -6.71816
               9       -6.3754      -3.81705
              10        0.8542      -0.96708
```

Output 2.13 displays a "variance summary" produced by PROC PLS as well as a listing of first ten values of the two PLS components (XSCR1 and XSCR2). In general, navigating and understanding the PROC PLS output was discussed in great detail by Tobias (1997a, 1997b), and there is little need for us to repeat that discussion here. Rather, our purpose is simply to extract and then access the PLS scores for purposes of performing either a formal or ad hoc classification analysis. However, it is relevant to see how well the original feature space has been summarized by these extracted components.

When ten PLS components are requested, the "variance summary" in Output 2.13 appears for the permeability data. Notice that 26.70% of the variability in the model effects (X-space) was summarized by the first PLS component and 43.36% was summarized by the first two, etc. It may take as many as 20 components to adequately summarize the descriptor space variability. The right-most columns have recorded the correlation between the X-space score and the categorical response. For a discriminant application, this is not meaningful since the group coding is arbitrary.

For purposes of this chapter, the X-space scores XSCR values are what we are primarily interested in (see Figure 2.10) since it is these new data that are then transferred to PROC DISCRIM for classification. These scores appear on the PROC PLS output as shown (for two components) below. The results of the classification, summarized in terms of each group's misclassification ("error") rate, for various numbers of PLS components, are displayed in Table 2.6. These error rates are the so-called "apparent" misclassification rates and are completely standard in discriminant analysis and can be read directly off the default SAS output. Both the error rates for a simple reclassification of the training set (called "re-substitution" rates on the SAS output) and the leave-one-out cross-validated rates (called "cross-validated" rates on the SAS output) are reported. In all cases the pooled covariance matrix was used for the (linear) classification.

Table 2.6 Error Rates for PLS and PCA Followed by PROC DISCRIM on Component Scores (bolded entries are the minimum (observed) misclassification rates)

| Number of components | Re-substitution method (Cross-validation method) | | | | | |
| | Total error rate | | 0-class error rate | | 1-class error rate | |
	PLS	PCA	PLS	PCA	PLS	PCA
2	0.3136	0.3390	0.2881	0.3220	0.3390	0.3559
	(0.3192)	(0.3418)	(0.2938)	(0.3277)	(0.3446)	(0.3559)
10	0.2345	0.2966	0.2599	0.2768	0.2090	0.3164
	(**0.2514**)	(0.3192)	(0.2655)	(0.2994)	(0.2373)	(0.3390)
20	**0.2090**	0.2885	0.2260	0.2768	0.1921	0.2881
	(0.2571)	(0.3390)	(0.2881)	(0.2994)	(0.2260)	(0.3785)
30	0.2401	0.2345	0.2260	0.2429	0.2542	0.2260
	(0.3079)	(**0.2853**)	(0.3051)	(0.2881)	(0.3107)	(0.2825)
40	0.2232	**0.2345**	0.2203	0.2712	0.2260	0.1977
	(0.3136)	(0.3220)	(0.3164)	(0.3616)	(0.3107)	(0.2825)
50	0.2429	0.2458	0.2316	0.2599	0.2542	0.2316
	(0.3333)	(0.3333)	(0.3333)	(0.3672)	(0.3333)	(0.3277)
PROC DISCRIM with quasi-inverse	**0.2373** (**0.3192**)		0.2316 (0.2938)		0.2429 (0.3446)	

The following assessments are evident from Table 2.6:

- The cross-validated estimates of total misclassification rates suggest about a 25-30% rate. This rate is fairly consistent with the non-cross-validated direct re-substitution method, although this latter method is overly optimistic, as expected.

- Practically stated, the permeable and non-permeable compounds are not well separated, even with quadratic boundaries, and the corresponding misclassification rates can be expected to be fairly high in practice if one were to use the classification rule that would result from this analysis.

- PROC DISCRIM with the quasi-inverse is recorded on the last line of the table. Notice that the cross-validated misclassification rates for this alternative appear to do no better, perhaps worse, than PLS-LDA.

- As expected, PLS-LDA does a better job (with the exception of 30 components) than does PCA followed by an LDA. This is not really a surprise, of course, since the PLS dimension reduction step basically involved maximizing among-groups differences.

2.5.6 Summary

In Section 2.5 we have briefly reviewed the formal sense in which an empirically obvious use of PLS for the purposes of discrimination is actually optimal. This formal connection is easily turned into a functional paradigm using PROC PLS. Direct connections to the theory developed by Barker and Rayens (2003) can be had by invoking the SIMPLS option in PROC PLS, but the connections would certainly hold (on an intuitive level) for other realizations of PLS as well.

Again, what one does with the PLS scores after they are obtained seems to vary widely by user. Some will want to then perform a formal discriminant analysis (as we did in the example), while others will consider the problem "worked" at this point and simply do ad hoc classification at the level of the scores, perhaps by identifying the sample group mean score that is closest to an unknown observation.

The real point to the work of Barker and Rayens (2003) is that when a classical discrimination is desired, but can't be performed owing to singular covariances (either within or pooled), then initially reducing the dimension of the problem by identifying optimal linear combinations of the original features is a rational and well-accepted approach. In fact, it is common to use PCA to accomplish this first step. One of the points made clear by Barker and Rayens (2003) is that PLS can always be expected to do a better job at this initial reduction than can PCA. It is an open question as to whether PLS used in this fashion would outperform the SAS use of a quasi-inverse. One of the advantages enjoyed by PLS-LDA (followed by PROC DISCRIM, perhaps) over using the quasi-inverse is that the sense in which these activities are optimal is well understood. It also would be interesting to know if employing the quasi-inverse leads to understated non-cross-validated misclassification rates in general, or if that is just a characteristic that is somehow specific to this data set. That issue is a matter for future research, however, and will not be discussed further here.

2.6   Summary

Both boosting and partial least squares for linear discriminant analysis can be easily implemented using widely available SAS modules. The theoretical properties and performance of each of these methods have been thoroughly researched by a number of authors. In this chapter, we have provided additional details about each method, have provided code to implement each method, and have illustrated their application on a typical drug discovery data set. Undoubtedly, as data structures become more complex, methods like these become extremely valuable tools for uncovering relationships between descriptors and response classification.

References

Barker, M., Rayens, W.S. (2003). "A partial least squares paradigm for discrimination." *Journal of Chemometrics.* 17, 166–173.

Bartlett, M.S. (1938). "Further aspects of the theory of multiple regression." *Proceedings of the Cambridge Philosophical Society.* 34, 33–40.

Bauer, E., Kohavi, R. (1999). "An empirical comparison of voting classification algorithms: Bagging, boosting, and variants." *Machine Learning.* 36, 105–142.

Breiman, L. (1996). "Bagging predictors." *Machine Learning.* 26, 123–140.

Breiman, L. (1998). "Arcing classifiers." *The Annals of Statistics.* 26, 801–824.

Breiman, L., Friedman, J., Olshen, R., Stone, C. (1984). *Classification and Regression Trees.* Wadsworth.

Frank, I., Friedman, J. (1993). "Statistical view of chemometric regression tools." *Technometrics.* 35, 109–135.

Freund, Y. (1995). "Boosting a weak learning algorithm by majority." *Information and Computation.* 121, 256–285.

Freund, Y., Mansour, Y., Schapire, R.E. (2001). "Why averaging classifiers can protect against overfitting." *Proceedings of the Eighth International workshop on Artificial Intelligence and Statistics.* 1–8.

Freund, Y., Schapire, R.E. (1996a). "Game theory, on-line prediction and boosting." *Proceedings of the Ninth Annual Conference on Computational Learning Theory.* 325–332.

Freund, Y., Schapire, R.E. (1996b). "Experiments with a new boosting algorithm." *Machine Learning: Proceedings of the Thirteenth International Conference.* 148–156.

Freund, Y., Schapire, R.E. (2000). "Discussion of the paper 'Additive logistic regression: A statistical view of boosting' by Jerome Friedman, Trevor Hastie and Robert Tibshirani." *The Annals of Statistics.* 38, 391–393.

Friedman, J. (1989). "Regularized discriminant analysis." *Journal of the American Statistical Association.* 405, 165–175.

Friedman, J., Hastie, T., Tibshirani, R. (2000). "Additive logistic regression: A statistical view of boosting." *The Annals of Statistics.* 38, 337–374.

Greene, T., Rayens, W.S. (1989). "Partially pooled covariance matrix estimation in discriminant analysis." *Communications in Statistics.* 18, 3679–3702.

Hastie, T. J., Buja, A., Tibshirani, R. (1995). "Penalized discriminant analysis." *Annals of Statistics.* 23, 73–102.

de Jong, S. (1993). "SIMPLS: An alternative approach to partial least squares regression." *Chemometrics and Intelligent Laboratory Systems.* 18, 251–263.

Kansy, M., Senner, F., Gubernator, K. (1998). "Physicochemical high throughput screening: Parallel artificial membrane permeation assay in the description of passive absorption processes." *J. Med. Chem.* 41, 1007–1010.

Kearns, M., Valiant, L.G. (1989). "Cryptographic limitations on learning Boolean formulae and finite automata." *Proceedings of the Twenty-First Annual ACM Symposium on Theory of Computing.* New York, NY: ACM Press.

Krieger, A., Long, C., Wyner, A. (2001). "Boosting noisy data." *Proceedings of the Eighteenth International Conference on Machine Learning.* 274–281.

Lavine, B.K., Davidson, C.E., Rayens, W.S. (2004). "Machine learning based pattern recognition applied to microarray data." *Combinatorial Chemistry and High Throughput Screening.* 7, 115–131.

Optiz, D., Maclin, R. (1999). "Popular ensemble methods: An empirical study." *Journal of Artificial Intelligence Research.* 11, 169–198.

Rayens, W.S., Greene, T. (1991). "Covariance pooling and stabilization for classification." *Computational Statistics and Data Analysis.* 11, 17–42.

Rayens, W.S. (1990). "A role for covariance stabilization in the construction of the classical mixture surface." *Journal of Chemometrics.* 4, 159–170.

Ripley, B.D. (1996). *Pattern Recognition and Neural Networks.* Cambridge: Cambridge University Press.

Schapire, R.E. (1990). "The strength of weak learnability." *Machine Learning.* 5, 197–227.

Schapire, R.E. (2002). "Advances in boosting." *Uncertainty in Artificial Intelligence: Proceedings of the Eighteenth Conference.* 446–452.

Schapire, R.E., Freund, Y., Bartlett, P., Lee, W.S. (1998). "Boosting the margin: A new explanation for the effectiveness of voting methods." *The Annals of Statistics.* 26, 1651–1686.

Tobias, R. (1997a). "An introduction to partial least squares regression." TS-509, SAS Institute Inc., Cary, NC. Available at http://www.sas.com/rnd/app/papers/pls.pdf.

Tobias, R. (1997b). "Examples using the PLS procedure." SAS Institute Inc., Cary, NC. Available at http://www.sas.com/rnd/app/papers/plsex.pdf.

Vapnik, V. (1996). *The Nature of Statistical Learning Theory.* New York: Springer-Verlag.

Wold, H. (1966). *Estimation of principal components and related models by iterative least squares. Multivariate Analysis.* New York: Academic Press.

Wold, H. (1981). *Soft modeling: The basic design and some extensions. Systems under indirect observation, causality-structure-prediction.* Amsterdam: North Holland.

Chapter 3

Model Building Techniques in Drug Discovery

Kimberly Crimin
Thomas Vidmar

This chapter focuses on techniques used in the early stages of drug discovery to optimize the model building process. The main topics discussed in this chapter are selecting training and test sets, selecting variables, and determining when the prediction of a new observation is valid. Statistical procedures that are used to build models are also reviewed. For each of the main topics we provide background, discuss a new procedure, and provide guidance on implementation in SAS. We apply these methods to a real drug discovery data set.

3.1 Introduction

Computational models are used in many stages of drug discovery to predict structure activity and property relationships. The models built are based on the principle that compounds with similar structures are expected to have similar biological activities (Golbraikh and Tropsha, 2002). In the early stage of drug discovery, high throughput screens are used to identify a compound's activity against a specific target. Computational models used at this stage need to predict a compound's activity but the model does not need to be interpretable. Once a series of compounds has been identified as being active against the biological target, then the goal is to optimize the compound in terms of structure properties such as solubility, permeability, etc. At this stage, it is useful to have a computational model that can be interpreted by the scientists so they understand the impact on the property when a particular feature of the structure is changed. For example, suppose solubility has an inverse relationship with a certain feature of the structure, such

Kimberly Crimin is Senior Principal Biostatistician II, Early Development, Wyeth, USA. Thomas Vidmar is Senior Director, Midwest Nonclinical Statistics, Pfizer, USA.

as number of rotatable bonds; if the goal is to increase solubility then the team knows that decreasing the number of rotatable bonds will increase solubility. In this chapter we focus on methods the statistician can use to optimize the model building process. As George Box (Box et al., 1978) once said "All models are wrong, but some are useful".

To build a model, a statistician will perform the following steps: divide the data into training and test sets, select the variables, and select the best statistical tool for prediction. There exists a vast body of literature on statistical procedures for model building, a few of which we discuss here. In this chapter, we focus on statistical techniques that can optimize the model building process independent of the statistical procedure used to build the model.

A computational model is evaluated on the ability to predict future observations. One measure of this performance is the prediction error of the model calculated on an independent test set; therefore, it is important that the test set be representative of the training set. Otherwise, the prediction error obtained from the test set will not be indicative of the model's performance.

Molecular descriptors describe geometric, topological, and electronic properties of the compound. The molecular descriptors used to build structure activity and property relationship models are often computer generated and many programs are available to generate molecular descriptors. Because of this, the modeler is often faced with hundreds, if not thousands, of descriptors. Since the molecular descriptors are based on attributes of the structure many of these descriptors are highly correlated. To build a model that is interpretable, the goal of variable selection is to choose descriptors that are highly predictive of the response and independent of one another. A computational model built with independent descriptors will be more interpretable than a model built with correlated descriptors.

When using a model to predict a response for a new set of data, it is important to know how different the new data are from the data used to train the model. If the new set of data is "far" from the data used to train and test the model, then one might consider retraining the model because the error in the prediction will be large.

In Section 3.2 we provide an example using a real drug discovery data set. In Section 3.3 we review methods for training and test set selection, and we discuss a novel technique that we found useful. In Section 3.4 we provide an overview of variable selection techniques and provide a new technique. In Section 3.5 we briefly review statistical procedures for model building. In Section 3.6 we discuss a simple procedure that can be used to determine if a new observation is in the descriptor space the model was trained on. And in Section 3.7 we use SAS Enterprise Miner to build a computational model.

To save space, some SAS code has been shortened and some output is not shown. The complete SAS code and data sets used in this book are available on the book's companion Web site at `http://support.sas.com/publishing/bbu/companion_site/60622.html`.

3.2 Example: Solubility Data

Solubility is an important physical-chemical property in drug discovery. In the simplest definition, the solubility of a solute is the maximum amount of the solute that can dissolve in a certain amount of solvent. Intrinsic solubility is the solubility of the neutral form of the salt; the salt has not separated into ions. Some of the factors affecting solubility are: temperature, molecular weight, number of hydrogen bonds, and polarity.

For a solvent to dissolve a solute, particles of the solvent must be able to separate the particles of the solute and occupy the intervening spaces. Water is a polar substance and polar solvents can generally dissolve solutes that are ionic. Dissolving takes place when the solvent is able to pull ions out of their crystal structure. Separation of ions by the action of a solvent is called dissociation; ions are dissociated by the water molecules and spread evenly throughout the solution.

Estimating solubility then becomes a bit more complicated when compounds have an ionized group because multiple forms of the compound exist with varying solubilities. The term *aqueous solubility* refers to the solubility of all forms of a compound. It is the sum of the solubility for the neutral compound plus the solubility of each ionized form of the compound; aqueous solubility is, therefore, a function of pH. Since most drugs are either weakly acidic or basic, it is important to use the intrinsic solubility when building a computational model to predict solubility.

The example that we use in this chapter is a solubility data set (SOLUBILITY data set can be found on the book's companion Web site). The solubilities were estimated using the old-fashioned method. An excessive amount of the compound was added to a flask of water; the flask was shaken and the amount of remaining compound measured. The amount of the compound dissolved is the solubility of that compound. In this data set, there are 171 compounds. To limit the complexity of the example, only one software package was used to generate the molecular descriptors. The goal is to build an interpretable computational model that predicts the solubility of new compounds.

3.3   Training and Test Set Selection

In order to build reliable and predictive models, care must be taken in selecting training and test sets. When the test set is representative of the training set, one can obtain an accurate estimate of the model's performance. Ideally, if there is sufficient data, the modeler can divide the data into three different data sets: training set, validation set, and test set. The training set is used to train different models. Then each of these models is applied to the validation set and the model with the minimum error is selected as the final model and applied to the test data set to estimate the prediction error of the model. More often than not, there is insufficient data to apply this technique.

Another method used to calculate the model prediction error is *cross-validation*. In this method, the data are divided into K parts, each approximately the same size. For each of the K parts, the model is trained on the remaining $K-1$ parts and then applied to the Kth part and the prediction error calculated. The final estimate of the model prediction error is the average prediction error over all K parts. The biggest question associated with this method is how to choose K. If K is chosen to be equal to the sample size, then the estimate of error will be unbiased, but the variance will be high. As K decreases, the variance decreases but the bias increases, so one needs to find the appropriate trade-off between bias and variance. See Hastie, Tibshirani, and Friedman (2001) for more information.

Another method which is similar to cross-validation is the *bootstrap method*. This method takes a bootstrap sample from the original data set and uses the data to train the model. Then, to calculate the model prediction error, the model is applied to either the original data set or to the data not included in the bootstrap sample. This procedure is repeated a large number of times and the final prediction error is the average prediction error over all bootstrap samples. The main issue with this method is whether the original data set is used to calculate the prediction error or only the observations not included in the bootstrap sample. When the original sample is used, the model will give relatively good predictions because the bootstrap sample (training set) is very similar to the original sample (test set). One solution to reduce this bias is to use the "0.632 estimator". For more information on this method see Efron and Tibshirani (1998).

3.3.1   Proposed Procedure for Selecting Training and Test Sets

The methods discussed above are easy to implement and will give adequate estimates of the prediction error provided there are no outlying points in either the response space or the descriptor space. Often in drug discovery data sets there are points of high influence or outliers, so excluding them from the training set is an important consideration. Therefore, we developed a training and test set selection procedure that limits the number of outliers and points of influence in the training set. In this section we provide an overview of the proposed method and apply it to the solubility data set.

Our goal is to develop a training and test set selection algorithm that accomplishes the following goals:

- Select a test set that is representative of the training set.
- Minimize the number of outliers and influential points in the training set.

Suppose a data set contains only one descriptor. Then it would be fairly easy to come up with an algorithm that meets the above goals. For instance, the modeler could use a two-dimensional plot to identify outliers and points of influence, assign these to the test set, and then take a random sample of the remaining points and assign them to the training set. To view the data set selection, the modeler could again use a two-dimensional plot and if the test set did not visually appear to be representative of the training set, the modeler could take another random sample.

The above procedure is simple and easy to explain to scientists and accomplishes the two goals when $p = 1$, but it is not easily extendable for $p > 1$ because the modeler would have to view $\binom{p+1}{2}$ two-dimensional projections to identify outliers and points of influence. Furthermore, selecting a representative sample in $p + 1$-dimensional space would be impossible unless the following were true: instead of using each descriptor, a summary measure for each observation in the descriptor space were used, for instance, the leverage value of each observation. Recall that the leverage value for the ith observation is the distance between the observation and the center of the descriptor space. By using the leverage values and the responses, we have reduced the space down to 2 from $p + 1$ and the algorithm discussed above (when $p = 1$) could be used to develop an algorithm for training and test set selection.

The basic algorithm for the proposed procedure is defined as follows:

1. Calculate the leverage value for each observation.
2. Bin the leverage values and the responses and then combine the bins into cells. Two possible ways to calculate the bin size are discussed below.
3. Potential outliers and points of influence are identified as points outside a $(1 - \alpha)\%$ confidence region around the responses and the leverage values. These points are assigned to the test set.
4. For each cell, take a random selection of the points within the cell and assign them to the training set; assign the other points to the test set.

To enhance the above algorithm, the user can select how the bin size is calculated. One option, referred to as the *normal bin width*, defines the bin width to be:

$$W = 3.49\sigma N^{-1/3},$$

where N is the number of samples and σ is the standard deviation of either the responses or the leverages. The second option, referred to as *robust bin width*, defines the bin width to be:

$$W = 2(IQR)N^{-1/3},$$

where IQR is the interquartile range of either the response or the leverage values and N is defined above. We have found that the robust bin width results in more cells being defined, and we suggest the normal bin width be used if the data are sparse. For discussions on the optimal bin size see Scott (1979).

After running the basic algorithm on a number of examples, we added two additional features to improve it. The user has an option of pre-allocating particular observations to either the training or test data sets. The other feature we added allows the user to specify the seed used to initialize the random number generator. This allows the user to get the same results on multiple runs.

3.3.2 TRAINTEST Module

The algorithm described above was implemented in a SAS/IML module (TRAINTEST module) that can be found on the book's companion Web site. There are seven inputs to this module:

- the matrix of descriptors
- the vector of responses
- the type of bins to use (normal or robust)
- the desired proportion of observations in the training set
- a vector indicating the data set each observation should be in (if the user doesn't want to pre-allocate any observations, then this vector should be initialized to a vector of zeros)
- the level of the confidence region around the leverage values and the response
- the seed

(If the seed is set to zero, then the module allows SAS to initialize the random number generator; SAS uses the system time for initialization.)

The TRAINTEST module produces three outputs plus a visual display of the training and test set selections. The three outputs are: a vector indicating the data set each observation belongs to, the vector of leverage values, and the observed proportion of observations in the training set. Finally, the module creates a SAS data set (PLOTINFO) that can be used to create a plot of leverage values versus responses. The PLOTTYPE variable in this data set serves as a label for individual observations: PLOTTYPE=1 if the observation is in the training set, PLOTTYPE=2 if the observation is in the test set, and PLOTTYPE=3 if the observation is an outlier.

3.3.3 Analysis of Solubility Data

The SOLUBILITY data set that can be found on the book's companion Web site has 171 observations (compounds) and 21 possible descriptors. The response variable, the log of the measured solubility of the compound, is continuous. The objective is to divide the data set into two data sets, one for training the model and the other for testing the model. Based on the size of the data set, we wanted approximately 75% of the observations to be allocated to the training set and we considered any point outside the 95% confidence region to be an outlier. Program 3.1 contains the SAS code to run the training and test set selection module on the solubility data.

Program 3.1 first reads in the SOLUBILITY data set and stores the descriptors into a matrix **X** and the response into a vector **y**. After that, it calls the TRAINTEST module and, based on the allocations returned from the module, the full data set is divided into the training and test sets.

Program 3.1 Training and test set selection in the solubility example

```
proc iml;
    * Read in the solubility data;
    use solubility;
    read all var ('x1':'x21') into x;
    read all var {y} into y;
```

```
* Initialize the input variables;
type=1;
prop=0.75;
level=0.95;
init=j(nrow(x),1,0);
seed=7692;
* Next two lines of code are not executed on the first run,;
* only the second run to allocate certain observations to the training set;
*outindi={109, 128, 139, 165};
*init[outindi,1]=1;
call traintest(ds,hi,obsprop,x,y,type,prop,init,level,seed);
print obsprop[format=5.3];
* Using data set definitions returned in DS, create two data sets;
trindi=loc(choose(ds=1,1,0))`;
tsindi=loc(choose(ds=2,1,0))`;
trx=x[trindi,]; try=y[trindi,1]; trxy=trx||try;
tsx=x[tsindi,]; tsy=y[tsindi,1]; tsxy=tsx||tsy;
* Number of observations in the train and test data set;
ntrain=nrow(trxy); ntest=nrow(tsxy);
print ntrain ntest;
create soltrain from trxy[colname=('x1':'x21'||'y')];
append from trxy;
close soltrain;
create soltest from tsxy[colname=('x1':'x21'||'y')];
append from tsxy;
close soltest;
quit;
* Vertical axis;
axis1 minor=none label=(angle=90 "Response")  order=(-7 to 5 by 2) width=1;
* Horizontal axis;
axis2 minor=none label=("Leverage") order=(0 to 0.8 by 0.2) width=1;
symbol1 i=none value=circle color=black height=5;
symbol2 i=none value=star color=black height=5;
symbol3 i=none value=dot color=black height=5;
proc gplot data=plotinfo;
    plot response*leverage=plottype/vaxis=axis1 haxis=axis2 frame nolegend;
    run;
    quit;
```

Output from first run of Program 3.1

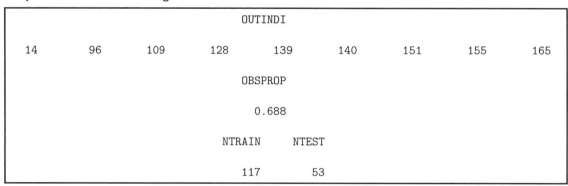

		OUTINDI						
14	96	109	128	139	140	151	155	165

OBSPROP

0.688

NTRAIN	NTEST
117	53

Output 3.1 lists the IDs of the outlying observations (OUTINDI), observed proportion (OBSPROP) and number of observations in the training and test sets (NTRAIN and NTEST) after the first run of Program 3.1. Figure 3.1 contains the plot of the points assigned to each data set.

Figure 3.1 Initial assignment of observations, training set (circle), test set (star) and outlier (dot)

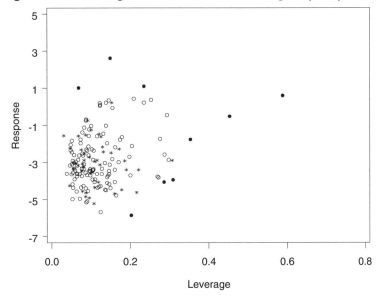

Output from second run of Program 3.1

	OUTINDI			
14	96	140	151	155

OBSPROP

0.712

NTRAIN	NTEST
121	49

The first run of the program allocated 117 observations (68.8%) to the training set and 53 observations to the test set. The proportion of the observations assigned to the training set was a bit lower than the desired proportion. Upon visual inspection of the data set definitions, Figure 3.1, it was decided to run the program again only this time allocating 4 observations identified as outliers to the training set (observations 109, 128, 139 and 165). Figure 3.2 contains the plot after the second run of Program 3.1 and the output shows the IDs of the outlying observations, observed proportion, and number of observations in each data set after the second run. After the second run, 121 observations (71.2%) were assigned to the training set and 49 observations to the test set.

3.4   Variable Selection

The data used to develop computational models can often be characterized by a few observations and a large number of measured variables, some of which are highly correlated. Traditional approaches to modeling these data are principle component regression (PCR) and partial least squares (PLS) (Frank and Friedman, 1993). The factors obtained from PCR and PLS are usually not interpretable, so if the goal is to develop an interpretable model, these are not as useful. By considering the loadings given to each

Figure 3.2 Final assignment of observations, training set (circle), test set (star) and outlier (dot)

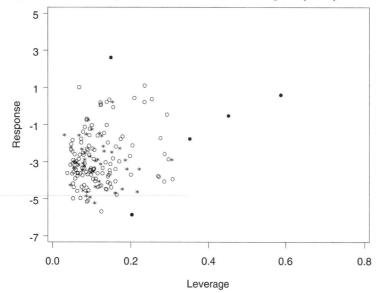

variable, these methods can be used to prune the variables, although the criteria for removing variables can be inconsistent between data sets.

Other popular variable selection procedures are tied to model selection and are often used when fitting multiple linear regression models, although they can be used with other models, such as a linear discriminant model. A few of the more popular procedures are discussed below. Shrinkage methods are also available to the modeler. Instead of selecting variables, shrinkage methods shrink the regression coefficients by minimizing a penalized error sum of squares, subject to a constraint. In *ridge regression*, a common shrinkage method, the L_2 norm is used as the constraint, and *Lasso*, another shrinkage method, the L_1 norm is used as the constraint.

Best K models examine all possible models and the top K models are selected for further investigation by the modeler. The top models are selected using a measure of fit such as adjusted R^2 (R_a^2) or Mallow's C_p. Since all possible models are examined, the sample size, n, must be strictly greater than the number of descriptors, p. Examining all possible models is computationally intensive and, realistically, this method should be used only if p is small. A rule of thumb is $p < 11$.

Best subset algorithms, such as *leaps and bounds* (Furnival and Wilson, 1974), do not require the examination of all possible subsets, but use an efficient algorithm to search for the K best models using a statistical criteria such as R_a^2 or Mallow's C_p. It is recommended that other variable selection procedures be used if $p > 60$.

Stepwise procedures, including forward selection and backward elimination, develop models at each step by adding (forward selection) or deleting (backward elimination) the variable that has the largest (forward selection) or smallest (backward elimination) F-value meeting a particular pre-determined limit. Stepwise procedures are commonly used to build computational models.

The variable selection methods mentioned above are dependent on the type of model being fit and often select variables for the final model which are highly correlated. This is not an issue unless the modeler is interested in an interpretable model. In the following section we present a variable selection method that is based on a stable numerical linear algebra technique. Our technique ranks the predictor variables in terms of importance, taking into account the correlation with the response and the predictor variables previously selected. Our technique deals with the collinearity problem by separating correlated variables during the ranking. Our procedure is a modification of the variable selection procedure proposed by Golub and Van Loan (1996) and is not dependent on the modeler pre-specifying the type of model.

3.4.1   Variable Selection Algorithm

The proposed variable selection algorithm is defined below:

1. Center the matrix of descriptors, **X**.
2. Center the vector of responses, **Y**.
3. Calculate $\hat{\mathbf{Y}}$, where each column is the projection of y onto x_i, where x_i' is the ith row of **X**.
4. Use the Singular-Value Decomposition to calculate the rank of $\hat{\mathbf{Y}}$.
5. Apply the QR decomposition with column-pivoting to $\hat{\mathbf{Y}}$.

The output from the QR decomposition algorithm when column pivoting is invoked is an ordering of the columns of $\hat{\mathbf{Y}}$. The number of descriptors to use in the model can be obtained from the magnitude of the singular values of the ordered descriptor matrix. This is similar to using a scree plot in PCR to determine the number of directions to use.

The QR decomposition of the matrix **X** is $X = QR$, where Q is an orthogonal matrix and R is upper triangular. The matrix **Q** if formed from the product of p Householder matrices, $\mathbf{H}_i$, $i = 1, \ldots, p$. A Householder matrix is an elementary reflector matrix. When column-pivoting is incorporated into the QR algorithm, prior to calculating each Householder matrix, the column with the largest norm is moved to the front of the matrix. Since the matrix on which the QR algorithm operates is $\hat{\mathbf{Y}}$, at each stage the column moved to the front of the matrix will correspond to the descriptor with the highest correlation with y but relatively independent of the previously selected descriptors, x_i. In other words, the first column selected will be the descriptor that has the highest correlation with the response. The second column selected will be the descriptor with the highest correlation with the response and the lowest correlation with the first descriptor selected. This continues until all the descriptors have been ordered.

3.4.2   ORDERVARS Module

The variable ordering procedure described in Section 3.4.1 was implemented in a SAS/IML module (ORDERVARS module) that can be found on the book's companion Web site. The inputs to this module are the matrix **X** of descriptors and the vector **y** of responses. The outputs are the variable ordering, the rank of X, and a scree plot of the singular values.

The ORDERVARS module starts by centering X and y and then calculates the matrix $\hat{\mathbf{Y}}$. The QR-algorithm implemented in SAS/IML (QR subroutine) is called. Column-pivoting is invoked in the QR subroutine by initializing the input vector ORD to all zeros. The ORDERVARS module also creates a SAS data set (SINGVALS) that stores the computed singular values.

3.4.3   Example: Normally Distributed Responses

As an illustrative example, suppose x is distributed $N_5(0, \Sigma)$, where

$$\Sigma = \begin{bmatrix} 1 & 0.75 & 0.05 & 0 & 0 \\ 0.75 & 1 & 0 & 0 & 0 \\ 0.05 & 0 & 1 & 0.98 & 0 \\ 0 & 0 & 0.98 & 1 & 0 \\ 0 & 0 & 0 & 0 & 1 \end{bmatrix}.$$

In this example, x_1 is correlated with x_2 and x_3 is highly correlated with x_4. Take 25 samples from the above distribution and define

$$y = 2x_1 + 1.5x_3 + x_5 + \epsilon,$$

where ϵ is distributed $N(0,1)$. The response is a linear function of three of the five descriptors, namely, x_1, x_3, x_5. A variable selection procedure should select 1, 3, 5 as the first three variables.

Program 3.2 provides the code to randomly generate the descriptors for this example and calls the ORDERVARS module on the simulated data. The program starts by initializing the covariance matrix, randomly generating the matrix of descriptors, and then calculating y. To randomly generate data from a multivariate distribution with mean μ and positive-definite covariance matrix Σ, the following transformation was used

$$x = \Sigma^{1/2}z + \mu,$$

where z is a $p \times 1$ vector of observations, z_i distributed $N(0,1)$, and the matrix $\Sigma^{1/2}$ is the square root matrix of Σ. The square root matrix of a positive definite matrix can be calculated from the spectral decomposition of Σ as:

$$\Sigma^{1/2} = CD^{1/2}C',$$

where $\mathbf{C}$ is the matrix of eigenvectors and $\mathbf{D}$ is a diagonal matrix whose elements are the square-root of the eigenvalues.

Program 3.2 Variable ordering in the simulated example

```
proc iml;
   * Initialize Sigma;
   call randseed(564);
   sig={1 0.75 0.05 0 0,
       0.75 1 0 0 0,
       0.05 0 1 0.98 0,
       0 0 0.98 1 0,
       0 0 0 0 1};
   * Calculate sigma^{1/2};
   call eigen(evals,evecs,sig);
   sig12=evecs*diag(evals##0.5)*evecs`;
   * Generate the matrix of descriptors, x;
   n=25;
   x=j(n,5,0);
   do i=1 to n by 1;
   d=j(1,5,.);
   call randgen(d,'normal',0,1);
   x[i,]=(sig12*d`)`;
   end;
   err=j(1,n,.);
   * Randomly generate errors and calculate y;
   call randgen(err,'normal',0,1);
   y=j(n,1,0);
   y=2*x[,1] + 1.5*x[,3] + x[,5] + err`;
   * Concatenate x and y into one matrix;
   xy=x||y;
   * Calculate the correlation matrix;
   cormat=j(6,6,.);
   cormat=corr(xy);
   print cormat[format=5.3];
```

```
* Save example data set;
create test from xy[colname={'x1' 'x2' 'x3' 'x4' 'x5' 'y'}];
append from xy;
close test;
use test;
read all var ('x1':'x5') into x;
read all var {y} into y;
* Call the variable selection procedure;
call ordervars(x,y,order,rank);
* Print rank and variable ordering *;
print rank;
print order;
quit;
```

Output from Program 3.2

```
                          CORMAT

         1.000  0.808  0.129  0.044  0.175  0.746
         0.808  1.000  0.014 -.037  0.062  0.511
         0.129  0.014  1.000  0.977  0.294  0.603
         0.044 -.037  0.977  1.000  0.308  0.550
         0.175  0.062  0.294  0.308  1.000  0.576
         0.746  0.511  0.603  0.550  0.576  1.000

                           RANK

                            5

                          ORDER

         1        3        5        2        4
```

After randomly generating the data, the correlation matrix is calculated for the matrix $[Xy]$. The correlation matrix is given in the Output 3.2. The last row of the correlation matrix corresponds to y. Looking at the correlations, we see that x_1 has the highest correlation with the response y and should be selected first by the variable ordering procedure. Output 3.2 also contains the variable ordering produced by the ORDERVARS module. The ordering of the variables is: x_1, x_3, x_5, x_2, and x_4.

3.4.4   Forward Selection

Let's compare the above results to the results obtained from forward selection. Program 3.3 contains the code to run forward selection in SAS using the REG procedure. The TEST data set used in this program was created in Program 3.2.

Program 3.3 Forward selection on simulated data

```
proc reg data=test;
    model y = x1 x2 x3 x4 x5/selection=forward;
    run;
```

Output from Program 3.3

```
              All variables have been entered into the model.

                      Summary of Forward Selection

          Variable    Number    Partial    Model
   Step   Entered     Vars In   R-Square   R-Square    C(p)     F Value   Pr > F

     1    x1            1        0.5572     0.5572    93.1599    28.94    <.0001
     2    x4            2        0.2678     0.8250    26.1153    33.67    <.0001
     3    x5            3        0.0956     0.9206     3.4650    25.29    <.0001
     4    x2            4        0.0036     0.9242     4.5439     0.94    0.3432
     5    x3            5        0.0021     0.9263     6.0000     0.54    0.4698
```

Output 3.3 contains the output from Program 3.3. Using forward selection, the variables are entered x_1, x_4, x_5, x_2, x_3, which is different from the variables used to generate y.

3.4.5   Variable Ordering in the Solubility Data

Program 3.4 executes the ORDERVARS module on the solubility training set. The solubility training and test sets (SOLTRAIN and SOLTEST data sets) were generated in Program 3.1. At the end of Program 3.4, the training and test data sets are created using the new variable ordering, and they are stored as SAS data sets (SOLTRUSE and SOLTSUSE data sets).

Program 3.4 Variable ordering in the solubility data

```
proc iml;
    use soltrain;
    read all var ('x1':'x21') into x;
    read all var {y} into y;
    * Call the variable ordering module;
    call ordervars(x,y,order,rank);
    print rank;
    print order;
    * Store the variable ordering;
    create solorder var{order};
    append;
    close solorder;
    * Create the training and test sets using the new variable order;
    * Store the training and test sets;
    trxy=x[,order[1,]]||y;
    create soltruse from trxy[colname=('x1':'x21'||'y')];
    append from trxy;
    close soltruse;
    use soltest;
    read all var ('x1':'x21') into tsx;
    read all var {y} into tsy;
    tsxy=tsx[,order[1,]]||tsy;
    create soltsuse from tsxy[colname=('x1':'x21'||'y')];
    append from tsxy;
    close soltsuse;
    * Number of observations in the train and test data set;
    ntrain=nrow(trxy); ntest=nrow(tsxy);
    print ntrain ntest;
    quit;
```

```
* Vertical axis;
axis1 minor=none label=(angle=90 "Singular value") order=(0 to 12000 by 2000) width=1;
* Horizontal axis;
axis2 minor=none label=("Index") order=(1 to 21 by 1) width=1;
symbol1 i=none value=dot color=black height=5;
proc gplot data=singvals;
    plot singularvalues*index/vaxis=axis1 haxis=axis2 frame;
    run;
    quit;
```

Figure 3.3 Singular values in the solubility data set

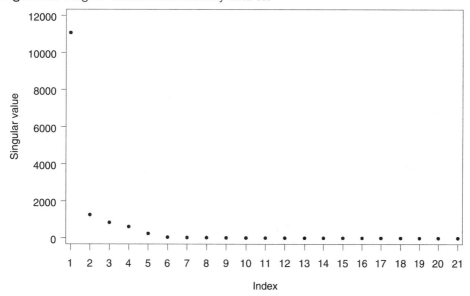

Output from Program 3.4

	RANK					
	21					
	ORDER					
	COL1	COL2	COL3	COL4	COL5	COL6	COL7
ROW1	8	13	3	1	21	9	19
	ORDER					
	COL8	COL9	COL10	COL11	COL12	COL13	COL14
ROW1	12	20	2	4	5	16	10
	ORDER					
	COL15	COL16	COL17	COL18	COL19	COL20	COL21
ROW1	6	17	18	7	14	15	11
	NTRAIN	NTEST				
	121	49				

Figure 3.3 is a plot of the singular values for each descriptor. Based on this plot it appears only five descriptors are needed to build a predictive model, but sequential regression should probably be used to determine the number of descriptors needed in the model. In sequential regression the variables can be added in order, provided that a certain statistical criterion is met. An example is provided in Section 3.5.1. Output 3.4 contains the variable ordering (ORDER) as well as the number of observations in the training and test data sets (NTRAIN and NTEST).

3.5   Statistical Procedures for Model Building

In this section we review a number of procedures that can be used to build statistical models. The potential model builder will need to obtain a substantial toolbox since no one statistical tool will work on every set of data. As a corollary to the previous statement, the nature of data will help determine the appropriate statistical model-building tool. See Two Crows Corporation (1999) for a review of model building procedures used to extract *predictive patterns* in data.

3.5.1   Multiple Linear Regression

Multiple linear regression has certainly been the workhorse for building predictive models for many years. It is appropriate for the situation where there are many more observations than descriptors and where the dependent variable is continuous. Models take the form of:

$$y = \beta_0 + \beta_1 x_1 + \beta_2 x_2 + \cdots + \beta_p x_p + e,$$

where y is the dependent variable, $x_1, \ldots, x_p$ are the independent variables, and e is the random error. If a linear combination of the descriptors can approximate the response, then this formulation works well and is simple to perform. Terms such as x_i^n or $x_i x_j$ are also allowed. PROC REG is appropriate and allows for variations of this simple procedure, such as when terms are added or deleted in a stepwise fashion. Program 3.3 illustrated the use of PROC REG with forward selection.

PROC REG can also be used interactively. After you specify a model in PROC REG with a RUN command, but not a QUIT command, you can use a variety of other commands interactively such as ADD, DELETE, PRINT. See the PROC REG documentation for more information. To demonstrate the utility of this, Program 3.5 provides the commands for running PROC REG interactively on the solubility training set. In Program 3.5, the initial model fit contains the five most important variables, $x_1, \ldots, x_5$ (they are included in the MODEL statement) and the VAR statement specifies the other variables that might be added to the model. The following commands add variables to the initial model with the ADD statement. The QUIT command ends PROC REG. The results for the final model are included in Output 3.5. The output contains, for the intercept and each of the 11 variables added to the model interactively, the parameter estimate, the standard error of the estimate, and the one-degree of freedom test.

Program 3.5 Using PROC REG interactively on the solubility data

```
proc reg data=soltruse;
    model y = x1 x2 x3 x4 x5;
    var x6 x7 x8 x9 x10 x11 x12 x13 x14;
    run;
    add x6 x7;
    print;
    run;
    add x8 x9;
    print;
    run;
    add x10 x11;
    print;
    run;
    quit;
```

Output from Program 3.5

		Parameter	Standard		
Variable	DF	Estimate	Error	t Value	Pr > |t|
Intercept	1	-3.40278	3.57476	-0.95	0.3433
x1	1	-0.00771	0.00198	-3.89	0.0002
x2	1	-0.00851	0.00162	-5.25	<.0001
x3	1	0.77612	0.22350	3.47	0.0007
x4	1	0.08929	0.04636	1.93	0.0567
x5	1	0.23927	0.11684	2.05	0.0430
x6	1	0.07884	0.03048	2.59	0.0110
x7	1	7.36932	3.37147	2.19	0.0310
x8	1	-0.00510	0.00126	-4.06	<.0001
x9	1	-0.40608	0.12531	-3.24	0.0016
x10	1	0.63875	0.29842	2.14	0.0345
x11	1	0.88649	0.20612	4.30	<.0001

Parameter Estimates

3.5.2 Logistic Regression

Logistic regression is a generalization of linear regression in that it is formulated to use a binary (0 or 1) dependent variable. It uses the log odds or logit transformation:

$$\log\left(\frac{P_i}{1 - P_i}\right) = \beta_0 + \beta_1 x_1 + \beta_2 x_2 + \cdots + \beta_p x_p + e,$$

where P_i is the probability of the event occurring. The model above assumes the log odds ratio is a linear function of the predictors. It is easy to see that similar pros and cons exist between logistic regression and multiple linear regression discussed earlier.

There are several SAS procedures that you can use to perform logistic regression such as the LOGISTIC, CATMOD, and GENMOD procedures.

3.5.3 Discriminant Analysis

Discriminant analysis is an extremely old statistical technique developed by R. A. Fisher in the 1930's and used to classify the famous Iris data set. This procedure determines the hyper-planes that separate or discriminate between the various classes in the data. It is a very simple technique to use and interpret since an observation is on one side of the plane or the other. However, there are assumptions such as normality of the descriptors, which can be problematic, and the fact that the boundaries separating the different classes are linear may reduce its discrimination ability.

There are several SAS procedures that you can use to perform discriminant analysis such as the DISCRIM, CANDISC, and STEPDISC procedures.

3.5.4 Generalized Additive Regression

The standard linear regression model is used quite often for building models. It is extremely easy to use but suffers from the fact that, in real life, many effects are not linear. The generalized additive regression model offers a flexible statistical technique that can accommodate factors that are not linear. Additionally, generalized additive regression is an interesting tool because it allows both continuous and discrete responses.

Recall that the standard linear regression model assumes the expected value of y has the linear form:

$$E(y|X) = f(x_1, \ldots, x_p) = \beta_0 + \beta_1 x_1 + \cdots + \beta_p x_p.$$

The additive model generalizes the linear model by modeling the expected value of y as

$$E(y|X) = s_0 + s_1(x_1) + s_2(x_2) + \cdots + s_p(x_p),$$

where $s_i, i = 1, \ldots, p$ are smooth functions. These function are estimated in a nonparametric fashion. Generalized additive models are able to fit both discrete and continuous responses by allowing for a link between $f(x_1, \ldots, x_p)$ and the expected value of y. Additional details may be found in the GAM procedure documentation.

This procedure has a lot of appeal and can be very useful. Since it is new to SAS, we will illustrate its use on the solubility training data set. Program 3.6 contains the SAS code to read the solubility training set and fit the model using five variables. The GAM procedure determines the values of smoothing parameters for each of the variables. For the solubility example, the SPLINE smoothing effect was chosen for each variable. The other options available are LOESS, SPLINE2, or PARAM. The LOESS option fits a local regression with the variable, the SPLINE2 option fits a bivariate thin-plate spline to two variables, and PARAM specifies a parametric variable; a smoothing function is not applied to the variable.

Program 3.6 Analysis of the solubility data using PROC GAM

```
proc gam data=soltruse;
    model y=spline(x1) spline(x2) spline(x3) spline(x4) spline(x5)/dist=normal;
    run;
    quit;
```

Output from Program 3.6

```
                         Regression Model Analysis
                           Parameter Estimates

                      Parameter      Standard
           Parameter   Estimate         Error   t Value   Pr > |t|

           Intercept   -0.73649       0.36152     -2.04     0.0442
           Linear(x1)  -0.00741       0.00118     -6.31     <.0001
           Linear(x2)  -0.00677       0.00158     -4.30     <.0001
           Linear(x3)   0.79513       0.19670      4.04     0.0001
           Linear(x4)   0.01799       0.03777      0.48     0.6349
           Linear(x5)   0.17723       0.10212      1.74     0.0857


                         Smoothing Model Analysis
                           Analysis of Deviance

                                      Sum of
           Source          DF        Squares   Chi-Square   Pr > ChiSq

           Spline(x1)   3.00000    14.958834     23.9431       <.0001
           Spline(x2)   3.00000     5.345832      8.5565       0.0358
           Spline(x3)   1.00000     0.038719      0.0620       0.8034
           Spline(x4)   3.00000     1.252530      2.0048       0.5714
           Spline(x5)   3.00000     7.541784     12.0714       0.0071
```

Output 3.6 provides the output from Program 3.6. From the output we can see that x_1, x_2, and x_3 have significant linear trends; p-values for the corresponding linear tests are significant at 5%. From the analysis of deviance table, the chi-square tests are significant for x_1, x_2, and x_5, indicating that these variables need to be smoothed.

A plot of the smoothing components with 95% confidence bands for each of the five descriptors can be created using the ODS GRAPHICS statement as shown below:

```
ods html;
ods graphics on;
proc gam data=soltruse plots(clm commonaxes);
    model y=spline(x1) spline(x2) spline(x3) spline(x4) spline(x5)/dist=normal;
    run;
    quit;
ods graphics off;
ods html close;
```

The plot generated using ODS GRAPHICS is not provided because it is not publication-quality and, due to the experimental nature of PROC GAM, the option to save the smoothed components to a SAS data set is not yet available.

3.6   Determining When a New Observation Is Not in a Training Set

Often in drug discovery, a computational model is built for a non-statistician, and the user relies on easily understood diagnostics and graphical representations to assess the accuracy of the prediction.

Once a model is in use, a common question asked of the statistician is, "How accurate is the prediction?" If the model is providing the user with a quantitative estimates, then the error associated with the prediction can be used to answer this question. This measure takes into account the distance between the new observation and the training set as well as the error in the model. If the model is providing the user with a categorical response and a linear discriminant model is used, then a bar chart of the probability that an observation is in each group can be provided to the user.

Another measure that can be provided to the user is the *leverage value*. Recall the leverage value is simply the distance between the new observation and the centroid of the training data set. If the leverage value of a new observation is greater than the maximum observed leverage value in the training set, then the prediction for the new observation will be unreliable.

Also, it is important for the modeler to keep track of the data sets the model is being applied to. If the data is far from the data the model was trained on, then the modeler should think about retraining the data sets so the predictions will be reliable.

Once the statistician has provided the model, along with easily understood diagnostics, to the team, the statistician should also monitor how the model is being used. If the chemistry space the model is being used on is different from the chemistry space the model was trained on, then the statistician should consider retraining the model to provide the team with a more useful model. This monitor can easily be done by keeping track of the leverage value from each prediction and by using a flag to indicate whether the prediction was an extrapolation or not.

As an example, let's determine if the observations in the solubility test set are included in the solubility training space. Program 3.7 contains the code to fit the selected model to the test set and store the residuals in a SAS data set (RESDS data set). This is accomplished using PROC REG.

Program 3.7 also contains SAS/IML code to calculate the maximum leverage value in the training set. The program reads in the test set, finds the leverage values for each observation and determines which, if any, observations are outside the training set space. If some observations in the test set are outside the space, the program calculates the prediction error excluding these observations.

Program 3.7 Determine if the test set is in the training set space

```
proc reg data=soltsuse;
    model y=x1 x2 x3 x4 x5 x6 x7 x8 x9 x10 x11;
    output out=resds r=yresids;
    run;
    quit;
proc iml;
    use soltruse;
    read all var ('x1':'x11') into x;
    * Calculate the maximum leverage value of the training set;
    xtx=x`*x;
    call eigen(evals,evecs,xtx);
    invxtx=evecs*diag(1/evals)*evecs`;
    himax=max(vecdiag(x*invxtx*x`));
    print himax[format=5.3];
```

```
* Read in the test set and calculate the test set leverage values;
* Determine which ones are in training set space;
use soltsuse;
read all var ('x1':'x11') into x;
hi=vecdiag(x*invxtx*x`);
indi=choose(hi>himax,1,0);
outindi=loc(indi);
print outindi;
use resds;
read all var{yresids} into e;
ein=e[loc(^indi),1];
err=sqrt(sum(e##2)/(nrow(e)-ncol(x)));
print err[format=5.3];
quit;
```

Output from Program 3.7

```
                            Analysis of Variance

                                Sum of          Mean
          Source        DF     Squares         Square    F Value    Pr > F

          Model         11     89.03923        8.09448     9.16     <.0001
          Error         37     32.69240        0.88358
          Corrected Total  48  121.73163

                                HIMAX

                                0.269

                               OUTINDI

                        6           40        41

                                 ERR

                               0.928
```

The output analysis of variance table is contained in Output 3.7. The prediction error of the model, which is the square root of the Mean Square Error, calculated from the test set is $\sqrt{0.8836} \approx 0.94$. The maximum leverage value in the training set is approximately 0.269, so any observation in the test set whose leverage value is greater than this is not in the training set space. There were three observations in the test set that are outside the training set space (observations 6, 40 and 41). Excluding these observations results in a prediction error for the model of 0.928.

3.7   Using SAS Enterprise Miner

As an alternative to the procedures presented in the previous sections, model building can be done using popular software, SAS Enterprise Miner. We will use SAS Enterprise Miner to build a model for the SOLUBILITY data set and make comparisons to the results presented in the previous sections.

To begin, we will use the basic SAS Enterprise Miner routines to input the solubility data using the drag and drop Input Data Source node. This particular node, the first one in Figure 3.4, allows the solubility data to be accessed by the software.

Figure 3.4 SAS Enterprise Miner flow diagram

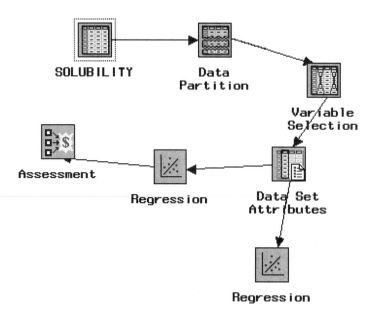

The second node in the Figure 3.4, the Data Partition node, provides an easy way to partition the data into training, test, and validation subsets. Table 3.1 below contains the settings used in this example. We excluded a validation set so the results could be compared to the results obtained in Example 3.3.3.

Table 3.1 Settings Used in the Variable Selection Node

Setting	value
Method	Simple random
Random seed	5164
Training percentage	75%
Validation percentage	0%
Test percentage	25%

The output from running the Data Partition node is not included, but the procedure selected 128 observations for the training set and 42 observations for the test set.

The Variable Selection node, the third in Figure 3.4, was used to select descriptors that are associated with the designated target variable, log-solubility. The rules used by this node are:

1. Compute the squared correlation for each descriptor with the target variable and then assign the rejected role to those descriptors that have a value less than the squared correlation criterion, 0.005.

2. Evaluate the remaining significant descriptors using a forward stepwise R^2 regression. Descriptors that have a stepwise R^2 improvement less than the threshold criterion, 0.0005, are assigned the rejected role.

3. For binary descriptors, perform a logistic regression using the predicted values output from the forward stepwise regression as the independent input.

The results from the variable selection process for the solubility data are shown below. It follows from the output that variables x_5, x_6, x_7, x_{11}, x_{15}, and x_{18} were rejected because of a low R^2 value.

Output from the variable selection process

Name	Role	Rejection Reason	% Missing	# of Levels
X1	input		0	12
X2	input		0	2
X3	input		0	3
X4	input		0	3
X5	rejected	Low R2 w/ target	0	11
X6	rejected	Low R2 w/ target	0	4
X7	rejected	Low R2 w/ target	0	5
X8	input		0	122
X9	input		0	127
X10	input		0	116
X11	rejected	Low R2 w/ target	0	117
X12	input		0	109
X13	input		0	71
X14	input		0	128
X15	rejected	Low R2 w/ target	0	8
X16	input		0	49
X17	input		0	125
X18	rejected	Low R2 w/ target	0	95
X19	input		0	128
X20	input		0	125
X21	input		0	126

The fourth node in Figure 3.4 is the Data Set Attributes node. This node allows the user to change certain features of the input data sets, which are the training and test sets after variable selection. This node is needed to re-specify the target variable, or the response. The target variable was specified after running the Input Data Source node but is no longer specified after the Variable Selection node.

The next two nodes in Figure 3.4 are for constructing a model to predict solubility. Two models were fit to the training set, one using all the variables selected in the third node and the second using forward selection on the variables selected by the software.

The output below displays the output from fitting a linear model using all the variables the Variable Selection node did not reject. The output contains the analysis of variance table, model fit statistics, and the parameter estimates along with the one-degree of freedom tests. The Root Mean Squared Error (RMSE) for this model is 0.6992. The p-values from the one-degree of freedom tests indicate that some of the variables are not needed in the model.

Output that shows solubility model performance

		Analysis of Variance			
Source	DF	Sum of Squares	Mean Square	F Value	Pr > F
Model	15	204.302215	13.620148	27.86	<.0001
Error	112	54.750576	0.488844		
Corrected Total	127	259.052791			

```
                    Model Fit Statistics

  R-Square          0.7887    Adj R-Sq          0.7603
  AIC              -76.7030   BIC              -70.1724
  SBC              -31.0705   C(p)              16.0000


               Analysis of Maximum Likelihood Estimates


                                     Standard
  Parameter    DF     Estimate        Error     t Value    Pr > |t|

  Intercept     1       4.0946        6.3716       0.64      0.5218
  x1            1       0.0973        0.0435       2.24      0.0273
  x16           1       0.1986        0.0631       3.15      0.0021
  x17           1     -18.6201        6.4189      -2.90      0.0045
  x19           1      -0.2266        5.7512      -0.04      0.9686
  x2            1       0.7320        0.2485       2.95      0.0039
  x20           1      -0.3103        0.1191      -2.60      0.0104
  x21           1      -0.2759        0.1118      -2.47      0.0151
  x4            1       0.6935        0.1936       3.58      0.0005
  x8            1       0.00276       0.00883      0.31      0.7554
  x9            1       0.3260        0.0984       3.31      0.0012
  x3            1       0.4099        0.2569       1.60      0.1135
  x12           1      -0.00290       0.00238     -1.22      0.2265
  x13           1      -0.00929       0.00322     -2.88      0.0047
  x14           1      -0.00695       0.00406     -1.71      0.0898
  x10           1      -0.00119       0.00229     -0.52      0.6043
```

The next output contains a subset of the output from the Regression node when forward selection is used. The output displays a summary of forward selection, the analysis of variance table, model fit statistics, and parameter estimates along with the one-degree of freedom tests. The RMSE for this model is 0.7673 which is slightly higher than the model using all the variables. Therefore, the trade-off is complexity versus parsimony.

We decided to use the simpler model to obtain the model prediction error, which is the prediction error of the test data set. The Assessment node, the last node in Figure 3.4, is used to calculate the model prediction error. The RMSE for the test set is 1.0268.

Output that shows solubility model performance using forward selection

```
             Summary of Forward Selection

         Effect             Number
  Step   Entered     DF       In     F Value    Pr > F

    1    x8           1        1       89.36    <.0001
    2    x16          1        2       94.15    <.0001
    3    x3           1        3       13.48    0.0004
    4    x2           1        4        4.28    0.0407


  The selected model, based on the CHOOSE=AIC criterion, is
  the model trained in Step 4. It consists of the following effects:

  Intercept  x16  x2  x8  x3
```

```
                      Analysis of Variance

                        Sum of
  Source          DF   Squares   Mean Square   F Value   Pr > F

  Model            4   183.700468    45.925117    74.97   <.0001
  Error          123    75.352323     0.612621
  Corrected Total 127   259.052791


                   Model Fit Statistics

  R-Square      0.7091    Adj R-Sq      0.6997
  AIC         -57.8215    BIC         -57.5483
  SBC         -43.5614    C(p)         35.6797


            Analysis of Maximum Likelihood Estimates

                                   Standard
  Parameter   DF   Estimate      Error   t Value   Pr > |t|

  Intercept    1    -0.0114     0.2261     -0.05     0.9599
  x16          1     0.3613     0.0362      9.97     <.0001
  x2           1     0.5384     0.2603      2.07     0.0407
  x8           1    -0.0187     0.00129   -14.48     <.0001
  x3           1     0.4819     0.1969      2.45     0.0158
```

3.8 Summary

This chapter described statistical techniques used in drug discovery to facilitate model building. The techniques are useful for accelerating the pace of moving potential drugs through the development process. Procedures provided in this chapter deal with partitioning relevant data into training and test sets, selecting potential variables to be used to construct the model, building the model, and making predictions.

The model building process described in the chapter is illustrated by using a real drug discovery data set to predict the solubility of various chemical compounds. The SAS code is provided so that the interested reader can use these procedures on their own sets of data.

References

Box, G.E.P., Hunter, J.S., Hunter, W.G. (1978). *Statistics for Experimenters: Design, Innovation, and Discovery.* Second Edition. New York: Wiley.

Breiman, L. (2001). "Statistical modeling: The two cultures." *Statistical Science.* 16, 199–231.

Efron, B., Tibshirani, R.J. (1998). *An Introduction to the Bootstrap.* Boca Raton, FL: CRC Press.

Frank, I.E., Friedman, J.H. (1993). "A statistical view of some chemometrics regression tools." *Technometrics.* 35, 109–148.

Furnival, G.M., Wilson, R.W. (1974). "Regression by leaps and bounds." *Technometrics.* 16, 499–511.

Golbraikh, A., Tropsha, A. (2002). "Predictive QSAR modeling based on diversity sampling of experimental datasets for the training and test set selection." *Journal of Computer-Aided Molecular Design.* 16, 357–369.

Golub, G.H., Van Loan, C.F. (1996). *Matrix Computations.* Third Edition. Baltimore, MD: The John Hopkins Press.

Hand, D. (1998). "Data mining: Statistics and more." *The American Statistician.* 52, 112–118.

Hastie, T., Tibshirani, R., Friedman, J. (2001). *The Elements of Statistical Learning: Data Mining, Inference and Prediction.* New York: Springer-Verlag.

Scott, D.W. (1979). "On optimal and data-based histograms." *Biometrika.* 66, 3, 605–610.

Two Crows Corporation (1999). *Introduction to Data Mining and Knowledge Discovery.* Third Edition. Available at www.twocrows.com.

Chapter 4

Statistical Considerations in Analytical Method Validation

Bruno Boulanger
Viswanath Devanaryan
Walthère Dewé
Wendell Smith

In the context of a pre-study validation of an analytical method, this chapter describes some advanced statistical methodologies available in the literature and provides SAS code to calculate the validation criteria. Both linear and nonlinear methods are described within the calibration framework. In terms of validation criteria, i.e., when the calculated concentrations are available, criteria based on measurement error are utilized, and useful graphical representations are proposed.

4.1 Introduction

In the pharmaceutical industry, analytical methods play a vital role in all experiments performed in the development of a drug product. If the quality of an analytical method is doubtful, then the whole set of decisions based on those measures is questionable.

Bruno Boulanger is Manager, European Early Phase Statistics, Eli Lilly and Company, Belgium. Viswanath Devanaryan is Associate Director, Biostatistics and Research Decision Sciences, Merck, USA. Walthère Dewé is Senior Research Scientist, Global Discovery and Development Statistics, Eli Lilly and Company, Belgium. Wendell Smith is Partner and Senior Statistician, Bowsher Brunelle Smith (B2S Consulting), USA.

Consequently, assessment of the quality of an analytical method is far more than a statistical challenge, it is a matter of good ethics and good business practices.

Many regulatory documents have been released in the pharmaceutical industry to address quality issues. These are primarily ICH and FDA documents. Those that are related to analytical and bioanalytical method validation (ICH, 1995, 1997; FDA, 2001) suggest that analytical methods must comply with specific acceptance criteria to be recognized as validated procedures. The primary aim of these documents is to require evidence that the analytical methods are suitable for their intended use. Unfortunately, discrepancies exist among these documents with respect to the definition of acceptance criteria, and limited guidance is provided for estimating the performance criteria.

In this chapter, background information will be provided on analytical method validation concepts, and apparent inconsistencies will be addressed from a statistical perspective. Statistical methods will be described for estimation of analytical performance parameters, and decision criteria will be illustrated that are consistent with the concept of a "good" analytical procedure. The impact of these methods on the design of experiments needed to obtain reliable estimates of the performance criteria will be considered. A major emphasis of this chapter will be the use of SAS programs to illustrate the computation of assay performance parameters, with limited discussion about the philosophy of the assay validation purpose or practices.

To save space, some SAS code has been shortened and some output is not shown. The complete SAS code and data sets used in this book are available on the book's companion Web site at `http://support.sas.com/publishing/bbu/companion_site/60622.html`.

4.1.1 Method Classification Based on Data Types

The ultimate goal of an analytical method or procedure is to measure accurately a quantity, such as the concentration of an analyte, or to measure a specific activity, as for example for a biomarker. However, many assays such as cell-based and enzyme activity biomarker assays may not be very sensitive, may lack precision, and/or may not offer definitive reference standards. Assays based on physicochemical (such as chromatographic methods) or biochemical (such as ligand binding assays) properties of an analyte assume that these quantifiable characteristics are reflective of the quantity, concentration, or biological activity of the analyte. For the purpose of analytical validation, we will follow the recently proposed classifications for assay data by Lee et al. (2003). These classifications, summarized below, provide a clear distinction with respect to the analytical validation practice and requirements.

Qualitative methods generate data which do not have a continuous proportionality relationship with the amount of analyte in a sample; the data are categorical in nature. Data may be nominal such as a present/absent call for a gene or gene product. Alternatively, data might be ordinal in nature, with discrete scoring scales (e.g., 1 to 5 or $-, +, + + +$) such as for immuno-histochemistry assays or Fluorescence In Situ Hybridization (FISH).

Quantitative methods are assays where the response signal has a continuous relationship with the quantity or activity of the analyte. These responses can therefore be described by a mathematical function. Inclusion of reference standards at discrete concentrations allows the quantification of sample responses by interpolation. The availability of a well-defined reference standard may be limited, or may not be representative of the in vivo presentation, so quantification may not be absolute. To that end, three types of quantitative methods have been defined:

- A *definitive quantitative assay* uses calibrators fit to a known model to provide absolute quantitative values for unknown samples. Typically, such assays are possible only where the analyte is not endogenous. An example of this is a small molecule drug.

- A *relative quantitative assay* is similar in approach, but generally involves the measurement of endogenously occurring analytes. In this case, even a "zero" or blank calibrator may contain some amount of analyte, and quantification can only be done relative to this "zero" level. Examples of this include immunoassays for cytokines such as sTNFRII, or gene expression assays, e.g., Reverse Transcriptase-Polymerase Chain Reaction (RT-PCR).

- A *quasi-quantitative assay* does not involve the use of calibrators, mostly due to the lack of suitable reference material, so the analytical result for a test sample is reported only in terms of the assay signal (e.g., optical density in ELISA).

This chapter deals with the assessment of definite and relative quantitative assays. A full discussion of quasi-quantitative and qualitative assays and statistical considerations thereof is beyond the scope of this chapter. A good reference on the analytical validation of a typical quasi-quantitative assay is a white paper on immunogenicity by Mire-Sluis et al. (2004).

4.1.2 Objective of an Analytical Method

The objective of a definite and relative quantitative analytical method is to be able to quantify as accurately as possible *each* of the unknown quantities that the laboratory will have to determine. In other words, what all analysts expect from an analytical procedure is that the difference between the measurement or observation (X) and the unknown "true value" μ_T of the test sample be small or inferior to an acceptance limit λ:

$$-\lambda < X - \mu_T < \lambda \Longleftrightarrow |X - \mu_T| < \lambda. \tag{4.1}$$

The acceptance limit λ can be different depending on the requirements of the analyst and the objective of the analytical procedure. The objective is linked to the requirements usually admitted by the practice (e.g., 1% or 2% on bulk, 5% on pharmaceutical specialties, 15% for biological samples. Acceptance limits vary in clinical applications depending on factors such as the physiological variability and the intent of use).

4.1.3 Objective of the Pre-Study Validation Phase

The aim of the pre-study validation phase is to generate information to guarantee that the analytical method will provide, in routine use, measurements close to the true value (DeSilva et al., 2003; Findlay, 2001; Hubert et al., 2004; Smith and Sittampalam, 1998; Finney, 1978) without being affected by other elements present in the sample. In other words, the validation phase should demonstrate that the inequality described in Equation (4.1) holds for a certain proportion of the sample population.

The difference between the measurement X and its true value is a sum of a systematic error (bias or trueness) and a random error (variance or precision). The true values of these parameters are unknown but they can be estimated based on the validation experiments. The reliability of these estimates depends on the adequacy of these experiments (design, size).

Consequently, the objective of the validation phase is to evaluate whether, given the estimates of bias and variance, the expected proportion of measures that will fall within the acceptance limits is greater than a predefined level, say, β:

$$E_{\widehat{\mu},\widehat{\sigma}}\left(P[|X - \mu_T| < \lambda|\widehat{\mu}_M, \widehat{\sigma}_M]\right) \geq \beta. \tag{4.2}$$

Although Equation (4.2) cannot be solved exactly within a frequentist framework, Section 4.6 will discuss approximate solutions that can be used in practice.

4.1.4 Classical Design in Pre-Study Validation

Experiments performed during pre-study validation are designed to mimic the processes and practices to be followed during routine application of a method. All aspects of the analytical method should be taken into account, such as the lot of a solvent, operator, preparation of samples, etc. If measurements generated under these "simulated" conditions are acceptable (see Section 4.6) then the method will be declared valid for routine use. Usually, two sets of samples will be prepared for simulating the real process: calibration and validation samples.

- Calibration samples (CS) must be prepared according to the protocol that will be followed during routine use, i.e., the same operational mode, the same number of concentration levels for the standard curve, and the same number of repetitions at each level.

- Validation samples (VS) must be prepared in the sample matrix when applicable. In the validation phase, they mimic the unknown samples that the analytical procedure will have to quantify in routine use. Each validation standard should be prepared independently, in order to have realistic estimates of the variance components.

The minimum design of a pre-study validation phase is at least two replicates per run or series in a minimum of three runs. However, it is highly recommended to consider at least six runs in order to have a good estimate of the between-run variance. The number of runs and replicates to perform at each concentration level to demonstrate that an analytical procedure is valid could be estimated (by simulations) and depends, of course, on the inherent but unknown properties of the analytical procedure itself. The more variable the method, the more experiments are necessary.

Table 4.1 displays the minimal sample size for r runs and s replicates per run for 10% acceptance limits (the table was computed via simulations, assuming a potential small bias of 2%). It is clear that the number of runs increases with increasing between-run variance. The higher number of runs can be compensated for by more replicates per runs, but this leads to a larger total number of experiments (rs). Also, as expected when the sum of bias (2%) and the within-run and between-run variances become greater than 10%, it becomes unlikely that the method will ever be validated for such acceptance limits. More development in the laboratory is required to achieve this objective. The reproducibility that requires between-laboratory experiments will not be discussed in this chapter.

Table 4.1 Minimal Sample Size for r Runs and s Replicates per Run for 10% Acceptance Limits

Between -run variance	Within-run variance									
	1%		2%		3%		4%		5%	
	r	s	r	s	r	s	r	s	r	s
1%	4	3	4	3	4	3	4	4	5	9
	5	3	5	3	5	3	5	4	6	7
2%	4	3	4	3	4	3	4	6	8	9
	5	3	5	3	5	3	5	6	9	7
3%	4	4	4	6	5	6	7	10		
	5	3	5	3	6	5	8	9		
4%	7	10	9	8						
	8	7	10	6						

4.1.5   Example: A Sandwich ELISA Assay

A sandwich ELISA assay, optimized by statistical design of experiments and validated at Lilly Research Laboratories, will be used throughout this chapter for illustration purposes. This data set is available on the book's companion Web site.

The objective of this assay was to quantify a protein used as a biomarker in neurological disease therapeutic research. The ELISA consisted of incubation of samples on plates pre-coated with a capture antibody specific to the protein of interest followed by immunological detection of the specific bound protein by an enzyme conjugate and measurement (optical density) of the colored product. In order to validate this assay, calibration standards and validation samples were prepared in appropriate matrices from stock protein solution via serial dilution and then tested in triplicate. The procedure was followed for four independent runs with two plates per run over four days.

4.2   Validation Criteria

The main validation criteria widely recommended by various regulatory documents (ICH, FDA, European Union) and commonly used in analytical laboratories are:

- Specificity-selectivity.
- Response function (calibration curve).
- Linearity.
- Precision (repeatability and intermediate precision).
- Accuracy (trueness).
- Measurement error (total error).
- Limit of detection (LOD).
- Limit of quantification (LOQ).
- Assay range.
- Sensitivity.

In addition, according to the domains concerned, other specific criteria can be required:

- Analyte stability.
- Recovery.
- Effect of the dilution.

A full validation is necessary for an analytical procedure to pass from the development phase to the phase of routine analysis. The validation step is not only necessary but also required at the time specifications (tests and acceptance limits) are set up for an active ingredient or a finished product.

The validation criteria mentioned above must be established, insofar as possible, in the same matrix as that of samples to be analyzed. Every new analytical procedure will have to be validated for each type of matrix (e.g., for each type of biological fluid and for each animal species). Nevertheless, the definition of a matrix depends on analyst responsibility. Some matrix regrouping, generally admitted by the profession for an application domain given, can be performed.

This section will focus only on the main criteria that apply to most, if not all, analytical methods and that must be adequately estimated and documented for ensuring compliance to regulations. The criteria that are very analytical, but that have no particular statistical content, will be defined here, but will not be discussed.

4.2.1 Specificity-Selectivity

The specificity of an analytical procedure is the ability to unequivocally assess the analyte in the presence of components that may be expected to be present. Usually, the analyst must demonstrate that the measured result is directly related to the analyte or product of interest and that other aspects in the sample, such as the matrix, do not interfere with the signal or measurement. For example, selectivity of a chromatographic method is verified typically by showing that the product of interest is clearly separated from all other products.

4.3   Response Function or Calibration Curve

The response function for an analytical procedure is the existing relationship, within a specified range, between the response (signal, e.g., area under the curve, peak height, absorption) and the concentration (quantity) of the analyte in the sample. The calibration curve should be described preferably by a simple monotonic response function that gives accurate measurements. Note that the response function is frequently confused with the linearity criteria. However, the latter criterion refers to the relationship between the quantity introduced and the quantity back-calculated from the calibration curve (see Section 4.4). Because of the confusion, it is common to see laboratory analysts try very hard to ensure that the response function is linear in the classical sense, i.e., a straight line. Not only is this not required, but it is often irrelevant and can lead to large errors in measured results (e.g., for ligand binding assays). A significant source of bias and imprecision in analytical measurements can be the choice of the statistical model for the calibration curve.

4.3.1   Computational Aspects

Statistical models for calibration curves can be either linear or nonlinear in their parameter(s). The choice between these two families of models will depend on the type of method and/or the range of concentrations of interest. If the range is very narrow, an unweighted linear model may suffice, while a larger range may require a more advanced and weighted model. High Performance Liquid Chromatography (HPLC) methods are usually linear while immunoassays are typically nonlinear. Weighting may be important for both methods because a common feature for many analytical methods is that the variance of the signal is a function of the level or quantity to be measured.

Methodologies for fitting linear and nonlinear models generally require different SAS procedures. For both model types, curves are fit by finding values for the model parameters that minimize the sum of squares of the distances between observations and the fitted curve. For linear models, parameter estimates can be derived analytically while this is not the case for many nonlinear models. Consequently, iterative procedures are often required to estimate the parameters of a nonlinear model. In this section, both linear and nonlinear models will be considered.

In case of heterogeneous variances of the signal across the concentration range, it is recommended that observations be weighted when fitting a curve. If observations are not weighted, an observation more distant to the curve than others has more influence on the curve fit. As a consequence, the curve fit may not be good where the variances are smaller. Weighting each term of the sum of squares is frequently used to solve this problem, where this can be viewed as minimizing the relative distances instead of minimizing the actual distances. When replicates are present at each concentration level, it is often better to fit the model to their average/median response values. Regardless of model type, it is assumed that all observations fit to a model are completely independent. In reality, replicates are often not independent for many analytical procedures because of the steps followed in preparation and analysis of samples. In such cases, replicates should not be used separately.

Models are typically applied on either a linear scale or log scale of the assay signal and/or the calibrator concentrations. The linear scale is used in case of homogeneous variance across the concentration range, and the log scale is often more appropriate when variance increases with increasing response.

4.3.2   Linear and Polynomial Models

Most commonly used types of polynomial models include simple linear regression (with or without an intercept) and quadratic regression models.

As an illustration, consider a subset of the ELISA_CS data set (Plate A, Series 1) that includes three replicate measurements at each of eight concentration levels (the ELISA_CS data set is provided on the book's web site). The selected measurements are included in the CALIB data set shown below in Program 4.1. The CONCENTRATION variable is the concentration value and the rep1, rep2, and rep3 variables represent the three replicates. Program 4.1 uses the MIXED procedure to fit a linear model to the data collected in the study. The model parameters are estimated using the restricted maximum likelihood method, which is equivalent to the ordinary least square method when the data are normally distributed. The SOLUTION option requests parameter estimates which are then saved to a SAS data set specified in the ODS statement. The OUTPREDM option is used to save the predicted signal values from the fitted linear model to another SAS data set. The WEIGHT statement is used to assign weights to the individual measurements (the W variable represents the weight). For example, the weights can be defined as the inverse of the signal level or the inverse of the squared signal. In this example, the weights are defined using the following formula: $w = 1/s^{1.3}$, where s is the signal level. The choice of the exponent (e.g., 1.3) depends on the relationship between the signal's variability and its average level.

Program 4.1 Fitting a simple linear regression model using PROC MIXED

```
data calib;
    set elisa_cs;
    if plate='A' and series=1;
    array y(3) rep1 rep2 rep3;
    do j=1 to 3;
        signal=y(j);
        w=1/signal**1.3;
    output;
    end;
proc mixed data=calib method=reml;
    model signal=concentration/solution outpredm=predict;
    weight w;
    ods output solutionf=parameter_estimates;
proc sort data=predict;
    by concentration;
axis1 minor=none label=(angle=90 'Signal') order=(0 to 4 by 1);
axis2 minor=none label=('Concentration') logbase=10 logstyle=expand;
symbol1 value=none i=join color=black line=1 width=3;
symbol2 value=dot i=none color=black height=5;
proc gplot data=predict;
    plot (pred signal)*concentration/overlay frame haxis=axis2 vaxis=axis1;
    run;
    quit;
```

Output from Program 4.1

	Solution for Fixed Effects				
Effect	Estimate	Standard Error	DF	t Value	Pr > |t|
Intercept	0.3684	0.02850	22	12.93	<.0001
concentration	0.005802	0.000339	22	17.12	<.0001

Output 4.1 lists the parameter estimates generated by PROC MIXED and associated *p*-values. Figure 4.1 displays the fitted regression line. It is obvious that the linear model provides a poor fit to the data.

Figure 4.1 Fit of the linear model

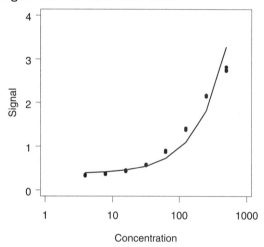

While Program 4.1 focuses on a simple linear regression model with an intercept, one can consider a linear regression model without an intercept, which is requested using the following MODEL statement:

```
model signal=concentration/solution noint;
```

or a quadratic regression model:

```
model signal=concentration concentration*concentration/solution;
```

In addition, the BY statement can be included in PROC MIXED to fit a calibration curve within each run.

4.3.3   Nonlinear Models (PROC NLIN)

As briefly mentioned in the introduction of this section, to fit a nonlinear model, one needs to rely on iterative methods. These methods begin with an initial set of parameter values for the model of interest and update the parameter values at each step in order to improve the fit. The iterative process is stopped when the fit can no longer be improved.

Nonlinear models frequently used in curve calibration include the 4-parameter logistic regression, 5-parameter logistic regression and power model. Consider, for example, the 4-parameter logistic regression model:

$$y = f(x) = \beta_1 + \frac{\beta_2 - \beta_1}{1 + (x/\beta_3)^{\beta_4}},$$

where β_1, β_2, β_3, and β_4 are the top asymptote, bottom asymptote, and concentration corresponding to half distance between β_1 and β_2, and the slope, respectively.

Both the NLIN and NLMIXED procedures can be used to fit such models. Except for the fact that they both optimize a function of interest, they do not work in the same manner. PROC NLIN fits nonlinear models by minimizing the error sum of squares and it can handle only models with fixed effects. PROC NLMIXED enables us to fit models with fixed and random effects by maximizing an approximation to the likelihood function integrated over the random effects. In this context, PROC NLMIXED is used only in models with fixed effect, and thus the problem of integration is avoided.

Program 4.2 relies on PROC NLIN to model the relationship between concentration and signal levels in the subset of the ELISA_CS data set. PROC NLIN supports several iterative methods for minimizing the error sum of squares (they can be specified using the METHOD statement). The more robust methods are GAUSS, NEWTON, and MARQUARDT. The default method is GAUSS, but the most commonly used is MARQUARDT.

Program 4.2 analyzes the CALIB data set created in Program 4.1. The initial values of the four parameters are specified in the PARAMETERS statement and the _WEIGHT_ statement is used to weight the observations when fitting the model.

Program 4.2 Fitting a 4-parameter logistic regression model using PROC NLIN

```
proc nlin data=calib method=marquardt outest=parameter_estimates;
    parameters top=3 bottom=0.2 c50=250 slope=1;
    model signal=top+(bottom-top)/(1+(concentration/c50)**slope);
    _weight_=w;
    output out=fitted_values predicted=pred;
proc sort data=fitted_values;
    by concentration;
axis1 minor=none label=(angle=90 'Signal') order=(0 to 4 by 1);
axis2 minor=none label=('Concentration') logbase=10 logstyle=expand;
symbol1 value=none i=join color=black line=1 width=3;
symbol2 value=dot i=none color=black height=5;
proc gplot data=fitted_values;
    plot (pred signal)*concentration/overlay frame haxis=axis2 vaxis=axis1;
    run;
    quit;
```

Output from Program 4.2

Source	DF	Sum of Squares	Mean Square	F Value	Approx Pr > F
Model	3	9.5904	3.1968	6181.88	<.0001
Error	20	0.0103	0.000517		
Corrected Total	23	9.6008			

Parameter	Estimate	Approx Std Error	Approximate 95% Confidence Limits	
top	3.6611	0.1186	3.4138	3.9085
bottom	0.3198	0.00681	0.3056	0.3340
c50	219.9	13.9168	190.9	249.0
slope	1.2611	0.0391	1.1795	1.3427

Output 4.2 shows that the overall F value is very large while the standard errors of the four parameter estimates are small. This suggests that the model's fit was excellent (see also Figure 4.2). It is important to note that one sometimes encounters convergence problems and it is helpful to examine the iteration steps included in the output. If the iterative algorithm does not converge, it is prudent to explore different sets of initial values that can be obtained by a visual inspection of the raw data.

Figure 4.2 Fit of the 4-parameter logistic regression model

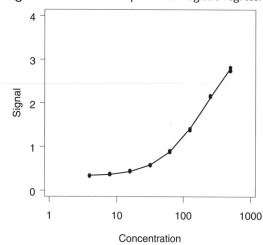

4.3.4   Nonlinear Models (PROC NLMIXED)

PROC NLMIXED supports a large number of iterative methods for fitting nonlinear models. Unfortunately, there is no general rule for choosing the most appropriate method. The choice is problem-dependent and, most of the time, one needs to select the iterative method by trial and error (see PROC NLMIXED documentation for general recommendations).

Note that the METHOD option in PROC NLMIXED does not specify the optimization method as in PROC NLIN, but rather the method for approximating the integral of the likelihood function over the random effects. The TECHNIQUE option is used in PROC NLMIXED to select the optimization method. Program 4.3 fits a 4-parameter logistic regression model to the concentration/signal data from the CALIB data set created in Program 4.1. The program uses the Newton-Raphson method with ridging as an optimization method (NRRIDG) in PROC NLMIXED. Other TECHNIQUE options include NEWRAP (Newton-Raphson optimization combining a line-search algorithm with ridging) or QUANEW (quasi-Newton). Since no random effects are included in this model, the METHOD option is not used in this example.

PROC NLMIXED does not allow the fitting of weighted regression models; however, it allows us to specify a variance function. The most popular variance function in nonlinear calibration is the power of the mean (O'Connell, Belanger, and Haaland, 1993). To specify this function in Program 4.3, we introduce the VAR and THETA parameters:

```
model signal~normal(expect,(expect**theta)*var);
```

Here VAR is the residual variance at baseline, and THETA defines the rate at which this variance changes with the predicted concentration (EXPECT variable). The

FITTED_VALUES data set contains the fitted value for each concentration value as well as the corresponding residual.

Program 4.3 Fitting a 4-parameter logistic regression model using PROC NLMIXED

```
proc nlmixed data=calib technique=nrridg;
    parms top=3 bottom=0.2 c50=250 slope=1 theta=1 var=0.0001;
    expect=top+(bottom-top)/(1+(concentration/c50)**slope);
    model signal~normal(expect,(expect**theta)*var);
    predict top+(bottom-top)/(1+(concentration/c50)**slope) out=fitted_values;
    run;
```

Output from Program 4.3

			Parameter Estimates					
Parameter	Estimate	Standard Error	DF	t Value	Pr > |t|	Alpha	Lower	Upper
top	3.6584	0.1213	24	30.16	<.0001	0.05	3.4081	3.9087
bottom	0.3205	0.006942	24	46.18	<.0001	0.05	0.3062	0.3349
c50	219.50	14.2383	24	15.42	<.0001	0.05	190.12	248.89
slope	1.2628	0.04151	24	30.42	<.0001	0.05	1.1771	1.3485
theta	1.2879	0.4150	24	3.10	0.0049	0.05	0.4313	2.1445
var	0.000434	0.000130	24	3.35	0.0027	0.05	0.000167	0.000701

Output 4.3 shows that the parameter estimates and their approximate standard errors produced by PROC NLMIXED are very close to those displayed in the PROC NLIN output (Output 4.2). This is partly due to the fact that the estimated THETA parameter (1.2879) is very close to the exponent used in the weighting scheme in PROC NLIN. The fitted calibration curve is very similar to the calibration curve displayed in Figure 4.2.

4.3.5   Precision Profile for Immuno-Assays

After a calibration curve and weighting model have been chosen, a precision profile may be employed to characterize the precision of the back-calculated concentrations for unknown test samples using this calibration curve. The precision profile is a plot of the coefficient-of-variation (CV) of the calibrated concentration versus the true concentration on a log scale. Ideally, the calculated standard error of the calibrated concentration must take into account both the variability in the calibration curve and variability in the assay response. Wald's method is generally recommended for computing these standard errors (Belanger, Davidian and Giltinan, 1996) and the resulting coefficient-of-variation is given by:

$$\mathrm{CV}(x_0) = \frac{100}{x_0}\left[\left(\frac{\partial f^{-1}(y_0,\widehat{\beta})}{\partial y}\right)\frac{\widehat{\sigma}^2 y_0^{2\widehat{\theta}}}{m} + \left(\frac{\partial f^{-1}(y_0,\widehat{\beta})}{\partial y}\right)'\Sigma(\widehat{\beta})\left(\frac{\partial f^{-1}(y_0,\widehat{\beta})}{\partial y}\right)\right]^{1/2},$$

where m is the number of replicates and $\Sigma(\widehat{\beta})$ is the covariance matrix of the parameter estimates $\widehat{\beta}$.

As an illustration, Program 4.4 computes a precision profile for a 5-parameter logistic model:

$$y = f(x) = \beta_1 + \frac{\beta_2 - \beta_1}{[1 + (x/\beta_3)^{\beta_4}]^{\gamma}}.$$

The γ parameter is known as the asymmetry factor and, when it is set to 1, this model is equivalent to a 4-parameter logistic model fitted in Program 4.3. In fact, Program 4.4 fits this special case of the 5-parameter logistic model by forcing the γ parameter to be 1 (note the restrictions on this parameter in the BOUNDS statement). A general 5-parameter logistic model can be obtained by removing these constraints. Further, Program 4.4 uses the IML procedure to calculate the precision profile (in general, it is easier to use the matrix language instead of DATA steps to calculate the two terms under the square root). The covariance matrix is extracted from PROC NLMIXED and is imported into PROC IML. The computed response profile is displayed in Figure 4.3.

Program 4.4 Computation of the precision profile for a 4-parameter logistic regression model using PROC NLMIXED

```
proc nlmixed data=calib technique=nrridg;
    parms top=3 bottom=0.2 c50=250 slope=1 theta=1 var=0.0001 g=1;
    bounds g>=1, g<=1, theta>=0, var>0; * Constraints on model parameters;
    expect=top+(bottom-top)/((1+(concentration/c50)**slope)**g);
    model signal~normal(expect,(expect**theta)*var);
    predict top+(bottom-top)/((1+(concentration/c50)**slope)**g) out=fitted_values;
    ods output ParameterEstimates=parm_est_repl;
    ods output CovMatParmEst=cov_parm;
* Data set containing parameter estimates;
data b;
    set parm_est_repl;
    where parameter in ('top','bottom','c50','slope');
    keep estimate;
* Data set containing the covariance matrix of the parameter estimates;
data covb;
    set cov_parm;
    where parameter in ('top','bottom','c50','slope');
    keep top bottom c50 slope;
proc sql noprint;
    select distinct estimate into: sigma_sq from parm_est_repl where parameter='var';
    select distinct estimate into: theta from parm_est_repl where parameter='theta';
    select min(concentration) into: minconc from calib where concentration>0;
    select max(concentration) into: maxconc from calib;
    run;
    quit;
%let m=3; * Number of replicates at each concentration level;
proc iml;
    * Import the data sets;
    use b; read all into b;
    use covb; read all into covb;
    * Initialize matrices;
    y=j(101,1,0);
    h=j(101,1,0);
    hy=j(101,1,0);
    hb=j(101,4,0);
    varx0=j(101,1,0);
    pp=j(101,1,0);
    top=b[1];
    bottom=b[2];
    c50=b[3];
    slope=b[4];
```

```
    * Calculate the precision profile;
    do i=1 to 101;
        h[i,1]=10**(log10(&minconc)+(i-1)*(log10(&maxconc)-log10(&minconc))/100);
        y[i,1]=top+(bottom-top)/(1+(h[i,1]/c50)**slope);
        hy[i,1]=h[i,1]*(top-bottom)/(slope*(bottom-y[i,1])*(y[i,1]-top));
        hb[i,1]=h[i,1]/(slope*(y[i,1]-top));
        hb[i,2]=h[i,1]/(slope*(bottom-y[i,1]));
        hb[i,3]=h[i,1]/c50;
        hb[i,4]=-h[i,1]*log((bottom-y[i,1])/(y[i,1]-top))/(slope**2);
        varx0[i,1]=((hy[i,1]**2)*&sigma_sq*(y[i,1]**(2*&theta))/&m)+hb[i,]*covb*hb[i,]`;
        pp[i,1]=100*sqrt(varx0[i,1])/h[i,1];
    end;
    create plot var{h hy y pp};
    append;
    quit;
axis1 minor=none label=(angle=90 'CV (%)') order=(0 to 30 by 10);
axis2 minor=none label=('Concentration') logbase=10 logstyle=expand;
symbol1 value=none i=join color=black line=1 width=3;
proc gplot data=plot;
    plot pp*h/frame haxis=axis2 vaxis=axis1 vref=20 lvref=34 href=4.2 lhref=34;
    run;
    quit;
```

Figure 4.3 Precision profile based on a 4-parameter logistic regression model

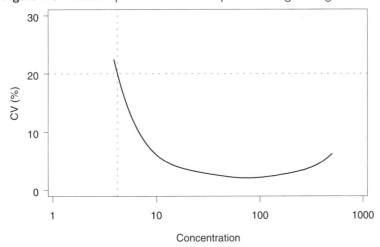

It can be seen in Figure 4.3 that, for concentration levels above 4 μM, the fitted model is a priori able to quantify with a precision better than 20% on a CV scale. Below this threshold, the variability of the back-calculated measurements explodes as expected because of the (low) asymptote effect. The precision achieves its maximum (approximately 2-3% on a CV scale) around the C50 value (220 μM).

The estimates of quantification limits from precision profiles are "optimistic" because they are based on only the calibration curve data themselves. These limits do not take into account matrix interference, cross-reactivity, operational factors, etc. However, these limits serve as a useful screening tool before beginning the pre-study validation exercise. Since the pre-study validation package encompasses several other sources of variability as well, if the quantification limits from a precision profile are not satisfactory, then almost definitely, the quantification limits derived from a rigorous pre-study validation package will not be satisfactory. In this case, it will be worth going back to the drawing board and further optimizing the assay protocol before proceeding to the pre-study validation phase.

4.3.6 Back-Calculated Quantities or Inverse Predictions

Once a calibration curve is fitted, concentrations of the samples of interest are calculated by inverting the estimated calibration function. In a pre-study validation, the calibration curves are fitted separately for each run and the validation samples are calculated using the calibration curve for the same run. The resulting data set consists of different concentration levels and, at each level, there are multiple runs and replicates within each run. Most of the time, the number of runs and the number of replicates are the same for all concentration levels. Inverse functions for widely used response functions are shown in Table 4.2.

For example, Program 4.5 calculates the concentrations of the validation samples (ELISA_VS data set available on the book's companion Web site) from the 4-parameter logistic regression model fitted by series and by plate as described in Program 4.3.

Program 4.5 Concentration calculation based on a 4-parameter logistic regression model

```
data calib;
    set elisa_cs;
    array y(3) rep1 rep2 rep3;
    do j=1 to 3;
        signal=y(j);
    output;
    end;
proc datasets nolist;
    delete parm_est;
%macro calib(series, plate);
    proc nlmixed data=calib technique=nrridg;
        parms top=3 bottom=0.2 c50=250 slope=1 theta=1 var=0.0001;
        expect=top+(bottom-top)/(1+(concentration/c50)**slope);
        model signal~normal(expect,(expect**(theta*2))*var);
        where series=&series and plate=&plate;
        ods output ParameterEstimates=parm_est_tmp;
    data parm_est_tmp;
        set parm_est_tmp;
        series=&series;
        plate=&plate;
    proc append data=parm_est_tmp base=parm_est;
        run;
%mend;
%calib(series=1, plate='A');
%calib(series=1, plate='B');
%calib(series=2, plate='A');
%calib(series=2, plate='B');
%calib(series=3, plate='A');
%calib(series=3, plate='B');
%calib(series=4, plate='A');
%calib(series=4, plate='B');
data valid;
    set elisa_vs;
    array y(3) rep1 rep2 rep3;
    do j=1 to 3;
        signal=y(j);
        output;
    end;
proc sort data=valid;
    by series plate;
```

```
proc transpose data=parm_est out=parm_est_t;
    var estimate;
    id parameter;
    idlabel parameter;
    by series plate;
data calc_conc;
    merge valid parm_est_t;
    by series plate;
    drop _name_ theta var;
data calc_conc;
    set calc_conc;
    if plate='A' then run=series;
    else run=series+4;
    calc_conc=c50*((((bottom-top)/(signal-top))-1)**(1/slope));
    run;
```

Table 4.2 Inverse Functions for Widely Used Response Functions

Response function	Back-calculated value x^* using the inverse function
$Y = \beta X$	$x^* = y/\widehat{\beta}$
$Y = \beta_0 + \beta_1 X$	$x^* = (y - \widehat{\beta}_0)/\widehat{\beta}_1$
$Y = \beta_0 + \beta_1 X + \beta_2 X^2$	$x^* = \left(-\widehat{\beta}_1 + \sqrt{\widehat{\beta}_1^2 - 4\widehat{\beta}_2(\widehat{\beta}_0 - y)}\right)/2\widehat{\beta}_2$
$Y = \beta_1 + \dfrac{\beta_2 - \beta_1}{1 + (X/\beta_3)^{\beta_4}}$	$x^* = \widehat{\beta}_3 \left(\dfrac{\widehat{\beta}_2 - \widehat{\beta}_1}{y - \widehat{\beta}_1} - 1\right)^{-1/\widehat{\beta}_4}$

4.4 Linearity

The linearity of an analytical procedure is defined in terms of its ability to obtain results directly proportional to the concentrations (quantities) of the analyte in the sample within a defined range (ICH, 1995). It is important to note that the linearity criterion is applied to the results, i.e., back-calculated quantities or concentrations, rather than the response signals or instrument response as a function of the dose or quantities.

To illustrate this concept, Program 4.6 performs a linearity assessment by comparing the original concentrations from the ELISA_VS data set to the corresponding back-calculated concentrations obtained in Program 4.5. This program uses the CALC_CONC data set created in Program 4.5. The results are displayed in Figure 4.4.

Program 4.6 Linearity assessment based on a 4-parameter logistic regression model

```
proc sql;
    create table linprof as
    select concentration, calc_conc, 't1' as type
            from calc_conc union
    select distinct concentration, concentration*0.7 as calc_conc, 't2' as type
            from calc_conc union
    select distinct concentration, concentration*1.3 as calc_conc, 't3' as type
            from calc_conc union
```

```
        select distinct concentration, concentration as calc_conc, 't4' as type
            from calc_conc;
    run;
    quit;
proc sort data=linprof;
    by type concentration;
axis1 label=("Nominal concentration") order=(0 to 600 by 200);
axis2 label=(angle=90 "Calculated concentration") order=(0 to 600 by 200);
symbol1 value=circle i=none color=black height=5;
symbol2 value=none i=join color=black line=34 width=3;
symbol3 value=none i=join color=black line=34 width=3;
symbol4 value=none i=join color=black line=1 width=3;
proc gplot data=linprof;
    plot calc_conc*concentration=type/haxis=axis1 vaxis=axis2 nolegend;
    run;
    quit;
```

Figure 4.4 Linearity assessment based on a 4-parameter logistic regression model

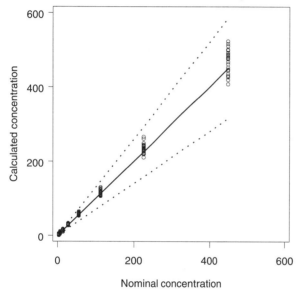

Figure 4.4 plots the raw concentrations versus the corresponding back-calculated concentrations. The solid line is the identity line and, if the method is reasonably unbiased, all data points are expected to cluster around this line. The two dotted lines represent the $[-30\%, 30\%]$ acceptance limits drawn on an absolute scale. If the precision of a method is satisfactory, most data points should lie between the two lines.

Figure 4.4 displays the linearity of the results for the data obtained using the calibration curve of Figure 4.2. The contrast between those two figures highlights clearly the contrast between the concepts of linearity of the results and response function of the signals. When chromatographic methods with narrow ranges are envisaged, both the response function and linearity will look linear but, except for this specific case, the two graphs will look different.

4.5 Accuracy and Precision

4.5.1 Accuracy (Trueness)

The accuracy (the preferred term is trueness) of an analytical procedure, according to ICH and related documents, expresses the closeness of agreement between the mean value obtained from a series of measurements and the value which is accepted either as a conventional true value or an accepted reference value (e.g., international standard, standard from a pharmacopoeia). It is a measure of the systematic error of test results obtained by the analytical method from its theoretical true/reference value. The measure of trueness is generally expressed in terms of recovery and absolute/relative bias.

Note that "accuracy" is synonymous with bias or trueness only within the pharmaceutical set of regulations covered by ICH (and related national documents implementing ICH Q2A and Q2B). Outside the pharmaceutical industry, e.g., in industries covered by the International Organization for Standardization or National Committee for Clinical Laboratory Standards guidelines (food, chemistry, clinical biology, and other industries), "accuracy" refers to the total error, i.e., the sum of trueness and precision.

4.5.2 Precision

The precision of an analytical procedure is defined by the closeness of agreement (usually expressed as standard deviation or coefficient of variation) between a series of measurements obtained from multiple sampling of the same homogeneous sample (independent assays) under the prescribed conditions. The term "independent results" means that the results are obtained and prepared the same way that unknown samples will be quantified and prepared.

Precision provides information on random errors and can be evaluated at three levels: repeatability, intermediate precision (within laboratory), and reproducibility (between laboratories). The precision represents the distribution of the random errors only and is not related to the true or specified value. A measure of precision is calculated from the standard deviation of the results.

Quantitative measures of precision depend in a critical manner on stipulated conditions. One can distinguish among the following three conditions:

- **Repeatability.** Repeatability expresses the precision under conditions where the results of independent assays are obtained by the same analytical procedure on identical samples in the same laboratory, with the same operator, using the same equipment during a short interval of time. It is estimated by the within-series variance component.

- **Intermediate precision.** Intermediate precision expresses the precision under conditions where the results of independent assays are obtained by the same analytical procedure on identical samples in the same laboratory, with different operators, using different equipment during a given time interval. It is estimated by the sum of within-series and between-series variance components. Intermediate precision is representative of the total random error for a single measurement within a laboratory, whatever the day or series.

- **Reproducibility.** Reproducibility expresses the precision under conditions where the results are obtained by the same analytical procedure on identical samples in different laboratories, with different operators, using different equipment. It is estimated by the sum of within-series, between-series and between-laboratories variance components.

4.5.3 Total Error or Measurement Error

The measurement error of an analytical procedure is related to the closeness of agreement between the value found and the value that is accepted either as a conventional true value or an accepted reference value. The closeness of agreement observed is based on the sum of the systematic and random errors; in other words, the total error linked to the result. Consequently, the measurement error is the expression of the sum of the trueness and precision, i.e., the total error.

As shown below, the observation X is a result of the true sample value μ_T, the method's Bias (estimated by the mean of many results) and Precision (estimated by the standard deviation or, in most cases, the Intermediate Precision). Equivalently, the difference between an observation X and the true value is the sum of the systematic and random errors, i.e., Total Error or Measurement Error.

$$X = \mu_T + \text{Bias} + \text{Precision}$$

$$\Leftrightarrow X - \mu_T = \text{Bias} + \text{Precision}$$

$$\Leftrightarrow X - \mu_T = \text{Total error}$$

$$\Leftrightarrow X - \mu_T = \text{Measurement Error}$$

4.5.4 Computational Aspects

As described in Section 4.1, the classical design of a pre-study validation consists in performing different runs with replicates in each run. Let's note p the number of runs, n_i the number of replicates in the ith run, $N = n_1 + \cdots + n_p$ the total number of measurements, and x_{ij} the jth measurement in the ith run.

Validation criteria such as accuracy and precision are estimated for each concentration level by statistical analysis of the back-calculated quantities. Computationally, a one-factor random effects Analysis of Variance (ANOVA) model is fit to the back-calculated values at each level with run as the random effects factor:

$$x_{ij} = \mu + \alpha_i + \varepsilon_{ij},$$

where μ is the mean of calculated concentrations, α_i and ε_{ij} are normally distributed with mean 0 and variances σ_B^2 and σ_W^2, respectively. Here σ_B^2 is the run-to-run variance and σ_W^2 is the within-run variance. The estimates of μ, σ_B^2, and σ_W^2 are given by:

$$\widehat{\mu} = \overline{x}_{..} = \frac{1}{N}\sum_{i=1}^{p} n_i \overline{x}_{i\cdot}, \quad \widehat{\sigma}_W^2 = \frac{1}{N-p}\sum_{i=1}^{p}\sum_{j=1}^{n_i}(x_{ij} - \overline{x}_{i\cdot})^2,$$

$$\widehat{\sigma}_B^2 = \frac{p-1}{N-\overline{n}}\left[\left(\frac{1}{p-1}\sum_{i=1}^{p}(\overline{x}_{i\cdot} - \overline{x}_{..})^2\right) - \widehat{\sigma}_W^2\right],$$

where $\overline{x}_{i\cdot} = n_i^{-1}\sum_{j=1}^{n_i} x_{ij}$ and $\overline{n} = N^{-1}\sum_{i=1}^{p} n_i$. By definition, the estimate of the within-run variance corresponds to the variance of repeatability and the sum of the within-run and run-to-run components corresponds to the intermediate precision. That is, the intermediate precision variance is $\sigma_B^2 + \sigma_W^2$.

The relative error (RE), the coefficient of variation of the intermediate precision (CV_{IP}), and the total error (TE) are calculated as follows:

$$\text{RE} = 100\frac{\widehat{\mu} - \mu}{\mu}, \quad \text{CV}_{IP} = 100\frac{\sqrt{\widehat{\sigma}_W^2 + \widehat{\sigma}_B^2}}{\mu}, \quad \text{TE} = \text{RE} + \text{CV}_{IP}.$$

Program 4.7 utilizes these formulas to compute the accuracy, precision, and total error performance parameters. In this example, each combination of series and plate is considered as a run. Again, this program uses the CALC_CONC data set created in Program 4.5.

Program 4.7 Computation of the accuracy and precision parameters

```
proc sort data=calc_conc;
    by concentration;
proc mixed data=calc_conc;
    class run;
    model calc_conc=/solution;
    random run;
    by concentration;
    ods output dimensions=dim;
    ods output solutionf=solf;
    ods output CovParms=var_comp;
proc sort data=calc_conc;
    by concentration run plate;
proc univariate data=calc_conc noprint;
    var calc_conc;
    by concentration run;
    output out=stat_by_conc_run n=n;
proc sql;
    create table stat_by_conc as
    select t1.concentration, cap_n, sum_n_sq/cap_n as n, n_run, mean,
        sqrt(sigma_w_sq) as sigma_w,
        100*sqrt(sigma_w_sq)/t1.concentration as cv_w,
        sqrt(sigma_b_sq) as sigma_b,
        100*sqrt(sigma_b_sq)/t1.concentration as cv_b,
        sqrt(sigma_w_sq+sigma_w_sq) as sigma_t,
        100*sqrt(sigma_w_sq+sigma_w_sq)/t1.concentration as cv_t
    from (select concentration, value as cap_n from dim where
        descr='Observations Used') as t1,
        (select concentration, estimate as mean from solf) as t2,
        (select concentration, sum(n**2) as
        sum_n_sq from stat_by_conc_run group by concentration) as t3,
        (select concentration, estimate as sigma_b_sq from var_comp
        where covparm='run') as t4,
        (select concentration, estimate as sigma_w_sq from var_comp
        where covparm='Residual') as t5,
        (select count(run) as n_run from (select distinct run from
        stat_by_conc_run)) as t6 where t1.concentration=t2.concentration=
        t3.concentration=t4.concentration=t5.concentration;
    run;
    quit;
* Calculation of validation criteria;
data stat_by_conc;
    set stat_by_conc;
    format mean re cv_w cv_b cv_t te 4.1 low_tol_lim_rel upp_tol_lim_rel 5.1;
    re=100*(mean-concentration)/concentration;
    te=abs(re)+cv_t;
    r=(sigma_b/sigma_w)**2;
    b=sqrt((r+1)/(n*r+1));
    ip=sigma_b**2+sigma_w**2;
    ddl=((r+1)**2)/(((r+1/n)**2)/(n_run-1)+(1-1/n)/(cap_n));
    low_tol_lim_abs=mean-tinv(0.975,ddl)*sqrt(ip*(1+1/(cap_n*b**2)));
    upp_tol_lim_abs=mean+tinv(0.975,ddl)*sqrt(ip*(1+1/(cap_n*b**2)));
    low_tol_lim_rel=(low_tol_lim_abs-concentration)*100/concentration;
    upp_tol_lim_rel=(upp_tol_lim_abs-concentration)*100/concentration;
```

```
     label concentration='Nominal*concentration'
           mean='Calculated*concentration'
           re='Relative*error (%)'
           cv_w='Within-run*CV (%)'
           cv_b='Run-to-run*CV (%)'
           cv_t='Intermediate*precision*CV (%)'
           te='Total*error(%)'
           low_tol_lim_rel='Upper 95%*tolerance*limit'
           upp_tol_lim_rel='95% TI*upper*limit';
proc print data=stat_by_conc noobs split='*';
     format mean re cv_w cv_b cv_t te 4.1 low_tol_lim_rel upp_tol_lim_rel 5.1;
     var concentration mean re cv_w cv_b cv_t te;
     run;
```

Output from Program 4.7

Nominal concentration	Calculated concentration	Relative error (%)	Within-run CV (%)	Run-to-run CV (%)	Intermediate precision CV (%)	Total error(%)
3.5	3.2	-7.3	37.7	23.0	53.3	60.6
7.0	8.2	16.6	14.7	18.5	20.8	37.5
14.1	15.5	9.8	12.7	8.8	18.0	27.8
28.0	30.0	7.3	4.3	5.2	6.1	13.4
56.0	57.9	3.3	2.7	5.3	3.9	7.2
113.0	114	1.0	2.4	6.2	3.4	4.3
225.0	235	4.6	3.7	4.5	5.2	9.8
450.0	469	4.1	5.5	5.0	7.7	11.9

Output 4.7 shows the estimates of the main accuracy, precision, and total error performance parameters of the method as a function of the estimated concentration. It is clear that, at low concentration levels (less than or equal to 14.1 μM), either the trueness or precision is not acceptable. This is well summarized by the total error that becomes smaller than 30% for concentration values greater than 14.1 μM. This already gives an idea about the capability of the method, but as it will be shown in the following section, the decision rule should be established on a more refined method that takes the notion of risk into consideration.

4.6 Decision Rule

It was mentioned in the Introduction that Equation (4.2), which describes the main objective of an analytical method, cannot be solved exactly. A simple way to resolve this problem and make a reliable decision, proposed by several authors (Hubert et al., 2004; Boulanger et al., 2000a, 2000b; Hoffman and Kringle, 2005), relies on computing the β-expectation tolerance intervals (Mee, 1984):

$$E_{\widehat{\mu}_M, \widehat{\sigma}_M} \left(P_X \left[\widehat{\mu}_M - k\widehat{\sigma}_M < X < \widehat{\mu}_M + k\widehat{\sigma}_M | \widehat{\mu}_M, \widehat{\sigma}_M \right] \right) = \beta,$$

where the k factor is determined so that the expected proportion of the population falling within the interval is equal to β. If the β-expectation tolerance interval is totally included within the acceptance limits $[-\lambda, +\lambda]$, i.e., if $\widehat{\mu}_M - k\widehat{\sigma}_M > -\lambda$ and $\widehat{\mu}_M + k\widehat{\sigma}_M < \lambda$, the expected proportion of measurements within the same acceptance limits is greater than or equal to β. Note that the opposite statement is not true, i.e., if either $\widehat{\mu}_M - k\widehat{\sigma}_M < -\lambda$ or $\widehat{\mu}_M + k\widehat{\sigma}_M > \lambda$, the expected proportion is not necessarily smaller than β.

Most of the time, an analytical procedure is intended to quantify over a range of quantities or concentrations. Consequently, during the validation phase, samples are prepared to adequately cover this range, and a β-expectation tolerance interval is calculated at each level.

A measurement error profile is obtained, on one hand, by connecting the lower limits and, on the other hand, by connecting the upper limits. A procedure is valid over a certain range of values if the measurement error profile is included within the acceptance limits $[-\lambda, +\lambda]$. This concept is illustrated in Figure 4.5 for a chromatographic bio-analytical method (Hubert et al., 1999). In the left panel of Figure 4.5, the measurement error profile is included within the 20% acceptance limits over the entire range of concentrations, and thus the method is valid to quantify concentration values over this range. However, in the right panel of Figure 4.5, the method can accurately quantify only concentration values greater than 300μM.

Figure 4.5 Measurement error profiles with observations, 20% acceptance limits (dotted lines), β-expectation tolerance limits (solid curves), and biases (dashed curves) expressed in terms of relative errors as a function of the concentration values for a chromatographic bio-analytical procedure

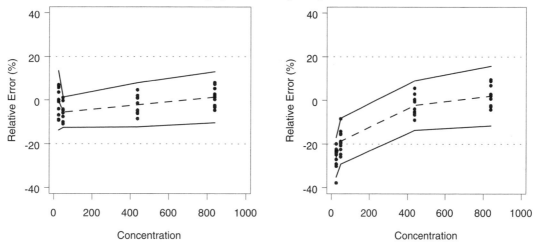

A measurement error profile gives the analyst a sense of what a procedure will be able to produce over the intended range. The interpretation of a measurement error profile is that it shows where $100\beta\%$ of the measures provided by this analytical method will lie, which is directly connected to the objective of the analytical method (produce measures close to the unknown true values).

In practice, the β-expectation tolerance interval is obtained as follows (on a relative scale):

$$\left[\mathrm{RE} - Q_t\left(v, \frac{1+\beta}{2}\right)\sqrt{1 + \frac{1}{pnB^2}}\mathrm{CV}_{IP}, \mathrm{RE} + Q_t\left(v, \frac{1+\beta}{2}\right)\sqrt{1 + \frac{1}{pnB^2}}\mathrm{CV}_{IP} \right],$$

where p is the number of runs, n is the number of replicates within each run, $Q_t(a,b)$ is the $100b\%$ quantile of the t distribution with a degrees of freedom, and

$$R = \frac{\hat{\sigma}_B^2}{\hat{\sigma}_W^2}, \quad B = \sqrt{\frac{R+1}{nR+1}}, \quad v = (R+1)^2 \left[\frac{(R+1/n)^2}{p-1} + \frac{1-1/n}{pn} \right]^{-1}.$$

Program 4.8 lists and plots the β-expectation tolerance limits for this study considered in Program 4.7 (the limits are included in the STAT_BY_CONC data set created by Program 4.7).

Program 4.8 Computation of the β-expectation tolerance limits

```
proc print data=stat_by_conc noobs split='*';
    var concentration mean low_tol_lim_rel upp_tol_lim_rel;
    run;
axis1 minor=none label=(angle=90 "Total error (%)") order=(0 to 70 by 10);
axis2 minor=none label=("Nominal concentration") logbase=10 logstyle=expand;
symbol1 value=dot i=join color=black line=1 width=3 height=3;
proc gplot data=stat_by_conc;
    format te 4.0;
    plot te*concentration/frame haxis=axis2 vaxis=axis1 vref=30 lvref=34;
    run;
    quit;
axis1 minor=none label=(angle=90 "Relative error (%)") order=(-120 to 100 by 20);
axis2 minor=none label=("Nominal concentration") logbase=10 logstyle=expand;
symbol1 value=dot i=join color=black line=1 width=3 height=3;
symbol2 value=none i=join color=black line=8 width=3;
symbol3 value=dot i=join color=black line=1 width=3 height=3;
proc gplot data=stat_by_conc;
    format low_tol_lim_rel re upp_tol_lim_rel 4.0;
    plot (low_tol_lim_rel re upp_tol_lim_rel)*concentration/overlay frame
        haxis=axis2 vaxis=axis1 vref=-30,30 lvref=34;
    run;
    quit;
```

Output from Program 4.8

Nominal concentration	Calculated concentration	Lower 95% tolerance limit	Upper 95% tolerance limit
3.5	3.2	-103	88.4
7.0	8.2	-37.1	70.4
14.1	15.5	-23.8	43.4
28.0	30.0	-8.1	22.7
56.0	57.9	-10.8	17.4
113.0	114	-15.0	16.9
225.0	235	-8.5	17.7
450.0	469	-12.2	20.5

Output 4.8 shows that the tolerance interval is fully included within the 30% acceptance limits for concentration levels strictly greater than 14.1 μM. It is also important to notice that once the tolerance interval is fully included within the acceptance limits, the classical validation criteria are also satisfied. That is, the accuracy and intermediate precision are better than the usually recommended 15% threshold. Figure 4.6 provides a graphical summary of the quantities produced by Program 4.8. In the right panel of Figure 4.6, the two horizontal (dotted) lines represent the 30% acceptance limits and are equivalent to the acceptance limits in Figure 4.4 on an absolute scale. The dashed curve represents the accuracy (relative error).

Program 4.9 computes the individual total and individual relative errors and plots them along with the total error and measurement error curves displayed in Figure 4.6. The resulting total error and measurement error profiles are shown in Figure 4.7.

Figure 4.6 Total error (left panel) and measurement error (right panel) profiles

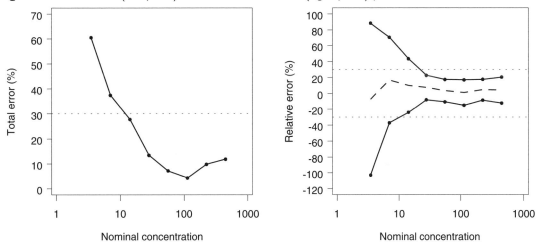

Program 4.9 Computation of individual total and individual relative errors

```
proc sql;
    create table te as
    select distinct 2 as type, calc_conc.concentration,
        abs(calc_conc-calc_conc.concentration)*100/calc_conc.concentration as te
    from calc_conc, stat_by_conc
    where calc_conc.concentration=stat_by_conc.concentration
    union
    select 1 as type, concentration, te from stat_by_conc;
    run;
    quit;
axis1 minor=none label=(angle=90 "Total error (%)") order=(0 to 100 by 20);
axis2 minor=none label=("Nominal concentration") logbase=10 logstyle=expand;
symbol1 value=dot i=join color=black line=1 width=3;
symbol2 value=circle i=none color=black height=5;
proc gplot data=te;
    plot te*concentration=type/vaxis=axis1 haxis=axis2 vref=30 lvref=34 nolegend;
    run;
    quit;
data me;
    set stat_by_conc;
    array y(3) low_tol_lim_rel re upp_tol_lim_rel;
    do j=1 to 3;
        me=y(j); type=j;
    output;
    end;
proc sql;
    create table me as
    select distinct 4 as type, calc_conc.concentration,
        (calc_conc-calc_conc.concentration)*100/calc_conc.concentration as me
    from calc_conc, stat_by_conc
    where calc_conc.concentration=stat_by_conc.concentration
    union
    select type, concentration, me from me;
    run;
    quit;
```

```
axis1 minor=none label=(angle=90 "Relative error (%)") order=(-120 to 100 by 20);
axis2 minor=none label=("Nominal concentration") logbase=10 logstyle=expand;
symbol1 value=dot i=join color=black line=1 width=3 height=5;
symbol2 value=none i=join color=black line=20 width=3;
symbol3 value=dot i=join color=black line=1 width=3 height=5;
symbol4 value=circle i=none color=black height=5;
proc gplot data=me;
    plot me*concentration=type/frame haxis=axis2 vaxis=axis1 nolegend vref=-30,30 lvref=34;
    run;
    quit;
```

Figure 4.7 demonstrates that most of the individual relative errors are within the tolerance limits. This highlights the fact that both the total error and measurement error profiles provide clues about how extreme results may be when produced by an analytical method. That is, most results will be within the calculated limits. The measurement error profile is preferred because it is predictive (with a specified risk) of future results obtained by the method, while the total error profile is more descriptive.

Figure 4.7 Total error (left panel) and measurement error (right panel) profiles with individual total and individual relative errors

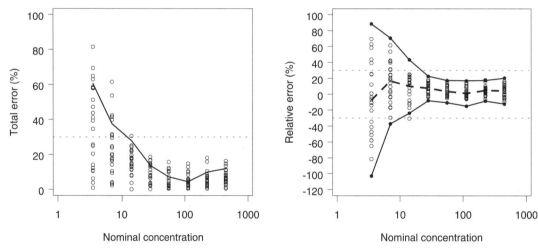

The choice of the criteria depends on the analyst's preference. If a certain amount of risk is associated with the decision, the measurement error profile should be used. Otherwise, the total error profile could be considered.

In conclusion, the measurement error profile is an easy-to-interpret tool that allows the analyst to check trueness, precision, and accuracy of the method. Moreover, it is statistically meaningful and objectively oriented. Note that the lower and upper limits of quantitation can be estimated using either total error and measurement error profiles by intersecting the profiles with the acceptance limits.

4.7 Limits of Quantification and Range of the Assay

The upper and lower limits of quantitation (ULOQ and LLOQ) of an analytical procedure are the lowest and highest amounts of the targeted substance in the sample that can be quantitatively determined under the prescribed experimental conditions. As a consequence, the range of an analytical procedure is the range between the lower and upper limits of quantitation for which the analytical procedure was demonstrated to have a suitable level of measurement error.

In practice, the information needed to establish the limits of quantitation and the associated range is already available in the measurement profile plot. As proposed by Hubert et al. (1999) and Hubert et al. (2004), the limits of quantitation are the most extreme (low, high) concentrations (quantities) at which the tolerance interval is still within the acceptance limits, should the tolerance limits cross the acceptance limits. If all tolerance limits lie within the acceptance limits, the limits of quantitation are defined as the most extreme quantities tested in the study.

Figure 4.8 Derivation of the limits of quantification and range of the assay using the measurement error profile

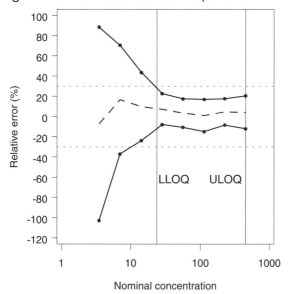

Figure 4.8 shows how the upper and lower limits of quantitation are defined. In this example, the lower limit is approximately 24 μM, and the upper limit is the maximum concentration investigated in this study, i.e., 450 μM.

4.8   Limit of Detection

The limit of detection (LOD) of an analytical procedure is the lowest amount of the targeted substance in the sample that can be detected reliably, but not necessarily quantified as an accurate value using the experimental conditions prescribed. A variety of methods to estimate the LOD have been proposed in the literature, generally based on the calibration information and estimates. None of them are really satisfactory and are sensitive to various assumptions related to the model, the design of experiment and modeling of heterogeneity of variances. Based on our experience, the "best" and most consistent estimate proposed for bioanalytical methods aiming at covering a large range of concentrations is based on dividing the lower limit of quantitation by 3 or 3.33. This is justified by the fact that it is largely accepted that the LOD is three times the noise of the signal and the LLOQ is ten times the same noise.

So, for the example used in Program 4.8, the LOD can be established around 8 μM. This is consistent with Figure 4.2. The 8 μM level appears to be the lowest concentration level producing a signal that is clearly above the signals in the lower asymptote.

4.9   Summary

Throughout this chapter we have proposed ways to understand, estimate, and interpret various criteria required for assessing the validity of an analytical method. These criteria must be reported, and the code used to compute the criteria must be documented according to best practices. Regardless of the complexity of computations and models, the objective of an analytical method (are the measurement errors acceptable?) should never be forgotten and should remain the primary focus. The information needed to make a decision is contained in the measurement error profile. All performance criteria—linearity, accuracy, precision, limits, measurement errors—can be assessed using a graphical profile and can be easily understood and interpreted by an analyst.

Another important remark related to the decision is that, if the tolerance intervals are within the acceptance limits, all of the required criteria are guaranteed to be met. The opposite, however, is not true. That is, even when all of the performance criteria are satisfied, the measurement errors will not necessarily be acceptable. Although it is common to assume that good methods will always produce good results and most regulatory documents were written in this spirit, it is important to remember that only the opposite statement holds true: good results can only be obtained with a good method.

4.10 Terminology

Analytical procedure (or method or assay). Written procedure which describes all the means and the operating procedures required to perform the analysis of the analyte. That is, field of application, principle and/or reactions, definitions, reactants, equipments, operating procedures, expression of results, suitability tests, test reports.

Analyte or activity. The analyte (or activity when relevant) is the matter of the analytical procedure. The analyte is a physical entity (e.g., water activity), chemical entity (e.g., active substance alone or in a pharmaceutical formulation, total lipids, aspartame, lead) or biological entity (e.g., atpmetric activity). In the case of quantitative analytical procedures of farm and food products, "analyte" is equivalent to "measurand".

Matrix. All constituents of the laboratory sample other than the analyte. A type of matrix is defined as a group of materials or of products considered by the analyst as having a homogeneous behavior with regard to the analytical method used.

Blank. Test performed in the absence of the matrix (blank reactant) or in a matrix without analyte (blank matrix). By extension, the instrument response in the absence of the analyte is used (blank instrumental).

Accepted reference value. An accepted reference value is a value used as a reference, agreed for a comparison and derived from:

- A theoretical or established value, based on scientific reasons.

- An assigned or certified value, based on experimental data from a national or international organization.

- A consensus value, based on a collaborative experimental work.

- The mathematical expectation of the (measurable) quantity, in cases where the previous points are not applicable, i.e., the mean of a specified population of determinations (cf. NF ISO 5725-1).

Calibration standard (or calibration sample). Calibration standards are samples of known concentrations, with or without matrix, that allow drawing the calibration curve.

Validation standard (or validation sample). Validation standards are samples reconstituted in the matrix or in any other reference material with true values set by consensus and used to validate the analytical procedure.

References

Belanger, B.A., Davidian, M., Giltinan, D.M. (1996). "The effect of variance function estimation on nonlinear calibration inference in immunoassay data." *Biometrics*. 52, 158–175.

Boulanger, B., Hubert, P., Chiap, P., Dewé, W. (2000a). "Objectives of pre-study validation and decision rules." Presented at AAPS APQ Open forum, Indianapolis, Indiana.

Boulanger, B., Hubert, P., Chiap, P., Dewé, W., Crommen, J. (2000b). "Analyse statistique des résultats de validation de méthodes chromatographiques." Presented at Journées du GMP, Bordeaux, France.

DeSilva, B., Smith, W., Weiner, R., Kelley, M., Smolec, J., Lee, B., Khan, M., Tacey, R., Hill, H., Celniker, A. (2003). "Recommendations for bioanalytical method validation of ligand-binding assays to support pharmacokinetic assessments of macromolecules." *Pharmaceutical Research.* 20, 1885–1900.

FDA (Food and Drug Administration). (2001). Guidance for industry: Bioanalytical methods validation.

Finney, D. (1978). *Statistical Methods in Biological Assays.* London: Charles Griffin.

Findlay, J.W.A., Smith, W.C., Lee, J.W., Nordblom, G.D., Das, I., Desilva, B.S., Khan, M.N., Bowsher, R.R. (2001). "Validation of immunoassays for bioanalysis: a pharmaceutical industry perspective." *Journal of Pharmaceutical and Biomedical Analysis.* 21, 1249–1273.

Hoffman, D., Kringle, R. (2005). "Two-sided tolerance intervals for balanced and unbalanced random effects models." *Journal of Biopharmaceutical Statistics.* 15, 283–293.

Hubert, P., Chiap, P., Crommen, J., Boulanger, B., Chapuzet, E., Mercier, N., Bervoas-Martin, S., Chevalier, P., Grandjean, D., Lagorce, P., Lallier, M., Laparra, M.C., Laurentie, M., Nivet, C. (1999). "The SFSTP guide on the validation of chromatographic methods for drug bioanalysis: from the Washington Conference to the laboratory." *Analytica Chimica Acta.* 391, 135–148.

Hubert, P. Nguyen-Huu, J.-J., Boulanger, B., Chapuzet, E., Chiap, P., Cohen, N., Compagnon, P.-A., Dewé, W., Feinberg, M., Lallier, M., Laurentie, M., Mercier, N., Muzard, G., Nivert, C., Valat, L. (2004). "Harmonization of strategies for the validation of quantitative analytical procedures. An SFSTP proposal Part I." *Journal of Pharmaceutical and Biomedical Analysis.* 36, 579–586.

ICH (International Conference on Harmonization). (1995). Guideline on validation of analytical procedures: definitions and terminology.

ICH (International Conference on Harmonization). (1997). Guideline on validation of analytical procedures: methodology.

International Organization for Standardization. ISO 5725-1:1994 *Accuracy (Trueness and Precision) of Measurement Methods and Results—Part 2: Basic Method for the Determination of Repeatability and Reproducibility of a Standard Measurement Method [Edition 1].*

Lee, J.W., Nordblom, G.D., Smith, W.C., Bowsher, R.R. (2003). "Validation of bioanalytical assays for novel biomarkers: Practical recommendations for clinical investigation of new drug entities." In *Biomarkers in Clinical Drug Development.* Bloom, J., Dean, R.A., eds. New York: Marcel Dekker, 119–148.

Mee, R.W. (1984). "β-expectation and β-content tolerance limits for balanced one-way ANOVA pandom model." *Technometrics.* 26, 251–254.

Mire-Sluis, A.R, Barrett, Y.C., Devanarayan, V., Koren, E., Liu, H., Maia, M., Parish, T., Scott, G., Shankar, G., Shores, E., Swanson, S.J., Taniguchi, G., Wierda, D., Zuckerman, L.A. (2004). "Recommendations for the design of immunoassays used in the detection of antibodies against biological products." *Journal of Immunological Methods.* 289, 1–16.

O'Connell, M.A., Belanger, B.A., Haaland P.D. (1993). "Calibration and assay development using the four-parameter logistic model." *Chemometrics and Intelligent Laboratory Systems.* 20, 97–114.

SAS Institute Inc. *SAS/STAT User's Guide.* Cary, NC: SAS Institute Inc.

Smith, W.C., Sittampalam, S.G. (1998). "Conceptual and statistical issues in the validation of analytical dilution assays for pharmaceutical applications." *Journal of Biopharmaceutical Statistics.* 8, 509–532.

Chapter 5

Some Statistical Considerations in Nonclinical Safety Assessment

Wherly Hoffman
Cindy Lee
Alan Chiang
Kevin Guo
Daniel Ness

Understanding the intrinsic toxicological properties of chemicals is fundamental to evaluating their safety. Such data are derived from toxicological studies that are required by national and international regulatory organizations. These data are used to determine the potential hazards and to gain an understanding of the potential risks to humans. This chapter provides a brief overview of the role of nonclinical safety assessment in drug development and covers several important statistical considerations in defining hazards. The topics include randomization, power evaluation, and data analysis. Examples are given throughout the chapter to illustrate the described statistical methods for each topic.

5.1 Overview of Nonclinical Safety Assessment

The goal of drug development is to identify safe and efficacious treatments for human diseases. Safety and efficacy are assessed through careful evaluation in animal models and *in vitro* systems (nonclinical setting) and in exposed human populations (clinical setting)

Wherly Hoffman is Head, Toxicology/Drug Disposition and Animal Health Statistics, Eli Lilly, USA. Cindy Lee is Associate Senior Statistical Analyst, Toxicology Statistics, Eli Lilly, USA. Alan Chiang is Senior Research Scientist, Toxicology Statistics, Eli Lilly, USA. Kevin Guo is Assistant Senior Statistician, Toxicology Statistics, Eli Lilly, USA. Daniel Ness is Research Advisor, Toxicology, Eli Lilly, USA.

throughout drug development and post-marketing. This chapter focuses on the design and analysis of studies that are integral to nonclinical safety assessment.

The objectives in nonclinical safety assessment are to define the toxicity profile of candidate drugs, estimate the margin of safety by understanding the relationship between toxic exposures and efficacious exposures, and provide a judgment on the likelihood that the animal findings can be extrapolated to humans. These assessments are important

- prior to candidate selection because they provide internal decision-makers with a probability of technical success for the new compound,
- during clinical development because they protect the safety of individuals in clinical trials, and
- during the post-marketing phase because they support alternative formulations or clinical indications or further investigation of newly discovered safety issues.

It is beyond the scope of this chapter to consider all questions asked in nonclinical studies; however, a few questions are particularly central to safety assessment. These include:

- What dose or plasma exposure of the experimental drug is not associated with any adverse outcome (i.e., the *no observed adverse effect level*, or NOAEL)?
- What dose or plasma exposure of drug does not result in any observed biological effect (i.e., the *no observed effect level*, or NOEL)?
- What is the *maximum tolerated dose* (MTD)?
- What is the nature of the dose-response relationship? For example, how steep is the dose-response curve as judged by the difference between the NOAEL and the MTD, or is the response monotonic or nonmonotonic?
- What are the target organs of toxicity?
- Are the toxicities monitorable? That is, are there antemortem measures that are predictive of tissue pathology?
- Are the toxicities reversible?

Most of the studies designed to answer the questions above are mandated by regulatory agencies worldwide. Guidance documents for pharmacology and toxicology are available electronically via the Food and Drug Administration (FDA), European Medicines Agency (EMEA), and the International Conference on Harmonization (ICH) Web sites. These guidance documents assist in determining the relevant single-dose, repeat-dose, genetic toxicity, safety pharmacology (e.g., central nervous system, cardiovascular, and respiratory) studies, immunotoxicity, reproductive/developmental, and carcinogenicity studies as well as the necessary studies to ensure adequate quality of the formulated material (e.g., impurity qualification). Other studies to examine more closely certain mechanisms of toxicity or compound-specific issues are conducted as needed.

5.2   Key Statistical Aspects of Toxicology Studies

The key statistical aspects of the design and analysis of toxicology studies to be discussed are randomization, power evaluation, and data analysis.

5.2.1   Randomization

Since proper randomization is the foundation of valid statistical inference, this chapter starts with the randomization methods for two commonly used designs in toxicology, namely, the parallel design and Latin square design.

Parallel designs include studies with one factor, two factors, one-factor with repeated measurements, and more. In a parallel design, separate groups of animals are assigned to combinations of factor levels. For example, in a one-factor design with four dose groups, there would be four different groups of animals, each receiving one of the four doses. For a two-factor design with four dose groups and two routes of delivery, there would be eight different groups of animals, each receiving one of the eight combinations of the four doses and two routes.

Latin square designs are efficient for incorporating three factors each with the same number of levels. For example, a basic 4×4 Latin square design can accommodate four animals, each receiving one of four treatments in four dosing periods. This design allows for the evaluation of the treatment, animal, and time effects. However, one has to assume that all two- or three-way interactions are not significant.

The random assignment of animals to the groups for the parallel and Latin square designs is discussed in Section 5.3.

5.2.2   Power Evaluation

Power evaluation characterizes the strength of an inferential test. It is a function of the size of the change to be detected, the variability, the Type I error rate, and the sample size. Although certain sample sizes in standard toxicology studies are commonly accepted and based, in part, on regulatory guidance, many toxicology studies targeted at special endpoints merit an assessment of the power and sample size. An example of this is the evaluation of QT prolongation in large animal toxicology studies with four treatment groups and three or four animals per group for each sex. The QT interval is a measure of the time between the start of the Q wave and the end of the T wave in the heart's electrical cycle. The details of the study design and statistical tests are described in Section 5.4. The evaluation of power is performed in a two-factor analysis of variance (ANOVA) framework by simulation.

5.2.3   Data Analysis

For each well-designed study, the statistical hypotheses and tests are defined in the protocol. The most commonly collected data for toxicology studies is body weight. A body weight change, either a gain or loss, is important for monitoring the well-being of the animal and toxicities of a compound. Section 5.5 describes statistical methods used in the analysis of body weight data in toxicology studies based on a one-factor ANOVA model with repeated measures.

To save space, some SAS code has been shortened and some output is not shown. The complete SAS code and data sets used in this book are available on the book's companion Web site at `http://support.sas.com/publishing/bbu/companion_site/60622.html`.

5.3   Randomization in Toxicology Studies

One of the most important elements of any research design is the concept of randomization. This concept is central to all types of toxicology research (Lin, 2001). Random assignment of experimental animals to different groups helps reduce the potential biases among the comparison groups and allows a valid interpretation of the research results. Most statistical packages can produce random numbers within a specified range, which can be used to assign experimental units to treatments. Some textbooks have tables of random numbers designed for this purpose.

There are generally two approaches to creating randomization tables in SAS. The first approach is based on a direct generation of randomization tables using a DATA step. The other approach relies on the PLAN procedure that provides different randomization layouts depending on the experiment. For a more detailed description of random allocation methods in pre-clinical and clinical studies, see Chapter 9, "Allocation in Randomized Clinical Trials".

5.3.1   Block Randomization in a Parallel Design

In this subsection, the randomized complete block design is discussed to illustrate the randomization approach. Consider designing a study to determine whether different dose levels of a compound affect the liver weight of a mouse after one week of treatment. If four dose levels, including a control, are evaluated and 48 mice are available for the study, then these mice would need to be assigned to the four dose groups. Since the liver weight is related to the body weight, a common practice is to assign the animals to treatments using a randomized complete block design with body weight stratification.

A *randomized complete block design* structure is any blocking scheme in which the number of experimental units within a block is a multiple of the number of treatments, and thus the experimental units can be assigned completely at random to a complete set of treatments in each block.

Program 5.1 creates a randomization table for the study using a DATA step. First, the program defines the allocation number for each animal based on the animal id and sorts the animals by their body weights. The random variable RAND is created using the RANUNI function that generates a random number from the uniform distribution on the $(0, 1)$ interval. The BLOCK variable assigns each animal to a block: the first four mice to Block 1 (the four lightest animals), the second four mice to Block 2, etc. The RANK procedure ranks the random numbers from the smallest to the largest within each block. The ranking of the random numbers within a block is used for treatment assignment.

Program 5.1 Randomization using a DATA step

```
data bodyweight;
    animal_id=_n_;
    input bw @@;
    datalines;
22.1 25.5 24.2 26.1 23.3 22.0 21.8 24.8 23.1 23.0 23.0 24.8
25.3 25.6 24.8 24.5 26.0 23.6 26.3 24.0 26.9 25.9 22.7 22.4
22.5 22.3 22.3 25.5 20.9 24.5 22.2 23.3 20.3 26.3 27.6 26.5
26.8 25.6 26.6 23.5 22.4 21.3 23.7 26.8 24.6 24.2 26.1 26.2
;
proc sort data=bodyweight;
    by bw;
data bdwt;
    set bodyweight;
    rand=ranuni(1202019);
    block=1+int((_n_-1)/4);
proc rank data=bdwt out=bwstrat (drop=rand);
    by block;
    var rand;
    ranks group;
proc sort data=bwstrat;
    by group animal_id;
proc print data=bwstrat noobs label;
    label animal_id="Animal ID"
          bw="Body weight"
          block="Block"
          group="Treatment group";
    run;
```

Partial output from Program 5.1

Animal ID	Body weight	Block	Treatment group
10	23.0	4	1
12	24.8	8	1
18	23.6	5	1
20	24.0	6	1
24	22.4	3	1
28	25.5	9	1
29	20.9	1	1
31	22.2	2	1
37	26.8	12	1
39	26.6	11	1
45	24.6	7	1
48	26.2	10	1
2	25.5	8	2
7	21.8	1	2
17	26.0	10	2

Output 5.1 displays the first 15 rows in the randomization table generated by Program 5.1.

5.3.2   Randomization in a Latin Square Design

The other popular design used in toxicology studies is the Latin square design. The *Latin square design* consists of blocking in two directions. For an experiment involving n treatments, n^2 experimental units are assigned into an $n \times n$ square in which the rows are called row blocks and the columns are called column blocks. Thus, the $n \times n$ arrangement of experimental units is blocked in two directions. In a Latin square design, the treatments are randomly assigned to animals in the square such that each treatment occurs once and only once in each row block and once and only once in each column block. See Box et al. (1978) for various arrangements of treatments into row and column blocks.

The %LATINSQ macro in Program 5.2 uses PROC PLAN to generate an $n \times n$ Latin square design. The macro includes two arguments: N is the dimension of the Latin square, and SEED is the seed for randomization.

The FACTORS statement in PROC PLAN specifies the row and column blocks of the design (ANIMAL_ID and PERIOD). The TREATMENTS statement specifies the treatment groups (GROUP). The OUTPUT statement saves the design generated to the specified data set (LATIN). Creating a randomization schedule for a Latin square design involves randomly permuting the row, column, and treatment values independently. To accomplish this, the association type of each factor is specified as RANDOM in the OUTPUT statement. The output is summarized using the TABULATE procedure.

Program 5.2 Randomization in an $n \times n$ Latin square design

```
%macro latinsq(n,seed);
proc plan seed=&seed;
    factors animal_id=&n ordered period=&n ordered/noprint;
    treatments group=&n cyclic;
    output out=latin period random animal_id random group random;
proc tabulate data=latin formchar='                           ';
    label animal_id="Animal ID" period="Period";
    keylabel sum=' ';
    class period animal_id;
    var group;
    table animal_id, period*(group=''*f=3.)/rts=8;
    run;
%mend latinsq;
%latinsq(n=4,seed=1034567);
```

Output from Program 5.2

```
            Period

          1   2   3   4

Animal ID

 1        1   3   4   2
 2        3   4   2   1
 3        2   1   3   4
 4        4   2   1   3
```

5.4   Power Evaluation in a Two-Factor Model for QT Interval

This section describes the process of estimating the power of a trend test in a two-factor ANOVA model. The application is for a general toxicology study in beagle dogs with a vehicle control (Dose 0) and three groups of increasing doses of a compound (Doses 1, 2, and 3). A sample size of three or four animals per sex per group is generally used. The purpose of the analysis is to evaluate the treatment effects on heart rate-corrected QT intervals by identifying the highest no observed effect dose level (NOEL). The power evaluation is important because the recent ICH S7B guideline (2004) recommends that the sensitivity and reproducibility of the *in vivo* test system be characterized. See also Chiang et al. (2004).

The QT interval of the electrocardiogram (ECG) is a common end-point for characterizing potential drug-associated delayed ventricular repolarization *in vivo* (ICH S7B, 2004). It has been conjectured that delayed ventricular repolarization caused by a compound may lead to serious ventricular tachyarrhythmias in humans (see, for example, Kinter, Siegl and Bass, 2004). Statistical analysis of QT interval data is complicated by the fact that the analysis is inversely correlated with heart rate (HR). Therefore, analysis of QT interval data generally includes an adjustment for the RR interval (the RR interval, expressed in seconds, is equal to 60 times the reciprocal of heart rate, i.e., RR=60/HR). Fridericia's formula ($QTc = QT/\sqrt[3]{RR}$, Fridericia, 1920) is commonly used to obtain the heart rate-corrected QT intervals in nonclinical evaluation.

5.4.1   Sequential Testing Method

At Eli Lilly and Company, QT intervals in a general toxicology study are collected at pre-specified time points both before and after treatment on selected dosing dates. QTc data at each time point are analyzed using a two-factor ANOVA model. Factors in the model include treatment, sex, and the interaction of those two factors. Effects associated with treatment and treatment-by-sex interaction are tested using an *F*-test at the 0.05 significance level. Monotonicity of dose response is examined by first testing for an interaction between the treatment linear trend and sex group at the 0.05 significance level. If this interaction is significant, a sequential trend test (Tukey et al., 1985) on treatment means is performed at the 0.05 significance level for each sex group and for the two sex groups combined. Otherwise, the sequential trend test is performed only for the combined group. For a detailed description of linear and other trend tests in dose-ranging studies, see Chapter 11, "Design and Analysis of Dose-Ranging Clinical Studies".

To define the sequential testing procedure in the two-factor ANOVA model, consider the following four tests:

Test A. An interaction between the treatment linear trend and sex.

Test B. The overall treatment linear trend for combined sexes.

Test C. The treatment linear trend for each sex.

Test D. Test C when Test A is significant or Test B when Test A is not significant.

Figure 5.1 Flow chart for analysis of two-factor ANOVA

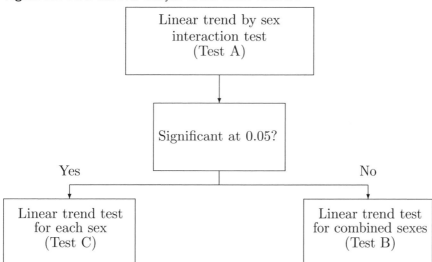

Figure 5.1 is a flow chart of the sequential testing procedure. Test D represents an overall assessment of the study and is of primary interest.

The sequential trend test by Tukey et al. (1985) is carried out as follows. Let μ_0 denote the mean QTc effect in the control group (Dose 0) and μ_1, μ_2, μ_3 denote the mean QTc effects at Doses 1, 2, and 3, respectively. The linear trend based on an ordinal dosing scale is examined by testing the following linear contrast:

$$\text{Contrast of four means } = -3\mu_0 - 1\mu_1 + 1\mu_2 + 3\mu_3.$$

This trend test is performed in a sequential fashion to identify the highest NOEL. If the linear contrast of the four means is not significant, one concludes that the high dose (Dose 3) is the NOEL and no further testing is performed. If the linear contrast is significant, the test procedure continues to assess the significance of a linear trend in the first three means using the following contrast:

$$\text{Contrast of three means} = -1\mu_0 + 0\mu_1 + 1\mu_2.$$

If this linear contrast is not significant, one concludes that the medium dose (Dose 2) is the NOEL and testing stops. However, if the test is significant, the test procedure continues to compare the control and low dose (Dose 1):

$$\text{Contrast of two means} = -1\mu_0 + 1\mu_1.$$

If this linear contrast is not significant, the low dose (Dose 1) is the NOEL. Otherwise, the NOEL is not established for this response variable.

5.4.2   Power Evaluation

We evaluate the statistical power of the sequential testing method via simulation. The following parameters are needed: the number of simulations, the sample size in each treatment group, the effect size in each treatment group, and the common variance.

The simulation study will use 2000 simulations with $n = 3$ per sex or $n = 4$ per sex in each treatment group. Let y_{ijk} denote the observation for the k-th animal in the i-th treatment group ($i = 0, 1, 2, 3$) and j-th sex group ($j = 1$ denotes males and $j = 2$ denotes females). Suppose that y_{ijk} is normally distributed with mean μ_{ij} and variance σ^2.

The dose-response profile is defined to be flat from Dose 0 up to Dose 2 and assumed to show an increase at Dose 3. In other words,

$$\mu_{0j} = \mu_{1j} = \mu_{2j}$$

for $j = 1, 2$. To allow for a sex difference at Dose 3, we make the following assumptions. The treatment effects in male and female animals at Dose 3 are given by

$$\mu_{31} = \mu_{01}(1 + \delta_1), \quad \mu_{32} = \mu_{02}(1 + \delta_2).$$

Here δ_1 is the relative treatment difference in the male animals at Dose 3 compared to Dose 0 and, similarly, δ_2 is the relative treatment difference in the female animals at Dose 3 compared to Dose 0. It is worth noting that the assumptions given above result in conservative power estimates. If drug effects were present at Doses 1 and 2, the power of the sequential testing method would be greater.

The control means in males and females as well as the variance of QTc interval were estimated from a historical database including pre-treatment QTc values from 91 male and 91 female beagle dogs:

$$\mu_{01} = 236.35, \quad \mu_{02} = 237.88, \quad \sigma^2 = 102.98.$$

Power simulations are performed under 12 scenarios defined by 6 combinations of δ_1 and δ_2 ($\delta_1 = 0.05$, $\delta_2 = 0$; $\delta_1 = 0.05$, $\delta_2 = 0.05$; $\delta_1 = 0.1$, $\delta_2 = 0$; $\delta_1 = 0.1$, $\delta_2 = 0.1$; $\delta_1 = 0.15$, $\delta_2 = 0$; $\delta_1 = 0.15$, $\delta_2 = 0.15$) and two values of the sample size per sex per treatment group ($n = 3$ and $n = 4$). Simulated data are generated by calling the %SIMULQT macro that can be found on the book's companion Web site. For example, the following call simulates QTc interval data for $\delta_1 = \delta_2 = 0.05$ and three animals per sex per treatment group (the parameters of the %SIMULQT macro are defined in the code available on the book's companion Web site):

```
%simulqt(n_sim=2000, avgmale=236.35, avgfemale=237.88, var=102.98,
    n=3, delta1=0.05, delta2=0.05, out=simul1, seed=2631);
```

Program 5.3 analyzes the simulated data using the MIXED procedure, with the ESTIMATE statements specifying the linear contrasts of treatment means. The *p*-values for Tests A, B, and C are saved in the TESTS data set. To evaluate the power of Test D, the FINALCOUNT data set converts the three *p*-values into binary variables (TESTA, TESTB, and TESTC) based on the 0.05 significance level. The TESTD variable is then defined as the sum of TESTA*TESTC and (1-TESTA)*TESTB. Finally, the binary variables are summarized using the MEANS procedure to compute the power of each test.

Program 5.3 Power evaluation in the two-factor model for QT interval

```
proc mixed data=simul1;
    by simul;
    class simul dose sex animal;
    model qtc=sex|dose;
    /* Test A */
    estimate 'Linear trend*sex'     dose*sex 3 -3 1 -1 -1 1 -3 3;
    /* Test B */
    estimate 'Combined trend test'  dose -3 -1 1 3;
    /* Test C */
    estimate 'Trend test in females' dose -3 -1 1 3
                                     dose*sex -3 0 -1 0 1 0 3 0;
    estimate 'Trend test in males'   dose -3 -1 1 3
                                     dose*sex 0 -3 0 -1 0 1 0 3;
    ods output estimates=tests;
data tests (keep=simul label probt);
    set tests;
    if probt<0.0001 then probt=0.0001;
proc transpose data=tests out=testfinal;
    by simul;
    var label probt;
data testfinal (drop=_name_ _label_);
    set testfinal;
    rename col1=flagp col2=combp col3=femalep col4=malep;
    if _name_='Probt';
data finalcount;
    set testfinal;
    retain testA testB testC testD 0;
    testA=(flagp<0.05);
    testB=(combp<0.05);
    testC=max((femalep<0.05),(malep<0.05));
    testD=testA*testC+(1-testA)*testB;
proc means data=finalcount noprint;
    var testA testB testC testD;
    output out=power;
data summary (keep=testA testB testC testD);
    set power;
    if _stat_='MEAN';
proc print data=summary noobs;
    format testA testB testC testD 5.3;
    run;
```

Output from Program 5.3

testA	testB	testC	testD
0.051	0.421	0.414	0.445

Output 5.3 displays the estimated power of the four tests. The power of the overall analysis (Test D) in this scenario ($\delta_1 = \delta_2 = 0.05$ and $n = 3$) is clearly too low (44.5%). The estimated powers for all 12 scenarios are summarized in Table 5.1. The power of the sequential testing procedure is above 95% when Dose 3 is expected to prolong the QTc interval by 10% in both male and female beagle dogs.

Table 5.1 Estimated Power of Sequential Testing Procedure in the Two-Factor Model for QT Interval

Treatment effect in males (δ_1)	Treatment effect in females (δ_2)	Sample size per sex per treatment group	
		$n = 3$	$n = 4$
0.05	0	23.2%	30.2%
0.05	0.05	44.5%	59.2%
0.1	0	63.4%	78.7%
0.1	0.1	95.7%	98.8%
0.15	0	93.2%	98.4%
0.15	0.15	100.0%	100.0%

Evaluation of Type I Error Rate in a Two-Factor ANOVA Model

Another important characteristic of the sequential testing method is the probability of a Type I error. The Type I error rates for Tests A and B are 5% and, under an additional assumption of independent multiple tests, the Type I error rate for Test C is $1 - (1 - 0.05)^2 = 9.75\%$. Although calculation of the Type I error rate for Test D is not as straightforward, it can be evaluated by simulation.

To estimate the Type I error rate associated with Test D, one needs to create a simulated data set under the global null hypothesis of no drug effect ($\delta_1 = \delta_2 = 0$) as shown below:

```
%simulqt(n_sim=2000, avgmale=236.35, avgfemale=237.88, var=102.98,
    n=3, delta1=0, delta2=0, out=simul1, seed=4641);
```

The Type I error rate of Test D is computed using Program 5.3.

Output from Program 5.3 (Computation of the Type I error rate)

testA	testB	testC	testD
0.054	0.054	0.098	0.084

Output 5.3 (Computation of the Type I error rate) shows the estimated Type I error probabilities of Tests A, B, C, and D. The estimated Type I error rate of the overall analysis (Test D) is 8.4%. Since this rate is greater than 5%, one can consider adjusting the significance levels for Tests A and B downward, i.e., carrying them out at a level that is lower than 0.05.

5.5 Statistical Analysis of a One-Factor Design with Repeated Measures

In this section, we discuss the methods for analyzing the body weight data from rodent and large animal toxicology studies. Since body weights of each animal are collected throughout the study, they are repeated measures data. The repeated measures analysis is a logical choice for large animal toxicology studies (Thakur, 2000) and rodent toxicology studies

(Hoffman et al., 2002) with data that are repeated in nature. For a typical four-treatment-group two-year carcinogenicity study, there could be as many as 60 rats per sex for each treatment group, each receiving a 0 (vehicle), low, mid, or high dose of a compound daily for the entire two years. The body weights of each animal are usually collected weekly up to 13 weeks and biweekly thereafter. The purpose of the statistical analysis is to evaluate compound-related effects on body weight. The statistical methods are described and the analysis of the data is carried out using SAS. For a detailed description of repeated measures models and issues arising in the analysis of incomplete data, we refer the reader to Chapter 12, "Analysis of Incomplete Data."

5.5.1 Objectives of Statistical Comparisons

The main interest of the analysis is to identify the NOEL on body weight in each sex by comparing treatment groups to the vehicle control. Treatment-related effects need to be interpreted with reference to the time intervals in which they were observed. The compound-related effects at a single time point are discussed first to help lay the foundation for evaluation of those effects within a time span.

5.5.2 Assessment of Compound-Related Effects at a Single Time Point

Compound-related effects can be assessed using the sequential trend test described in Section 5.4.1. The trend test is performed to evaluate the monotonic dose-response relationship in group means. However, if researchers are also interested in detecting statistically significant differences between the lower doses and the control in the absence of a high dose effect, then the Dunnett test (Dunnett, 1964) at the significance level of 0.05 can be applied as a secondary test. To avoid an inflated Type I error rate by routinely performing Dunnett's t-test in the absence of a high dose effect, researchers can first perform an F-test at a lower significance level, 0.01. Dunnett's t-test is performed only if the F-test is significant. If the effects detected at lower doses in the absence of a high dose effect would be dismissed as not being dose-responsive, then neither the F-test nor Dunnett's t-test would be considered.

5.5.3 Assessment of Compound-Related Effects within a Time Span

The evaluation of compound effects can be expanded from a single time point to the entire time span of the experiment (Hoffman et al., 2002). As the duration of a study increases, so does the number of body weight measurements. For example, in a two-year carcinogenicity study there would be more than 50 body weight measurements for each animal surviving to the end of the study. As one expects the growth pattern to change as animals mature, body weights are averaged accordingly to capture the characteristics. The calculation of average body weights in selected analysis intervals and the handling of missing data due to death are discussed in this section. The inclusion of baseline and the selection of a covariance structure for each animal across time are detailed first. After that, statistical methods for evaluation of monotonic and nonmonotonic dose-response relationships are explained.

Calculation of Interval-Averaged Body Weights

Analysis intervals for averaging body weights that are used in rodent and large animal studies at Eli Lilly and Company are listed in Table 5.2. Statistical analysis of rodent body weights collected in the first 3 months of the rapid growth phase is performed in one analysis. After 3 months, the growth of a rodent slows down and enters into the maintenance phase. Body weights beyond 3 months are evaluated in the second statistical analysis. The baseline body weight of each animal is included in both analyses as a covariate. Large animal studies are typically 2 weeks, or 1, 3, 6, 9, or 12 months in duration. Evaluation of large animal body weights collected in the first 6 months is

Table 5.2 Analysis Intervals for Rodent and Large Animal Studies

Time[a]	Rodent studies	Large animal studies
Weeks 1–5	Every week	Every week
Weeks 6–14	Every 2 weeks	Every 4 weeks
Weeks 15–26	Every 4 weeks	Every 4 weeks
Weeks 27–104[b]	Every 14 weeks	Every 4 weeks

[a] Approximate time intervals which may vary from study to study.
[b] Duration of dog studies is generally up to 52 weeks.

performed in one statistical analysis, while additional body weights collected beyond 6 months are analyzed in a second analysis.

Interval-averaged body weights for rodents are derived using the %ONEWEEK and %WEEKINT macros provided on the book's companion Web site. The INDDATA data set includes body weight measurements, collection days and general study information. The first few weeks of the study are called the initial period when no averaging is required. Weekly body weights are saved as INT_BW for those weeks starting from the study day that the first weekly data are recorded, WKLY_FIRST, to the study day that the last weekly data are recorded, WKLY_END. This is done using the %ONEWEEK macro. After the initial period, body weights in the selected intervals are averaged and saved as INT_BW in the macro. To obtain a representative body weight for the selected interval, the initial body weight in the current interval, which is the last body weight of the previous interval, is also included in the calculation. The %WEEKINT macro computes INT_BW by specifying the number of time points for averaging (NUM_TIMEPTS), the start day (INT_START), and the end day (INT_END) of the interval. This macro pulls the last observation from the previous interval into the calculation of INT_BW for the current interval.

As an illustration, Program 5.4 applies the macros to derive five weekly body weights and four biweekly interval-averaged body weights post dosing as INT_BW. These are appended to the base data set, ALL_INTERVAL. ALL_INTERVAL contains the weekly body weights for Days -1, 6, 13, 20, 27, and 34, and biweekly interval-averaged body weights for Days 48, 62, 76, and 90.

Program 5.4 Calculation of interval-averaged body weights

```
%oneweek(wkly_first=-1,wkly_end=34);
%weekint(num_timepts=2,int_start=41,int_end=90);
```

Partial output of ALL_INTERVAL from Program 5.4

```
Gender    Phase        Day    Animal    Group     BW     INTVL_BW

Male      Treatment     -1     1001       1      132.5    132.5
Male      Treatment      6     1001       1      166.5    166.5
Male      Treatment     13     1001       1      194.8    194.8
Male      Treatment     20     1001       1      212.1    212.1
Male      Treatment     27     1001       1      246.3    246.3
Male      Treatment     34     1001       1      255.5    255.5
Male      Treatment     41     1001       1      277.3     NC
Male      Treatment     48     1001       1      294.7    275.8
Male      Treatment     55     1001       1      308.8     NC
Male      Treatment     62     1001       1      317.3    306.9
Male      Treatment     69     1001       1      330.8     NC
Male      Treatment     76     1001       1      328.3    325.5
Male      Treatment     83     1001       1      339.0     NC
Male      Treatment     90     1001       1      339.8    335.7
```

Female	Treatment	-1	2051	2	101.8	101.8
Female	Treatment	6	2051	2	116.5	116.5
Female	Treatment	13	2051	2	127.5	127.5
Female	Treatment	20	2051	2	145.5	145.5
Female	Treatment	27	2051	2	152.1	152.1
Female	Treatment	34	2051	2	156.0	156
Female	Treatment	41	2051	2	162.6	NC
Female	Treatment	48	2051	2	171.9	163.5
Female	Treatment	55	2051	2	178.6	NC
Female	Treatment	62	2051	2	189.5	180
Female	Treatment	69	2051	2	196.5	NC
Female	Treatment	76	2051	2	197.8	194.6
Female	Treatment	83	2051	2	201.9	NC
Female	Treatment	90	2051	2	204.2	201.3

Handling of Missing Data Due to Death

As animals age, their mortality rate also increases. It is not unusual to see 50% of the animals die before the end of a carcinogenicity study. Missing data due to death are different from missing data due to other reasons. When averaging body weights for an animal in a selected time interval, a missing value will be assigned in the former case while an average would be calculated based on all available body weights in the latter case. All body weights up to the time interval of death are included in the statistical analysis. Body weights that are collected prior to death but that are not included in the statistical analysis would be incorporated into the overall assessment of mortality data.

Program 5.5 demonstrates how to properly handle the missing data. For each animal, a survival index (SURVIVAL_DAY) is created to indicate the last day of non-missing observations. This index is used to identify if a missing body weight is due to death or other reasons. If this survival index day falls within an interval, the derived interval body weight (INT_BW) is set to missing for that animal. Otherwise, the program averages all available non-missing body weights.

The following arguments are used in Program 5.5:

- SURVIVAL_DAY is the last day of non-missing body weights.
- START is the first day of an interval (e.g., Day 48).
- END is the last day of an interval (e.g., Day 69).

Program 5.5 Handling of missing data

```
data survival_index;
    set all;
    if bw ne .;
proc sort data=survival_index;
    by animal gender;
data survival_index;
    set survival_index;
    by animal gender day;
    if last.animal;
    survival_day=day;
    keep animal gender survival_day;
data week&end;
    merge week&end(in=in1) survival_index(in=in2);
    by animal;
    if in1 and in2;
    if &start<=survival_day<&end then int_bw=.;
    run;
```

The output of Program 5.5 can be summarized into four possible scenarios. To illustrate, Table 5.3 includes both the weekly body weights and the derived interval body weights. For the first month, INT_BW values were captured directly from the five weekly body weights. INT_BW ending on Day 48 was computed by averaging the body weights collected on Days 34, 41, and 48. For the last time interval that ended on Day 62, INT_BW is derived from the average of three time points: Days 48, 55, and 62. The following scenarios are presented in Table 5.3:

Scenario 1. Animal 1001 did not have any missing values.

Scenario 2. Animal 1002 had a missing value on Day 55. Hence INT_BW for Day 62 was computed by averaging the body weights from Days 48 and 62.

Scenario 3. Animal 1003 had two missing values on Days 41 and 48. Hence INT_BW for Day 48 was computed from Day 34, and INT_BW for Day 62 was computed by averaging the body weights from Days 55 and 62.

Scenario 4. Animal 1004 is assumed dead right after Day 48 since there were no data collected beyond Day 48. Therefore, INT_BW for the last interval was set to be missing.

Table 5.3 Examples of INT_BW Computation and Missing Data Handling

Animal ID	Parameter	Study day								
		6	13	20	27	34	41	48	55	62
1001	BW	166.5	194.8	212.2	246.3	277.3	294.7	308.8	317.3	330.8
	INT_BW	166.5	194.8	212.2	246.3	277.3		**293.6**		**319.0**
1002	BW	166.5	194.8	212.2	246.3	277.3	294.7	308.8	.	330.8
	INT_BW	166.5	194.8	212.2	246.3	277.3		**293.6**		**319.8**
1003	BW	166.5	194.8	212.2	246.3	277.3	.	.	317.3	330.8
	INT_BW	166.5	194.8	212.2	246.3	277.3		**277.3**		**324.1**
1004	BW	166.5	194.8	212.2	246.3	277.3	294.7	308.8	.	.
	INT_BW	166.5	194.8	212.2	246.3	277.3		**293.6**		.

5.5.4 Selection of the Covariance Structure for Each Animal

To account for the initial differences in baseline body weight, it is important to include it as a covariate in the statistical model. In addition, to account for the correlation among repeated measures from the same animal, one needs to select an appropriate covariance structure for each animal. As a default covariance structure, consider repeated measures data from a split-plot design with a treatments, b time points, and c animals. The split-plot model (Aldworth and Hoffman, 2002) with treatment as the whole-plot factor, time as the subplot factor, and baseline as a covariate is

$$y_{ijk} = \mu + \alpha_i + \gamma(x_{ik} - \bar{x}_{..}) + d_{k(i)} + \beta_j + (\alpha\beta)_{ij} + e_{ijk},$$

where $i = 1, \ldots, a$, $j = 1, \ldots, b$ and $k = 1, \ldots, c$. In this equation, y is the body weight, μ is the grand mean, α is the treatment effect, γ is the regression coefficient for the covariate x and the overall mean of the covariate $\bar{x}$, β is the time effect, $\alpha\beta$ is the treatment by time interaction, d is the random error for the animal nested in the treatment, and e is random error at each time point.

The covariance matrix for an animal in this model has the variance components structure (VC). The covariance terms in VC are nonnegative. If negative correlations are allowed in the covariance structure of VC, the matrix will have a compound symmetric structure (CS).

In addition, the variances of an animal at different time points could be heterogeneous. Therefore, one can consider other covariance structures, including homogeneous variance components (VC), heterogeneous variance components (UN(1)), homogeneous compound symmetry (CS), heterogeneous compound symmetry (CSH) and spatial power (SP[POW]).

For the UN(1) and SP(POW) structures, both REPEATED and RANDOM statements are included in the MIXED procedure. An animal is specified as the subject in both statements and the intercept is specified in the RANDOM statement as a random effect.

For the CS and CSH structures, only the REPEATED statement is included while for the VC structure, only the RANDOM statement is included. The finite-sample corrected Akaike's Information Criterion (Keselman et al., 1998) can be used for selecting the covariance structure. The Kenward and Roger method (Kenward and Roger, 1997) is generally recommended for the denominator degrees of freedom. The PROC MIXED syntax for each covariance structure discussed above follows:

Homogeneous variance components (VC)

```
proc mixed data=one;
    class trt time animal;
    id trt time animal;
    model body_wt=trt time trt*time baseline/ddfm=kenwardroger solution;
    random int/type=vc subject=animal s;
```

Heterogeneous variance components (UN(1))

```
proc mixed data=one;
    class trt time animal;
    id trt time animal;
    model body_wt=trt time trt*time baseline/ddfm=kenwardroger solution;
    repeated time/type=un(1) subject=animal r;
    random int/subject=animal s;
```

Note that TYPE=UN(1) can be replaced with TYPE=VC GROUP=TIME.

Homogeneous compound symmetry (CS)

```
proc mixed data=one;
    class trt time animal;
    id trt time animal;
    model body_wt=trt time trt*time baseline/ddfm=kenwardroger solution;
    repeated time/type=cs subject=animal r;
```

Heterogeneous compound symmetry (CSH)

```
proc mixed data=one;
    class trt time animal;
    id trt time animal;
    model body_wt=trt time trt*time baseline/ddfm=kenwardroger solution;
    repeated time/type=csh subject=animal r;
```

Spatial power (SP[POW])

```
proc mixed data=one;
    class trt time animal;
    id trt time animal;
    model body_wt =trt time trt*time baseline/ddfm=kenwardroger solution;
    repeated time/type=sp(pow) (time) subject=animal r;
    random int/subject=animal s;
```

Evaluation of Monotonic and Nonmonotonic Dose-Response Relationships

A repeated measures analysis of variance is performed to assess compound-related effects across time. Treatment and time are entered in the statistical model as the two main factors. The interaction between treatment and time is also included in the model. Based on the significance of the interaction, compound effects are assessed either at each time point when the interaction is significant, or on results pooled across the entire time span when the interaction is not significant. This approach allows researchers to detect the start and end of compound-related effects.

The assessment of compound-related effects for repeated measures analysis of variance is carried out in the same manner as for single-time measurements. The additional dimension of time requires more detailed specifications for the comparisons of treatment groups at selected time points. The basic concept of evaluating a monotonic dose-response relationship using the sequential trend test by Tukey, Ciminera and Heyse (1985) and supplementing it with a pairwise to control test for a nonmonotonic dose-response relationship remains the overall strategy.

To test for a monotonic dose-response relationship in treatment, one first needs to evaluate the interaction between time and treatment by performing the following three interaction tests:

- Test 1 (linear trend in treatment by time at the 0.01 significance level). This test evaluates the similarity of the monotonic dose-response relationship across time. Consider a slope for each dose-response relationship being evaluated. The test checks for equality of these slopes at all time points.

- Test 2 (linear trend in treatment by linear trend in time at the 0.05 level). This test checks if the slopes defined above change consistently across time, e.g., if they continue to rise or continue to drop.

- Test 3 (linear trend in treatment by quadratic trend in time at the 0.05 level). This test checks if the slopes change in a quadratic fashion, i.e., rising or dropping in one direction to a point and then changing directions for the rest of the time span.

Tests 2 and 3 help researchers understand the treatment related dose-response effects within a time span. The interaction between the linear trend in treatment and time is a more general catch-all test to capture those scenarios when the slopes of the monotonic dose-response vary significantly across time and the profile of the slopes may be more complicated than linear or quadratic.

Following the strategy for handling monotonic and nonmonotonic treatment effects at a single time point, Test 1 is carried out at the 0.01 level and Tests 2 and 3 are carried out at the 0.05 level. If any of the three interaction tests is statistically significant at the respective significance level, contrasts need to be specified to examine the treatment effects either on the results pooled across time, or at each time point in addition.

In order to evaluate the monotonic dose-response relationship in treatment and the treatment by time interaction terms, several ESTIMATE and CONTRAST statements need to be specified. The %REPMEAS49 macro provided on the book's companion Web site is an example of proper ESTIMATE and CONTRAST parameters in a study design of four groups with nine time points post dosing. The doses are measured on an ordinal scale.

If no monotonic dose-response relationship is detected in treatment and a lower dose effect in the absence of a high dose effect may be meaningful, one can evaluate a nonmonotonic dose-response relationship. In the one-factor design, this is accomplished by conducting an F-test at a lower significance level, 0.01, and carrying out Dunnett's t-test at the 0.05 level only after the F-test is shown to be significant.

In the repeated measures analysis based on PROC MIXED, Dunnett's t-test would compare all combinations of treatment and time back to the first time point of the control group. This is not the same as comparing the treated group means back to the control at each time point. Therefore, Dunnett's t-test is replaced with the Bonferroni-adjusted t-test at the 0.05 level. The Bonferroni-adjusted t-test is preceded by two F-tests carried out at the 0.01 level for (1) the treatment by time interaction and (2) the treatment main effect:

- If the treatment by time interaction is significant, the Bonferroni-adjusted t-test will be applied to the treatment means for each time point.
- If the treatment main effect alone is significant, the Bonferroni-adjusted t-test will be applied only to the treatment means pooled across all time points.
- If neither is significant, no further testing is performed.

Note that the Bonferroni-adjusted t-test can be carried out using the %BONF_ADJ macro given on the book's companion Web site.

SAS Module for the Evaluation of Monotonic and Nonmonotonic Dose-Response Relationships

A comprehensive and flexible SAS module was developed by the authors to carry out the statistical analysis for the evaluation of monotonic and nonmonotonic dose-response relationships in a repeated measures ANOVA framework. In this macro-parameter-driven module, users can specify the names of the analysis variables and covariates (if any), levels of time factor, the covariance structure, denominator degrees of freedom method (DDFM), and an option of yes/no for inference tests. As an illustration, SAS code provided on the book's companion Web site uses the macro %PRF1FRM to analyze the INDDATA data set. Results from the analysis of the INDDATA data set using the module are given in Tables 5.4 and 5.5.

5.6   Summary

The growing complexity of guideline studies, and the increasing number of measurements required therein, present major challenges to nonclinical scientists and statisticians. Central to defining hazards and subsequently assessing risk is a clear understanding of the study design and data characteristics. This chapter briefly defines statistical aspects of toxicology studies with examples implemented using SAS. We have described randomization schemes for two commonly used designs in toxicology, namely, the parallel design and Latin square design. We have presented the statistical power evaluation of the heart-rate corrected QT intervals from a large animal toxicology study. We have also discussed the statistical analysis of treatment-related effects on body weights from general toxicology studies. From creating randomization schemes, calculating the power of a test, performing statistical analysis and finally to reporting results in nicely formatted tables, SAS has been an indispensable tool for nonclinical statisticians in the pharmaceutical industry.

Table 5.4 Summary Table for Body Weight Data Using a Repeated Measures Analysis of Variance, Days 6–48

Repeated Measures ANOVA : Compound - TEST_STUDY Study - int_parms_BW_test
Analysis Variable : Interval BW (g)
Baseline (Day -1) was used as covariate
GENDER = Female

Group	Statistics	Baseline	Overall	Day 6	Day 13	Day 20	Day 27	Day 34	Day 34–48
1	Mean	100.39	159.22	119.51	130.98	145.51	153.62	160.55	168.32
	SD	4.34	NA	5.23	5.22	7.55	8.11	7.41	8.09
	N	15	15	15	15	15	15	15	15
	LSM	NA	159.27	119.56	131.03	145.56	153.67	160.60	168.36
	LSM s.e.	NA	1.42	1.66	1.66	1.66	1.66	1.66	1.66
2	Mean	100.25	162.23	118.12	131.64	147.95	156.39	162.15	170.54
	SD	4.36	NA	5.32	6.00	5.94	7.42	7.40	7.67
	N	15	15	15	15	15	15	15	15
	Mean: % Chg from Cntrl	0	2	-1	1	2	2	1	1
	LSM	NA	162.41	118.30	131.82	148.13	156.57	162.33	170.72
	LSM s.e.	NA	1.42	1.66	1.66	1.66	1.66	1.66	1.66
	Trend p-val#	NT	0.124	NT	NT	NT	NT	0.462	0.317
3	Mean	100.49	152.19	117.95	129.25	142.09	151.23	155.19	160.28
	SD	4.60	NA	5.27	5.03	5.79	5.92	7.03	7.21
	N	15	15	15	15	15	15	15	15
	Mean: % Chg from Cntrl	0	-4	-1	-1	-2	-2	-3	-5
	LSM	NA	152.14	117.90	129.20	142.03	151.17	155.13	160.22
	LSM s.e.	NA	1.42	1.66	1.66	1.66	1.66	1.66	1.66
	Trend p-val#	NT	0.001*	NT	NT	0.135	NT	0.022*	0.001*
4	Mean	100.61	154.34	117.62	129.28	142.68	151.04	155.15	161.36
	SD	4.27	NA	5.88	6.01	7.24	7.75	7.84	7.20
	N	15	15	15	15	15	15	15	15
	Mean: % Chg from Cntrl	0	-3	-2	-1	-2	-2	-3	-4
	LSM	NA	154.17	117.45	129.11	142.51	150.87	154.97	161.18
	LSM s.e.	NA	1.42	1.66	1.66	1.66	1.66	1.66	1.66
	Trend p-val#	NT	<.001*	0.364	0.260	0.042*	0.065	0.002*	<.001*
ALL	Trt F-test p-val++		<.001*						
INTN	Trt*Time p-val++		<.001*						
	LinTrt*Time p-val++		<.001*						
	LinTrt*LinTime p-val+		<.001*						
	LinTrt*QdrTime p-val+		0.698						

\# : Level of significance tested = .05; Two-sided test.   NT : Not tested.
++ : Level of significance tested = .01.   NA : Not applicable.
\+ : Level of significance tested = .05.   KENWARDROGER was used for the DDFM.
*   : Statistically significant.   CS covariance structure over time was selected for the model.

Table 5.5 Summary Table for Body Weight Data Using a Repeated Measures Analysis of Variance, Days 48–90

Repeated Measures ANOVA : Compound - TEST_STUDY Study - int_parms_BW_test
Analysis Variable : Interval BW (g)
Baseline (Day -1) was used as covariate
GENDER = Female

Group	Statistics	Day 48–62	Day 62–76	Day 76–90
1	Mean	178.38	185.52	190.61
	SD	9.99	11.50	12.82
	N	15	15	15
	LSM	178.43	185.56	190.67
	LSM s.e.	1.66	1.66	1.66
2	Mean	182.91	192.63	197.70
	SD	8.72	9.37	9.26
	N	15	15	15
	Mean: % Chg from Cntrl	3	4	4
	LSM	183.09	192.82	197.88
	LSM s.e.	1.66	1.66	1.66
	Trend p-val#	0.049*	0.003*	0.003*
3	Mean	166.55	171.59	175.60
	SD	8.08	8.86	8.87
	N	15	15	15
	Mean: % Chg from Cntrl	-7	-8	-8
	LSM	166.49	171.54	175.54
	LSM s.e.	1.66	1.66	1.66
	Trend p-val#	<.001*	<.001*	<.001*
4	Mean	170.20	178.28	183.45
	SD	7.21	7.98	8.06
	N	15	15	15
	Mean: % Chg from Cntrl	-5	-4	-4
	LSM	170.03	178.11	183.27
	LSM s.e.	1.66	1.66	1.66
	Trend p-val#	<.001*	<.001*	<.001*
ALL	Trt F-test p-val++			
INTN	Trt*Time p-val++			
	LinTrt*Time p-val++			
	LinTrt*LinTime p-val+			
	LinTrt*QdrTime p-val+			

# : Level of significance tested = .05; Two-sided test.	NT : Not tested.
++ : Level of significance tested = .01.	NA : Not applicable.
+ : Level of significance tested = .05.	KENWARDROGER was used for the DDFM.
* : Statistically significant.	CS covariance structure over time was selected for the model.

Acknowledgments

The authors wish to thank Dr. Gheorghe Doros and Mr. James Hoffman for reviewing and refining the chapter.

References

Aldworth, J., Hoffman, W.P. (2002). "Split-plot model with covariate: A cautionary tale." *The American Statistician.* 56, 284–289.

Box, G.E.P., Hunter, W.G., Hunter, J.S. (1978). "Designs with more than one blocking variable." *Statistics for experimenters.* New York: Wiley. 245–280.

Chiang, A.Y., Smith, W.C., Main, B.W., Sarazan, R.D. (2004). "Statistical power analysis for hemodynamic cardiovascular safety pharmacology studies in beagle dogs." *Journal of Pharmacological and Toxicological Methods.* 50, 121–130.

Dunnett, C.W. (1964). "New tables for multiple comparisons with a control." *Biometrics.* 20, 482–491.

Fridericia, L.S. (1920). "Die Systolendauer im Elecktrokardiogramm bei normalen Menschen und bei Herzkranken." *Acta Medica Scandinavia.* 53, 469–486.

Hoffman, W.P., Ness, D.K., van Lier, R.B. (2002). "Analysis of rodent growth data in toxicology studies." *Toxicological Sciences.* 66, 313–319.

International Conference on Harmonization of Pharmaceuticals for Human Use (ICH). (2004). S7B: "The nonclinical evaluation of the potential for delayed Ventricular repolarization (QT interval prolongation) by Human Pharmaceuticals. Step 2 Revision." Available at `http://www.ich.org`.

Kenward, M.G., Roger, J.H. (1997). "Small sample inference for fixed effects from restricted maximum likelihood." *Biometrics.* 53, 983–997.

Keselman, H.J., Algina, J., Kowalchuk, R.K., Wolfinger, R.D. (1998). "A comparison of two approaches for selecting covariance structures in the analysis of repeated measurements." *Communications in Statistics. Simulation and Computation.* 27, 591–604.

Kinter, L.B., Siegl, P.K.S., Bass, A.S. (2004). "New preclinical guidelines on drug effects on ventricular repolarization: Safety pharmacology comes of age." *Journal of Pharmacological and Toxicological Methods.* 49, 153–158.

Lin, K.K. (2001). Guidance for industry statistical aspects of the design, analysis, and interpretation of chronic rodent carcinogenicity studies of pharmaceuticals (draft guidance online). *Rockville: US Department of Health and Human Services, Food and Drug Administration, Center for Drug Evaluation and Research (CDER).* Available at `http://www.fda.gov/cder/guidance/815dft.pdf`.

Thakur, A.K. (2000). "Statistical issues in large animal toxicology." *International Journal of Toxicology.* 19, 133–140.

Tukey, J.W., Ciminera, J.L., Heyse, J.F. (1985). "Testing the statistical certainty of a response to increasing doses of a drug." *Biometrics.* 41, 295–301.

Chapter 6

Nonparametric Methods in Pharmaceutical Statistics

Paul Juneau

Nonparametric, or distribution-free, statistical methods are very useful in the setting of pharmaceutical research. These methods afford data analysts the ability to relax some of the assumptions typically made by their Gaussian (normality-based) analogues. In some settings (e.g., drug discovery investigations), these assumptions may not be verifiable due to small sample sizes. In others, where larger sample sizes are employed (e.g., clinical trial settings), the assumption of a Gaussian (normal) distribution is not met because of the presence of heavy-tails in measurement response or a large degree of skewness. This chapter covers two settings found commonly in pharmaceutical research (two-sample setting and one-way layout) and discusses sample size determination in a nonparametric sense. The introduced statistical methods are illustrated using examples from drug discovery studies and clinical trials.

6.1 Introduction

Why are nonparametric statistical methods important to pharmaceutical research? To really address this question, we first must begin by looking at the nature of pharmaceutical research. The goal of pharmaceutical research is to discover and develop new medicines that are able to cure or moderate various disease symptoms or conditions. The *disease response* (i.e., physiological trauma) can vary greatly from individual to individual because of a complex inherent genetic variability. The corresponding statistical phenomenon associated with this genetic variation to disease response is that measurements collected on such a cohort can be skewed by the extreme response of a small portion of the individuals under consideration.

Consider, as an example, the search for a new anti-inflammatory agent. Inflammatory response (e.g., swelling) can vary greatly in untreated subjects. Some individuals can have a very extreme response to a biological insult that produces an inflammatory response,

Paul Juneau is Associate Director, Nonclinical Statistics, Pfizer, USA.

causing the distribution of those responses to be very "heavy" or "long-tailed" in the upper direction (e.g., a great deal of swelling). The common summary statistic for continuous responses in such a setting, the mean, is greatly influenced by the presence of a small number of high responses and "over-represents" the location of the untreated subjects' distribution (we will see that a similar phenomenon can also occur in the study of novel antibacterial drugs).

A similar phenomenon can occur, but this time in treated subjects, in the study of the toxicology of new agents. The genetic variability of individuals can vary greatly as the body responds to higher and higher levels of a compound. Some subjects in a high dose group of a toxicology study may exhibit very high or very low responses that skew the distribution. In both cases (the study of efficacy and toxicology), it becomes apparent that traditional Gaussian (normal) statistical methods may fail because of the required assumption of symmetry. Nonparametric, or distribution-free, statistical methods liberate us from the need to assume symmetry in response. We no longer impose an assumption on our inference that may not be verifiable (because of small sample sizes) or observable.

This chapter is an introduction to popular nonparametric statistical methods in the analysis of studies in the pharmaceutical industry. It provides a review of some new statistical methods for inference and an introduction to some fairly infrequently used techniques that have existed "below the radar" for several years.

Section 6.2 will be devoted to the two-sample setting, exploring both the equal and unequal dispersion cases. Section 6.3 will discuss aspects of the one-way layout and associated multiple comparison procedures. Section 6.4 will introduce one approach to sample size determination in a "nonparametric" sense, using data from pilot studies. Each section also contains a mixture of SAS/STAT procedures and some original macros to perform elementary nonparametric data analyses.

Throughout this chapter, the discussion of topics will be motivated by providing some real or simulated examples of data from studies in medical research. In either situation, the purpose of the example is not to make a medical claim (efficacy or safety of an anonymous compound) but to illustrate a feature of data critical to the outcome of the statistical inference.

To save space, some SAS code has been shortened and some output is not shown. The complete SAS code and data sets used in this book are available on the book's companion Web site at http://support.sas.com/publishing/bbu/companion_site/60622.html.

6.2   Two Independent Samples Setting

An experiment involving two independent groups is one of the most elementary types of investigations in scientific research, yet is no less valuable than many other more complicated multifactor or multilevel experiments. Typically, subjects are randomly assigned to one of two groups, a treatment is applied, and a measurement is collected on all subjects in both groups. It is then usually the desire of the investigators to ascertain whether sufficient evidence exists to declare the two groups to be different. Although many are familiar enough with statistics to perform two sample comparisons, often, key assumptions are swept under the rug, as it were, and as a result, errors in inference sometimes occur.

This section will begin with a review of the assumptions necessary to perform a two-sample comparison and then discuss a common means to compare the location parameters of two groups (Section 6.2.2). Section 6.2.3 will discuss one possible solution for the problem of unequal spread (heteroscedasticity) between two groups. One important false notion about the rank transform and its effect on groups with unequal dispersion will also be discussed. This section will include examples of data from two clinical trials.

6.2.1   A Review of the Two-Sample Setting

We will begin this section with a slight bit of mathematical formality. Suppose that we have two random samples, denoted $X_1, \ldots, X_{n_1}$ and $Y_1, \ldots, Y_{n_2}$. We assume that the X's and Y's come from a distribution, say, F. Typically, the two-sample problem is characterized by a desire to perform an inference about a *shift in location* between the distributions of the two samples. Suppose $F_X = F(\mu)$ is the distribution of the X's (with location parameter μ) and $F_Y = F(\mu + \delta)$ is the distribution of the Y's, shifted by δ. What can be said about a measure δ, where $F_X(\mu) = F_Y(\mu + \delta)$? Is δ "reasonably close" to zero or not? If the X's follow a Gaussian (normal) distribution, say, $N(\mu_1, \sigma^2)$, and the Y's are distributed from another Gaussian distribution, say, $N(\mu_2, \sigma^2)$, we could think of the problem as considering whether $\mu_1 = \mu_2 + \delta$. In this classic "normal" case, it is very important to note that we assume that the variance of the two Gaussian distributions is "reasonably" the same (often referred to as a common σ^2). If latter is not true, we are working under the conditions of a famous setting in statistics, the Behrens-Fisher problem, and we need to make some changes to our inferential procedures. In either case, some form of a Student's t-statistic may be used to compare the locations (means) of the two distributions.

Do the measurements we collect come from a Gaussian distribution? In drug discovery settings in the pharmaceutical industry, sample sizes in some investigations can be very small (total sample size for a study less than 12) and, thus, it might be very difficult to establish the form of a measurement's distribution. Moreover, as little is known in the discovery stages of research about the impact of a compound on a complicated mammalian physiology, a seemingly outlying or extreme value may in fact be part of the tail of a highly skewed distribution. With small sample sizes and limited background information, it really is anyone's best guess as to the parametric form of a measurement's distribution.

What if the distributions are not Gaussian? The standard Student's t-test will perform at the designated level of statistical significance. In most cases, departures from a Gaussian distribution do not greatly affect the operating characteristics of inferential procedures if the two distributions have mild skewness or kurtosis. However, moderate to high levels of skewness can influence the power of a comparison of two location parameters (Bickel and Docksum, 1977) and, thus, influence our ability to identify potentially efficacious or toxic compounds.

Shortly, we will look at an example of a two-sample setting where the data are sufficiently skewed to warrant the introduction of a distribution-free test that serves as a competitor to the standard Student's two-sample t-test. First, however, it is necessary to describe a setting where such data could be found: asthma treatment research.

EXAMPLE: Asthma Clinical Trial

Asthma is a disease characterized by an obstruction of airflow in the lungs, and thus referred to as an obstructive ventilatory defect (National Asthma Council Australia, 2002). One of the chief measurements used to characterize the degree of disease is the Forced Expiratory Volume in 1 second, or FEV1. A patient will exhale into a device (a spirometer) that will measure his or her volume of expired air.

Physicians working on developing new drugs for treatment of asthma are interested in a measure that characterizes the degree of reversibility of the airflow obstruction. Thus, a patient's baseline FEV1 will be measured before treatment (at baseline). Typically, 10-15 minutes after the administration of a treatment (e.g., a beta$_2$ agonist bronchodilator), FEV1 will be measured a second time. The percent improvement in FEV1 from baseline is used to measure the effectiveness of a new drug:

$$\text{Percent improvement} = \frac{\text{Post-treatment FEV1} - \text{Baseline FEV1}}{\text{Baseline FEV1}} 100\%.$$

A medically important improvement is typically at least 12%; however, not all researchers in the field accept this level.

Consider a simple trial to establish the efficacy of a new treatment for asthma. Seven hundred patients are randomized to one of two groups. Patients randomized to the first group had baseline FEV1 measurements collected, then received a placebo treatment and, at 10 minutes post-treatment, had a second FEV1 measurement taken. The procedure was similar for the second group of patients, except that after measurement at baseline, they received a novel treatment. The percent changes in FEV1 collected in this study are included in the SPIRO data set that can be found on the book's companion Web site. Table 6.1 provides a summary of the results in the two treatment groups and Figure 6.1 displays a box plot summary of percent changes in FEV1.

Table 6.1 Summary of Percent Changes in FEV1 in the Asthma Clinical Trial

Group	Sample size	Mean	Standard deviation	Median	Interquartile range
Placebo	347	9.58	30.89	14.00	37.55
Treated	343	13.75	20.08	14.14	29.16

Figure 6.1 Results of the asthma clinical trial

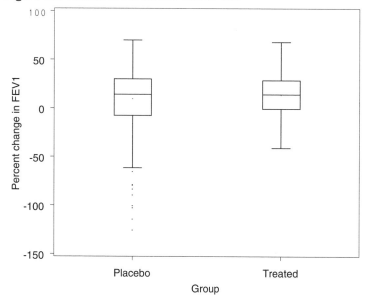

The box plots and summary statistics provide some interesting insight into the results of the study. First, it appears that for a small number of patients (10 total) a full percent change score was unavailable. Also, it appears that the placebo group included some patients whose results were dramatically worse between the initial baseline FEV1 and the later post-treatment measurement (note the large number of outliers represented by dots outside the fence in the box plot). These results produce a long downward tail in the distribution of change from baseline values and contribute to skewing of the distribution. This also can be observed by the difference of 4 percentage points between the mean and median for this group. The level of dispersion, or spread, of the two groups is not identical; however, it is not greatly different. How do we compare the location parameters of two distributions in the presence of equal, or similar, dispersion?

6.2.2   Comparing the Location of Two Independent Samples I: Similar Dispersion

Returning to the notation introduced previously in Section 6.2.1:

- Group 1: $X_1, \ldots, X_{n_1}$ with $X_i \sim F(\mu)$, $i = 1, \ldots, n_1$.
- Group 2: $Y_1, \ldots, Y_{n_2}$ with $Y_j \sim F(\mu + \delta)$, $j = 1, \ldots, n_2$.

Here $F(\lambda)$ is a continuous distribution with location parameter, λ.

Wilcoxon Rank Sum Test

The inference of interest is $H_0 : \delta = 0$ versus $H_A : \delta > 0$. This null hypothesis can be tested using the Wilcoxon Rank Sum test (Wilcoxon, 1945). To carry out this test, we begin by performing a joint ranking of the X's and Y's from the smallest value to the largest. If two or more values are tied, their ranks are replaced by the average of the tied ranks. The Wilcoxon, W, is simply the sum of the corresponding ranks for the Y's, i.e., $W = \sum_{j=1}^{n_2} R_j$, where R_j is the rank of Y_j in the combined sample. We reject $H_0 : \delta = 0$ at an α level if $W \geq c_\alpha$, for a cutoff value, c_α, that corresponds to the null distribution of W.

If the minimum of n_1 and n_2 is sufficiently large, an approximate procedure may be employed. The Wilcoxon Rank Sum statistic specified above requires some modifications so that the Central Limit Theorem will hold and the statistic will be asymptotically normal:

$$W_L = \left(W - \frac{n_1(n_2 + n_1 + 1)}{2} \right) \Big/ \sqrt{\text{Var}(W_T)}.$$

What is the form of the variance of the W statistics for tied observations, i.e., $\text{Var}(W_T)$? Suppose that for all $n_1 + n_2$ observations, tied groups of size t_v $(1, 2, \ldots, v, \ldots, \xi)$ exist. The variance in the denominator of the statistic is as follows (Hollander and Wolfe, 1999):

$$\text{Var}(W_T) = \frac{n_1 n_2}{12} \left(n_1 + n_2 + 1 - \frac{\sum_{v=1}^{\xi}(t_v - 1)t_v(t_v + 1)}{(n_1 + n_2)(n_1 + n_2 - 1)} \right).$$

The large sample version of the one-sided hypothesis test is then: Reject $H_0 : \delta = 0$ at an α level if $W_L \geq z_\alpha$, where z_α is the $100(1 - \alpha)$ quantile from a standard normal distribution.

Before we embark on a discussion of the lower-tailed and two-sided Wilcoxon Rank Sum tests, let's return to the asthma clinical trial example. Program 6.1 uses the NPAR1WAY procedure to perform the Wilcoxon Rank Sum test using the SPIRO data set. The Wilcoxon Rank Sum test is requested by the WILCOXON option. PROC NPAR1WAY automatically uses tied ranks and performs the appropriate adjustment to the test statistic based upon the large sample form of the statistic. The ODS statement is included to select the relevant portion of the procedure's output. The CLASS statement identifies the classification or grouping variable. The VAR statement identifies the response variable (percent change in FEV1).

Program 6.1 Wilcoxon Rank Sum test in the asthma trial example

```
proc npar1way data=spiro wilcoxon;
    ods select WilcoxonTest;
    class group;
    var fevpc;
    run;
```

Output from Program 6.1

```
                    Wilcoxon Two-Sample Test

            Statistic              120620.0000

            Normal Approximation
            Z                          0.8071
            One-Sided Pr >  Z          0.2098
            Two-Sided Pr > |Z|         0.4196


            t Approximation
            One-Sided Pr >  Z          0.2099
            Two-Sided Pr > |Z|         0.4199


           Z includes a continuity correction of 0.5.
```

We conclude from Output 6.1 that a statistically significant improvement from baseline FEV1 did not exist at the 0.05 level of statistical significance (one-tailed p-value based on the normal approximation is 0.2098).

Student's Two-Sample t-Test

What about other methods of data analysis? Commonly, in the face of skewed data, analysts will use some form of transformation. A commonly used transformation is the natural logarithm transformation applied to each of the responses. In the example illustrated previously, a log or power transformation will not work, as negative values exist in the data. Suppose we chose to ignore the long tail of the placebo group's distribution and performed a standard Student's two-sample t-test (Student, 1908). Program 6.2 performs Student's two-sample t-test using PROC TTEST. The code to perform a Student's two-sample t-test is very similar to the PROC NPAR1WAY code. As in Program 6.1, the ODS statement is included to select the relevant portion of the procedure's output.

Program 6.2 Student's two-sample t-test in the asthma trial example

```
proc ttest data=spiro;
    ods select ttests;
    class group;
    var fevpc;
    run;
```

Output from Program 6.2

		T-Tests			
Variable	Method	Variances	DF	t Value	Pr > |t|
fevpc	Pooled	Equal	688	-2.10	0.0363
fevpc	Satterthwaite	Unequal	595	-2.10	0.0360

If we select the two-tailed p-value from Output 6.2 (assuming unequal variances in the two groups) and adjust it so that it reflects a one-tailed t-test in ($p = 0.036/2 = 0.018$), we will infer that the mean FEV1 percent change score was statistically significantly increased by the new treatment. The long tail of extreme low responses in the control group influenced this conclusion. The extreme negative FEV1 percent change scores did not adversely affect the Wilcoxon Rank Sum test.

Lower- and Two-Tailed Versions of the Wilcoxon Rank Sum Test

We will now move on to complete this section with a brief discussion of the lower-tailed and two-tailed versions of the Wilcoxon Rank Sum test. If the inference of interest is $H_0 : \delta = 0$ versus $H_A : \delta < 0$, we will reject the null hypothesis at level α if

$$W \leq n_1(n_2 + n_1 + 1) - c_\alpha,$$

where c_α is the same cutoff mentioned in the first part of this section for the upper-tailed test. Similarly, for the two-tailed test, if the testing problem is $H_0 : \delta = 0$ versus $H_A : \delta \neq 0$, the null hypothesis is rejected if

$$W \leq n_1(n_2 + n_1 + 1) - c_{\alpha/2} \text{ or } W \geq c_{\alpha/2}.$$

For the large-sample approximate procedures, we will reject the null hypothesis at level α if $W_L \leq z_\alpha$ (upper-tailed test) or $|W_L| \geq z_{\alpha/2}$ (two-tailed test), where W_L is the large-sample Wilcoxon statistic (Hollander and Wolfe, 1999).

6.2.3 Comparing the Location of Two Independent Samples II: Comparisons in the Presence of Unequal Dispersion

EXAMPLE: Antibacterial Clinical Trial

Let's begin our discussion by examining a very simple clinical trial design from clinical antibacterial research. Bacteria-resistant infections are becoming a serious public health threat in many parts of the world (World Health Organization, 2002). Consider an agent that is believed to show some effect against bacteria in an infection that is now thought to be resistant to most common antibacterial agents. Infected patients are randomized to one of two groups: a placebo group or a new treatment group. Originally, 400 subjects were randomized to one of the two arms of the study. Forty-eight hours after receiving treatment, 5–10 milliliters of blood were drawn aseptically from each subject to determine his or her blood bacterial count (BBC) in colony-forming units/ml of blood (CFU/ml). Due to some clinical complications during the study, the resultant sample sizes were 184 and 109 (new therapy and placebo, respectively). The blood bacterial count data collected in the study are included in the BBC data set that can be found on the book's companion Web site.

Table 6.2 Summary of Blood Bacterial Count Data in the Antibacterial Clinical Trial

Group	Sample size	Mean	Standard deviation	Median	Interquartile range
Placebo	184	60.5	92.8	8.0	88.0
Treated	109	19.5	38.4	8.0	12.0

Table 6.2 and Figure 6.2 provide a summary of the blood bacterial count data in the antibacterial clinical trial. A first inspection of the box plots in Figure 6.2 suggests some rather peculiar behavior in the placebo subjects: the distribution of this group appears to have a very long and significant tail upwards away from zero. Moreover, the data are quite skewed (as evidenced by the difference between the mean and median in the summary statistics). Note that despite the rather large difference in the two means, the medians are the same. A final feature worth examining is the seemingly large difference in dispersion between the two groups.

A rather naïve way to approach a comparison of the two location parameters would begin with a simple application of a Student's two-sample *t*-test based on the TTEST procedure (Program 6.3).

Figure 6.2 Results of the antibacterial clinical trial

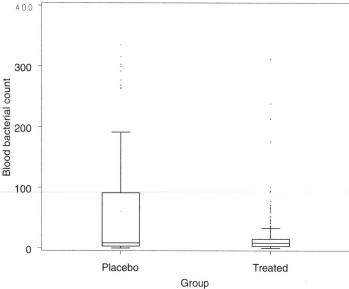

Program 6.3 Student's two-sample *t*-test in the antibacterial trial example

```
proc ttest data=bbc;
    ods select ttests;
    class group;
    var bbc;
    run;
```

Output from Program 6.3

T-Tests					
Variable	Method	Variances	DF	t Value	Pr > |t|
bbc	Pooled	Equal	291	5.28	<.0001
bbc	Satterthwaite	Unequal	130	4.40	<.0001

Output 6.3 shows that the two- and one-tailed *p*-values (assuming unequal variances) are highly significant ($p < 0.0001$). It will be unwise to be very happy at this statistically significant result (we have evidence that a statistically significant decrease in the treatment mean exists) and more prudent to reflect on the apparent lack of symmetry in the data. Given the observed asymmetry, one can decide that perhaps an independent two-sample nonparametric test might be more appropriate in this setting. Program 6.4 performs the Wilcoxon Rank Sum test on the BBC data set.

Program 6.4 Wilcoxon Rank Sum test in the antibacterial trial example

```
proc npar1way data=bbc wilcoxon;
    ods select WilcoxonTest;
    class group;
    var bbc;
    run;
```

Output from Program 6.4

```
                         Wilcoxon Two-Sample Test

              Statistic                      17223.0000

              Normal Approximation
              Z                                  1.7149
              One-Sided Pr >  Z                  0.0432
              Two-Sided Pr > |Z|                 0.0864

              t Approximation
              One-Sided Pr >  Z                  0.0437
              Two-Sided Pr > |Z|                 0.0874

              Z includes a continuity correction of 0.5.
```

The one-tailed p-value (based on the normal approximation) displayed in Output 6.4 is significant ($p = 0.0432$). Once again, we appear to have a statistically significant difference (this time, it is a difference between the two medians); however, this conclusion appears to be at odds with the descriptive statistics in Table 6.2 which indicate that the two groups have a median of 8.0.

The problem with the second analysis, although it accounts for a lack of symmetry in the two distributions, is that it does not account for a difference in dispersion or scale. One of the chief assumptions of the Wilcoxon Rank Sum test is that the two samples have the same level of dispersion. This feature of our study could be the cause of the seemingly strange statistical results. It appears that a difference exists in the dispersion of the two distributions, yet does a more objective analytical means exist to substantiate this belief?

Comparison of Two Distributions' Dispersion

One means to compare the dispersion or spread of two distributions, assuming that the medians are identical, is to use the Ansari-Bradley test (Ansari and Bradley, 1960). Inference about the equality of two distributions' variation is based upon the ratio of the two distributions' scale parameters. If X comes from a distribution $F(\mu, \varphi_1)$, with location parameter μ and scale parameter φ_1, and Y comes from a distribution $F(\mu, \varphi_2)$, with location parameter μ and scale parameter φ_2, we are interested in $\phi = \varphi_1/\varphi_2$. If ϕ is close to unity, evidence does not exist to declare the two distributions to have differing amounts of variation. Otherwise, one distribution is declared to have greater dispersion than the other. Note that the Ansari-Bradley test assumes that the two distributions have a common location parameter, μ.

What if we are working in an area of research where we really do not have knowledge about the assumption of a common location parameter, μ? A way around making assumptions in this setting is to transform each of the n_1 X values and the n_2 Y values by subtracting each group's median from the original raw data:

$$X_i^* = X_i - m_X, \quad i = 1, \dots, n_1, \quad Y_j^* = Y_j - m_Y, \quad j = 1, \dots, n_2,$$

where m_X and m_Y are the medians of the X's and Y's, respectively. To construct the Ansari-Bradley statistic, we begin by jointly ordering the X^*'s and Y^*'s from smallest to largest. The ranking procedure is slightly different from most nonparametric techniques. Table 6.3 shows how the ranking procedure works for the Ansari-Bradley test. The ranks are assigned from the "outward edges, inward," for the ordered values. Ties are resolved as before in the calculation of the Wilcoxon Rank Sum test (see Section 6.2.2).

Table 6.3 Ranking Procedure for the Calculation of the Ansari-Bradley Statistic

Step	Procedure
0	Order $X_1^*, X_2^*, \ldots, X_{n_1}^*, Y_1^*, Y_2^*, \ldots, Y_{n_2}^*$ from smallest to largest, label these values O_i for $i = 1, \ldots, n_1 + n_2$
1	Assign O_1 and $O_{n_1+n_2}$ rank= 1
2	Assign O_2 and $O_{n_1+n_2-1}$ rank= 2
3	Assign O_3 and $O_{n_1+n_2-2}$ rank= 3
...	...

For the $n_1 + n_2$ ranked values, the Ansari-Bradley statistic is calculated by adding up the n_2 ranks of the Y^* values: $S = \sum_{j=1}^{n_2} R_j$, where R_j is the rank of Y_j^* in the combined sample.

The null hypothesis $H_0 : \phi = 1$ is rejected in favor of $H_A : \phi \neq 1$ at level $\alpha = \alpha_1 + \alpha_2$ if $S \geq c_{\alpha_1}$ or $S \leq c_{1-\alpha_2} - 1$, where c_{α_1} and $c_{1-\alpha_2}$ are quantiles from the null distribution of the Ansari-Bradley statistic with $\alpha_1 + \alpha_2 = \alpha$. If $n_1 = n_2$, it is reasonable to pick $\alpha_1 = \alpha_2 = \alpha/2$. A corresponding large sample approximate test exists as well as corrections to the test statistic for the presence of ties. The interested reader is encouraged to read a discussion of these items in Hollander and Wolfe (1999, Chapter 5).

Let's now return to the antibacterial trial example. It would be interesting to test whether the scale parameters of the two distributions are different. If evidence exists that the dispersion of the two distributions is different, one of the principal assumptions of the Wilcoxon Rank Sum test is violated.

The Ansari-Bradley test may be carried out easily with PROC NPAR1WAY (Program 6.5). The first step in performing the version of the Ansari-Bradley test described above is to determine each sample's median (this is accomplished by using the UNIVARIATE procedure). After that, each value is adjusted by its location (median) and the Ansari-Bradley statistic is calculated from the adjusted values using PROC NPAR1WAY with the AB option.

Program 6.5 Ansari-Bradley test in the antibacterial trial example

```
proc univariate data=bbc;
    class group;
    var bbc;
    output out=medianbbc median=median;
proc sort data=bbc;
    by group;
proc sort data=medianbbc;
    by group;
data adjust;
    merge bbc medianbbc;
    by group;
    bbc_star=bbc-median;
proc npar1way data=adjust ab;
    ods select ABAnalysis;
    class group;
    var bbc_star;
    run;
```

Output from Program 6.5

```
                    Ansari-Bradley One-Way Analysis

          Chi-Square                      9.8686
          DF                                   1
          Pr > Chi-Square                 0.0017
```

Output 6.5 lists the Ansari-Bradley test statistic ($S = 9.8686$) and associated p-value based on a normal approximation ($p = 0.0017$). Note that the sample size is sufficiently large to warrant the use of the large sample approximate test. At the 0.05 level of statistical significance, we can declare that a statistically significant difference exists between the two groups with respect to the scale parameter or, equivalently, dispersion of the blood bacterial count data.

Fligner-Policello Approach to the Behrens-Fisher Problem

It is clear from Output 6.5 that one of the chief assumptions of the Wilcoxon Rank Sum test is violated because the two distributions have differing amounts of variation. Does a statistical inferential procedure exist that will compare two group medians in the face of unequal spread?

Fligner and Policello (Fligner and Policello, 1981) suggested the following test to compare two location parameters from distributions with differing amounts of variation. The statistic is based upon a quantity called a *placement*. A placement is the number of values in one sample strictly less than a given value of a second sample. Consider samples $X_1, \ldots, X_{n_1}$ and $Y_1, \ldots, Y_{n_2}$. For a value X_i, $i = 1, \ldots, n_1$, its corresponding placement, P_i, is the number of values from $Y_1, \ldots, Y_{n_2}$ less than X_i. Similarly, for a value Y_j, $j = 1, \ldots, n_2$, its placement, Q_j, is the number of values from the first sample less than Y_j. The next steps involve calculating quantities similar to those required for a standard Student's two-sample t-test. To perform the hypothesis test

$$H_0 : \delta = 0 \text{ when } \varphi_1 \neq \varphi_2 \text{ versus } H_A : \delta > 0 \text{ when } \varphi_1 \neq \varphi_2,$$

we need to compute the average placements, $\bar{P} = n_1^{-1} \sum_{i=1}^{n_1} P_i$ and $\bar{Q} = n_2^{-1} \sum_{j=1}^{n_2} Q_j$, as well as the variances:

$$V_1 = \sum_{i=1}^{n_1} (P_i - \bar{P})^2, \quad V_2 = \sum_{j=1}^{n_2} (Q_j - \bar{Q})^2.$$

The Fligner-Policello test statistic is of the form:

$$F = \frac{n_1 \bar{P} - n_2 \bar{Q}}{2\sqrt{V_1 + V_2 + \bar{P}\bar{Q}}}.$$

We reject H_0 in favor of H_A at level α if $F \geq c_\alpha$, where c_α is the cutoff from the null distribution of the Fligner-Policello statistic. By the Central Limit Theorem, $F \sim N(0, 1)$, c_α may be conveniently replaced with the quantile from a standard normal distribution, z_α, when n_1 and n_2 are large. This gives us the large-sample versions of the lower and two-tailed tests, respectively:

- Reject $H_0 : \delta = 0$ (when $\varphi_1 \neq \varphi_2$) in favor of $H_A : \delta < 0$ (when $\varphi_1 \neq \varphi_2$) if $F \leq -z_\alpha$.
- Reject $H_0 : \delta = 0$ (when $\varphi_1 \neq \varphi_2$) in favor of $H_A : \delta \neq 0$ (when $\varphi_1 \neq \varphi_2$) if $|F| \geq z_{\alpha/2}$.

If ties exist in the data, one needs to adjust the Fligner-Policello statistic defined above. The adjustment occurs at the level of the placements. When calculating one set of the placements, for example, to compute P_i, we count the number of the Y's less than X_i and add half of the values equal to X_i. The same is true for the other set of placements. The balance of the method remains unchanged.

%FPSTATISTIC Macro for Performing the Comparison of Two Location Parameters in the Absence of Unequal Scale

The %FPSTATISTIC macro, available on the book's companion Web site, performs the large sample version of the Fligner-Policello test. To invoke the macro, the user needs to specify three parameters:

- DATASET is the name of the data set.
- VAR is the name of the response variable.
- GROUPVAR is the name of the grouping (classification) variable for the macro. This variable must have only two levels.

The macro begins by determining the sample size for each level of the grouping variable. After the macro determines the two group sample sizes, it begins to calculate the placements for each group including an adjustment for ties. The algorithm involves a loop that iteratively compares values and assigns an indicator variable with the values 1 (greater than response in other group), 0.5 (tied or equal values) or 0 (for less than response in other group). These indicators are added to form the placements for the first group with n_1 observations. The same procedure is followed for the n_2 placements of the second group.

The final portion of the macro uses the MEANS procedure to calculate the sum of the placements, mean placements, and corrected sums of squares for the placements for each level of the grouping variable. The PROBNORM function is then used to determine the p-values. The macro outputs three versions of the large sample Fligner-Policello test, two-sided, upper and lower (the results are saved in the FPSTATISTIC data set).

Program 6.6 calls the %FPSTATISTIC macro to perform a comparison of the blood bacterial count data in the placebo and treated groups. Note that the location parameters for the two levels of the classification variable are compared in reverse alphanumeric sort order. Thus, for the BBC data set, the macro will test whether the location parameter in the treated group is greater than the location parameter in the placebo group. In the antibacterial trial example, we are interested in knowing if the treatment lowered the blood bacterial count. Thus we need to focus on the one-sided, lower-tailed test.

Program 6.6 Fligner-Policello test in the antibacterial trial example

```
%fpstatistic(dataset=bbc,var=bbc,groupvar=group);
proc print data=fpstatistic noobs label split="*";
    format uhat 6.3 p2sided up1sided low1sided 5.3;
    var n1 n2 uhat p2sided up1sided low1sided;
    run;
```

Output from Program 6.6

Sample size group 1	Sample size group 2	FP statistic	Two-sided p-value	Upper-tailed p-value	Lower-tailed p-value
109	184	-1.615	0.106	0.947	0.053

Output 6.6 shows that the one-sided, lower-tailed p-value is not significant at a 0.05 level ($p = 0.053$). Thus the Fligner-Policello test fails to reject the null. We conclude that

there is not enough evidence to declare the blood bacterial counts to be statistically significantly different between the two treatment groups. The results of this analysis are much more in agreement with the summary statistics presented earlier in Table 6.2.

Role of Data Transformations

It is very common for data analysts to employ basic transformations on data so that the transformed data will have desirable properties. Two common means of transforming data are the use of logarithms and rank-transformations. Logarithmic transformations are not always useful, as zeros or negative values may exist in data sets. Moreover, applying a logarithmic transformation to data, followed by performing a two-sample Student's t-test, does not always test the desired hypothesis comparing the locations of two samples' distributions (see Zhou et al., 1997, for a discussion of this practice).

Some folklore has evolved among practitioners of statistical methods in the field with respect to the rank-transformations. A common misconception is that the rank-transform eliminates the need to adjust for unequal dispersion, i.e., the rank-transformation corrects for the problem of unequal variation. As illustrated previously, the results of performing a rank-transformation and a large-sample approximate Wilcoxon Rank Sum test may produce a counter-intuitive inference. The author would recommend the application of the Fligner-Policello method in an analysis of two independent samples where the distributions are believed to differ in scale.

Although, on the surface, the analysis of the two independent samples seems trivial, the setting still requires the care necessary for any data analysis. The features of the data must be understood and matched appropriately to the assumptions of the statistical method(s) employed. This is a theme that will be reiterated more times throughout this chapter.

Now that we have examined a few aspects of the two independent samples case, it is time to advance to the setting of k independent samples. The next section will address several nonparametric approaches to analyzing data in this common experimental setting, largely emphasizing examples from drug discovery research.

6.2.4   Summary

The two-sample problem is not unlike other problems in statistical inference: careful examination of the assumptions is required to execute an appropriate analysis. Nonparametric statistics methods afford techniques for handling the comparison of two distributions' locations in both the case of similar dispersion (Wilcoxon Rank Sum test) and differing dispersion (Fligner-Policello test). In the case of the former, the corresponding analysis may be carried out with PROC NPAR1WAY. In the latter case, the analysis may be executed with a SAS macro created by the author (%FPSTATISTIC macro).

6.3   The One-Way Layout

The one-way layout is a commonly used experimental design in biological research. A recent Web search of the archival scientific literature (Science Direct) covering the domain of published pharmacological, toxicological, and pharmaceutical science research from 2000 to first quarter 2005 produced 3,671 hits pertaining to the application of a one-way layout as the means to address a specific scientific hypothesis. The one-way layout is a powerful means to study basic scientific questions because of its simplicity. Recall that in a one-way layout, experimental units (e.g., subjects) are randomly assigned to two or more treatment groups. The treatment groups could consist of two or more distinct treatments or increasing doses of a single agent (as in the case of later stage drug discovery studies or many standard toxicology investigations). Each unit receives its treatment and a continuous measurement is collected. Formally, consider a study (drug discovery experiment or clinical trial) with k

treatment groups and n_i units/subjects in the ith group. Let X_{ij} be the response of the jth experiment unit ($j = 1, \ldots, n_i$) to the ith treatment ($i = 1, \ldots, k$). The total sample size for the study is $N = n_1 + \ldots + n_k$. Let τ_i be the effect of the ith treatment group. Then, we can express the relationship between the treatment and response with a simple linear model:

$$X_{ij} = \tau_i + \varepsilon_{ij}. \tag{6.1}$$

where ε_{ij} represents the measurement error associated with the jth experimental unit (subject) assigned to the ith group (or treatment). We will assume that the ε_{ij}'s are independent and identically distributed by virtue of the original random assignment of units to treatments.

EXAMPLE: Dopamine Experiment

In an effort to make the discussion more concrete, consider the following simple experiment as an example (Juneau, 2004). Samples of PC12 cells were randomized to one of three groups. The PC12 cells were cultured in one of three media: the first medium was infected with a particular strain of bacteria postulated to be associated with the development of Parkinson's disease. The second group used a medium cultured with a second strain of bacteria. The third group of cells was cultured in a normal uninfected medium. All cells were incubated for 24 hours and harvested to determine the dopamine concentration. Due to some unanticipated circumstances, some samples were lost during processing and the resultant sample sizes were unequal at the end of the study. The dopamine concentration data from this study are included in the DOPAMINE data set provided on the book's companion Web site.

From the formal statement above, X_{ij} would represent the dopamine concentration response of the jth experiment unit ($j = 1, \ldots, n_i$) to the ith treatment ($i = 1, 2, 3$), where treatment 1 (control group) sample size is $n_1 = 14$, treatment 2 (Strain I) sample size is $n_2 = 7$, and treatment 3 (Strain II) sample size is $n_3 = 10$.

Typically, researchers will be interested in making decisions about the τ_i's: "Does evidence exist to declare the response of at least one of the treatment groups to be different from the others?" Mathematically, one could ask: "Is it the case that $\tau_1 = \tau_2 = \ldots = \tau_k$ or does evidence exist to declare at least one $\tau_i \neq \tau_l$ for $i \neq l$?" For the dopamine experiment illustrated previously, this question would translate into "How does the presence of one of the three treatments affect dopamine concentration?" Once again, one could express this question mathematically: "Is it the case that $\tau_1 = \tau_2 = \tau_3$ or that at least one difference exists among all pairs of τ_1, τ_2, or τ_3?"

In some sense, evidence to reject the idea that the treatment effects are all the same is not very informative. Typically, such conclusions do not establish differences in activity among several novel agents. Investigators will often be more interested in examining questions with alternatives designed for the comparison of one or more of the treatment effects against a designated group (e.g., a control group) or elucidating where treatment effects exist among the k treatments (e.g., pair-wise comparisons). This section will be devoted to the examination of various alternative hypotheses that arise in typical pharmaceutical investigations where the one-way layout is the design of choice.

Section 6.3.1 will address the general alternative hypothesis of unequal treatment effects associated with the Kruskal-Wallis test. Examples will be demonstrated with PROC NPAR1WAY. Section 6.3.2 will deal with nonparametric multiple comparison procedures in the one-way layout when an investigator is interested in comparing all treatment groups against a single designated (control) group or performing all simultaneous pair-wise comparisons of treatment effects. Section 6.3.2 will introduce an original SAS macro code to perform the desired statistical tests.

6.3.1   Testing a General Alternative for k Groups: The Kruskal-Wallis Test

Table 6.4 and Figure 6.3 summarize the dopamine concentration response data collected in the dopamine experiment. Notice the slight asymmetry in the distribution patterns of the control and Strain I-infected samples. The control group has two responses that seem extreme relative to the rest of the measurements.

Table 6.4 Summary of the Dopamine Concentration Response Data in the Dopamine Experiment

Group	Sample size	Mean	Standard deviation	Median	Interquartile range
Control	14	100.00	15.14	95.24	13.71
Strain I	7	63.12	28.65	54.29	46.38
Strain II	10	78.04	29.81	72.98	45.12

Figure 6.3 Results of the dopamine experiment

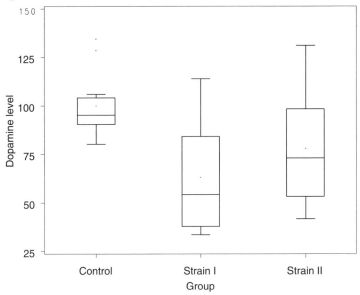

A standard approach to comparing these three treatments would be to begin with a classical linear model relating the dopamine response to the treatment:

$$X_{ij} = \mu + \tau_i + \eta_{ij},$$

where for $i = 1$ (controls), $j = 1, \ldots, 14$; $i = 2$ (Strain I), $j = 1, \ldots, 7$; $i = 3$ (Strain II), $j = 1, \ldots, 10$, η_{ij} are independent and identically distributed Gaussian errors ($N(0, \sigma^2)$), and μ represents the overall population dopamine concentration mean response. This model could then be fit with the GLM procedure. Let's examine a plot of the density of the residuals after fitting the dopamine concentration data with a standard linear model based upon analysis of variance assumptions.

The two curves superimposed in Figure 6.4 allow us to make an interesting conclusion. The dashed curve represents a Gaussian (normal) distribution fitted to the residuals. A solid kernel smooth curve fit to the same residuals suggests that the residuals are asymmetrical (this is also demonstrated by the separation of the means and medians in Table 6.4).

It appears by the evidence presented above that the normality assumption of the η_{ij}'s is in question. The typical parametric model and inferences based upon the analysis of variance are thus inappropriate for the evaluation of this experiment. A more appropriate

Figure 6.4 Plot of residuals after fitting a traditional analysis of variance-type model in the dopamine experiment

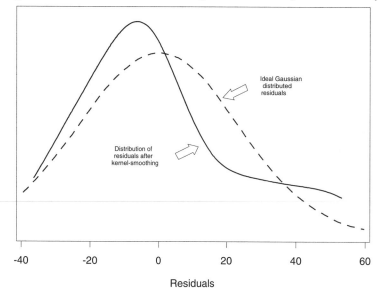

analysis would be based upon the model introduced earlier in this section and the Kruskal-Wallis test (Kruskal and Wallis, 1952).

Consider the model introduced at the beginning of Section 6.3.1 that relates the response of an experimental unit to the effect of the ith treatment τ_i. The Kruskal-Wallis test is used to test the following null hypothesis against a very general alternative:

$$H_0 : \tau_1 = \tau_2 = \cdots = \tau_k$$

versus

$$H_A : \text{ at least one } \tau_i \neq \tau_l \text{ for } 1 \leq i \leq k, \quad 1 \leq l \leq k, \quad i \neq l.$$

In the dopamine study, $H_0 : \tau_1 = \tau_2 = \tau_3$ is tested versus the alternative that states that at least two out of three τ's are different from each other.

The test statistic for the Kruskal-Wallis test is constructed by ranking all N observations from the smallest value to the largest. Tied values are assigned the average rank. The mean rank is then calculated for each group. Denote the mean rank for the ith group by $R_{\bar{i}}$. The Kruskal-Wallis test statistic, K, is then defined by

$$K = \frac{12T}{N(N+1)} \sum_{i=1}^{k} n_i \left(R_{\bar{i}} - \frac{N+1}{2} \right)^2,$$

where T is a correction factor for tied values. If for all N values, ξ groups of size t_v $(1, 2, \ldots, v, \ldots, \xi)$ exist, the correction factor T is expressed as (Hollander and Wolfe, 1999)

$$T = 1 - \frac{1}{N^3 - N} \sum_{v=1}^{\xi} (t_v^3 - t_v).$$

H_0 is rejected in favor of H_A at level α, if $K \geq c_\alpha$.

The cutoff value for the test, c_α, comes in two varieties. The first is the large sample approximate test, which is based upon a chi-squared cutoff. The degrees of freedom for this test are $k - 1$, i.e., $\chi^2_{k-1}(\alpha)$. The other type of cutoff for this test is the exact version, based upon the permutation distribution of the ranks. PROC NPAR1WAY has had the capability to perform the exact version of the test since SAS Version 6 via the EXACT statement. The formulation of the Kruskal-Wallis statistic and test stated above is not truly an exact α-level test, but only an approximate α-level test. Fortunately, if ties exist in the data, PROC NPAR1WAY will perform the exact α-level test because it determines the permutation distribution of the tied ranks when it determines the p-value for the analysis.

Large-Sample Kruskal-Wallis Test

Program 6.7 performs the large-sample Kruskal-Wallis test in the dopamine experiment using PROC NPAR1WAY. The WILCOXON option is used in the procedure to specify Wilcoxon-type scores to perform the Kruskal-Wallis test.

Program 6.7 Large-sample Kruskal-Wallis test in the dopamine experiment

```
proc npar1way data=dopamine wilcoxon;
   class group;
   var dopamine;
   run;
```

Output from Program 6.7

```
                    Kruskal-Wallis Test

                Chi-Square          9.0710
                DF                       2
                Pr > Chi-Square     0.0107
```

Output 6.7 shows that the p-value produced by the large-sample Kruskal-Wallis test is 0.0107 and thus we would reject the null hypothesis of equal dopamine response for all three groups in favor of the alternative that at least one difference exists among the three sets of treated samples at $\alpha = 0.05$.

Exact Kruskal-Wallis Test

If we examine the DOPAMINE data set, we notice that several tied responses exist for the dopamine response. Tied values are a reality in many real experiments because of limitations on measurement precision, rounding off of values by scientists, or various other reasons. Let's take advantage of the EXACT statement in PROC NPAR1WAY and see if the results change dramatically (Program 6.8). The program is virtually identical to Program 6.7, with the exception of the insertion of the EXACT statement with a WILCOXON option.

Program 6.8 Exact Kruskal-Wallis test in the dopamine experiment

```
proc npar1way data=dopamine wilcoxon;
   class group;
   var dopamine;
   exact wilcoxon;
   run;
```

Output from Program 6.8

```
                        Kruskal-Wallis Test

              Chi-Square                      9.0710
              DF                                   2
              Asymptotic Pr > Chi-Square     0.0107
              Exact Pr >= Chi-Square         0.0074
```

Output 6.8 lists the exact p-value ($p = 0.0074$), which is also significant at a 0.05 level. It is important to mention that running the exact Kruskal-Wallis test is *very resource-intensive*. The real time run for this short program and relatively small data set was about three hours long. From the author's perspective, the additional resources required may not justify the implementation of the exact test in this particular setting. This may not always be the case, however. Let's consider another experiment.

EXAMPLE: Hyperplastic Alveolar Nodule (HAN) Study

Samples of four types of cell subsets are prepared for a study of the progression of mouse mammary preneoplastic hyperplastic alveolar nodule (HAN) line C4 to carcinoma. Hoechst fluorescence is measured as a response for each sample. The data from this study are included in the HAN data set given on the book's companion Web site.

Let's assume that the small sample size for this study is enough to warrant the selection of a nonparametric statistical method (see Section 6.1). We also assume that the investigator has a very wide alternative of interest: the investigator is interested in evidence of any difference in response. Does the choice of an exact test or asymptotic test matter? Program 6.9 carries out the large-sample and exact Kruskal-Wallis tests in the HAN study.

Program 6.9 Large-sample and exact Kruskal-Wallis tests in the HAN study

```
proc npar1way data=han wilcoxon;
    class celltype;
    var fluor;
    exact wilcoxon;
    run;
```

Output from Program 6.9

```
                        Kruskal-Wallis Test

              Chi-Square                      6.7315
              DF                                   3
              Asymptotic Pr > Chi-Square     0.0810
              Exact Pr >= Chi-Square         0.0383
```

Output 6.9 lists the large-sample and exact p-values produced by the Kruskal-Wallis tests. It appears that if the investigator were interested in using the traditional 0.05 level of statistical significance, he or she might be disappointed with the results of the asymptotic test ($p = 0.0810 > 0.05$), or elated at the conclusion reached by the exact Kruskal-Wallis analysis ($p = 0.0383 < 0.05$). The point of this illustration is that small sample sizes can influence the results of a testing procedure and should be considered before an analysis strategy is selected. The next section offers some rules of thumb about group size and about when the asymptotic procedures typically perform reasonably.

6.3.2   Nonparametric Multiple Comparison Procedures for Specific Alternatives

A common objective of many pharmaceutical research investigations with a one-way layout is to compare treatments in a pair-wise fashion. Investigators are typically interested in comparing all treatment responses against one another or in comparing several agents against a designated group (e.g., an inactive control group). Both of these settings involve the simultaneous comparison of several group location parameters (say, means in the parametric case, medians in the nonparametric). It is also often desirable to preserve the family-wise error rate in this inference; i.e., the probability of declaring at least one treatment to be different from another, when in fact, it is not.

If we return to the dopamine experiment first presented in Section 6.3.1, we can see an example of a discovery biology study where investigators might be interested in performing all pair-wise comparisons of treatment groups. The investigators would certainly be interested in comparing the typical response in the Strain I and Strain II infected samples with that of the untreated media samples (control); however, it might be of scientific consequence to know if a difference exists between the two sets of infected samples as well.

Dunn's Procedure for All Pair-Wise Comparisons

If an investigator is interested in comparing the location parameters of k experimental groups simultaneously and preserving the family-wise error rate, he or she could use an approach suggested by Dunn (Dunn, 1964) for the linear model introduced previously (6.1), i.e., conclude that $\tau_i \neq \tau_l$ if

$$|R_{\bar{i}} - R_{\bar{l}}| > z_{\alpha/k(k-1)} \sqrt{\frac{N(N+1)}{12}\left(\frac{1}{n_i} + \frac{1}{n_l}\right)},$$

where $R_{\bar{i}}$ is the mean of the joint ranks for the ith group, $R_{\bar{c}}$ is the mean of the joint ranks for the lth group, and n_i and n_l are sample sizes in the two groups, respectively, N is the total sample size, k is the total number of groups (in the dopamine experiment, $k = 3$) and $z_{\alpha/k(k-1)}$ is the $100\alpha/[k(k-1)]$th upper quantile from a standard Gaussian distribution.

Recall that the *joint ranking* of a data set is determined by ranking all of the N observations together from smallest to largest.

Dunn's procedure offers the following advantages:

- The symmetry assumption, which is often difficult to assess in drug discovery settings with small sample sizes, may be relaxed or ignored.

- Equal sample sizes are not required.

- Relatively small total sample sizes may be analyzed with this technique, i.e., three groups with five experimental units/group or more than three groups with four units/group (see Lehman, 1975).

%DUNN Macro

A simple set of macros can be constructed in SAS to perform Dunn's procedure for all pair-wise comparisons. The author composed a simple SAS macro to perform Dunn's procedure as part of a presentation for the 2004 PharmaSUG meeting in San Diego (Juneau, 2004). The %DUNN macro was designed to imitate PROC NPAR1WAY, with respect to ease of use and input of information, and to produce output quite similar to SAS' very popular procedure, GLM, using the MEANS statement and the CLDIFF option. The macro can be found on the book's companion Web site.

The %DUNN macro consists of a body of code containing one embedded macro (%GROUPS). The embedded macro determines the number of groups present and assigns that value to a macro variable (NGRPS). If a group in the data set does not contain at least one response value (i.e., all values are missing), it will not be included in the analysis. The embedded macro also creates one global macro variable that contains the group labels (GRPVEC) for the levels of the class variable. The main body of the SAS code determines summary statistics (e.g., average ranks, sample sizes, etc.). This information is employed to calculate the pair-wise test statistics. The corresponding cutoff for the test statistic is calculated with the PROBIT function. Results are then printed out using the PRINT procedure.

Program 6.10 carries out Dunn's test in the dopamine experiment by invoking the %DUNN macro. The first parameter of the macro is the input SAS data set name (DOPAMINE), the second the classification or grouping variable (GROUP), the third the name of the response variable in the input data set (DOPAMINE), and the fourth macro parameter is the overall family-wise error rate (0.05).

Program 6.10 Large-sample and exact Kruskal-Wallis tests in the HAN study

```
%dunn(dopamine,group,dopamine,0.05);
```

Output from Program 6.10

```
              Large sample approximation multiple comparison procedure
                           designed for unbalanced data
         3 groups: Control StrainI StrainII (respective sample sizes: 14 7 10)
                                  Alpha = 0.05


                                   Difference
                                       in           Cutoff
           Comparison        Group      average       at           Significance
             number      comparisons    ranks     alpha=0.05      difference = **


               1         Control-StrainI   11.8571    10.0759            **
               2         Control-StrainI    7.6429     9.0121
               3         StrainI-StrainI    4.2143    10.7266
```

Output 6.10 displays the output of the %DUNN macro. The %DUNN macro produces TITLE statements that state the number of class levels, a list of each of the class levels with non-missing values, and the corresponding group sample sizes. Moreover, the macro generates output that contains the relevant test statistic for each comparison (a function of the average ranks/class level), the corresponding cutoff for the chosen level of family-wise error, and a symbol indicating whether the results of the statistical inference are statistically significant. From this analysis, one would conclude that a statistically significant difference existed between the median dopamine levels in the controls and samples treated with the first strain of bacteria.

Dwass-Steel-Critchlow-Fligner Procedure for All Pair-Wise Comparisons

The Dwass-Steel-Critchlow-Fligner (DSCF) procedure is another popular form of simultaneous nonparametric inference in the one-way layout for all pair-wise comparisons (Hollander and Wolfe, 1999). If a pharmaceutical researcher is interested in comparing the location parameters of k experimental groups $(\tau_1, \ldots, \tau_k)$ simultaneously, and preserving the family-wise error rate, he or she could use an approach suggested by Dwass (Dwass, 1960) and Steel (Steel, 1960) for the linear model previously introduced (6.1).

The procedure begins by calculating the $k(k-1)/2$ pairs of Wilcoxon Rank Sum statistics, W_{il} (Wilcoxon, 1945) for each pair, i and l $(1 \leq i \leq l \leq k)$. The Wilcoxon statistics *should* include an adjustment for tied values. Conclude that $\tau_i \neq \tau_l$ if $|D_{il}| > q_\alpha$, where

$$D_{il} = \sqrt{\frac{2}{\text{Var}(W_{il})}} \left(W_{il} - \frac{\min(n_i, n_l)(n_i + n_l + 1)}{2} \right),$$

$\text{Var}(W_{il})$ is the tie-adjusted variance for the Wilcoxon statistic, and q_α is the 100αth upper quantile from the Studentized Range distribution (Harter, 1960).

Suppose that for all $n_i + n_l$ observations in the comparison, ξ tied groups of size t_v $(1, 2, \ldots, v, \ldots, \xi)$ exist. The variance in the denominator of the statistic is given by (Hollander and Wolfe, 1999)

$$\frac{n_i n_l}{24} \left(n_i + n_l + 1 - \frac{\sum_{v=1}^{\xi}(t_v - 1)t_v(t_v + 1)}{(n_i + n_l)(n_i + n_l - 1)} \right).$$

%DSCF Macro

An analysis using the DSCF method may be conducted using a SAS macro called %DSCF created by the author for a 2004 PharmaSUG presentation (Juneau, 2004). The call of the %DSCF macro is as follows:

```
%dscf(dataset,group,response,alpha);
```

The first parameter of the macro is the input data set name (DATASET), the second the classification or grouping variable (GROUP), the third the name of the response variable in the input data set (RESPONSE), and the fourth macro parameter is the overall family-wise error rate (ALPHA).

The %DSCF macro consists of a body of code containing one embedded macro (%GROUPS). The embedded macro determines the number of groups present (NGRPS) as in the %DUNN macro. If a group in the input data set does not contain at least one response value, it will be excluded from the analysis. The embedded macro also creates two global macro variables that contain the group labels (GRPVEC) for the levels of the class variable and information about the sample size for each group (NVEC). The main body of the code calculates the necessary summary statistics (e.g., Wilcoxon Rank Sum test statistics) and the number of ties present in each pair-wise comparison. This information is then used to calculate the pair-wise test statistics. The cutoff for the test statistic is calculated with the PROBMC function (using the Studentized Range distribution). As the macro iterates between all pair-wise comparisons it concatenates successive results in a data set called STAT. The final results are then printed out with PROC PRINT.

EXAMPLE: Cardiovascular Discovery Study

Subjects were randomized to one of four treatment groups: three active agents and one vehicle control group. The goal of the experiment was to determine whether the agents could affect triglyceride level (in mg/dl) relative to the vehicle controls and to determine whether evidence existed to declare one agent different from another with respect to triglyceride response. Each subject was treated and, after a fixed post-treatment period, the blood triglyceride level was measured for each subject.

The data gathered in the study are included in the TRIG data set provided on the book's companion Web site. The results of the experiment are displayed in Table 6.5 and Figure 6.5 with box plots and summary statistics. Note the balanced sample sizes. The DSCF multiple comparison method works optimally under settings with equal sample sizes per group.

The triglyceride data can be analyzed using the %DSCF macro designed to perform the desired simultaneous pair-wise nonparametric comparisons of all treatments (see Program 6.11).

Table 6.5 Summary of Triglyceride Data in the Cardiovascular Discovery Study

Group	Sample size	Mean	Standard deviation	Median	Interquartile range
Vehicle	5	99.8	16.71	105.0	20.00
Treatment A	5	78.2	29.75	71.0	26.00
Treatment B	5	49.0	17.33	42.0	24.00
Treatment C	5	59.0	15.00	62.0	26.00

Figure 6.5 Results of the cardiovascular discovery study

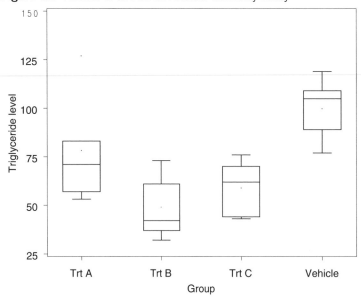

Program 6.11 Dwass-Steel-Critchlow-Fligner (DSCF) method in the cardiovascular discovery study

```
%dscf(trig,group,trig,0.05);
```

Output from Program 6.11

```
              Large sample approximation multiple comparison procedure
                            or all treatment pairs
                          based upon pairwise rankings
       4 groups: TreatmentA TreatmentB TreatmentC Vehicle (respective sample sizes: 5 5 5 5)
                              Alpha = 0.05


                                           Test statistic   Cutoff
            Group          Comparison    Test      absolute     at        Significant
          comparisons        number    statistic    value   alpha=0.05  difference = **


    TreatmentA - TreatmentB     1       -2.21565    2.21565    3.63316
    TreatmentA - TreatmentC     2       -1.62481    1.62481    3.63316
    TreatmentA - Vehicle        3        1.92023    1.92023    3.63316
    TreatmentB - TreatmentC     4        1.92023    1.92023    3.63316
    TreatmentB - Vehicle        5        3.69274    3.69274    3.63316         **
    TreatmentC - Vehicle        6        3.69274    3.69274    3.63316         **
```

The output of the %DSCF macro in Output 6.11 includes TITLE statements that state the number of class levels, a list of each of the class levels with non-missing values and the corresponding group sample sizes. The macro also lists the relevant test statistic for each comparison (a function of the Wilcoxon statistic), the corresponding cutoff for the chosen level of family-wise error, and a symbol indicating whether the results of the statistical inference are statistically significant. We conclude from Output 6.11 that a statistically significant difference existed between the median triglyceride levels in the controls and samples treated with Treatments B and C. Statistically significant differences did not exist amongst the three active agents (Treatments A, B, and C).

Comparison of Dunn and Dwass-Steel-Critchlow-Fligner Methods

Is the DSCF method really necessary, when the multiple comparison procedure suggested by Dunn is a more general case (unequal sample sizes)? The triglyceride data from the cardiovascular discovery study used previously as an example for the application of the DSCF method are analyzed in Program 6.12 with Dunn's method for all pair-wise comparisons.

Program 6.12 Dunn method in the cardiovascular discovery study

```
%dunn(trig,group,trig,0.05);
```

Output from Program 6.12

```
               Large sample approximation multiple comparison procedure
                          designed for unbalanced data
   4 groups: TreatmentA TreatmentB TreatmentC Vehicle (respective sample sizes: 5 5 5 5)
                               Alpha = 0.05


                                   Difference
                                       in          Cutoff
   Comparison          Group        average          at         Significance
     number        comparisons       ranks      alpha=0.05    difference = **


         1        TreatmentA-TreatmentB      6.6        9.87145
         2        TreatmentA-TreatmentC      3.6        9.87145
         3        TreatmentA-Vehicle         5.0        9.87145
         4        TreatmentB-TreatmentC      3.0        9.87145
         5        TreatmentB-Vehicle        11.6        9.87145           **
         6        TreatmentC-Vehicle         8.6        9.87145
```

It is instructive to compare Output 6.11 (DSCF method) and Output 6.12 (Dunn method). Recall that in Output 6.11, Treatment C was also declared to be statistically significantly different from the vehicle control group for the median triglyceride response. A comparison of the properties of these two methods may shed some light on the reason for the differing inferential conclusions. First, the Dunn method uses a Bonferroni-like correction to the family-wise error rate (Miller, 1981) and might be a bit too conservative. Second, the Dunn method employs joint ranking, and thus the comparison of two groups is highly influenced by the behavior of other groups in the experiment as the data are initially ranked over the entire experiment (Hollander and Wolfe, 1999). The balanced sample sizes for all groups also suggest that the DSCF method might be the most appropriate technique to employ for all pair-wise nonparametric comparisons.

Simultaneous Nonparametric Inference in the One-Way Layout for All Group Comparisons with a Designated Control Group

It is often the case that investigators are interested in comparing the location parameters of treatments against the location parameter of a designated group. This designated group could be a control group consisting of the response of subjects or units completely untreated by any form of active agent. These are often referred to as untreated control subjects or units. More commonly, a group of experimental units or subjects treated with the vehicle, or medium for delivery of the compound, is the designated group of interest. Such subjects or units are referred to as forming the *vehicle control group*. A third type of control group is a set of subjects or units treated with an agent known to provoke a response that is the same as one desired by proponents of one of the compounds under examination. Such a collection of subjects or units is called a *positive control group*.

Other forms of designated control groups exist. The point, however, is that if pharmaceutical researchers are interested in comparing the location parameters of several experimental groups $(\tau_1, \ldots, \tau_k)$ to a designated group (τ_c), simultaneously and preserving the family-wise error rate, they can use one of two approaches originally suggested by Dunn (Dunn, 1964) or Miller (Miller, 1966) for the linear model (6.1). These approaches are described below.

Dunn's Method for All Group Comparisons to a Designated Control Group (Unequal Sample Sizes)

Using the notation first introduced in the beginning of Section 6.3, suppose that $n_i \neq n_j$ for at least one (i, j) pair $(1 \leq i < j \leq k)$ of treatments. We could conclude that $\tau_i \neq \tau_c$ if

$$|R_{\bar{i}} - R_{\bar{c}}| > z_{\alpha/2(k-1)} \sqrt{\frac{N(N+1)}{12} \left(\frac{1}{n_i} + \frac{1}{n_c}\right)},$$

where $R_{\bar{i}}$ is the mean of the joint ranks for the group i, $R_{\bar{c}}$ is the mean of the joint ranks for the control group c, and n_i and n_c are sample sizes for group i and the control group, c, respectively, N is the total sample size, $k - 1$ is the total number of comparisons desired, and $z_{\alpha/2(k-1)}$ is the $100\alpha/[2(k-1)]$th upper quantile from a standard Gaussian distribution.

Miller's Method for All Group Comparisons to a Designated Control Group (Equal Sample Sizes)

Using the notation introduced above, assume that $n_i = n_j$ for all (i, j) pairs $(1 \leq i < j \leq k)$ of treatments. Then τ_i is declared different from τ_c if

$$|R_{\bar{i}} - R_{\bar{c}}| > |M_{\alpha,k-1,\infty}| \sqrt{\frac{N(N+1)}{12} \left(\frac{1}{n_i} + \frac{1}{n_c}\right)},$$

where $R_{\bar{i}}$, $R_{\bar{c}}$, n_i, n_c, N, K are defined as above and $|M_{\alpha,k-1,\infty}|$ is the αth quantile from the Studentized Maximum Modulus distribution (Pillai and Ramachandran, 1954).

%NPARMCC Macro

All nonparametric pair-wise comparisons to a designated control group in a one-way layout can be performed with the %NPARMCC macro that can be found on the book's companion Web site. The macro consists of a body of code containing one embedded macro (%GROUPS). The embedded macro determines the number of groups present (NGRPS) as in the %DUNN and %DSCF macros. If a group in the SAS data set does not contain at least one response value, it will be excluded from the analysis. The embedded macro also creates one global macro variable that contains the group labels (GRPVEC) for the levels of the class variable. The main body of the macro determines the necessary summary statistics (e.g., average ranks, sample sizes, etc.). This information is then employed to calculate the pair-wise test statistics. The cutoff for the test statistic is calculated with the PROBIT function for Dunn's method and the PROBMC function (with the Studentized Maximum Modulus distribution) for Miller's method. The results are then printed out with PROC PRINT.

EXAMPLE: Thrombosis Experiment

An experiment was designed to study the effect of increasing the dose of a novel agent on activated clotting time (ACT). Subjects were randomized to one of four groups: a vehicle control group, a low dose group, a medium dose group, and a high dose group. About 200 minutes after receiving treatment, the ACT for each subject was measured (in seconds). The data from this experiment are contained in the THROMB data set (available on the book's companion Web site) and its results are shown in Table 6.6 and Figure 6.6 (as before, with box plots and summary statistics).

Table 6.6 Summary of Activated Clotting Times in the Thrombosis Experiment

Group	Sample size	Mean	Standard deviation	Median	Interquartile range
Control	5	64.2	8.11	63.0	6.00
Low dose	5	98.4	17.47	106.0	14.00
Middle dose	5	91.8	14.06	98.0	16.00
High dose	5	156.4	9.40	153.0	2.00

Figure 6.6 Results of the thrombosis experiment

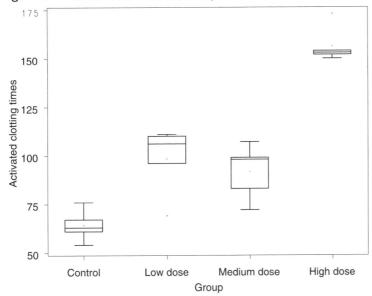

The thrombosis experiment data can be analyzed using the %NPARMCC macro designed to perform the desired simultaneous pair-wise nonparametric comparisons of all groups against a designated control group. Program 6.13 applies the %NPARMCC macro to the THROMB data.

Program 6.13 Dunn's and Miller's methods in the thrombosis experiment

```
%nparmcc(thromb,group,Control,act,0.05);
```

Output from Program 6.13

```
            Large sample approximation multiple comparison procedure
          for all treatment compared against a control (control group = Control)
        4 groups: Control HighDose LowDose MedDose (respective sample sizes: 5 5 5 5)
                                  Alpha = 0.05


                           Difference   Dunn     Miller
                               in       cutoff   cutoff   Significant  Significant
               Group      Comparison   average     at       at        difference   difference
             comparisons    number      ranks   alpha=0.05 alpha=0.05 Dunn's = *D* Miller's = *D*


         HighDose-Control     1          14.6    8.95745   8.93410      *D*           *M*
         LowDose-Control      2           7.6    8.95745   8.93410
         MedDose-Control      3           6.2    8.95745   8.93410
```

The output of the %NPARMCC macro displayed in Output 6.13 includes a list of the class levels with non-missing values, corresponding group sample sizes, and relevant test statistics for each comparison (the difference in the mean ranks), the corresponding cutoff for the chosen level of family-wise error, and a symbol indicating whether the results of the statistical inference are statistically significant by Dunn's method (*D*) and Miller's method (*M*). We conclude from Output 6.13 that a statistically significant difference existed between the median activated clotting time in the controls and samples treated with a high dose of the new agent. As the experiment consisted of balanced sample sizes, the conclusion presented by the Miller approach would be considered the appropriate one to report.

Comparison of Dunn's and Miller's Methods for All Pair-Wise Comparisons Versus a Designated Control Group

It is interesting to compare the behavior of Dunn's and Miller's respective methods using a data set that is not balanced with respect to sample size. Suppose that we conducted a similar thrombosis experiment as described previously a second time, yet this time with unequal sample sizes. The experimental results are slightly different (the control and low dose groups have different sample sizes than in the original thrombosis experiment). The results are summarized in Table 6.7 and Figure 6.7. The data from this study are included in the THROMB2 data set available on the book's companion Web site.

Table 6.7 Summary of Activated Clotting Times in the Thrombosis Experiment with Unequal Sample Sizes

Group	Sample size	Mean	Standard deviation	Median	Interquartile range
Control	9	62.2	11.10	63.0	6.00
Low dose	8	100.0	20.87	98.5	26.00
Middle dose	5	96.4	18.37	106.0	24.00
High dose	5	144.4	31.63	153.0	2.00

The data from the thrombosis experiment with unequal sample sizes were once again analyzed using the %NPARMCC macro (Program 6.14).

Program 6.14 Dunn's and Miller's methods in the thrombosis experiment with unequal sample sizes

```
%nparmcc(thromb2,group,Control,act,0.05);
```

Figure 6.7 Results of the thrombosis experiment with unequal sample sizes

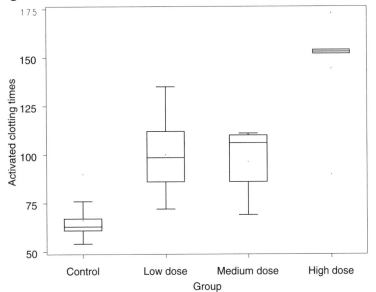

Output from Program 6.14

```
              Large sample approximation multiple comparison procedure
            for all treatment compared against a control (control group = Control)
          4 groups: Control HighDose LowDose MedDose (respective sample sizes: 9 5 8 5)
                                    Alpha = 0.05


                           Difference    Dunn       Miller
                               in        cutoff     cutoff     Significant   Significant
           Group          Comparison   average       at         at        difference    difference
           comparisons      number      ranks     alpha=0.05  alpha=0.05  Dunn's = *D*  Miller's = *D*


           HighDose-Control    1        17.4667     9.42126    10.5710        *D*           *M*
           LowDose-Control     2        10.2917     8.20747     9.2091        *D*           *M*
           MedDose-Control     3        10.1667     9.42126    10.5710        *D*
```

Output 6.14 displays the cutoffs computed by the %NPARMCC macro in the modified thrombosis experiment data set. The results of the analysis are quite interesting. Note that Miller's method *does not* declare the median response for the medium dose group to be statistically significantly different from the control group, while Dunn's method *does* declare this result to be statistically significant. The difference in the inferential conclusions can be attributed to the fact that Dunn's method was derived to handle the unequal sample size setting, whereas the Miller method, using the Studentized Maximum Modulus cutoff, works optimally under conditions of equal sample size (Hochberg and Tamhane, 1987).

Summarization of the Nonparametric Multiple Comparison Methods

Table 6.8 summarizes the nonparametric methods presented in this section and provides advice on the method to use given the study design of a particular investigation.

Table 6.8 A Summary of the Nonparametric Multiple Comparison Methods Discussed in Section 6.4

Desired comparison	Study design feature	Nonparametric method	SAS macro to perform analysis
All pair-wise comparison	Equal group sample sizes	Dwass-Steel-Critchlow-Fligner method	%DSCF
	Unequal group sample sizes	Dunn's method	%Dunn
Comparisons with a designated control group	Equal group sample sizes	Miller's method	%NPARMCC
	Unequal group sample sizes	Dunn's method	%NPARMCC

6.3.3   Summary

The one-way layout is commonly used in scientific investigations in the pharmaceutical industry. For asymmetrical response data, nonparametric statistical methods offer data analysis methods for a general alternatives of at least one change in location (Kruskal-Wallis test) or more specific, such as all pair-wise comparisons of treatment group locations (Dunn and Dwass-Steel-Critchlow-Fligner methods) or comparisons with a designated control group (Dunn and Miller methods). All techniques can be useful in settings where a small number of responses in a group skew responses or adversely affect the mean response. As was the case for the two-sample setting, it is important to study data properties (i.e., balanced vs. unbalanced sample sizes) before selecting a statistical method.

6.4   Power Determination in a Purely Nonparametric Sense

The title of this section must seem a bit of an enigma. What does the expression "in a purely nonparametric sense" really mean? Let's begin by reviewing some basic concepts in experimental design and power in the familiar parametric sense. We will then demonstrate a nonparametric approach to the same setting. Lastly, we will compare and contrast the two approaches to highlight important features.

Consider the design of a simple two-group experiment. Subjects or samples are randomized to one of two groups and receive a treatment. After treatment, a measurement is collected and the investigators wish to see if evidence exists to declare the responses of the two groups to be "different".

What does it mean for two treatments to be different? Typically, the investigators are interested in finding a statistically significant effect at some level (say, 0.05) with some reasonable degree of certainty (say, at least 80% of the time). Using some common symbolism from mathematical statistics, the latter expression may be written as:

$$\text{Test } H_0 : \mu_1 - \mu_2 = 0 \text{ vesus } H_A : \mu_1 - \mu_2 = \delta \text{ at some level } \alpha,$$

where:

- $\alpha = P(\text{Conclude } \mu_1 - \mu_2 > 0 | (\mu_1 - \mu_2) = 0)$,
- $1 - \beta = P(\text{Conclude } \mu_1 - \mu_2 > 0 | (\mu_1 - \mu_2) = \delta)$.

In other words, we are interested in finding a statistically significant difference at an α level with $1 - \beta$ power when the true difference is δ. If we make the standard assumptions of two independent and identically distributed Gaussian samples ($N(\mu_1, \sigma^2)$ and $N(\mu_2, \sigma^2)$), respectively), we can perform such a comparison via a two-sample Student's t-test, i.e., reject H_0 in favor of H_A if

$$t = \frac{\bar{X}_1 - \bar{X}_2}{\sqrt{2s^2/n}} > T_{2(n-1)}(\alpha),$$

where $\bar{X}_1$ and $\bar{X}_2$ correspond to the sample means of Treatment Groups 1 and 2, respectively, s^2 is the pooled variance estimator, and $T_{2(n-1)}(\alpha)$ is a quantile from a standard Student's t-distribution with $2(n-1)$ degrees of freedom.

Naturally, the investigators are interested in the total number of experimental units or subjects, $2n$, required to conduct the experiment. The sample size per group, n, may be determined iteratively, using a standard inequality:

$$n \geq \frac{2s^2(T_{2(n-1)}(\alpha) + T_{2(n-1)}(\beta))^2}{\delta^2},$$

where $T_{2(n-1)}(\alpha)$ and $T_{2(n-1)}(\beta)$ are quantiles from a standard Student's t-distribution with $2(n-1)$ degrees of freedom.

Note the key features of this method of sample size determination are

- The person using this method needs to feel fairly confident that his or her measure comes from a distribution that is Gaussian (normal).

- He or she must have a reasonable idea of what the treatment effect should be.

- He or she must also have a reasonable estimate of the spread or dispersion of the desired effect (s^2).

It is generally difficult for scientific investigators (based on the author's personal experience in statistical consulting) to provide information about a measurement's distribution. In fact, it is sometimes very difficult in some scientific disciplines (e.g., toxicology) for investigators to predict the size of effect that will be elicited by a treatment a priori.

One possible way to solve this problem is to conduct a small pilot study to get some initial intuition about a measurement's properties. Such a study will provide investigators with insight into the size of treatment effect that is achievable, as well as some preliminary ideas about the measurement's properties, such as its pattern of dispersion. However, will such a study address the fundamental question of distribution (e.g., Gaussian distribution) when the sample size is small?

A distribution-free, or nonparametric, approach to sample size estimation frees scientists from guessing about the distribution of measurement or treatment effect. One common way to estimate a sample size in a study with two independent groups, suggested by Collings and Hamilton (1988), requires that investigators perform a small pilot study. Investigators then use the pilot study's results to refine their estimates of achievable treatment effect. The scientists must also provide an estimate of their desired statistical significance level; however, they do not have to have any estimate of the dispersion or variation of the measurement at the time of the sample size determination. This information can be indirectly gleaned from the pilot study. For a given pair of sample sizes, say, n_1 and n_2, scientists determine the level of power associated with the comparison.

6.4.1   Power Calculations in a Two-Sample Case

How does the Collings-Hamilton technique work? Suppose the pilot study consists of data from two groups, say, $X_1, \ldots, X_{n_1}$ and $Y_1, \ldots, Y_{n_2}$. For the moment, let's assume that $n_1 = n_2 = n$ and thus the total pilot sample size is $2n$ observations. Suppose we are interested in determining the power of a comparison of two treatment groups' location parameters, one shifted by an amount δ from the other, at the α-level of statistical significance, with a total sample size of $2m$ (m samples per treatment group). We begin by making a random selection (with replacement) of size $2m$ from the pilot study data, i.e., $X_1, \ldots, X_{n_1}, Y_1, \ldots, Y_{n_2}$. Denote the first m randomly sampled observations by $R_1, \ldots, R_m$. For the second set of randomly selected observations (denoted by $R_{m+1}, \ldots, R_{2m}$), add the change in location, δ, to each of the m values. Denote the resulting values $R_{m+1} + \delta, \ldots, R_{2m} + \delta$ by $S_1, \ldots, S_m$. Now compare $R_1, \ldots, R_m$ with $S_1, \ldots, S_m$ via a form of two-sample location comparison test, with a statistic, say, W, at level α. Create an indicator variable, I, and let $I = 1$ if $W \geq c_\alpha$ (c_α represents the α-level cutoff of the test) and 0 otherwise.

We can repeat the process B times (for some large value of B, i.e., $B > 1000$), recording the value of the indicator variable I as I_b for the bth repetition of the process ($b = 1, \ldots, B$). An estimate of the power, $\pi(\alpha, \beta, m)$, of the test conducted at the α-level of significance, for shift of size and sample size m per group is

$$\pi(\alpha, \beta, m) = \frac{1}{B} \sum_{b=1}^{B} I_b.$$

Collings and Hamilton went on to show in their paper that the described procedure can be improved in the following way. Consider $X_1, \ldots, X_n$ and randomly sample from the group $2m$ times. For the second half of the $2m$ terms, add the shift value δ to each term: $X_{m+1} + \delta, \ldots, X_{2m} + \delta$. Perform the pair-wise comparison at level α between $X_1, \ldots, X_m$ and $X_{m+1} + \delta, \ldots, X_{2m} + \delta$, record the result via an indicator variable, and repeat the procedure B times, as described previously. The same procedure is carried out for $Y_1, \ldots, Y_n$. Let $\pi_X(\alpha, \beta, m)$ be the power calculated for the X_i's and $\pi_Y(\alpha, \beta, m)$ be the power calculated for the Y_i's. The power for the comparison of $X_1, \ldots, X_n$ and $Y_1, \ldots, Y_n$ with respect to a shift δ, at level α, for total sample size $2m$ is:

$$\pi(\alpha, \beta, m) = \frac{1}{2}(\pi_X(\alpha, \beta, m) + \pi_Y(\alpha, \beta, m)).$$

Earlier in our discussion we assumed $n_1 = n_2 = n$. This condition is not necessary. If $n_1 \neq n_2$, this approach can be modified by using a weighted mean of the two power values; see Collings and Hamilton (1988) for details.

6.4.2 Power Calculations in a k-Sample Case

Mahoney and Magel (1996) extended the results of Collings and Hamilton to cover the case of k independent samples and their comparison by the Kruskal-Wallis test. The procedure is similar in spirit to that described previously. Consider the testing problem defined in Section 6.3, i.e., consider a study with k treatment groups and n_i units/subjects in the ith group. In the two-sample case, we talk about a single shift in location, δ. For a k-sample one-way layout, we have to talk about a k-dimensional vector, $\tilde{\delta} = (\delta_1, \ldots, \delta_k)$, where δ_i is the shift-change in location of the ith group. Without loss of generality, we assume that the first delta component is zero, i.e., $\delta_1 = 0$. As was the case in the two-sample setting, we again need to randomly sample (with replacement) from each of the groups. For the ith group, we sample km observations. We then apply the k location shifts to each of the km observations as follows. The first m observations are unshifted because $\delta_1 = 0$. The second set of m observations are then shifted by δ_2, the third set by δ_3, etc. Mahoney and Magel point out that, essentially, we now have k random samples of size m, only differing by a shift in location and thus we can perform a Kruskal-Wallis test at level α. If the test is statistically significant, we let $I_1 = 1$, otherwise $I_1 = 0$. We repeat the process B times, for a large value of B, recording the value of I_b for the bth repetition of the process. The power for the ith group is calculated as before, i.e., $\pi_i(\alpha, \beta, m) = B^{-1} \sum_{b=1}^{B} I_b$. The outlined procedure is repeated for the remaining $k - 1$ groups and the power for the Kruskal-Wallis test is then simply the average power determined k times:

$$\pi(\alpha, \beta, m) = \frac{1}{k} \sum_{i=1}^{k} \pi_i(\alpha, \beta, m).$$

6.4.3   Comparison of Parametric and Nonparametric Approaches

Now is probably a good time to compare and contrast the parametric approach to sample size determination in a two-sample setting with the nonparametric approach. Notice we require some similar pieces of information in both approaches: we require knowledge about the level of the test, α, the desired effect, δ, and the number of observations per group, say, m. For the parametric comparison, we assume a common form of distribution for both groups, with an assumed known and identical scale parameter σ, and known location parameters (means) μ_1 and μ_2, respectively. The nonparametric approach requires a small pilot data set consisting of two groups without any assumptions about the common distributional form. For pharmaceutical researchers, the nonparametric approach may require more work because it requires *data* before the power may be calculated. However, the benefits of the procedure may greatly outweigh the expenditure of resources for the pilot.

First, investigators get to see the distribution of real data, as opposed to imagining what it would look like, conditional on a value of location and scale and assuming some pre-specified mathematically convenient form. Often investigators will use the summary statistics from a previously published manuscript to determine the power in the parametric approach for their investigation. Were the originally published data distributed normally? Were they symmetrically distributed, for that matter? We really do not know by looking at the summary statistics alone in the manuscript, nor can we learn much more if the authors made a poor choice of data visualization to summarize the paper's findings. We only know for certain about the raw data's features if we can see all of the data used in the original manuscript. It has been the author's experience that such data are difficult to get, as scientific writers change institutional affiliations and are subsequently difficult to locate years after they publish their research. Moreover, some are unwilling to share raw data, or unable to share it because of the proprietary nature of their research or because the data are stored in an inconvenient form (e.g., in several spreadsheets or laboratory notebooks).

Second, the nonparametric approach is intuitive, given that a scientist understands how the nonparametric test works. Power corresponds to the probability of, in some sense, finding a true and context meaningful difference, δ, at the α-level of significance for a given sample size. By employing the calculations described above, scientists can see where the value of power comes from without having to understand the mathematical statistical properties of the Gaussian (normal) distribution.

%KWSS Macro

How do we perform a nonparametric power analysis in SAS? The Mahoney-Magel generalization of the Collings-Hamilton approach to power determination in the k-sample setting may be accomplished with a SAS macro called %KWSS provided on the book's companion Web site. The %KWSS macro consists of a body of code containing one embedded macro (%BOOTSTRAP). The macro requires seven input parameters:

- IDSN is the name of an input SAS data set.
- GROUP is the grouping or classification variable.
- VAR is the response measurement variable.
- SHIFTLIST is the shift vector, $\widetilde{\delta} = (\delta_1, \ldots, \delta_k)$, for the test.

- SAMPLE_SIZE is the sample size for power determination.
- ITERATION_NO is the number of re-sampled data sets.
- ALPHA is the desired level of statistical significance.

The macro first removes any observations that contain a missing response. It then determines the number of groups in the input data set IDSN and creates a list of these groups with the macro variable GRPVEC. The data are then subset by the various levels of the classification variable. The embedded %BOOTSTRAP macro then performs the required sampling (with replacement) of size SAMPLE_SIZE*GROUP for each level of GROUP. The location shifts specified in SHIFTLIST are applied to the re-sampled data by GROUP. If the number of location shifts specified in SHIFTLIST does not equal the number of groups or classes in the input data set, the program has a fail-safe that will terminate its execution. The macro then performs a Kruskal-Wallis test by PROC NPAR1WAY for each level of GROUP. The number of statistically significant (at level ALPHA) results is recorded as a fraction of the total number of re-sampled data sets (ITERATION_NO) with PROC FREQ. For the GROUP levels of power determined, PROC MEANS determines the *mean power* and the result is reported via PROC PRINT.

To illustrate the usage of %KWSS, Program 6.15 estimates the power of the Kruskal-Wallis test based on simulated data from a four-group experiment with 10 observations/group. The response variable is assumed to follow a gamma distribution. Suppose that we are interested in using this pilot study information to power a study, also with four groups, but with a sample size allocation of 20 subjects/group. Suppose that we are interested in seeing a statistically significant ($p < 0.05$) shift in the location parameters, $\tilde{\delta} = (0, 1.5, 1.5, 1.5)$. The macro call is specified as follows:

```
%kwss(one,group,x,0 1.5 1.5 1.5,20,5000,0.05)
```

The first parameter is the name of the input data set (ONE) containing a response variable (X) and grouping or classification variable (GROUP) of interest (the second and third macro parameters, respectively). The fourth parameter is the list of location shifts. The fifth parameter is the desired sample size/group. The sixth parameter is the number of iterations (bootstrapped samples) per group of size $4 \times 20 = 80$. The final parameter is the desired level of statistical significance for this four-sample test.

As a large number of bootstrap samples are requested for this power calculation, Program 6.15 will run for a considerable amount of time (about 20 minutes on the author's PC using an interactive SAS session). The NONOTES system option is specified before the invocation of the %BOOTSTRAP macro. Using this option prevents the annoying situation of a filled-up log in a PC SAS session that requires the user to empty or save the log in a file.

Program 6.15 Nonparametric power calculation using the %KWSS macro

```
data one;
    do i=1 to 10;
        x=5+rangam(1,0.5); group="A"; output;
        x=7+rangam(1,0.5); group="B"; output;
        x=7.2+rangam(1,0.5); group="C"; output;
        x=7.4+rangam(1,0.5); group="D"; output;
        drop i;
    end;
    run;
%kwss(one,group,x,0 1.5 1.5 1.5,20,5000,0.05)
```

Output from Program 6.15

```
The Mahoney-Magel Generalization of the Collings-Hamilton Approach    1
     For Power Estimation with Location Shifts 0 1.5 1.5 1.5
            of 4 Groups (A B C D) of Sample Size 20
      Compared at the 0.05-level of Statistical Significance


       (Power Based on a Bootstrap Conducted 5000 Times)


                           POWER

                          99.945
```

Output 6.15 shows the estimated power of the Kruskal-Wallis test (99.945%). As was the case for other macros presented in this chapter, the titles of the output inform the user about the information used to estimate the power (location shifts, number of groups, sample size, etc.).

6.4.4   Summary

The approach to power estimation illustrated in this section differs from the classical approach where an investigator is required to provide estimates of treatment effect and variation. In novel experimental investigations with limited resources, it may be difficult to provide such estimates without executing a pilot study first. The approach, suggested first by Collings and Hamilton (1988) and later generalized by Mahoney and Magel (1996), is a reasonable approach to estimation of the power of an inference comparing location parameters when estimates of treatment effect and variation are difficult to obtain. The author has developed a macro to estimate power using the approach suggested by Mahoney and Magel (the %KWSS macro).

Acknowledgments

The author is grateful to the following individuals for their contributions to this work: Dr. Dianne Camp (William Beaumont Research Institute) for her careful review and editorial suggestions; Drs. Robert Abel and Cathie Spino (Pfizer Global Research and Development - Michigan Laboratories: Ann Arbor Campus) for their assistance with the clinical trial examples; Ms Alice Birnbaum (Project Director, Axio Research) for her invaluable feedback on the multiple comparison macros; and Dr. Thomas Vidmar (Senior Director and Site Head, Midwest Nonclinical Statistics) for his support during this endeavor.

References

Ansari, A.R., Bradley, R.A. (1960). "Rank-sum tests for dispersion." *Annals of Mathematical Statistics*. 31, 1174–1189.

Bickel, P.J. Doksum, K.A. (1977). *Mathematical Statistics: Basic Ideas and Selected Topics*. Oakland: Holden Day.

Collings, B.J., Hamilton, M.A. (1988). "Estimating the power of the two-sample Wilcoxon test for location shift." *Biometrics*. 44, 847–860.

Dunn, O.J. (1964). "Multiple comparisons using rank sums." *Technometrics*. 6, 241–252.

Dwass, M. (1960). "Some *k*-sample rank-order statistics." *Contributions to Probability and Statistics*. I. Olkin, S.G. Ghurye, H. Hoeffding, W.G. Madow, H.B. Mann, eds. Palo Alto: Stanford University Press, 198–202.

Fligner, M.A., Policello, G.E. (1981). "Robust rank procedures for the Behrens-Fisher problem." *Journal of the American Statistical Association*. 76, 162–174.

Harter, H.L. (1960). "Tables of range and studentized range." *Annals of Mathematical Statistics.* 31, 1122–1147.

Hochberg, Y., Tamhane, A.C. (1987). *Multiple Comparison Procedures.* New York: Wiley.

Hollander, M., Wolfe, D.A. (1999). *Nonparametric Statistical Methods.* Second Edition. New York: Wiley.

Juneau, P. (2004). "Simultaneous nonparametric inference in a one-way layout using the SAS System." *Proceedings of the PharmaSUG 2004 Annual Meeting.* Available at `http://www.pharmasug.org/content/view/64/`.

Kruskal, W.H., Wallis, W.A. (1952). "Use of ranks in one-criterion variance analysis." *Journal of the American Statistical Association.* 47, 583–621.

Lehman, E.L. "Nonparametrics: Statistical Methods Based upon Ranks." New York: Holden-Day, 1975, pp. 206–207.

Mahoney, M., Magel, R. (1996). "Estimation of the power of the Kruskal-Wallis test." *Biometrical Journal.* 38, 613–630).

Miller, R.G., Jr. (1966). *Simultaneous Statistical Inference.* First edition. New York: Springer-Verlag.

Miller, R.G., Jr. (1981). *Simultaneous Statistical Inference.* Second edition. New York: Springer-Verlag.

National Asthma Council Australia. *Asthma Management Handbook 2002.* Available at `http://www.nationalasthma.org.au/html/management/amh/amh003.asp`.

Pillai, K.C.S., Ramachandran, K.V. (1954). "On the distribution of the ratio of the ith observation in an ordered sample from a normal population to an independent estimate of the standard deviation." *Annals of Mathematical Statistics.* 25, 565–572.

Steel, R.G.D. (1960). "A rank sum test for comparing all pairs of treatments." *Technometrics.* 2, 197–207.

Student. (1908). "The probable error of a mean." *Biometrika.* 6, 1–25.

Wilcoxon, F. (1945). "Individual comparisons by ranking methods." *Biometrics.* 1, 80–83.

World Health Organization Fact Sheet No. 194 (revised January 2002). Available at `http://www.who.int/mediacentre/factsheets/fs194/en/`.

Zhou, X-H, Gao, S., Hui, S. L. (1997). "Methods for comparing the means of two independent log-normal samples." *Biometrics.* 53, 1129–1135.

Chapter 7

Optimal Design of Experiments in Pharmaceutical Applications

Valerii Fedorov
Robert Gagnon
Sergei Leonov
Yuehui Wu

In this chapter we discuss optimal experimental designs for nonlinear models arising in various pharmaceutical applications and present a short survey of optimal design methods and numerical algorithms. We provide SAS code to implement optimal design algorithms for several examples:

- quantal models such as logistic models for analyzing success or failure in dose-response studies
- multi-parameter continuous logistic models in bioassays or pharmacodynamic studies, including models with unknown parameters in variance
- beta regression model

Valerii Fedorov is Group Director, Research Statistics Unit, Biomedical Data Sciences, GlaxoSmithKline Pharmaceuticals, USA. Robert Gagnon is Associate Director, Biomedical Data Sciences, GlaxoSmithKline Pharmaceuticals, USA. Sergei Leonov is Director, Research Statistics Unit, Biomedical Data Sciences, GlaxoSmithKline Pharmaceuticals, USA. Yuehui Wu is Senior Statistician, Research Statistics Unit, Biomedical Data Sciences, GlaxoSmithKline Pharmaceuticals, USA.

- models with multiple responses, such as models that measure both efficacy and safety in dose response studies and pharmacokinetic models with multiple samples per subject
- models with cost constraints.

For all examples, we use a first-order optimization algorithm in the space of information matrices. A short survey of other software tools for constructing optimal model-based designs is provided.

7.1 Optimal Design Problem

We start this chapter with the description of the general optimal design problems and concepts.

7.1.1 General Model

Consider a vector of observations $Y = \{y_1, \ldots, y_N\}$ and assume it follows a general parametric model. The joint probability density function of Y depends on x and θ, where x is the independent, or *design*, variable and $\theta = (\theta_1, \ldots, \theta_m)$ is the vector of unknown model parameters. The design variable x is chosen, or controlled, by researchers to obtain the best estimates of the unknown parameters.

This general model can be applied to a wide variety of problems arising in clinical and pre-clinical studies. The examples considered in this chapter are described below.

Dose-Response Studies

Dose-response models arise in clinical trials, either with a binary outcome (e.g., success-failure or dead-alive in toxicology studies; see Section 7.2) or continuous response (e.g., studies of pain medications when patients mark their pain level on a visual analog scale; see Section 7.4). In these examples, x represents the dose of a drug administered to the patient. Figure 7.1 illustrates the dependence of the probability of the success $\pi(x)$ at the dose x for a two-parameter logistic model. Dotted horizontal lines correspond to the probabilities of success for the two optimal doses x_1^* and x_2^*. See Section 7.2 for details.

Figure 7.1 Optimal design points (vertical lines) in a two-parameter logistic model

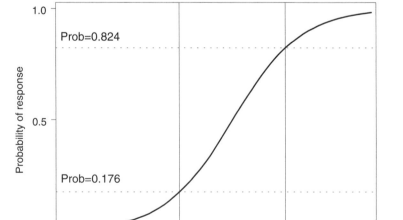

Bioassay Studies

Multi-parameter logistic models, sometimes referred to as the E_{max} or Hill models, are widely used in bioassay studies. Examples include models that relate the concentration of an experimental drug to the percentage or number of surviving cells in cell-based assays or models that quantify the concentration of antigens or antibodies in enzyme-linked immunosorbent assays (ELISA). In this context, the design variable x represents the drug concentration level; see Sections 7.2 and 7.3 for details.

As an illustration, Figure 7.2 plots a four-parameter logistic model that describes the number of surviving bad cells in a cell-based assay versus drug concentration x on the log-scale. The concentration-effect relationship is negative. This is expected since an increase in the drug concentration usually reduces the number of cells.

Figure 7.2 Four-parameter logistic model

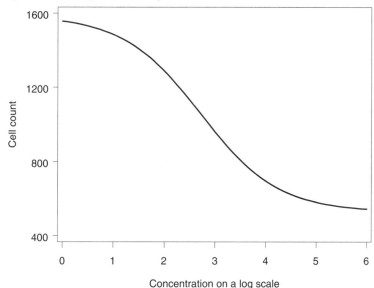

Concentration on a log scale

Two-Drug Combination Studies

Section 7.6.3 discusses two-drug combination studies in which drugs (e.g., Drug A and Drug B) are administered simultaneously and their effect on the patient's outcome is analyzed. In this case, the design variable x is two-dimensional, i.e., $x = (x_1, x_2)$, where x_1 is the dose of Drug A and x_2 is the dose of Drug B.

Clinical Pharmacokinetic Studies

Multiple blood samples are taken in virtually all clinical pharmacokinetic (PK) studies and the collected data are analyzed by means of various PK compartmental models. This analysis leads to quite sophisticated nonlinear mixed effects models which are discussed in Section 7.8. In these models x is a k-dimensional vector that represents a collection (sequence) of sampling times for a particular patient.

Cost-Based Designs

In the previous example (PK studies) it is quite obvious that each extra sample provides additional information. On the other hand, the number of samples that may be drawn from each patient is restricted because of blood volume limitations and other logistic and ethical reasons. Moreover, the analysis of each sample is associated with monetary cost. Thus, it makes sense to incorporate costs in the design. One potential approach for the construction of cost-based designs is discussed in Section 7.9.

SAS Code and Data Sets

To save space, some SAS code has been shortened and some output is not shown. The complete SAS code and data sets used in this book are available on the book's companion Web site at `http://support.sas.com/publishing/bbu/companion_site/60622.html`.

7.1.2 Information Matrix, Maximum Likelihood Estimates, and Optimality Criteria

For any parametric model, one of the main goals is to obtain the most precise estimates of unknown parameters θ. For example, in the context of a dose-response study, if clinical researchers can select doses x_i, in some admissible range, and number of patients n_i on each dose, then, given the total number of observations N, the question is how to allocate those doses and patients to obtain the best estimates of unknown parameters. The quality of estimators is traditionally measured by their variance-covariance matrix.

To introduce the theory underlying optimal design of experiments, we will assume that n_i is the number of independent observations y_{ij} that are made at design point x_i and that $N = n_1 + n_2 + \ldots + n_n$ is the total number of observations, $j = 1, \ldots, n_i$, $i = 1, \ldots, n$. Let the probability density function of y_{ij} be $p(y_{ij}|x_i, \theta)$. For example, in the context of a dose-response study, x_i is the ith dose of the drug and y_{ij} is the response of the jth patient on dose x_i.

Most of the examples discussed in this chapter deal with the case of a single measurement per patient, i.e., both x_i and y_{ij} are scalars (single dose and single response) and y_{ij} and y_{i_1,j_1} are independent if $i \neq i_1$ or $j \neq j_1$. The exceptions are in Section 7.7 (two binary responses for the same patient, such as efficacy and toxicity, at a given dose x_i) and Sections 7.8 and 7.9 that consider the problem of designing experiments with multiple PK samples over time for the same patient, i.e., a design point is a sequence $x_i = (x_i^1, x_i^2, \ldots, x_i^k)$ of k sampling times and $y_{ij} = (y_{ij}^1, \ldots, y_{ij}^k)$ is a vector of k observations. Note that in the latter case the elements of vector y_{ij} are, in general, correlated because they correspond to measurements on the same patient. However, similar to the case of a single measurement per patient, y_{ij} and y_{i_1,j_1} are still independent if $i \neq i_1$ or $j \neq j_1$.

Information Matrix

Let $\mu(x, \theta)$ be an $m \times m$ information matrix of a single observation at point x,

$$\mu(x, \theta) = E\left[\frac{\partial \log p(Y|x,\theta)}{\partial \theta} \frac{\partial \log p(Y|x,\theta)}{\partial \theta^T}\right], \tag{7.1}$$

where the expectation is taken with respect to the distribution of Y. The Fisher information matrix of the N experiments can be written as

$$M_N(\theta) = \sum_{i=1}^{n} n_i \mu(x_i, \theta). \tag{7.2}$$

Further, let $M(\xi, \theta)$ be a normalized information matrix,

$$M(\xi, \theta) = \frac{1}{N} M_N(\theta) = \sum_{i=1}^{n} w_i \mu(x_i, \theta), \quad w_i = \frac{n_i}{N}. \tag{7.3}$$

The collection $\xi = \{x_i, w_i\}$ is called a normalized (continuous or approximate) design with $w_1 + \ldots + w_n = 1$. In this setting N may be viewed as a resource available to researchers; see Section 7.9 for a different normalization.

Maximum Likelihood Estimates

The maximum likelihood estimate (MLE) of θ is given by

$$\widehat{\theta} = \arg\min_{\theta} \prod_{i=1}^{n}\prod_{j=1}^{n_i} p(y_{ij}|x_i,\theta) = \arg\min_{\theta} \sum_{i=1}^{n}\sum_{j=1}^{n_i} \log p(y_{ij}|x_i,\theta),$$

It is well known that the variance-covariance matrix of $\widehat{\theta}$ for large samples is approximated by

$$\mathrm{Var}(\widehat{\theta}) \approx M_N^{-1}(\theta) = \frac{1}{N}M^{-1}(\xi,\theta) = \frac{1}{N}D(\xi,\theta), \tag{7.4}$$

where $D(\xi,\theta) = M^{-1}(\xi,\theta)$ is the normalized variance-covariance matrix (Rao, 1973, Chapter 5). Throughout this chapter we assume that the normalized information matrix $M(\xi,\theta)$ is not singular and thus its inverse exists.

Optimality Criteria

In the convex design theory, it is standard to minimize various functionals depending on $D(\xi,\theta)$,

$$\xi^* = \arg\min_{\xi} \Psi[D(\xi,\theta)], \tag{7.5}$$

where Ψ is a selected functional (criterion of optimality). The optimization is performed with respect to designs ξ,

$$\xi = \{w_i, \quad 0 \le w_i \le 1, \quad \sum_{i=1}^{n} w_i = 1; \quad x_i \in \mathcal{X}\},$$

where $\mathcal{X}$ is a design region, e.g., admissible doses in a dose-response study, and the weights, $w_1,\ldots,w_n$, are continuous variables. The use of continuous weights leads to the concept of approximate optimal designs; the word *approximate* should draw attention to the fact that the solution of (7.5) does not generally give integer values $n_i = Nw_i$. However, this solution is often acceptable, in particular when the total number of observations N is relatively large and we round $n_i = Nw_i$ to the nearest integer while keeping $n_1 + \ldots + n_n = N$ (Pukelsheim, 1993).

The following optimality criteria are among the most popular ones:

***D*-optimality.** $\Psi = \ln|D(\xi,\theta)|$, where $|D|$ denotes the determinant of D. This criterion is often called a *generalized variance criterion* since the volume of the confidence ellipsoid for θ is proportional to $|D(\xi,\theta)|^{1/2}$; see Fedorov and Hackl (1997, Chapter 2).

***A*-optimality.** $\Psi = \mathrm{tr}[AD(\xi,\theta)]$, where A is an $m \times m$ non-negative definite matrix (*utility matrix*) and $\mathrm{tr}(D)$ denotes the trace, or sum of diagonal elements, of D. For example, if $A = m^{-1}I_m$, where I_m is an $m \times m$ identity matrix, the A-criterion is based on the average variance of the parameter estimates:

$$\Psi = m^{-1}\sum_{i=1}^{m}\mathrm{Var}(\widehat{\theta}_i).$$

***c*-optimality.** $\Psi = c^T D(\xi,\theta)c$, where c is an m-dimensional vector. A c-optimal design minimizes the variance of a linear combination of the model parameters, i.e., $c^T\theta$.

***E*-optimality.** $\Psi = \lambda_{\min}[M(\xi,\theta)] = \lambda_{\max}[D(\xi,\theta)]$, where $\lambda_{\min}(M)$ and $\lambda_{\max}(M)$ are minimal and maximal eigenvalues of M, respectively. Note that the length of the largest principal axis of the confidence ellipsoid is $\lambda_{\min}^{-1/2}[M(\xi,\theta)]$.

I-optimality. $\Psi = \int_{\mathcal{X}} \text{tr}[\mu(x,\theta)D(\xi,\theta)]dx$. In linear regression models, $\text{tr}[\mu(x,\theta)D(\xi,\theta)]$ represents the variance of the predicted response.

Minimax optimality. $\Psi = \max_{x \in \mathcal{X}} \text{tr}[\mu(x,\theta)D(\xi,\theta)]$.

For other criteria of optimality, see Fedorov (1972), Silvey (1980), or Pukelsheim (1993).

It is important to note that the D-, A-, and E-optimality criteria may be ordered for any design since

$$|D|^{1/m} \leq m^{-1}\text{tr}(D) \leq \lambda_{\max}(D), \quad D = D(\xi,\theta),$$

see Fedorov and Hackl (1997, Chapter 2). Also, the disadvantage of c-optimal designs is that they are often singular, i.e., the corresponding variance-covariance matrix is degenerate.

D-Optimality Criterion

In this chapter we concentrate on the D-optimality criterion (note that the SAS programs described in this chapter can also be used to construct A-optimal designs). D-optimal designs are popular among theoretical and applied researchers due to the following considerations:

- D-optimal designs minimize the volume of the asymptotic confidence region for θ. This property is easy to explain to practitioners in various fields.

- D-optimal designs are invariant with respect to non-degenerate transformations of parameters (e.g., changes of the parameter scale).

- D-optimal designs are attractive in practice because they often perform well according to other optimality criteria; see Atkinson and Donev (1992, Chapter 11) or Fedorov and Hackl (1997).

7.1.3   Locally Optimal Designs

It is easy to see from the definition of the information matrix $M(\xi,\theta)$ that, in general, optimal designs depend on the unknown parameter vector θ. This assumption leads to the concept of *locally optimal* designs that are defined as follows: first we need to specify a preliminary estimate of θ, e.g., $\tilde{\theta}$, and then solve the optimization problem for the given $\tilde{\theta}$ (Chernoff, 1953, Fedorov, 1972).

It is worth noting that in linear parametric models the information matrix does *not* depend on θ, which greatly simplifies the problem of computing optimal designs. Consider, for example, a linear regression model

$$y_{ij} = \theta_1 f_1(x_i) + \ldots + \theta_m f_m(x_i) + \varepsilon_{ij}, \quad i = 1,\ldots,n, \quad j = 1,\ldots,n_i$$

with normally distributed residuals, i.e., $\varepsilon_{ij} \sim N(0, \sigma^2(x_i))$. Here $f(x) = [f_1(x),\ldots,f_m(x)]^T$ is a vector of pre-defined *basis* functions. The information matrix is given by (Fedorov and Hackl, 1997)

$$\mu(x) = \sigma^{-2}(x_i)f(x)f^T(x)$$

and thus we can construct designs that will be optimal for any value of θ. In non-linear models, we need to select a value of θ first; however, once this value is fixed, the construction of an optimal design is absolutely the same as for linear problems.

Obviously, the quality of locally optimal designs may be poor if the preliminary estimate $\tilde{\theta}$ significantly differs from the true value of θ. We mention in passing that this problem can be tackled by using various techniques which include minimax designs (Fedorov and Hackl,

1997), Bayesian designs (Chaloner and Verdinelli, 1995), adaptive designs (Box and Hunter, 1965; Fedorov, 1972, Chapter 4; Zacks, 1996, Fedorov and Leonov, 2005, Section 5.6). While minimax and Bayesian approaches take into account prior uncertainties, they lead to optimization problems which are computationally more demanding than the construction of locally optimal designs. Instead of preliminary estimates of unknown parameters, we must provide an uncertainty set for the minimax approach and a prior distribution for Bayesian designs. The latter task is often based on a subjective judgment. Locally optimal designs serve as a reference point for other candidate designs, and sensitivity analysis with respect to parameter values is always required to validate the properties of a particular optimal design; see a discussion in Section 7.3.1.

In this chapter we concentrate on the construction of locally optimal designs.

7.1.4   Equivalence Theorem

Many theoretical results and numerical algorithms of the optimal experimental design theory rely on the following important theorem (Kiefer and Wolfowitz, 1960; Fedorov, 1972, White, 1973):

Generalized equivalence theorem. A design ξ^* is locally D-optimal if and only if

$$\psi(x, \xi^*, \theta) = \mathrm{tr}\left[\mu(x, \theta) M^{-1}(\xi^*, \theta)\right] \leq m, \tag{7.6}$$

where m is the number of model parameters. Similarly, a design ξ^* is locally A-optimal if and only if

$$\psi(x, \xi^*, \theta) = \mathrm{tr}\left[\mu(x, \theta) M^{-1}(\xi^*, \theta) A M^{-1}(\xi^*, \theta)\right] \leq \mathrm{tr}\left[A M^{-1}(\xi^*, \theta)\right]. \tag{7.7}$$

The equality in (7.6) and (7.7) is attained at the support points of the optimal design ξ^*.

The $\psi(x, \xi, \theta)$ function is termed the *sensitivity function* of the corresponding criterion. The sensitivity function helps identify design points that provide the most information with respect to the chosen optimality criterion (Fedorov and Hackl, 1997, Section 2.4). For instance, if we consider dose-response studies, optimal designs can be constructed iteratively by choosing doses x^* that maximize the sensitivity function $\psi(x, \xi, \theta)$ over the admissible range of doses.

As shown in the next subsection, the general formulas (7.6) and (7.7) form a basis of numerical procedures for constructing D- and A-optimal designs.

7.1.5   Computation of Optimal Designs

Any numerical procedure for constructing optimal designs requires two key elements:

- The information matrix $\mu(x, \theta)$ or, equivalently, sensitivity function $\psi(x, \xi, \theta)$.
- The design region $\mathcal{X}$ (the set of admissible design points).

In all examples discussed in this chapter (except for the examples considered in Sections 7.8 and 7.9), we define the design region as a compact set, but search for optimal points on a pre-defined discrete grid. This grid can be rather fine in order to guarantee that the resulting design is close to the optimal one.

The main idea behind the general nonlinear design algorithm is that, at each step of the algorithm, the sensitivity function $\psi(x, \xi, \theta)$ is maximized over the design region to determine the best new support point (forward step) and then minimized over the support points of the current design to remove the worst point in the current design (backward step). See Fedorov and Hackl (1997) or Atkinson and Donev (1992) for details.

To define a design algorithm, let $\xi_s = \{X_{is}, w_{is}\}$, $i = 1, \ldots, n_s$, be the design at Step s. Here $\{X_{is}\}$ is the vector of support points in the current design and w_{is} is the vector of weights assigned to X_{is}. The iterative algorithm is of the following form:

$$\xi_{s+1} = (1 - \alpha_s)\xi_s + \alpha_s \xi(X_s),$$

where $\xi(X)$ is a one-point design supported on point X.

> **Forward step.** At Step s, a point X_s^+ that maximizes $\psi(x, \xi, \theta)$ over $x \in \mathcal{X}$ is added to the design ξ_s with weight $\alpha_s = \gamma_s$, where $\gamma_s = 1/(n_0 + s)$ and n_0 is the number of points in the initial design.
>
> **Backward step.** After that, a point X_s^- that minimizes $\psi(x, \xi, \theta)$ over all support points in the current design is deleted from the design with weight

$$\alpha_s = \left\{ \begin{array}{ll} -\gamma_s, & w_s \geq \gamma_s, \\ -w_s/(1 - w_s), & w_s < \gamma_s. \end{array} \right.$$

In general, the user can change γ_s to $c_1/(n_0 + c_2 s)$, where c_1 and c_2 are two constants. The default values of the constants are $c_1 = c_2 = 1$.

In Section 7.2 we provide a detailed description of a SAS macro that implements the described optimal design algorithm for quantal dose-response models. However, once the information matrix $\mu(x, \theta)$ or the sensitivity function $\psi(x, \xi, \theta)$ is specified, the same technique can be used to generate optimal designs for any other model.

7.1.6   Existing Software

Computer algorithms for generating optimal designs have existed for quite some time (Fedorov, 1972; Wynn, 1970). These algorithms have been implemented in many commercially available software systems. Additionally, software written by academic groups is available. In general, the commercially available systems implement methods for problems such as simple linear regression and factorial designs, but not for more complex models such as nonlinear regression, beta regression, or population pharmacokinetics. Academic groups have developed and made available programs for the latter cases.

On the commercial side, SAS offers both SAS/QC (the OPTEX procedure) and JMP. The OPTEX procedure supports A- and D-optimal designs for simple regression models. JMP generates D-optimal factorial designs, as do software packages such as Statistica. In pharmacokinetic applications, the ADAPT II program developed at the University of Southern California implements c- and D-optimal and partially optimal design generation for individual pharmacokinetic models; however, it does not support more complex population pharmacokinetic models (D'Argenio and Schumitzky, 1997). Ogungbenro et al. (2005) implemented in Matlab the classical exchange algorithm for population PK experiments using D-optimality (note that the exchange algorithm improves the initial design with respect to selected optimality criterion but, in general, does not converge to the optimal design). Retout and Mentré (2003) developed extensive implementation of population pharmacokinetic designs in S-Plus.

In this chapter we discuss SAS/IML implementation of the general algorithm (which cannot be executed in PROC OPTEX) with applications to several widely used nonlinear models. We also describe optimal design algorithms for population pharmacokinetic models analogous to the S-Plus programs of Retout and Mentré (2003).

7.1.7 Structure of SAS Programs

There are five components in the programs:

- inputs
- establishment of the design region
- calculation of the information matrix
- using the optimal design algorithm
- outputs.

Only the optimal design algorithm component is unchanged for any model; the other components are model-specific and need to be modified accordingly. In any model, calculation of the information matrix is the critical (and most computationally intensive) step.

7.2 Quantal Dose-Response Models

Quantal, or binary, models arise in various pharmaceutical applications, such as clinical trials with a binary outcome, toxicology studies, and quantal bioassays.

7.2.1 General Model

In a quantal model, a binary response variable Y depends on the dose x,

$$Y = Y(x) = \begin{cases} 1, & \text{a response is present at dose } x, \\ 0, & \text{no response,} \end{cases}$$

and the probability of observing a response is modeled as

$$P\{Y = 1 | x\} = \eta(x, \theta),$$

where $0 \leq \eta(x, \theta) \leq 1$ is a given function and θ is a vector of m unknown parameters. It is often assumed that

$$\eta(x, \theta) = \pi(z),$$

where z is a linear combination of pre-defined functions of the dose x, i.e.,

$$z = \theta_1 f_1(x) + \ldots + \theta_m f_m(x).$$

In many applications, $\pi(z)$ is selected as a probability distribution function with a continuous derivative (see Fedorov and Leonov, 2001). Among the most popular choices of this function are:

Logistic (logit) model. The $\pi(z)$ function is a logistic function

$$\pi(z) = e^z/(1 + e^z).$$

Probit model. $\pi(z)$ is a standard normal cumulative distribution function,

$$\pi(z) = \frac{1}{\sqrt{2\pi}} \int_{-\infty}^{z} e^{-u^2/2} du.$$

In practice, when properly normalized, the two models lead to virtually identical results (Cramer, 1991, Section 2.3; Finney, 1971, p. 98).

In the quantal dose-response model the probability density function of Y at a point x is given by

$$p(Y|x,\theta) = \eta(x,\theta)^Y \left[1 - \eta(x,\theta)\right]^{1-Y}$$

and thus the information matrix of a single observation is equal to

$$\mu(x,\theta) = \frac{[\pi'(z)]^2}{\pi(z)[1 - \pi(z)]} f(x) f^T(x),$$

(Wu, 1988, Torsney and Musrati, 1993) where $f(x) = [f_1(x), \ldots, f_m(x)]^T$ and thus the variance-covariance matrix of the MLE, $\widehat{\theta}$, can be approximated by $D(\xi,\theta)/N$; see (7.4).

The quantal logistic model provides an example where the statement of the generalized equivalence theorem, (7.6) and (7.7), admits a more "numerically-friendly" presentation. For example, the information matrix for a single observation can be written as $\mu(x,\theta) = g(x,\theta) g^T(x,\theta)$, where

$$g(x,\theta) = \frac{\pi'(z) f(x)}{\sqrt{\pi(z)[1 - \pi(z)]}},$$

which, together with the matrix identity $\operatorname{tr}(AB) = \operatorname{tr}(BA)$, leads to the following presentation of the sensitivity functions for the quantal model:

$$D\text{-criterion: } \psi(x,\xi,\theta) = g(x,\theta) M^{-1}(\xi,\theta) g^T(x,\theta), \tag{7.8}$$

$$A\text{-criterion: } \psi(x,\xi,\theta) = g(x,\theta) M^{-1}(\xi,\theta) A M^{-1}(\xi,\theta) g^T(x,\theta). \tag{7.9}$$

7.2.2   Example 1: A Two-Parameter Logistic Model

In this subsection we will take a closer look at the logistic model with two unknown parameters:

$$P\{Y = 1|x\} = e^z/(1 + e^z), \quad z = \theta_1 + \theta_2 x,$$

where θ_1 and θ_2 are the intercept and slope parameters, respectively.

It can be shown that D-optimal designs associated with the described model are two-point designs, with half the measurements at each dose, i.e., $w_1 = w_2 = 0.5$ (White, 1975). It is interesting to note that, when the dose range is sufficiently wide, D-optimal designs are uniquely defined in the z-space and the optimal design points correspond to certain response probabilities. Specifically, the optimal design points on the z scale are $z_{opt} = \pm 1.543$ and the corresponding response probabilities are given by

$$\pi(-1.543) = 0.176, \quad \pi(1.543) = 0.824,$$

where $\pi(z) = e^z/(1 + e^z)$. Thus, if x_1^* and x_2^* are the two optimal doses corresponding to a D-optimal design, then

$$\theta_1 + \theta_2 x_1^* = -1.543, \quad \theta_1 + \theta_2 x_2^* = 1.543.$$

It is also worth pointing out that, if x_p is a dose level that causes a particular response in $100p\%$ of subjects, i.e., $\eta(x_p,\theta) = p$, the normalized variance of the MLE $\widehat{x}_p$ in the two-parameter logistic model is a special case of the c-optimality criterion and can be written as

$$\Psi = \operatorname{Var}(\widehat{x}_p) = c_p^T D(\xi,\theta) c_p = \operatorname{tr}\left[\left(c_p c_p^T\right) D(\xi,\theta)\right]$$

(Wu, 1988) with $c_p = f(x_p)/\theta_2$. For a discussion of c- and A-optimal designs for binary models, see Ford, Torsney and Wu (1992), Sitter and Wu (1993).

Illustration

Assume that the true values of the two parameters in the logistic model are $\theta_1 = 1$ and $\theta_2 = 3$. Assume also that the admissible dose range (design region) is given by $\mathcal{X} = [x_{\min}, x_{\max}]$ with $x_{\min} = -1$ and $x_{\max} = 1$. First of all, note that

$$1 + 3x_{\min} < -1.543, \quad 1 + 3x_{\max} > 1.543$$

and therefore the D-optimal dose levels lie inside the design region. The optimal doses, (x_1^*, x_2^*), are found from

$$1 + 3x_1^* = -1.543, \quad 1 + 3x_2^* = 1.543.$$

It is easy to verify that $x_1^* = -0.848$ and $x_2^* = 0.181$. The associated weights are $w_1 = w_2 = 0.5$.

Computation of the D-Optimal Design

This subsection introduces the %OptimalDesign1 macro that implements the optimal design algorithm in the univariate case (a single design variable x) and illustrates the process of computing a D-optimal design for the two-parameter logistic model described above. The advantage of using this simple model is that we can easily check individual elements of the resulting optimal design, including the information matrix, optimal doses, and weights.

The %OptimalDesign1 macro supports two main components of the optimal design algorithm:

- calculation of the information matrix (%deriv and %infod macros)
- implementation of the forward and backward steps of the algorithm (%doptimal macro)

The first component (information matrix) in this list is model-specific, whereas the second component (algorithm) does not depend on the chosen model. Note that the first component is the critical (and most computationally intensive) step in this macro.

Program 7.1 invokes the %OptimalDesign1 macro to compute a D-optimal design for the two-parameter logistic model. To save space, the complete SAS code is provided on the book's companion Web site. This subsection focuses on the user-specified parameters that define the design problem and set up the optimal design algorithm.

The first four macro variables in Program 7.1 define the following design parameters:

- POINTS is a vector of doses included in the initial design. In this case, the initial design contains four dose levels evenly spaced across the dose range $\mathcal{X} = [-1, 1]$.
- WEIGHTS defines the weights of the four doses in the initial design (the doses are equally weighted).
- GRID defines the grid points in the optimal design algorithm. The grid consists of 201 equally spaced points in the dose range. The DO function in SAS/IML creates a list of equally spaced points. In this case, DO produces the following row vector, $\{-1, -0.99, -0.98, \ldots, 0.98, 0.99, 1\}$.
- PARAMETER is a vector of model parameters. The true values of θ_1 and θ_2 are assumed to be 1 and 3, respectively.

The other five macro variables define the algorithm properties of the optimal design algorithm:

- CONVC is the convergence criterion. It determines when the iterative procedure terminates.

- MAXIMIT is the maximum number of iterations. A typical value of MAXIMIT is 200. A greater value is recommended when the final design is expected to contain a large number of points.

- CONST1 and CONST2 are used in the calculation of weights in the optimal design algorithm (Section 7.1.5). By default, these constants are set to 1. The user can change these values to facilitate the convergence of the algorithm in models with a large number of parameters.

- CMERGE is the *merging constant* in the optimal design algorithm that influences the process of merging design points. The default value of CMERGE is 3 and a larger value should be considered if the final design includes points with very small weights.

To calculate the information matrix $\mu(x, \theta)$ and the sensitivity function $\psi(x, \xi, \theta)$, we need to specify the $g(x, \theta)$ function introduced in Section 7.2.1. This function is defined in the %deriv macro. The MATRIX variable is the vector that contains the values of x, and the vector representing the $g(x, \theta)$ function is stored in the DERIVATIVE variable. In this case, the $g(x, \theta)$ function is given by

$$g(x, \theta) = \frac{e^{z/2}}{1 + e^z} f(x), \quad f(x) = (1, \; x)^T.$$

Program 7.1 *D*-optimal design for the two-parameter logistic model (Design parameters, algorithm parameters and $g(x, \theta)$ function)

```
* Design parameters;
%let points={-1 -0.333 0.333 1};
%let weights={0.25 0.25 0.25 0.25};
%let grid=do(-1,1,0.01);
%let parameter={1 3};
* Algorithm parameters;
%let convc=1e-7;
%let maximit=200;
%let const1=1;
%let const2=1;
%let cmerge=3;
* G function;
%macro deriv(matrix,derivative);
    nm=nrow(&matrix);
    one=j(nm,1,1);
    fm=one||&matrix;
    gc=exp(0.5*fm*t(parameter))/(1+exp(fm*t(parameter)));
    &derivative=gc#fm;
%mend deriv;
* Optimal design algorithm;
%OptimalDesign1;
```

Output from Program 7.1

```
                         Initial design

              Weight              X

               0.250           -1.000
               0.250           -0.333
               0.250            0.333
               0.250            1.000
```

```
Optimal design

Weight          X

0.501       -0.850
0.499        0.180


Determinants of the variance-covariance matrices

INITIAL     OPTIMAL

271.1        179.6


Variance-covariance matrix, initial design

COL1        COL2

11.0         9.2
 9.2        32.4


Variance-covariance matrix, optimal design

COL1        COL2

9.8          8.7
8.7         26.0
```

The output produced by Program 7.1 includes the following data sets (Output 7.1):

- INITIAL data set (design points and associated weights in the initial design).
- OPTIMAL data set (design points and associated weights in the D-optimal design).
- DDET data set (determinants of the variance-covariance matrix $D(\xi, \theta)$ for the initial and optimal designs).
- DINITIAL data set (variance-covariance matrix $D(\xi, \theta)$ for the initial design).
- DOPTIMAL data set (variance-covariance matrix $D(\xi, \theta)$ for the optimal design).

The program also creates two plots (Figure 7.3):

- Plot of the initial and optimal designs.
- Plot of the sensitivity functions associated with the initial and optimal designs.

Output 7.1 shows that the optimal doses are $x_1^* = -0.85$ and $x_2^* = 0.18$ with equal weights (see also the left panel in Figure 7.3). This example illustrates a well-known theoretical result that if a D-optimal design is supported at m points for a model with m unknown parameters, then the support points have equal weights, $w_i = 1/m$, $i = 1, \ldots, m$; see Fedorov (1972, Corollary to Theorem 2.3.1). It is easy to check that the optimal doses are very close to the doses we computed earlier from the equation $z_{opt} = \pm 1.543$.

We can also see from Output 7.1 that the determinant of the optimal variance-covariance matrix is 179.6 compared to the initial value of 271.1. There was also an improvement in the variance of the parameter estimates. The variance of $\widehat{\theta}_1$ dropped from 11.0 to 9.8 and the variance of $\widehat{\theta}_2$ decreased from 32.4 to 26.0.

The right panel in Figure 7.3 shows that the equivalence theorem serves as an excellent diagnostic tool in optimization problems. The sensitivity function of the D-optimal design hits the reference line $m = 2$ at the optimal doses, i.e., at $x = -0.85$ and $x = 0.18$. For the

Figure 7.3 Left panel: Initial (open circles) and optimal (closed circles) designs. Right panel: Sensitivity functions for the initial (dashed curve) and optimal (solid curve) designs.

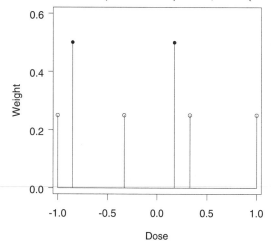 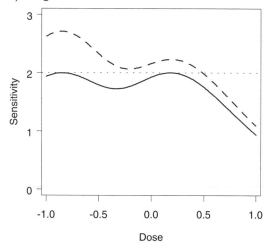

D-criterion, the sensitivity function is identical to the normalized variance of prediction. Thus, since the sensitivity function of the initial design is greater than $m = 2$ at the left end of the dose range, we conclude that

- The initial design is not D-optimal.
- More measurements should be placed at the left end of the design region in order to reduce the variance of prediction or, equivalently, push the sensitivity function down below the reference line defined by the equivalence theorem ($m = 2$).

7.2.3   Example 2: A Two-Parameter Logistic Model with a Narrow Dose Range

In Example 1, we considered the case when the admissible dose range was fairly wide, $\mathcal{X} = [-1, 1]$. If the dose range is not sufficiently wide, at least one of the D-optimal points may coincide with the boundary and we will no longer be able to use the simple rule introduced earlier in this section ($z_{opt} = \pm 1.543$).

As an illustration, consider a narrower dose range, $\mathcal{X} = [0, 1]$. To compute the D-optimal design for the two-parameter logistic model in this case, all we need to do is to modify the lower boundary of the dose range in Program 7.1. Program 7.2 derives the D-optimal design for the modified dose range (complete SAS code is given on the book's companion Web site).

Program 7.2 D-optimal design for the two-parameter logistic model with a narrow dose range (Algorithm parameters and $g(x, \theta)$ function are identical to those defined in Program 7.1)

```
* Design parameters;
%let points={0 0.333 0.667 1};
%let weights={0.25 0.25 0.25 0.25};
%let grid=do(0,1,0.025);
%let parameter={1 3};
```

Output 7.2

```
                        Initial design

                   Weight          X

                   0.250        0.000
                   0.250        0.333
                   0.250        0.667
                   0.250        1.000


                   Optimal design

                   Weight          X

                   0.500        0.000
                   0.500        0.725
```

Output 7.2 shows that the resulting D-optimal design is still a two-point design with equal weights $w_1 = w_2 = 0.5$; however, the optimal doses have changed. One dose is now located at the boundary of the region, $x_1^* = 0$, and the other dose has shifted to the right, $x_2^* = 0.725$ (see also Figure 7.4). As was pointed out in Section 7.1.5, optimal designs depend not only on the information matrix (which is identical to the information matrix in Example 1) but also on the dose range or, in general, the design region.

Figure 7.4 Left panel: Initial (open circles) and optimal (closed circles) designs. Right panel: Sensitivity functions for the initial (dashed curve) and optimal (solid curve) designs.

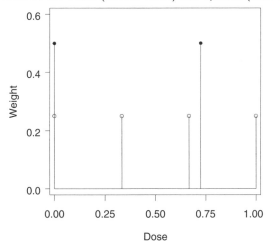
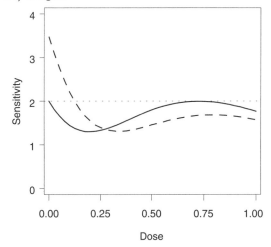

7.3   Nonlinear Regression Models with a Continuous Response

In this section we first formulate the optimal design problem for general nonlinear regression models and then consider a popular dose-response model with a continuous response variable (four-parameter logistic, or $E_{\max}$, model).

7.3.1   General Nonlinear Regression Model

Consider a general nonlinear regression model that is used in the analysis of concentration- or dose-response curves in a large number of clinical and pre-clinical studies. The model describes the relationship between a continuous response variable, Y, and design variable x (dose or concentration level):

$$E(y_{ij}|x_i;\theta) = \eta(x_i,\theta), \quad i = 1,\ldots,n, \quad j = 1,\ldots,n_i.$$

Here y_{ij}'s are often assumed to be independent observations with $\mathrm{Var}(y_{ij}|x_i;\theta) = \sigma^2(x_i)$, i.e., the variance of the response variable varies across doses or concentration levels. Also, x_i is the design variable that assumes values in a certain region, $\mathcal{X}$. Lastly, θ is an m-dimensional vector of unknown parameters.

When the response variable Y is normally distributed, the variance matrix of the MLE, $\widehat{\theta}$, can be approximated by $D(\xi,\theta)/N$, see (7.4), and the information matrix of a single observation is

$$\mu(x_i,\theta) = g(x_i,\theta)g^T(x_i,\theta), \quad g(x_i,\theta) = \frac{f(x_i,\theta)}{\sigma(x_i)},$$

where $f(x,\theta)$ is a vector of partial derivatives of the response function $\eta(x,\theta)$,

$$f(x,\theta) = \left[\frac{\partial\eta(x,\theta)}{\partial\theta_1}, \frac{\partial\eta(x,\theta)}{\partial\theta_2}, \ldots, \frac{\partial\eta(x,\theta)}{\partial\theta_m}\right]^T. \tag{7.10}$$

Now, as far as optimal experimental design is concerned, the sequence of steps that follow is exactly the same as for quantal dose-response models that are considered in Section 7.2:

1. Select a preliminary parameter estimate, $\widetilde{\theta}$.

2. Select an optimality criterion, $\Psi = \Psi[D(\xi,\widetilde{\theta})]$, e.g., the D-optimality criterion.

3. Construct a locally optimal design $\xi^*(\widetilde{\theta})$ for the selected criterion.

It is recommended that sensitivity analysis be performed, i.e., steps 1 through 3 must be repeated for slightly different $\widetilde{\theta}_s$, $s = 1,\ldots,N_s$, to verify robustness of the design $\xi^*(\widetilde{\theta})$; see Atkinson and Fedorov (1988).

Since we know that $\mu(x_i,\theta) = g(x_i,\theta)g^T(x_i,\theta)$, the process of constructing D- or A-optimal designs for the general nonlinear regression model will be virtually identical to the process of finding optimal designs for quantal dose-response models (described in detail in Section 7.2). We need only to redefine the $g(x,\theta)$ function that determines the information matrix $\mu(x,\theta)$ and sensitivity functions in (7.6) and (7.7) for the D- and A-optimality criteria, respectively.

7.3.2 Example 3: A Four-Parameter Logistic Model with a Continuous Response

The four-parameter logistic, or E_{max}, model, is used in many pharmaceutical applications, including bioassays (of which ELISA, or enzyme-linked immunosorbent assay, and cell-based assays are popular examples). See Finney (1976), Karpinski (1990), Hedayat et al. (1997), Källén and Larsson (1999) for details and references.

This example comes from a study of a compound that inhibits the proliferation of bone marrow erythroleukemia cells in a cell-based assay (Downing, Fedorov and Leonov, 2001; Fedorov and Leonov, 2005). The logistic model used in the study is defined as follows:

$$E(Y|x,\theta) = \eta(x,\theta) = \theta_3 + \frac{\theta_4 - \theta_3}{1 + (x/\theta_1)^{\theta_2}},$$

where Y is a continuous outcome variable (number of cells in the assay) and x is the design variable (concentration level). Further, θ_1 is often denoted as ED_{50} and represents the concentration at which the median response $(\theta_3 + \theta_4)/2$ is attained, θ_2 is a slope parameter, θ_4 and θ_3 are the lower and upper asymptotes (minimal and maximal responses).

As an aside note, the four-parameter logistic model provides an example of a *partially* nonlinear model since θ_3 and θ_4 enter the response function in a linear fashion. This linear entry implies that the D-optimal design does not depend on values of θ_3 and θ_4 (Hill, 1980).

Computation of the D-Optimal Design

The vector of unknown parameters θ was estimated by modeling the data collected from a 96-well plate assay, with six repeated measurements at each of the ten concentrations, $\{500, 250, 125, \ldots, 500/2^9\}$ ng/ml, which represents a two-fold serial dilution design with $N = 60$ and weights $w_i = 1/10$. The estimated parameters are

$$\theta = (15.03, 1.31, 530, 1587)^T.$$

The design region will be set to

$$\mathcal{X} = [\ln(500/2^9), \ln(500)] = [-0.024, 6.215]$$

on a log-scale. The x values will be exponentiated to compute the response function $\eta(x, \theta)$.

Program 7.3 constructs the D-optimal design for the introduced four-parameter logistic model by calling the %OptimalDesign1 macro. As in Programs 7.1 and 7.2, we need to define the following parameters and pass them to the macro:

- The initial design is based on ten equally spaced and equally weighted log-transformed concentration levels within the design region (POINTS and WEIGHTS variables).
- The grid consists of 801 equally spaced points in $\mathcal{X} = [-0.024, 6.215]$.
- The true values of the model parameters are specified using the PARAMETER variable.
- The algorithm parameters are identical to those used in Programs 7.1 and 7.2.
- To specify the $g(x, \theta)$ function, we need to compute a vector of partial derivatives. The derivatives are defined in the %deriv macro. The MATRIX variable contains the values of x. The DER1, DER2, DER3, and DER4 variables represent the derivatives of the response function with respect to θ_1, θ_2, θ_3, and θ_4, respectively, and the DERIVATIVE variable contains the vector of partial derivatives.

Also, note that in order to improve the stability of the iterative procedure, the merging constant (CMERGE) needs to be increased to 12.

As before, we will concentrate on the user-defined parameters. The complete SAS code can be found on the book's companion Web site.

Program 7.3 D-optimal design for the four-parameter logistic model with a continuous response (Design parameters, algorithm parameters and $g(x, \theta)$ function)

```
* Design parameters;
%let points=do(-0.024,6.215,6.239/9);
%let weights=repeat(0.1,1,10);
%let grid=do(-0.024,6.215,6.239/800);
%let parameter={15.03 1.31 530 1587};
* Algorithm parameters;
%let convc=1e-9;
%let maximit=1000;
%let const1=1;
%let const2=1;
%let cmerge=12;
```

```
* G function;
%macro deriv(matrix,derivative);
    nm=nrow(&matrix);
    one=j(nm,1,1);
    dera=(parameter[4]-parameter[3])/(1+(exp(&matrix[,1])/parameter[1])
        ##parameter[2])##2#(parameter[2]/parameter[1])#(exp(&matrix[,1])
        /parameter[1])##parameter[2];
    derb=-(parameter[4]-parameter[3])/(1+(exp(&matrix[,1])/parameter[1])
        ##parameter[2])##2#(exp(&matrix[,1])/parameter[1])##parameter[2]
        #log(exp(&matrix[,1])/parameter[1]);
    derc=1-1/(1+(exp(&matrix[,1])/parameter[1])##parameter[2]);
    derd=1/(1+(exp(&matrix[,1])/parameter[1])##parameter[2]);
    &derivative=dera||derb||derc||derd;
%mend deriv;
%OptimalDesign1;
```

Output from Program 7.3

```
                        Initial design

              Weight            X

               0.100          -0.024
               0.100           0.669
               0.100           1.362
               0.100           2.056
               0.100           2.749
               0.100           3.442
               0.100           4.135
               0.100           4.829
               0.100           5.522
               0.100           6.215

                        Optimal design

              Weight            X

               0.250          -0.024
               0.250           2.027
               0.250           3.540
               0.250           6.215

        Determinants of the variance-covariance matrices

                   INITIAL       OPTIMAL

                  0.000031      0.000015

        Variance-covariance matrix, initial design

             COL1      COL2      COL3      COL4

            0.019     0.000    -0.103    -0.283
            0.000     0.000     0.027    -0.041
           -0.103     0.027     6.745    -3.599
           -0.283    -0.041    -3.599    12.639
```

```
Variance-covariance matrix, optimal design

            COL1      COL2      COL3      COL4

           0.015     0.000    -0.118    -0.181
           0.000     0.000     0.013    -0.019
          -0.118     0.013     5.086    -1.065
          -0.181    -0.019    -1.065     7.087
```

Output 7.3 shows that the D-optimal design includes two points on the boundary of the design region ($x_1^* = -0.024$ and $x_4^* = 6.215$) and two in the middle of the design region ($x_2^* = 2.027$ and $x_3^* = 3.540$). The optimal weights are equal to $1/m = 0.25$ (compare Output 7.1); see the left panel in Figure 7.5. As in Examples 1 and 2 given in Section 7.2, the sensitivity function is equal to the number of the model parameters, $m = 4$, at the optimal concentrations (see the right panel in Figure 7.5).

Figure 7.5 Left panel: Initial (open circles) and optimal (closed circles) designs. Right panel: Sensitivity functions for the initial (dashed curve) and optimal (solid curve) designs.

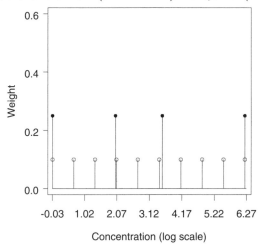 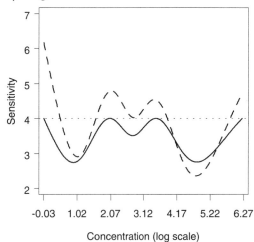

Output 7.3 also helps compare characteristics of the initial and optimal designs. The determinant of the final variance-covariance matrix is 1.5×10^{-5}, compared to the value of 3.1×10^{-5} for the initial design. The variance of individual parameter estimates has improved, too. For example, the variance of θ_1 has dropped from 0.019 to 0.015. Parameter θ_1, or ED_{50}, is usually the most important for practitioners. This example illustrates a good performance of a D-optimal design with respect to other criteria of optimality, like c-criterion for the estimation of θ_1.

7.4 Regression Models with Unknown Parameters in the Variance Function

In various pharmaceutical applications it is often assumed that the variance of the error term, σ^2, depends not only on the control variable x, but also on the unknown parameter θ, i.e., $\sigma_i^2 = S(x_i, \theta)$.

Consider, for instance, the four-parameter logistic model introduced in Section 7.2. The variance of the response variable, Y, may depend on its mean, as in the following power model (Finney, 1976; Karpinski, 1990; Hedayat et al., 1997):

$$\text{Var}(Y|x, \theta) = S(x, \theta) \sim \eta^\delta(x, \theta), \quad \delta > 0,$$

In cell-based assay studies, the response variable is a cell count and is often assumed to follow a Poisson distribution. Under this assumption, one can consider a power variance model with $\delta = 1$.

Another possible generalization of the nonlinear regression models considered in Section 7.2 is a regression model with multiple correlated responses. Multiple correlated responses are encountered in clinical trials with several endpoints or several objectives, e.g., simultaneous investigation of efficacy and toxicity in dose response studies (Heise and Myers, 1996; Fan and Chaloner, 2001).

7.4.1   General Model

Let the observed response Y be a normally distributed k-dimensional vector with

$$E[Y|x] = \eta(x, \theta), \quad \text{Var}[Y|x] = S(x, \theta), \tag{7.11}$$

where $\eta(x, \theta) = (\eta_1(x, \theta), \ldots, \eta_k(x, \theta))^T$ is a vector of pre-defined functions and $S(x, \theta)$ is a $k \times k$ positive definite matrix. It can be shown that the Fisher information matrix of a single observation, $\mu(x, \theta)$, is the sum of two terms; see Magnus and Neudecker, 1988, Chapter 6, or Muirhead, 1982, Chapter 1:

$$\mu(x, \theta) = \mu^R(x, \theta) + \mu^V(x, \theta), \tag{7.12}$$

$$\mu_{ij}^R(x, \theta) = \frac{\partial \eta^T(x, \theta)}{\partial \theta_i} S^{-1}(x, \theta) \frac{\partial \eta(x, \theta)}{\partial \theta_j}, \quad i, j = 1, \ldots, m,$$

$$\mu_{ij}^V(x, \theta) = \frac{1}{2} \text{tr}\left[S^{-1}(x, \theta) \frac{\partial S(x, \theta)}{\partial \theta_i} S^{-1}(x, \theta) \frac{\partial S(x, \theta)}{\partial \theta_j} \right],$$

where m is the number of unknown parameters. In the univariate case ($k = 1$), $\mu(x, \theta)$ permits the following factorization:

$$\mu(x, \theta) = g_R(x, \theta) g_R^T(x, \theta) + g_V(x, \theta) g_V^T(x, \theta), \tag{7.13}$$

where

$$g_R(x, \theta) = \frac{f(x, \theta)}{\sqrt{S(x, \theta)}}, \quad g_V(x, \theta) = \frac{h(x, \theta)}{\sqrt{2} S(x, \theta)},$$

$f(x, \theta)$ is the vector of partial derivatives of $\eta(x, \theta)$ as in (7.10) and $h(x, \theta)$ is the vector of partial derivatives of $S(x, \theta)$ with respect to θ, i.e.,

$$h(x, \theta) = \left[\frac{\partial S(x, \theta)}{\partial \theta_1}, \frac{\partial S(x, \theta)}{\partial \theta_2}, \ldots, \frac{\partial S(x, \theta)}{\partial \theta_m} \right]^T. \tag{7.14}$$

It is important to note that, since the information matrix $\mu(x, \theta)$ consists of two terms, its rank may be greater than 1 even in the univariate case. This may lead to the situation when the number of support points in the optimal design is less than the number of unknown parameters, i.e., less than m (see Downing, Fedorov and Leonov (2001) or Fan and Chaloner (2001) for examples). This cannot happen in regression models with a single response variable and a known variance since in this case the information matrix of the design would become singular.

Sensitivity Functions

Using classic optimal design techniques, it can be shown that the equivalence theorem (7.6), (7.7) remains valid for multivariate models in which the covariance matrix depends on unknown parameters (Atkinson and Cook, 1995; Fedorov, Gagnon and Leonov, 2002). In the univariate case, sensitivity functions for the D- and A-optimality criteria can be presented as the sum of two terms and can be factorized too. For example, the sensitivity function for the D-optimality criterion can be written as

$$\psi(x,\xi,\theta) = g_R(x,\theta)M^{-1}(\xi,\theta)g_R^T(x,\theta) + g_V(x,\theta)M^{-1}(\xi,\theta)g_V^T(x,\theta).$$

7.4.2   Example 4: A Four-Parameter Logistic Model with a Power Variance Model

This example is concerned with the computation of the D-optimal design in a four-parameter logistic model that extends the logistic model considered in Example 3 (see Section 7.3). Assume that Y is a normally distributed response variable with

$$E[Y|x] = \eta(x,\theta) = \theta_3 + \frac{\theta_4 - \theta_3}{1 + (x/\theta_1)^{\theta_2}},$$

$$\text{Var}[Y|x] = S(x,\theta) = \delta_1 \eta^{\delta_2}(x,\theta),$$

where δ_1 and δ_2 are positive parameters.

Parameters $\theta_1 - \theta_4$ are selected as in Example 3, i.e., $\theta = (15.03, 1.31, 530, 1587)^T$ and $\delta_1 = 0.5$, $\delta_2 = 1$. The combined vector of model parameters is $(15.03, 1.31, 530, 1587, 0.5, 1)^T$.

Computation of the D-Optimal Design

Program 7.4 computes the D-optimal design for the four-parameter logistic model specified above. This program is conceptually similar to Program 7.3. We need only to make the following changes:

- Add the two variance parameters to the PARAMETER variable.

- Change the method of calculating the derivatives since the analytic form of the derivatives is rather complicated. The %deriv macro computes $g_R(x,\theta)$ numerically using the NLPFDD function. Since the information matrix and sensitivity function also depend on $g_V(x,\theta)$, we introduce a new macro (%derivs) that calculates this function $g_V(x,\theta)$ (it is also derived using a numerical approximation).

- Since the information matrix is now the sum of two terms, the %infod macro needs to be modified as well.

- As in Program 7.3, the value of the merging constant (CMERGE) is increased to improve the stability of the iterative procedure in this complex optimal design problem.

The complete version of Program 7.4 is provided on the book's companion Web site.

Program 7.4 D-optimal design for the four-parameter logistic model with a continuous response and unknown parameters in the variance function (Design and algorithm parameters)

```
* Design parameters;
%let points=do(-0.024,6.215,6.239/7);
%let weights=repeat(0.125,1,8);
%let grid=do(-0.024,6.215,6.239/400);
%let parameter={15.03 1.31 530 1587 0.5 1};
* Number of parameters;
%let paran=6;
```

```
* Algorithm parameters;
%let convc=1e-9;
%let maximit=1000;
%let const1=1;
%let const2=1.2;
%let cmerge=7.5;
%OptimalDesign1;
```

Output from Program 7.4

```
                        Initial design

                    Weight          X

                    0.125       -0.024
                    0.125        0.867
                    0.125        1.759
                    0.125        2.650
                    0.125        3.541
                    0.125        4.432
                    0.125        5.324
                    0.125        6.215

                        Optimal design

                    Weight          X

                    0.294       -0.024
                    0.213        1.972
                    0.236        3.563
                    0.257        6.215
```

Output 7.4 lists the initial and optimal designs. Other parameters of the two designs, e.g., the variance-covariance matrices and their determinants, can be obtained by printing out the DINITIAL, DOPTIMAL, and DDET data sets. It was emphasized above that, in models with unknown parameters in the variance function, the number of support points in optimal designs can be less than the number of model parameters. Indeed, Output 7.4 shows that, although there are two more unknown parameters in this model compared to the model with a constant variance (Example 3 of Section 7.3), the D-optimal design is still a four-point design. The weights are no longer equal to $1/4$, and are equal to $(0.294, 0.213, 0.236, 0.257)$ for the four design points $(-0.024, 1.972, 3.563, 6.215)$ on the log-scale.

Figure 7.6 provides a visual comparison of the initial and optimal designs (left panel) and also plots the sensitivity functions associated with the two designs (right panel).

7.5 Models with a Bounded Response (Beta Models)

The Beta regression model provides another example where the information matrix can be calculated in a closed form.

7.5.1   General Model

In clinical trials, investigators often need to deal with ordinal variables containing many categories. The Beta regression model has been shown to be a good choice to analyze this type of response; see Wu, Fedorov and Propert (2005).

Figure 7.6 Left panel: Initial (open circles) and optimal (closed circles) designs. Right panel: Sensitivity functions for the initial (dashed curve) and optimal (solid curve) designs.

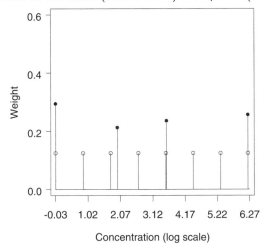
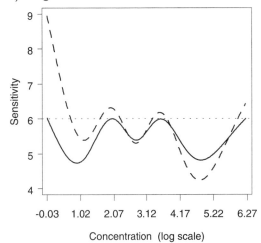

In this section we consider an example from a randomized, double-blind, placebo-controlled, parallel group clinical trial. Each patient was randomized to one of four treatment arms, with 0, 2, 4, and 8 dose units, respectively. The observations were taken at baseline and the end of the study at 6 months. The response is a severity measure score that varies between 0 and 1 on the normalized scale, with hundreds of levels. Higher scores indicate higher degrees of severity of the illness.

After investigating the data on the descriptive level, we observed that the median score has a slightly decreasing trend when dose increases. But the difference may not be big enough to be statistically and/or clinically significant. However, the 8 dose units group has a smaller variance than the other groups. Even among the other three treatment groups, the estimated population variances are different from one group to another. These facts indicate that, although the drug did not reduce the overall mean response, it might be effective in some subjects.

Let y_{ij} denote the response rate from patient j under dose level x_i and assume that y_{ij} follows the Beta distribution with parameters $p(x_i)$ and $q(x_i)$. We can model $p(x_i)$ and $q(x_i)$ as a function of the dose x:

$$\ln p(x) = \alpha^T f(x) \text{ and } \ln q(x) = \beta^T \phi(x), \tag{7.15}$$

Note that

$$E(y_{ij}) = \frac{p(x_i)}{p(x_i) + q(x_i)}, \quad \mathrm{Var}(y_{ij}) = \frac{p(x_i)q(x_i)}{(p(x_i) + q(x_i))^2(p(x_i) + q(x_i) + 1)^2},$$

where $B(p, q) = \Gamma(p)\Gamma(q)/\Gamma(p + q)$ and $\Gamma(p) = \int_0^\infty t^{p-1}e^{-t}dt$ is the Gamma function. Let $\theta = (\alpha^T, \beta^T)$, denote $p(x_i)$ and $q(x_i)$ as p_i and q_i, respectively, and consider the Digamma or Trigamma functions (Abramowitz and Stegun, 1972):

$$\Psi_\Gamma(v) = \frac{d \ln \Gamma(v)}{dv}, \quad \Psi'(v) = \frac{d\Psi_\Gamma(v)}{dv}.$$

The information matrix $\mu(x_i, \theta)$ of a single observation at point x_i is given by

$$\begin{pmatrix} [\Psi'(p_i) - \Psi'(p_i + q_i)]p_i^2 f(x_i)f^T(x_i) & -\Psi'(p_i + q_i)p_i q_i f(x_i)\phi^T(x_i) \\ -\Psi'(p_i + q_i)p_i q_i \phi(x_i)f^T(x_i) & [\Psi'(p_i) - \Psi'(p_i + q_i)]p_i^2 \phi(x_i)\phi^T(x_i) \end{pmatrix}. \tag{7.16}$$

The information matrix of N experiments, with n_i independent observations at dose x_i, admits presentation (7.2).

Here we consider a simple example with $\alpha = (\alpha_1, \alpha_2)^T$, $\beta = (\beta_1, \beta_2)^T$, and $f(x) = \phi(x) = (1, x)^T$. In that case the individual information matrix may be written as follows:

$$\mu(x_i, \theta) = \begin{pmatrix} [\Psi'(p_i) - \Psi'(p_i + q_i)]p_i^2 & -\Psi'(p_i + q_i)p_i q_i \\ -\Psi'(p_i + q_i)p_i q_i & [\Psi'(p_i) - \Psi'(p_i + q_i)]p_i^2 \end{pmatrix} \otimes \begin{pmatrix} 1 & x_i \\ x_i & x_i^2 \end{pmatrix}, \quad (7.17)$$

where $\otimes$ is the Kronecker product.

Figure 7.7 displays the fitted Beta densities for each treatment group with

$$\theta = (\alpha_1, \alpha_2, \beta_1, \beta_2)^T = (4, -0.49, 3.9, 0.15)^T.$$

The figure shows that, as the dose level increases, the mean response becomes smaller and thus we see evidence of a drug effect. Note also that the variance of the response variable clearly changes with the average response level and the highest dose group has the smallest variance.

Figure 7.7 Distribution of simulated responses in each treatment group

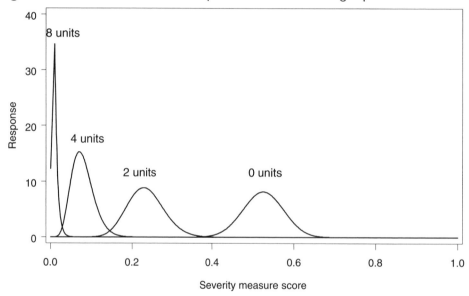

Unlike quantal dose-response models discussed in Section 7.2 or continuous logistic models from Sections 7.3–7.4, the presentation (7.16) for the information matrix of the Beta regression model cannot be factorized. As a result, the general formulas (7.6) and (7.7) have to be implemented in this case.

7.5.2   Example 5: Beta Regression Model

To construct a D-optimal design for the clinical trial example discussed above, we use the following Beta regression model:

$$\ln p(x) = 4 - 0.49x, \quad \ln q(x) = 3.9 + 0.15x.$$

Program 7.5 computes a D-optimal design for the Beta regression model by calling the %OptimalDesign2 macro. The initial design contains 6 equally spaced doses between 0 and 8 dose units, i.e., $\mathcal{X} = [0, 8]$. The true value of the parameter vector is $\theta = (4, -0.49, 3.9, 0.15)^T$.

The overall optimal design algorithm in this problem remains the same; however, unlike the previous examples, the information matrix is calculated directly, without using partial derivatives of the response function. Therefore, in this example (and subsequent examples in this chapter) the user does not need to specify the %deriv1 or %deriv2 macros. The direct calculation of the information matrix relies on the %Info, %infoele and %indinfo macros. The analytical form of the information matrix is specified in the %info macro. The %infoele macro is needed to store the individual information matrices based on the values of covariates on the grid. Elements of the information matrices are stored in different vectors based on the position in the matrix. Note that the information matrix is a symmetric matrix so only the upper (lower) triangular matrix is needed when storing it. For example, the ELEO14 vector contains the elements in the first row, fourth column of all the information matrices. The %indinfo macro outputs the individual information matrix for a given candidate point (note that the MDERIV macro variable is not used in this program and can be set to any value). The complete SAS code for the updated macros is provided on the book's companion Web site.

Program 7.5 *D*-optimal design for the beta regression model (Design and algorithm parameters)

```
* Design parameters;
%let points=do(0,8,8/5);
%let weights=repeat(1/6,1,6);
%let grid=do(0,8,8/200);
%let parameter={4 -0.49 3.9 0.15};
* Number of parameters;
%let paran=4;
* Algorithm parameters;
%let convc=1e-9;
%let maximit=1000;
%let const1=2;
%let const2=1;
%let cmerge=5;
%OptimalDesign2;
```

Output from Program 7.5

```
                      Initial design

                 Weight          X

                  0.167       0.000
                  0.167       1.600
                  0.167       3.200
                  0.167       4.800
                  0.167       6.400
                  0.167       8.000

                 Optimal design

                 Weight          X

                  0.491       0.000
                  0.071       4.280
                  0.438       8.000
```

Output 7.5 shows the 6-point initial and 3-point optimal designs. Note that most of the weight in this *D*-optimal design is assigned to the two points on the boundaries of the dose range (placebo and 8 dose units). The middle point receives very little weight (only 7% of the patients will be allocated to this dose). Figure 7.8 displays the initial and optimal designs and their sensitivity functions.

Figure 7.8 Left panel: Initial (open circles) and optimal (closed circles) designs. Right panel: Sensitivity functions for the initial (dashed curve) and optimal (solid curve) designs.

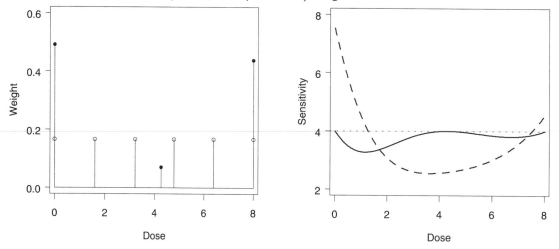

7.6 Models with a Bounded Response (Logit Link)

The Beta regression model takes care of the bounded variable naturally (by definition). In this section we introduce an alternative way to model the dose-response trial example from Section 7.5, using the linear regression model framework. After a simple transformation, the responses are not bounded and can be treated as normally distributed variables. Furthermore, to address the relationship between variance and dose level, one can model the variance as a function of the dose.

7.6.1 One-Dimensional Case

The model with a single drug can be written as

$$\ln \frac{B - Y}{Y - A} = \theta_1 + \theta_2 x + \sigma^2(x, \theta)\varepsilon,$$

where the variance function is defined as follows

$$\sigma^2(x, \theta) = \exp(\theta_3 + \theta_4 x).$$

Further, Y is the response varying from A to B, x denotes the dose, and the error term is normally distributed, i.e., $\varepsilon \sim N(0, 1)$.

In this example we take $A = 0$ and $B = 1$. Note that this model provides another example of models with unknown parameters in the variance function (compare with (7.11)). Thus formulas (7.13) and (7.14) can be applied here, with

$$f(x, \theta) = (1, x, 0, 0)^T, \quad h(x, \theta) = (0, 0, 1/\sqrt{2}, x/\sqrt{2})^T.$$

The information matrix of a single observation is

$$\mu(x, \theta) = \begin{pmatrix} \frac{1}{\exp(\theta_3 + \theta_4 x)} & 0 \\ 0 & 0.5 \end{pmatrix} \otimes \begin{pmatrix} 1 & x \\ x & x^2 \end{pmatrix}.$$

It follows from this equation that an optimal design will depend only on the values of θ_3 and θ_4.

7.6.2   Example 6: A Bounded-Response Model with a Logit Link and Unknown Parameters in the Variance Function

Program 7.6 calls the %OptimalDesign2 macro to find the D-optimal design for the bounded-response model defined earlier in this subsection. We need to change the %infoele macro that computes the information matrix and remove the %devpsi macro. The other macros do not change compared to Example 5. The complete SAS code for Program 7.6 can be found on the book's companion Web site.

Note that, as in Program 7.5, the initial design includes six equally spaced points in the dose region $\mathcal{X} = [0, 8]$. The parameter verctor is given by $\theta = (1, -0.5, 1, -0.6)^T$ and the fitted normal densities for each treatment group are displayed in Figure 7.9 (the densities are shown on a normal scale, i.e., $\ln(1-y)/y$). As in Figure 7.7, we see that, as the dose level increases, the mean response (disease severity) decreases and the variance decreases as well.

Figure 7.9 Distribution of the response variable in each treatment group

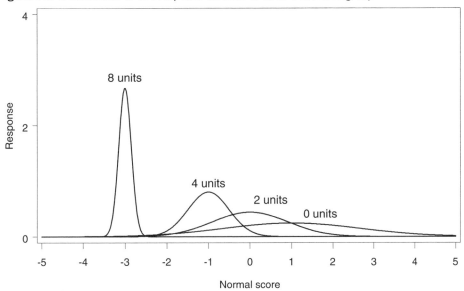

Program 7.6 D-optimal design for the bounded-response model with a logit link and unknown parameters in the variance function (Design and algorithm parameters)

```
* Design parameters;
%let points=do(0,8,8/5);
%let weights=repeat(1/6,1,6);
%let grid=do(0,8,8/400);
%let parameter={-1 0.5 -1 -0.6};
* Number of parameters;
%let paran=4;
* Algorithm parameters;
%let convc=1e-9;
%let maximit=1000;
%let const1=1;
%let const2=1;
%let cmerge=3;
%OptimalDesign2;
```

Output from Program 7.6

```
                            Initial design

                    Weight          X

                    0.167        0.000
                    0.167        1.600
                    0.167        3.200
                    0.167        4.800
                    0.167        6.400
                    0.167        8.000


                            Optimal design

                    Weight          X

                    0.307        0.000
                    0.228        4.720
                    0.465        8.000
```

Output 7.6 displays the initial and *D*-optimal designs. The optimal design is a three-point design with two points on the boundaries and one design point inside the design region. Comparing this optimal design with the design shown in Output 7.5, it is easy to see that the middle dose (4.720) is not much different from the optimal middle dose computed from the Beta regression model (4.280); however, it receives a much greater weight. The initial and optimal designs as well as the associated sensitivity functions are depicted in Figure 7.10.

Figure 7.10 Left panel: Initial (open circles) and optimal (closed circles) designs. Right panel: Sensitivity functions for the initial (dashed curve) and optimal (solid curve) designs.

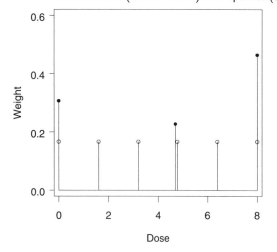 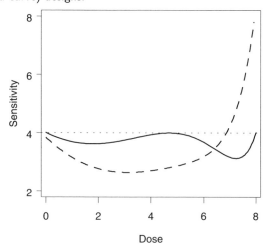

7.6.3   Models with a Bounded Response (Combination of Two Drugs)

Under the same dose-response set-up as in Section 7.5, we now consider the case of a two-drug combination: Drug A and Drug B. The daily dose of each drug ranges from 0.1 mg to 10 mg. Due to safety restrictions in drug development, often it is not safe to assign patients to the treatment with both drugs at the highest dose. In this example, we require that the total dose of the two drugs cannot exceed 10.1 mg. Note that the resulting design region is not square, it is a triangle.

Let x and z denote the daily doses of Drugs A and B, respectively. The corresponding model is

$$\ln\frac{1-y}{y} = \theta_1 + \theta_2 x + \theta_3 z + \theta_4 xz + \theta_5 x^2 + \theta_6 z^2 + \sigma^2(x,\theta,z)\varepsilon,$$

where the variance function is given by

$$\sigma^2(x,\theta,z) = \exp(\theta_7 + \theta_8 x + \theta_9 z).$$

It can be shown that the information matrix of a single observation is

$$\mu(x_i, z_j, \theta) = \begin{pmatrix} \exp(-\theta_7 - \theta_8 x_i - \theta_9 z_j) f_1 f_1^T & 0 \\ 0 & 0.5 f_2 f_2^T \end{pmatrix},$$

where $i = 1, \ldots, n$, $j = 1, \ldots, n$,

$$f_1 = (1,\ x_i,\ z_j,\ x_i,\ z_j,\ x_i^2,\ z_j^2)^T, \quad f_2 = (1,\ x_i,\ z_j)^T.$$

Here n is the number of possible doses of each drug. Note that, similar to the one-dimensional case considered in Example 6, the optimal design depends only on the parameters of the variance function, i.e., θ_7, θ_8 and θ_9.

7.6.4   Example 7: Two Design Variables (Two Drugs)

Program 7.7 computes the D-optimal design for the introduced model. This program calls the %OptimalDesign3 macro that supports a two-dimensional version of the optimal design algorithm defined in Section 7.1.5. The design parameters are specified in the %DesignParameters macro and then they are passed to the %OptimalDesign3 macro. The initial design (POINTS and WEIGHTS variables) includes ten equally weighted points that are evenly distributed across the triangular design region:

$$0.1 \le x \le 10, \quad 0.1 \le z \le 10, \quad x + z \le 10.1.$$

Program 7.7 D-optimal design for the bounded-response model with two design variables (Design and algorithm parameters)

```
%macro DesignParameters;
    * Initial design: Design points;
    * Four equally spaced doses for each drug;
    dosen=4;
    * Total number of points;
    dosem=dosen*(dosen+1)/2;
    points=j(dosem,2,0);
    k=0;
    do i=1 to dosen;
        do j=1 to dosen-i+1;
            k=k+1;
            points[k,1]=0.1+(i-1)*9.9/(dosen-1);
            points[k,2]=0.1+(j-1)*9.9/(dosen-1);
        end;
    end;
    * Initial design: Weights;
    * Equally weighted points;
    weights=t(repeat(1/10,1,10));
```

```
    * Grid;
    * 26 equally spaced points on each axis;
    gridn=26;
    gridm=gridn*(gridn+1)/2;
    grid=j(gridm,2,0);
    k=0;
    do i=1 to gridn;
        do j=1 to gridn-i+1;
            k=k+1;
            grid[k,1]=0.1+(i-1)*9.9/(gridn-1);
            grid[k,2]=0.1+(j-1)*9.9/(gridn-1);
        end;
    end;
    * Parameter values;
    para={1 1 1 1 1 1 1 0.6 0.4};
%mend DesignParameters;
* Number of parameters;
%let paran=9;
* Algorithm parameters;
%let convc=1e-9;
%let maximit=500;
%let const1=1;
%let const2=1;
%let cmerge=3;
%OptimalDesign3;
```

Output from Program 7.7

Initial design

Weight	Drug A	Drug B
0.10	0.10	0.10
0.10	0.10	3.40
0.10	0.10	6.70
0.10	0.10	10.0
0.10	3.40	0.10
0.10	3.40	3.40
0.10	3.40	6.70
0.10	6.70	0.10
0.10	6.70	3.40
0.10	10.0	0.10

Optimal design

Weight	Drug A	Drug B
0.17	0.10	0.10
0.14	0.10	2.87
0.20	0.10	10.0
0.14	2.48	0.10
0.14	3.66	5.64
0.21	10.0	0.10

Figure 7.11 Left panel: two-dimensional grid for the optimal design problem involving a combination therapy (dots) and initial design (open circles). Right panel: optimal design (closed circles)

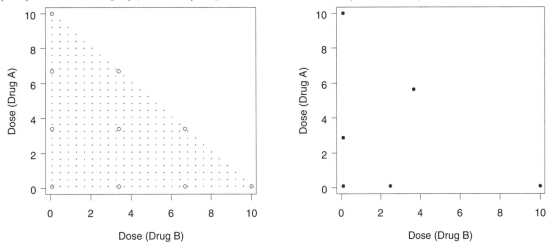

The grid (GRID variable) consists of 351 points in the design region. Figure 7.11 (left panel) depicts the selected grid (dots) and initial design (open circles). The parameter vector (PARAMETER variable) is given by

$$\theta = (1, 1, 1, 1, 1, 1, 1, 0.6, 0.4).$$

The information matrix and sensitivity function are defined using the %infoele, %indinfo, and %info macros. The complete version of Program 7.7 is given on the book's companion Web site.

Output 7.7 displays the initial and *D*-optimal designs. The right panel of Figure 7.11 provides a graphical summary of the optimal design (closed circles). It is worth noting that five of the six points in the optimal design lie on the boundary of the design region. The corresponding combination therapies involve the minimum dose (0.1 mg) of at least one of the two drugs. There is only one "non-trivial" combination with 3.7 mg of Drug A and 5.6 mg of Drug B.

7.7 Bivariate Probit Models for Correlated Binary Responses

In clinical trial analysis, the toxicity and efficacy responses usually occur together and it may be useful to assess them together. However, in practice the assessment of toxicity and efficacy is sometimes separated. For example, determining the maximum tolerated dose (MTD) is based on toxicity alone and then efficacy is evaluated in Phase II trials over the predetermined dose range. Obviously, the fact that the two responses from the same patient are correlated will introduce complexity into the analysis. But if we study these outcomes simultaneously, more information will be gained for future trials and treatment effects will be understood more thoroughly. In drug-response relationship, the correlation between efficacy and toxicity can be negative or positive depending on the therapeutical area. Two commonly used models, the Gumbel model (Kotz et al., 2000) and the bivariate binary Cox model (Cox, 1970), have been introduced to incorporate the two dependent outcomes, toxicity and efficacy, when both of them are dichotomous. In those two models, we need to model the probabilities of different outcome combinations separately. When both outcomes are binary, the total number of combinations is four, but when outcomes contain more than two categories, the number of unknown parameters may increase dramatically. Here we propose a bivariate probit model which incorporates the correlated responses naturally via the correlation structure of the underlying bivariate normal distribution; see Lesaffre and Molenberghs (1991). When the number of responses is more than two, the multivariate probit model may be used in a similar fashion.

Let $Y = \{0 \text{ or } 1\}$ denote the efficacy response and $U = \{0 \text{ or } 1\}$ denote the toxicity response in a clinical trial. Here 0 indicates no response and 1 indicates a response. Let d denote the dose of a drug and

$$p_{yu}(x) = P(Y = y, U = u|d = x), \quad y, u = 0 \text{ or } 1.$$

Assume that Z_1 and Z_2 follow bivariate normal distribution with zero mean and variance-covariance matrix

$$\Sigma = \begin{pmatrix} 1 & \rho \\ \rho & 1 \end{pmatrix},$$

where ρ may be interpreted as the correlation measure between toxicity and efficacy for the same patient. This set-up may be viewed as a standardization of the observed responses Y and U since under the natural scale the mean and variance vary from study to study. After simple transformation, the correlated responses follow the standard bivariate normal distribution,

$$p_{11} = F(\theta_1^T f_1, \theta_2^T f_2) = \int_{-\infty}^{\theta_1^T f_1} \int_{-\infty}^{\theta_2^T f_2} \frac{1}{2\pi|\Sigma|^{1/2}} \exp(-\frac{1}{2}Z^T \Sigma^{-1} Z) dz_1 dz_2,$$

where θ_1, θ_2 are unknown parameters and $f_1(x)$ and $f_2(x)$ contain the covariates of interest. In this section, we will study a simple linear model defined as follows (refer to Fedorov, Dragalin, and Wu, 2006).

$$\theta_1^T f_1 = \theta_{11} + \theta_{12}x, \quad \theta_2^T f_2 = \theta_{21} + \theta_{22}x.$$

In this case, the efficacy and toxicity response rates ($p_{1.}$ and $p_{.1}$, respectively) can be expressed as the marginals of the bivariate normal distribution,

$$p_{1.} = \Phi(\theta_1^T f_1), \quad p_{.1} = \Phi(\theta_2^T f_2),$$

where $\Phi(z)$ is the cumulative distribution function of the standard normal distribution. Note that p_{11}, $p_{1.}$, and $p_{.1}$ uniquely define the joint distribution of Y and U, i.e., $p_{10} = p_{1.} - p_{11}$, $p_{01} = p_{.1} - p_{11}$ and $p_{00} = 1 - p_{1.} - p_{.1} + p_{11}$.

Assume that the $\{y_i, u_i\}$'s are independent for different i's. Then the likelihood function for $\{Y, U\}$ is given by

$$L(Y, U|\theta) = \prod_{i=1}^{N} y_i u_i \log p_{11} + y_i(1 - u_i) \log p_{01} + (1 - y_i)(1 - u_i) \log p_{00}.$$

The information matrix of a single observation is

$$\begin{bmatrix} [C_1 C_2] \otimes f \\ \varphi_2 \quad -\varphi_2 \quad -\varphi_2 \end{bmatrix} (P - pp^T)^{-1} \begin{bmatrix} [C_1 C_2] \otimes f \\ \varphi_2 \quad -\varphi_2 \quad -\varphi_2 \end{bmatrix}^T,$$

where

$$C_1 = \begin{pmatrix} \varphi(\theta_1^T f) & 0 \\ 0 & \varphi(\theta_2^T f) \end{pmatrix}, C_2 = \begin{pmatrix} F(u_1) & 1 - F(u_1) & -F(u_1) \\ F(u_2) & -F(u_2) & 1 - F(u_2) \end{pmatrix},$$

$$u_1 = \frac{\theta_2^T f_2 - \rho \theta_1^T f_1}{\sqrt{1 - \rho^2}}, u_2 = \frac{\theta_1^T f_1 - \rho \theta_2^T f_2}{\sqrt{1 - \rho^2}},$$

$$P = \begin{pmatrix} p_{11} & 0 & 0 \\ 0 & p_{10} & 0 \\ 0 & 0 & p_{01} \end{pmatrix}, \quad p = (p_{11} \ p_{10} \ p_{01})^T,$$

$\varphi(v)$ denotes the probability density function of the standard normal distribution, $\varphi_2 = f(\theta_1^T f, \theta_2^T f, \rho)$ denotes the probability density function of bivariate normal distribution with mean $\theta_1^T f_1$ and $\theta_2^T f_2$, variance 1 and correlation coefficient ρ.

7.7.1   Example 8: A Bivariate Probit Model for Correlated Binary Responses (Efficacy and Toxicity)

Consider the problem of constructing the D-optimal design for the following model:

$$\theta_1^T f_1 = -0.9 + 10x, \quad \theta_2^T f_2 = -1.2 + 1.6x, \quad \rho = 0.5.$$

Here x denotes the dose of the experimental drug with the dose range given by $\mathcal{X} = [0, 1]$. Figure 7.12 depicts the response probabilities for

$$\theta = (-0.9, 10, -1.2, 1.6)^T \text{ and } \rho = 0.5.$$

This setting represents a scenario which is often encountered in clinical trial applications. The probabilities of efficacy and toxicity responses, $p_{.1}$ and $p_{1.}$, are both increasing as the dose increases. On the other hand, probability of "positive" response p_{10} (i.e., probability of positive efficacy and no toxicity) increases at the beginning. Then, at a certain dose level it begins to decrease (the probability of having a positive efficacy response without any side effects is low at high dose levels).

Figure 7.12 Bivariate probit model, probability of positive response (solid line), probability of efficacy response (dashed line), and probability of toxicity response (dotted line)

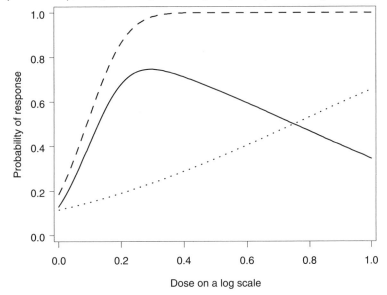

Program 7.8 constructs the D-optimal design for the described bivariate probit model by calling the %OptimalDesign4 macro. The initial design includes four equally spaced doses in the dose range $\mathcal{X} = [0, 1]$ (POINTS variable), and the doses are assumed to be equally weighted (WEIGHTS variable). The grid consists of 401 equally spaced points (GRID variable), and the PARAMETER variable specifies the true values of θ and ρ. The information matrix and sensitivity function are computed using the %infoele, %info, and %infod macros. The complete version of Program 7.8 is provided on the book's companion Web site.

Program 7.8 *D*-optimal design for the bivariate probit model with correlated binary responses (Design and algorithm parameters)

```
* Design parameters;
%let points={0 0.333 0.667 1};
%let weights={0.25 0.25 0.25 0.25};
%let grid=do(0,1,1/400);
%let parameter={-0.9 10 -1.2 1.6 0.5};
* Number of parameters;
%let paran=5;
* Algorithm parameters;
%let convc=1e-9;
%let maximit=1000;
%let const1=1;
%let const2=1;
%let cmerge=5;
%OptimalDesign4;
```

Output from Program 7.8

```
                        Initial design

                    Weight          X

                    0.250       0.000
                    0.250       0.333
                    0.250       0.667
                    0.250       1.000

                    Optimal design

                    Weight          X

                    0.451       0.000
                    0.356       0.180
                    0.194       1.000
```

Output 7.8 displays the initial and *D*-optimal designs. The locally *D*-optimal design is a three-point design with unequal weights. Two optimal doses are on the boundaries of the dose range ($x_1^* = 0$, $x_2^* = 1$) and the other optimal dose lies in the middle ($x_3^* = 0.18$). Almost half of the patients are assigned to the lowest dose and about 20% of patients are assigned to the highest dose. Figure 7.13 depicts the initial and optimal designs as well as the sensitivity funtions.

7.8 Pharmacokinetic Models with Multiple Measurements per Patient

In this section we discuss a clinical pharmacokinetic (PK) study where multiple blood samples are taken for each enrolled patient. This setup leads to nonlinear mixed effects regression models with multiple responses.

7.8.1 Two-Compartment Pharmacokinetic Model

Consider a clinical trial in which the drug was administered as a bolus input D_0 at the beginning of the study (at time $x = 0$) and then bolus inputs D_i were administered at 12- or 24-hour intervals until 72 hours after the first dose. To describe the concentration of

Figure 7.13 Left panel: Initial (open circles) and optimal (closed circles) designs. Right panel: Sensitivity functions for the initial (dashed curve) and optimal (solid curve) designs.

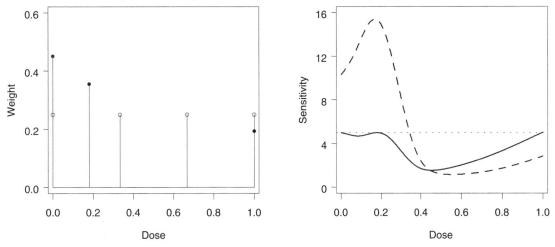

drug at time x, a two-compartment model was used with standard parameterization (volume of distribution V, transfer rate constants K_{12}, K_{21}, and K_{10}), see Figure 7.14.

Figure 7.14 Diagram of the two-compartment model

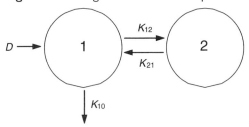

Measurements of drug concentration y_{ij} were taken from the central compartment (compartment 1 in Fig. 7.14) at times x_j for each patient,

$$y_{ij} = \frac{g_1(x_j)}{V_i} + z_{ij}, \quad i = 1, \dots, N, \quad j = 1, \dots, k, \tag{7.18}$$

$g_1(x_j) = g_1(x_j, \gamma_i)$ is the amount of the drug in the central compartment at time x_j for patient i, $\gamma_i = (V_i, K_{12,i}, K_{21,i}, K_{10,i})$ are individual PK parameters of patient i, and z_{ij} are measurement errors. The g_1 function and population model are discussed in more detail below.

In the trial under consideration, $k = 16$ samples were taken from each patient at times $x_j \in \mathcal{X}$,

$$x_j \in \mathcal{X} = \{5, 15, 30, 45 \text{ min}; 1, 2, 3, 4, 5, 6, 12, 24, 36, 48, 72, 144 \text{ h}\}.$$

Obviously, each extra sample provides additional information and increases the precision of parameter estimates. However, the number of samples that may be drawn from each patient is restricted because of blood volume limitations and other logistical and ethical reasons. Moreover, the analysis of each sample is associated with monetary costs. Therefore, it is reasonable to take the cost of drawing samples into account. If $X = (x_1, ..., x_k)$ is a collection of sampling times for a particular patient, we can consider the cost of the sampling sequence X (denoted by $c(X)$).

In this section we will consider the following questions:

1. Given restrictions on the number of samples for each patient, what are the optimal sampling times (i.e., how many samples and at which times)?

2. If not all 16 samples are taken, what would be the loss of information/precision?

3. If costs are taken into account, what is the optimal sampling scheme? Cost-based designs will be discussed in detail in Section 7.9.

By an *optimal sampling scheme* we mean a sequence which allows us to obtain the best precision of parameter estimates in terms of their variance-covariance matrix and the selected optimality criterion. First, we need to define the regression model with multiple responses, see (7.11). In this and next sections, X will denote a k-dimensional vector of sampling times and Y will be a k-dimensional vector of measurements at times X.

7.8.2   Regression Model

A two-compartment model described by the following system of linear differential equations was used to model the data:

$$\begin{cases} \dot{q}_1(x) = & -(K_{12} + K_{10})q_1(x) & +K_{21}q_2(x) \\ \dot{q}_2(x) = & +K_{12}q_1(x) & -K_{21}q_2(x), \end{cases} \tag{7.19}$$

for $x \in [t_i, t_{i+1})$ with initial conditions

$$q_1(t_i) = q_1(t_i - 0) + D_i, \quad q_1(0) = D_0, \quad q_2(0) = 0,$$

where $q_1(x)$ and $q_2(x)$ are amounts of the drug at time x in the central and peripheral compartments, respectively; t_i is a time of i-th bolus input, $t_0=0$; D_i is the amount of the drug administered at t_i.

The solution of the system (7.19) is a sum of exponential functions which depend on parameters (Gibaldi and Perrier, 1982, Appendix B):

$$\gamma = (V, K_{12}, K_{21}, K_{10})^T, \quad q(x) = [q_1(x, \gamma), q_2(x, \gamma)].$$

In population modeling it is often assumed that the γ_i parameters for patient i are independently sampled from the normal population with

$$E(\gamma_i) = \gamma^0, \text{Var}(\gamma_i) = \Lambda, \quad i = 1, \dots, N, \tag{7.20}$$

where the m_γ-dimensional vector γ^0 and $(m_\gamma \times m_\gamma)$ matrix Λ are usually referred to as the "population", or "global", parameters; see Gagnon and Leonov (2005) for a discussion on other population distributions.

To fit the model, it was assumed that the variance of the error term z_{ij} in (7.18) depends on the model parameters through the mean response,

$$z_{ij} = \varepsilon_{1,ij} + \varepsilon_{2,ij} \frac{q_1(x_j)}{V_i}, \tag{7.21}$$

where $\varepsilon_{1,ij}$ and $\varepsilon_{2,ij}$ are independent, identically distributed random variables with zero mean and variance σ_1^2 and σ_2^2, respectively.

Let $\theta = (\gamma^0, \Lambda, \sigma_1^2, \sigma_2^2)^T$ denote a combined vector of model parameters and k_i denote the number of samples taken for patient i. Let

$$X_i = (x_{1i}, x_{2i}, \dots, x_{k_i,i})^T \text{ and } Y_i = (y_{1i}, y_{2i}, \dots, y_{k_i,i})^T$$

be sampling times and measured concentrations, respectively, for patient i. Further, $\eta(x_{ji}, \gamma_i) = q_1(x_{ji}, \gamma_i)/V_i$, $\eta(x_{ji}, \theta) = q_1(x_{ji}, \gamma^0)/V$ and

$$\eta(X_i, \theta) = [\eta(x_{1i}, \theta), \ldots, \eta(x_{k_i, i}, \theta)]^T,$$

where $j = 1, 2, \ldots, k_i$.

If F is a $(k_i \times m_\gamma)$ matrix of partial derivatives of function $\eta(X_i, \theta)$ with respect to parameters γ_0,

$$F = F(X_i, \gamma^0) = \left[\frac{\partial \eta(X_i, \theta)}{\partial \gamma_\alpha} \right] \Big|_{\gamma = \gamma^0},$$

then, using the first order approximation together with (7.20) and (7.21), one obtains the following approximation of the variance-covariance matrix $S(X_i, \theta)$ for Y_i,

$$S(X_i, \theta) \simeq F \Lambda F^T + \sigma_2^2 \text{Diag}[\eta(X_i, \theta)\eta^T(X_i, \theta) + F\Lambda F^T] + \sigma_1^2 I_{k_i}, \tag{7.22}$$

where $\text{Diag}(A)$ denotes a diagonal matrix with elements a_{ii} on the diagonal; see also Fedorov, Gagnon and Leonov (2002; Section 4). If the γ parameters are assumed to be log-normally distributed, then Λ on the right-hand side of (7.22) has to be replaced with

$$\Lambda_1 = \text{Diag}(\gamma^0)\Lambda \text{Diag}(\gamma^0),$$

for details, see Gagnon and Leonov (2005). Therefore, for any sequence X_i, the response vector $\eta(X_i, \theta)$ and variance-covariance matrix $S(X_i, \theta)$ are defined, and we can use general formula (7.12) to calculate the information matrix $\mu(X_i, \theta)$.

7.8.3   Example 9: Pharmacokinetic Model without Cost Constraints

Consider a problem of selecting an optimal sampling scheme with a single bolus input at time $x = 0$. SAS code for generating optimal designs for the general case of multiple bolus inputs is available (Gagnon and Leonov, 2005), but the complexity of this code is beyond the scope of the chapter.

To fit the data, the K_{12} and K_{21} parameters were treated as constants ($K_{12} = 0.400$, $K_{21} = 0.345$), i.e., they were not accounted for while calculating the information matrix $\mu(x, \theta)$. Therefore, the combined vector of unknown parameters θ was defined as

$$\theta = \left(\gamma_1^0, \gamma_2^0; \lambda_{11}, \lambda_{22}, \lambda_{12}; \sigma_1^2, \sigma_2^2 \right),$$

where

$$\gamma_1^0 = CL, \quad \gamma_2^0 = V, \quad \lambda_{11} = \text{Var}(CL), \quad \lambda_{12} = \text{Cov}(CL, V), \quad \lambda_{22} = \text{Var}(V),$$

CL is the plasma clearance, and $K_{10} = CL/V$. The total number of unknown parameters is $m = 7$.

To construct locally optimal designs, we obtained the parameter estimates $\hat{\theta}$ using the NONMEM software (Beal and Sheiner, 1992) based on 16 samples (from the design region $\mathcal{X}$) for each of 27 subjects. The resulting estimate is given by

$$\hat{\theta} = (0.211, 5.50; 0.0365, 0.0949, 0.0443; 8060, 0.0213)^T.$$

Drug concentrations were expressed in μg/L, elimination rate constant K_{10} in L/h and volume V in L.

The PK modeling problem considered in this section greatly increases the complexity of the SAS implementation of D-optimal designs. Program 7.9 relies on a series of SAS macros and SAS/IML modules to solve the optimal selection problem formulated earlier in this section. Referring to Section 7.1.7, Components 1, 2, and 3 of the general optimal design algorithm require extensive changes. These changes are summarized below.

- In order to generate the grid, the set of sampling times $\mathcal{X}$ is defined in the CAND data set and the number of time points r in the final design is specified in the KS data set (in this case, we are looking for 8-point designs). Note that r ranges between 1 and k, where k is the total number of sampling times ($k = 16$ in this case). Any number of sets of size r can be applied to the optimization algorithm; however, for designs with the standard information matrix normalization, the design with the largest r will be D-optimal. This will not (necessarily) be true for cost-normalized designs (see Section 7.9).

- The problem of selecting candidate points is a $C(k, r)$ problem, where $C(k, r)$ is a binomial coefficient, $C(k, r) = k!/[r!(k - r)!]$. Sets of size r candidate points can become very large. For example, for $k = 16$ and $r = 8$, the number of possible 8-point sampling schemes is $C(16, 8) = 12,870$ which means that 12,870 information matrices need to be obtained. Note that in this example $m = 7$ and thus each information matrix is a 7×7 matrix. In order to store the information matrices for all possible 8-point designs, SAS needs to store $12,870 \times 7 = 90,090$ rows (and 7 columns) of data. Hence the data storage requirements for large designs can become formidable. Once the CAND and KS data sets are specified, Program 7.9 will generate the set of possible candidate points using the SAS/IML module TS. Also, the initial design is specified in the SAMPLE data set (note that all points in the initial design must be elements of the design region $\mathcal{X}$ and must be of length r, as specified in the KS dataset).

- The set of parameters has been defined as $\theta = (\gamma_1^0, \gamma_2^0, \lambda_{11}, \lambda_{22}, \lambda_{12}, \sigma_1^2, \sigma_2^2)^T$. Since we are dealing with a mixed effects model, the γ and λ parameters in the vector θ must be identified. This is accomplished by ordering the parameters in a specific way. In general, the parameters in θ *must* be entered as follows: vector γ, vector λ, and residual variance(s). The number of components in γ is defined by the NF macro variable. The ordering of the random effects (the λ's) is also highly important. Parameters along the main diagonal of Λ should be entered first: $\lambda_{11}, \lambda_{22}$. The off-diagonal element (λ_{12}) should be entered next. The Λ matrix is set up using the SAS/IML module LSETUP. In the general case, when Λ is a $b \times b$ matrix, the components of Λ need to be ordered as follows: diagonal elements, $\lambda_{11}, \lambda_{22}, \ldots, \lambda_{bb}$, followed by off-diagonal elements $\lambda_{12}, \lambda_{13}, \ldots, \lambda_{1b}, \lambda_{23}, \lambda_{24}, \lambda_{2b}, \ldots$, etc.

- The PK function, $\eta(x, \theta)$, derived from (7.19), is specified using the SAS/IML module FPARA. The ordering of the γ's in the vector of parameters must match the ordering in the FPARA module. As shown in Program 7.9, CL (plasma clearance) should be entered into θ first, and V (volume of distribution) should be entered second. In the current version, the equations in (7.19) must be solved, and the solution must be entered in the FPARA module.

- A computationally difficult issue is the calculation of derivatives for (7.12). Note that first- and second-order partial derivatives are required; see equations (7.12), (7.22). In the previous examples derivatives were obtained by either entering the analytical form or by using the NLPFDD function in SAS/IML. Here, we find it more practical to calculate the derivatives numerically using the following approximation

$$\partial\eta(x, \theta) = \frac{\eta(x, \theta + h) - \eta(x, \theta - h)}{2h},$$

where the value of h is a macro variable and is entered by the user ($h = 0.001$ in this example). The information matrix from (7.12) is computed via the SAS/IML module MOLD, which calls six other modules (HIGHERD, HIGHERM, FPLUS, FMINUS, FTHETA and JACOB2).

The complete SAS code for this example is provided on the book's companion Web site.

Program 7.9 *D*-optimal design for the pharmacokinetic model (Design and algorithm parameters)

```
* Design parameters;
%let h=0.001;  * Delta for finite difference derivative approximation;
%let paran=7;  * Number of parameters in the model;
%let nf=2;     * Number of fixed effect parameters;
%let cost=1;   * Cost function (1, no cost function,
                             2, user-specified function);
* Algorithm parameters;
%let convc=1e-9;
%let maximit=1000;
%let const1=2;
%let const2=1;
%let cmerge=5;
* PK parameters;
data para;
    input CL V vCL vV covCLV m s;
    datalines;
    0.211 5.50 0.0365 0.0949 0.0443 0.0213 8060
    ;
* All candidate points;
data cand;
    input x @@;
    datalines;
    0.083 0.25 0.5 0.75 1 2 3 4 5 6 12 24 36 48 72 144
    ;
* Number of time points in the final design;
data ks;
    input r @@;
    datalines;
    8
    ;
* Initial design;
data sample;
    input x1 x2 x3 x4 x5 x6 x7 x8 w @@;
    datalines;
    0.083 0.5 1 4 12 24 72 144 1.0
    ;
    run;
```

Output from Program 7.9

```
            Determinant of the covariance matrix D (initial design)

                                  IDED

                                0.0000136

             Determinant of the covariance matrix D (final design)

                                  DETD

                                8.4959E-6

                              Optimal design

       COL1    COL2    COL3    COL4    COL5    COL6    COL7    COL8

      0.083    0.25    0.5    0.75     36      48      72      144
```

Output 7.9 lists the optimal sampling times included in a D-optimal 8-point design (5 min, 15 min, 30 min, 45 min, 36 h, 48 h, 72 h and 144 h). The optimal design differs from the initial design in that it puts more weight on earlier time points; e.g., the optimal design includes the 15-min and 45-min which were not in the initial design. Note also that an application of the D-optimal algorithm reduced the determinant of the optimal variance-covariance matrix to 8.49×10^{-6} (from 1.36×10^{-5}).

7.9 Models with Cost Constraints

Traditionally, when normalized designs are discussed, the normalization factor is equal to the number of experiments N; see (7.3). In this section we will consider cost-normalized designs. Each measurement at point X_i is assumed to be associated with a cost $c(X_i)$, and a restriction on the total cost is given by

$$\sum_{i=1}^{n} n_i c(X_i) \leq C.$$

In this case it is quite natural to normalize the information matrix by the total cost C and introduce

$$M_C(\xi, \theta) = \frac{M_N(\theta)}{C} = \sum_i w_i \tilde{\mu}(X_i, \theta), \text{ with } w_i = \frac{n_i c(X_i)}{C}, \tilde{\mu}(X, \theta) = \frac{\mu(X, \theta)}{c(X)}. \quad (7.23)$$

Once the cost function $c(X)$ is defined, we can introduce a cost-based design $\xi_C = \{w_i, X_i\}$ and use exactly the same techniques of constructing continuous optimal designs for various optimality criteria as described in the previous sections,

$$\xi^* = \arg\min_{\xi} \Psi[D_C(\xi, \theta)], \text{ where } D_C(\xi, \theta) = M_C^{-1}(\xi, \theta) \text{ and } \xi = \{w_i, X_i\}.$$

As usual, to obtain counts n_i, values $\tilde{n}_i = w_i C / c(X_i)$ have to be rounded to the nearest integers n_i subject to $\sum_i n_i c(X_i) \leq C$.

We believe that the introduction of cost constraints and the alternative normalization (7.23) makes the construction of optimal designs for models with multiple responses more sound. It allows for a meaningful comparison of "points" X with distinct number of responses.

7.9.1   Example 10: Pharmacokinetic Model with Cost Constraints

The example considered in this section is the example discussed in Section 7.8.3 with cost functions added. Program 7.10 computes a D-optimal design by invoking the same SAS macros and SAS/IML modules as Program 7.9. The computational tools for calculating the cost-normalized information matrices are built into the macros and modules; we need only to specify the cost function and set the COST variable to 2 (to indicate that a user-defined cost function will be provided).

The cost function must be entered by the user into the SAS/IML module MCOST. In Program 7.10, we select a linear cost function with two components, the CV variable is associated with an overall cost of the study (i.e., the cost of a patient visit), and CS is the cost of obtaining/analyzing a single sample:

```
* Cost module;
start mcost(t);
   * t is vector sampling times;
   Cv = 1;
   Cs = 0.3;
   C = Cv + sum(Cs[1:nrow(t[loc(t>0)])]);
   return(C);
finish(mcost);
```

In general, the cost function can have almost any form.

The design region is defined as follows

$$X_1 = \left\{ X = (t_{i_1}, \ldots, t_{i_r}), \quad t_{i_j} \in \mathcal{X}, \quad j = 1, 2, \ldots, r, \quad 3 \leq r \leq 5 \right\},$$

that is, we allow any combination of r sampling times for each patient from the original sequence $\mathcal{X}, 3 \leq r \leq 5$. The values of r are specified in the KS data set. Note that a total of 6,748 information matrices (47,236 rows and 7 columns of data) are calculated and stored in this example.

Program 7.10 D-optimal design for the pharmacokinetic model with cost constraints

```
* Design parameters;
%let h=0.001;   * Delta for finite difference derivative approximation;
%let paran=7;   * Number of parameters in the model;
%let nf=2;      * Number of fixed effect parameters;
%let cost=2;    * Cost function (1, no cost function, 2, user-specified function);
* Algorithm parameters;
%let convc=1e-9;
%let maximit=1000;
%let const1=2;
%let const2=1;
%let cmerge=5;
* PK parameters;
data para;
    input CL V vCL vV covCLV m s;
    datalines;
    0.211 5.50 0.0365 0.0949 0.0443 0.0213 8060
    ;
* All candidate points;
data cand;
    input x @@;
    datalines;
    0.083 0.25 0.5 0.75 1 2 3 4 5 6 12 24 36 48 72 144
    ;
```

```
* Number of time points in the final design;
data ks;
    input r @@;
    datalines;
    3 4 5
    ;
* Initial design;
data sample;
    input x1 x2 x3 x4 x5 w @@;
    datalines;
    0.083 0.5 4 24 144 1.0
    ;
    run;
```

Output from Program 7.10

```
          Determinant of the covariance matrix D (initial design)

                                IDED

                              0.231426

          Determinant of the covariance matrix D (optimal design)

                                DETD

                             0.0209139

                            Optimal design

          Obs     COL1     COL2     COL3     COL4     COL5

           1      0.083    0.25      48       72      144
           2      0.083    0.25      72      144

                           Optimal weights

                          Obs       W

                           1      0.59125
                           2      0.40875
```

Output 7.10 displays the D-optimal design based on the linear cost funtion defined in the MCOST module. The D-optimal design is a collection of two sampling sequences,

$$\xi^* = \{X^{*1} = (5, 15 \text{ min}, \ 48, 72, 144 \text{ h}), \quad X^{*2} = (5, 15 \text{ min}, \ 72, 144 \text{ h})\},$$

with weights $w_1 = 0.59$ and $w_2 = 0.41$, respectively. This example shows that once costs are taken into account, sampling sequences with a smaller number of samples may become optimal.

7.10 Summary

Design optimization is a critical aspect of experimentation; this is particularly true in pharmaceutical applications. There are many barriers to routine application of optimal design theory. Among these is the lack of software. For the mathematically simplest cases, namely linear models where the parameters are independent of the variance function, many software packages are available for finding optimal designs (see Section 7.1.6). However, as models increase in complexity, the availability of software for solving optimal design problems drops off dramatically and pharmaceutical researchers often find it frustrating. Some academic labs have created software that supports selected optimal designs (see Section 7.1.6) but a general software package accessible to researchers in the pharmaceutical industry is not currently available. To this end, we have created a series of powerful modules that are based on SAS software for constructing optimal experimental designs for a large number of popular non-linear models. Although we focus on *D*-optimal designs, the same algorithm can be used for constructing other types of optimal designs, in particular *A*-optimal designs after minor changes.

In this chapter, we provide a brief introduction to optimal design theory, and we also discuss optimal experimental designs for nonlinear models arising in various pharmaceutical applications. We provide several examples, and the SAS code for executing these examples is included. The examples increase in complexity as the chapter moves forward. The first example generates *D*-optimal designs for quantal models; subsequent examples, in order, generate *D*-optimal designs for continuous logistic models, logistic regression models with unknown parameters in the variance function, the beta regression model, models with binary response, bivariate probit models for correlated binary responses, and pharmacokinetic models with multiple measurements per patient, with or without cost constraints. The most important hurdle in contructing these designs is the computation of the Fisher information matrix, and storage/retrieval of these matrices and associated design points. As designs increase in complexity, computation time is increased in order to handle the necessary calculations and storage/retrieval.

References

Abramowitz, M., Stegun, I. A. (1972). *Handbook of Mathematical Functions with Formulas, Graphs and Mathematical Tables*. 9th edition. New York: Dover.

Agresti, A. (1990). *Categorical Data Analysis*. New York: Wiley.

Anderson, J.A. (1984). "Regression and ordered categorical variables." *Journal of Royal Statistical Society. Series B*. 46, 1–30.

Ashby, D., Pocock, S.J., Shaper, A.G. (1986). "Ordered polytomous regression: An example relating serum biochemistry and haematology to alcohol consumption." *Applied Statistics*. 35, 289–301.

Atkinson, A.C., Cook, R.D. (1995). "*D*-optimum designs for heteroscedastic linear models." *Journal of the American Statistical Association*. 90, 204–212.

Atkinson, A.C., Donev, A.N. (1992). *Optimum Experimental Design*. Oxford: Clarendon Press.

Atkinson, A.C., Fedorov, V.V. (1988). "The optimum design of experiments in the presence of uncontrolled variability and prior information." In *Optimal Design and Analysis of Experiments*. Dodge, Y., Fedorov, V.V., Wynn, H. (editors). Amsterdam: North-Holland. 327–344.

Beal, S.L., Sheiner, L.B. (1992). *NONMEM User's Guide*. NONMEM Project Group. San Francisco: University of California.

Bezeau, M., Endrenyi, L. (1986). "Design of experiments for the precise estimation of dose-response parameters: the Hill equation." *Journal of Theoretical Biology*. 123, 415–430.

Box, G.E.P., Hunter, W.G. (1965). "Sequential design of experiments for nonlinear models." *Proceedings of IBM Scientific Computing Symposium*. Korth, J.J. (editor). White Plains, New York: IBM. 113–137.

Chaloner, K., Verdinelli, I. (1995). "Bayesian experimental design: a review." *Statistical Science*. 10, 273–304.

Chernoff, H. (1953). "Locally optimal designs for estimating parameters." *The Annals of Mathematical Statistics*. 24, 586–602.

Cox, D.R. (1970). *The Analysis of Binary Data*. London: Chapman and Hall.

Cramer, J.S. (1991). *The LOGIT Model: An Introduction for Economists.* London: Edward Arnold.

D'Argenio, D.Z., Schumitzky, A. (1997). *ADAPT II Users Guide: Pharmacokinetic/Pharmacodynamic Systems Analysis Software.* Biomedical Simulations Resource. Los Angeles: University of Southern California.

Downing, D.J., Fedorov, V.V., Leonov, S.L. (2001). "Extracting information from the variance function: optimal design." In *MODA6–Advances in Model-Oriented Design and Analysis.* Atkinson, A.C., Hackl, P., Müller, W.G. (editors). Heidelberg: Physica-Verlag. 45–52.

Dragalin, V., Fedorov V. (2006). "Adaptive designs for dose-finding based on efficacy-toxicity response," *Journal of Statistical Planning and Inference.* 136, 1800–1823.

Dragalin, V., Fedorov, V.V., Wu, Y. (2006). "Optimal design for bivariate probit model." GSK BDS Technical Report 2006-01. Collegeville, PA: GlaxoSmithKline. http://www.biometrics.com/downloads/TR_2006_01.pdf

Fan, S.K, Chaloner, K. (2001). "Optimal design for a continuation-ratio model." In *MODA6–Advances in Model-Oriented Design and Analysis.* Atkinson, A.C., Hackl, P., Müller, W.G. (editors). Heidelberg: Physica-Verlag. 77–85.

Fedorov, V.V. (1972). *Theory of Optimal Experiment.* New York: Academic Press.

Fedorov, V.V., Gagnon, R.C., Leonov, S.L. (2002). "Design of experiments with unknown parameters in variance." *Applied Stochastic Models in Business and Industry.* 18, 207–218.

Fedorov, V.V., Hackl, P. (1997). *Model-Oriented Design of Experiments.* New York: Springer-Verlag.

Fedorov, V.V., Leonov, S.L. (2001). "Optimal design of dose response experiments: a model-oriented approach." *Drug Information Journal.* 35, 1373–1383.

Fedorov, V., Leonov, S., (2005). "Response-driven designs in drug development." In *Applied Optimal Designs.* Wong, W.K., Berger, M. (editors). Chichester: Wiley. 103–136.

Finney, D.J. (1971). *Probit Analysis.* Third Edition. Cambridge: Cambridge University Press.

Finney, D.J. (1976). "Radioligand assay." *Biometrics.* 32, 721–740.

Ford, I., Torsney, B., Wu, C.F.J. (1992). "The use of a canonical form in the construction of locally optimal designs for non-linear problems." *Journal of the Royal Statistical Society. Series B.* 54, 569–583.

Gagnon, R., Leonov, S. (2005). "Optimal population designs for PK models with serial sampling." *Journal of Biopharmaceutical Statistics.* 15, 143–163.

Gibaldi, M., Perrier, D. (1982). *Pharmacokinetics.* Second edition. New York: Marcel Dekker.

Hedayat, A.S., Yan, B., Pezutto, J.M. (1997). "Modeling and identifying optimum designs for fitting dose-response curves based on raw optical density data." *Journal of American Statistical Association.* 92, 1132–1140.

Heise, M.A., Myers, R.H. (1996). "Optimal designs for bivariate logistic regression." *Biometrics.* 52, 613–624.

Hill, P.D.H. (1980). "*D*-optimal designs for partially nonlinear regression models." *Technometrics.* 22, 275–276.

Källén, A., Larsson, P. (1999). "Dose response studies: how do we make them conclusive?" *Statistics in Medicine.* 18, 629–641.

Karpinski, K.F. (1990). "Optimality assessment in the enzyme-linked immunosorbent assay (ELISA)." *Biometrics.* 46, 381–390.

Kiefer, J., Wolfowitz, J. (1960). "The equivalence of two extremum problems." *Canadian Journal of Mathematics.* 12, 363–366.

Kotz, S., Balakrishnan, N., Johnson, N.L. (2000). *Continuous Multivariate Distributions.* Second Edition. Volume 1. New York: Wiley.

Lesaffre, E., Molenberghs, G. (1991). "Multivariate probit analysis: A neglected procedure in medical statistics." *Statistics in Medicine.* 10, 1391–1403.

Magnus, J.R., Neudecker, H. (1988). *Matrix Differential Calculus with Applications in Statistics and Econometrics.* New York: Wiley.

McCullagh, P., Nelder, J. (1989). *Generalized Linear Models.* Second edition. New York: Chapman and Hall.

Mentré, F., Mallet, A., Baccar, D. (1997). "Optimal design in random-effects regression models." *Biometrika.* 84, 429–442.

Muirhead, R. (1982). *Aspects of Multivariate Statistical Theory.* New York: Wiley.

Ogungbenro, K., Graham, G., Gueorguieva, I., Aarons, L. (2005). "The use of a modified Fedorov exchange algorithm to optimise sampling times for population pharmacokinetic experiments." *Computer Methods and Programs in Biomedicine.* 80, 115–125.

Pukelsheim, F. (1993). *Optimal Design of Experiments.* New York: Wiley.

Rao, C.R. (1973). *Linear Statistical Inference and Its Applications.* Second edition. New York: Wiley.

Retout, S., Mentré, F. (2003). "Optimization of individual and population designs using S-Plus." *Journal of Pharmacokinetics and Pharmacodynamics.* 30, 417–443.

Silvey, S.D. (1980). *Optimal Design.* London: Chapman and Hall.

Sitter, R.R., Wu, C.F.J. (1993). "Optimal designs for binary response experiments: Fieller, *D*, and *A* criteria." *Scandinavian Journal of Statistics.* 20, 329–341.

Torsney, B., Musrati, A.K. (1993). "On the construction of optimal designs with applications to binary response and to weighted regression models." In *Model-Oriented Data Analysis.* Müller, W.G., Wynn, H.P., Zhigljavsky, A.A. (editors). Heidelberg: Physica-Verlag. 37–52.

White, L.V. (1973). "An extension of the general equivalence theorem to nonlinear models." *Biometrika.* 60, 345–348.

White, L.V. (1975). *The optimal design of experiments for estimation in nonlinear model.* Ph. D. thesis, University of London.

Wu, C.F.J. (1988). "Optimal design for percentile estimation of a quantal-response curve." In *Optimal Design and Analysis of Experiments.* Dodge, Y., Fedorov, V.V., Wynn, H.P. (editors). North Holland: Elsevier. 213–223.

Wu, Y., Fedorov, V.V., Propert, K. J. (2005). "Optimal design for dose response using beta distributed responses." *Journal of Biopharmaceutical Statistics.* 15, 753–771.

Wynn, H.P. (1970). "The sequential generation of *D*-optimal experimental designs." *Annals of Mathematical Statistics.* 41, 1055–1064.

Zacks, S. (1996). "Adaptive designs for parametric models." In *Design and Analysis of Experiments. Handbook of Statistics.* Volume 13. Ghosh, S., Rao, C.R. (editors). Amsterdam: Elsevier. 151–180.

Chapter 8

Analysis of Human Pharmacokinetic Data

Scott Patterson
Brian Smith

This chapter introduces pharmacokinetic terminology and also describes the collection, measurement, and statistical assessment of pharmacokinetic data obtained in Phase I and clinical pharmacology studies. It focuses on the following two topics that play a key role in pharmacokinetic analysis: testing for bioequivalence and assessing dose linearity. The chapter establishes a basic framework, and the examples demonstrate how to use restricted maximum likelihood models in the MIXED procedure to examine the most common pharmacokinetic data of practical interest.

8.1   Introduction

When a tablet of drug is taken orally, in general, it reaches the stomach and begins to disintegrate and is absorbed (A). When dissolved into solution in the stomach acid, the drug is passed on to the small intestine (Rowland and Tozer, 1980). At this point, some of the drug will pass through and be eliminated (E) from the body. Some will be metabolized (M) into a different substance in the intestine, and some drug will be picked up by the walls of the intestine and distributed (D) into the body. This last bit of drug substance passes through the liver first, where it is also often metabolized (M). The drug substance that remains then passes through the liver and reaches the bloodstream, where it is circulated throughout the body.

Pharmacokinetics (PK) is the study of absorption, distribution, metabolism, and excretion (ADME) properties (Atkinson et al., 2001). Following oral administration, the drug is held to undergo these four stages prior to being completely eliminated from the body. PK is also the study of what the body does to a drug, (as opposed to what a drug does to the body).

Measurement is central to PK, and the most common method is to measure by means of blood sampling how much drug substance has been put into the body (i.e., dose) relative to

---

Scott Patterson is Director, Statistical Sciences, GlaxoSmithKline Pharmaceuticals, USA. Brian Smith is Principal Biostatistician II, Medical Sciences Biostatistics, Amgen, USA.

how much drug reached the systemic circulation. For bioequivalence trials (U.S. Food and Drug Administration (FDA), 2003), at least 12 samples are collected over time following dosing with an additional sample collected prior to dosing.

As the drug is absorbed and distributed, the plasma concentration rises and reaches a maximum (called the C_{max} or maximum concentration). Plasma levels then decline until the body completely eliminates the drug from the body. The overall exposure to drug is measured by computing the area under the plasma concentration curve (AUC). AUC is derived in general by computing the area under each time interval and adding up all the areas. Other common summary measures are:

- T_{max} (time of maximum concentration),
- $T_{1/2}$ (half-life of drug substance)

More details of techniques used in the derivation of AUC may be found in Yeh and Kwan (1978). C_{max} and T_{max} are derived by inspection of the data, and $T_{1/2}$ is estimated by linear regression in accordance with the description of Rowland and Tozer (1980).

In this chapter, we will concentrate on the modeling of AUC and C_{max} using PROC MIXED (note that, in general, the methods described are applicable to other summary measures). It is acknowledged that AUC and C_{max} are log-normally distributed (Crow and Shimizu, 1988) for the purposes of this discussion in accordance with the findings of Westlake (1986), Lacey (1995), Lacey et al. (1997), and Julious and Debarnot (2000). Thus AUC and C_{max} data are analyzed under ln (natural-logarithmic) transformation (Box and Cox, 1964).

Analysis of ln-transformed AUC and C_{max} data in clinical pharmacology follows the general principles of analysis of repeated-measures, cross-over data using mixed effect linear models (Jones and Kenward, 2003; Wellek, 2003; Senn, 2002; Chow and Liu, 2000; Vonesh and Chinchilli, 1997; Milliken and Johnson, 1992). To illustrate this assumption, consider a study with n subjects and p periods where an observation y (either AUC or C_{max}, etc.) is observed for each subject in each period. Y may be expressed as a pn-dimensional response vector. Then, in matrix notation, this model can be expressed as

$$Y = X\beta + Z_1 u + Z_2 e,$$

where $X\beta$ is the usual design and fixed effects matrix, u is multivariate normal with expectation 0 and variance-covariance matrix Ω, i.e.,

$$u \sim MVN(0, \Omega),$$

and

$$e \sim MVN(0, \Lambda),$$

and u is independent of e. Subjects are assumed to be independent, and Z_1 and Z_2 are design matrices used to construct the variance-covariance structure as appropriate to the study design and desired variance components. The RANDOM and REPEATED statements in PROC MIXED are included as appropriate to calculate the desired variance-covariance structure.

Restricted maximum likelihood modeling (REML, Patterson, 1950; Patterson and Thompson, 1971) is applied to calculate unbiased variance estimates and the degrees of freedom (Kenward and Roger, 1997) are used to calculate appropriate degrees of freedom for any estimates or contrasts of interest.

Nonparametric statistical analysis of cross-over data will not be discussed further in this chapter but can be accomplished using SAS. Readers interested in such techniques should see Jones and Kenward (2003), Wellek (2003), Chow and Liu (2000) and Hauschke et al. (1990) for more details.

Other topics of general interest to readers of this chapter include *Population Pharmacokinetics* (FDA Guidance, 1999) and *Exposure-Response Modeling* (FDA Guidance, 2003). SAS is not generally used to support such nonlinear mixed-effect analyses, but is used in data management support (which will not be discussed here). It is theoretically possible to do such nonlinear mixed-effects modeling using the NLMIXED procedure, and we refer readers interested in such analyses to Atkinson et al. (2001), Sheiner et al. (1989), Sheiner (1997), Sheiner and Steimer (2000), and Machado et al. (1999) for more information on these topics.

To save space, some SAS code has been shortened and some output is not shown. The complete SAS code and data sets used in this book are available on the book's companion Web site at `http://support.sas.com/publishing/bbu/companion_site/60622.html`

8.2   Bioequivalence Testing

Bioequivalence testing is performed to provide evidence that a new formulation of drug substance (e.g., a new tablet) is equivalent *in vivo* to an existing formulation of drug product. Such testing is used by manufacturers of new chemical entities when making changes to a formulation of drug product used in confirmatory clinical trials, and additionally is used by the generic pharmaceutical industry to secure market access at patent expiration of an existing marketed product.

The new formulation (T) is generally studied relative to the existing formulation (R) in a randomized, open-label, cross-over trial (Jones and Kenward, 2003; Wellek, 2003; Senn, 2002; Chow and Liu, 2000). The structure of testing the question of bioequivalence is:

$$H_{01}: \quad \mu_T - \mu_R \leq -\ln 1.25,$$

$$H_{02}: \quad \mu_T - \mu_R \geq \ln 1.25,$$

where μ_T is the adjusted mean for the test formulation and μ_R is the adjusted mean for the reference formulation on the natural log scale. The factor $\ln 1.25$ was chosen by regulatory agencies (FDA, 1992; Barrett et al., 2000) and each one-sided test is performed at a 5% significance level without adjustment for multiplicity (Hauck et al., 1995).

Both tests must reject the null hypotheses for AUC and $C_{\max}$ in order for bioequivalence to be declared. This testing procedure is referred to as the TOST (two-one sided testing, see Schuirmann, 1987) assessment of average bioequivalence and constitutes the current FDA standard approach. See the FDA guidance page, `http://www.fda.gov/cder/guidance/` for more details. In practice, a 90% confidence interval for $\mu_T - \mu_R$ is derived for the ln-transformed AUC and $C_{\max}$, using a model appropriate to the study design. If both intervals fall completely within the interval $(-\ln 1.25, \ln 1.25)$, bioequivalence is concluded. This procedure is designated the *average bioequivalence* (ABE) because only the test and reference means are compared. A summary of other approaches and issues of interest in bioequivalence testing may be found in Hauck and Anderson (1992), Anderson and Hauck (1996), Benet (1999) and Zariffa and Patterson (2001). This type of approach is termed a *confirmatory* bioequivalence assessment for the purposes of this chapter.

This type of approach to data analysis is applied to clinical PK data in a variety of other setting in *exploratory* clinical pharmacology research. Examples include drug-drug interaction trials, food effect assessment and comparison of rate and extent of bioavailability in renal-impaired and hepatic insufficient populations. See the FDA guidance page, `http://www.fda.gov/cder/guidance/` for details. In these studies, the range of plausible values as expressed by a confidence interval is used to assess the degree of equivalence or comparability, depending upon the setting. A TOST need not necessarily be performed. Under such an approach, confidence level (Type I error rate) is termed *consumer* or *regulator* risk, i.e., the risk of the regulator agency in making an incorrect decision. Though often a pre-specified equivalence limit is difficult or impossible to define prior to study initiation, thus inhibiting the ability of study sponsors to adequately ensure adequate power to demonstrate equivalence if such is desired, power is of less concern when assessing the results of such studies than the confidence level. This approach to statistical inference gives regulator agencies an easy standard under which to assess the results of such exploratory pharmacokinetic studies.

We now turn to an example of a confirmatory bioequivalence trial.

EXAMPLE: Two-period cross-over study with test and reference formulations

Consider the following 2×2 cross-over study where a test formulation was compared to a reference formulation in approximately 50 normal healthy volunteer subjects.

The AUC data set, reproduced with permission from Patterson (2001), summarizes AUC values derived for the test and reference periods (the data set can be found on the book's companion Web site). The SUBJECT variable is the subject's ID number, the SEQUENCE variable codes the treatment sequence (RT denotes "'reference-test" and TR denotes "test-reference") and, lastly, the AUCT and AUCR variables contain AUC values when given Test and Reference formulations, respectively, in [ng/mL/h].

Program 8.1 uses PROC MIXED to fit a random-intercept model (Jones and Kenward, 2003) to the AUC data set once the AUC data are formatted by subject, sequence, period, and formulation (program not shown).

Program 8.1 Comparison of AUC values in the two-period cross-over study with test and reference formulations

```
data lnauc(keep=subject sequence period formula lnauc);
    set auc;
    lnauc=log(auc);
proc mixed data=lnauc method=reml;
    class sequence subject period formula;
    model lnauc=sequence period formula/ddfm=kenwardroger;
    random subject(sequence);
    estimate 'T-R' formula -1 1/cl alpha=0.10;
    run;
```

Output from Program 8.1

Cov Parm	Estimate
subject(sequence)	1.5921
Residual	0.1991

Type 3 Tests of Fixed Effects

Effect	Num DF	Den DF	F Value	Pr > F
sequence	1	44.8	0.12	0.7254
period	1	43.1	0.37	0.5463
formula	1	43.1	0.92	0.3424

```
                     Estimates

                  Standard
  Label   Estimate    Error     DF    t Value    Pr > |t|

  T-R      0.09023   0.09399    43.1     0.96       0.3424


  Label    Alpha     Lower     Upper

  T-R      0.1     -0.06776    0.2482
```

Output 8.1 displays the estimated difference between the test and reference formulations in the ln-transformed AUC with a 90% confidence interval. The 90% confidence interval is $(-0.0678, 0.2482)$. As insufficient information is present to reject H_{02}, bioequivalence was not demonstrated. No significant differences in sequence ($p = 0.7254$), period ($p = 0.5463$), and formulation ($p = 0.3424$) were detected. Between-subject variance in ln-AUC was 1.5921, and within-subject variance was 0.1991.

Alternatively we can use the following RANDOM statement to explicitly specify a random-intercept model:

```
    random int/subject=subject(sequence);
```

Specifying the RANDOM statement results in equivalent findings to the above.

These are the models required by FDA (1992) for the assessment of bioequivalence using a 2×2 cross-over design. Note that the Huyhn-Feldt condition (Hinkelmann and Kempthorne, 1994) is applied to the variance components using this model.

We may also use the REPEATED statement in PROC MIXED if the total variance (between plus within subject variation) of test and reference formulations and their covariance are of interest in a 2×2 cross-over study:

```
proc mixed data=lnauc method=reml;
    class sequence subject period formula;
    model lnauc=sequence period formula/ddfm=kenwardroger;
    repeated formula/type=un subject=subject(sequence);
    estimate 'T-R' formula -1 1/cl alpha=0.10;
    run;
```

Results of this last statement may differ slightly from the random-intercept model above when data are missing. It is unlikely, but not impossible, that missing data can affect inference.

8.2.1   Alternative Designs in Bioequivalence Testing

The FDA guidance also allows for alternative designs to demonstrate bioequivalence. One such design is the replicate cross-over design where each subject receives each formulation twice, with each administration being separated by a wash-out period of five half-lives.

EXAMPLE: A replicate cross-over study

Consider the following replicate design cross-over study where test and reference formulations were adminstered in a randomized, four-period cross-over design in approximately 36 normal healthy volunteer subjects.

The RAUC data summarizes AUC values collected in the test and reference periods (the data set can be found on the book's companion Web site). The SUBJECT variable is the subject's ID number, SEQUENCE variable codes the treatment sequence (ABBA denotes "test-reference-reference-test" and BAAB denotes "reference-test-test-reference") and, lastly, AUC1TEST, AUC2TEST, AUC1REF, AUC2REF variables contain AUC values when given Test and Reference formulations in the first and second administrations, respectively, in [ng/mL/h].

The FDA guidance requires that the construction of the confidence interval for $\mu_T - \mu_R$ be appropriate to the study design. In the case of a replicate design model, this recommendation results in a requirement to specify a factor-analytic variance-covariance structure for the between-subject variance components and calculation of within-subject variances for each formulation; see Patterson and Jones (2002) for more details.

Program 8.2 uses PROC MIXED to fit a factor-analytic variance-covariance structure using the FA0(2) option to the ln-transformed AUC data by subject, sequence, period, and formulation. See FDA Guidance (2001) for a description of why this option is used.

Program 8.2 Analysis of AUC data from a replicate cross-over study with test and reference formulations

```
data lnauc(keep=subject sequence period formula lnauc);
    set rauc;
    lnauc=log(auc);
proc mixed data=lnauc method=reml;
    class sequence subject period formula;
    model lnauc=sequence period formula/ddfm=kenwardroger;
    random formula/type=FA0(2) subject=subject;
    repeated/group=formula subject=subject;
    estimate 'T-R' formula 1 -1/cl alpha=0.1;
    ods output estimates=test;
data test;
    set test;
    lowerb=exp(lower); * Lower bound on original scale;
    upperb=exp(upper); * Upper bound on original scale;
proc print data=test noobs;
    var lowerb upperb;
    run;
```

Output from Program 8.2

```
Covariance Parameter Estimates

Cov Parm      Subject      Group        Estimate

FA(1,1)       subject                     0.5540
FA(2,1)       subject                     0.5542
FA(2,2)       subject                     1.24E-17
Residual      subject      formula A      0.09851
Residual      subject      formula B      0.1110


Type 3 Tests of Fixed Effects

              Num      Den
Effect        DF       DF      F Value     Pr > F

sequence       1       35.6      0.09       0.7706
period         3       106       2.28       0.0835
formula        1       106       7.68       0.0066
```

```
                        Estimates

                     Standard
Label      Estimate     Error        DF    t Value    Pr > |t|

T-R        -0.1449      0.05227      106    -2.77      0.0066


Label      Alpha        Lower        Upper

T-R        0.1          -0.2316      -0.05816


lowerb     upperb

0.79324    0.94350
```

Output 8.2 lists the findings. For ln-transformed AUC, a confidence interval of $(-0.2316, -0.0582)$ is found. As insufficient information is present to reject H_{01} in this case, bioequivalence was not demonstrated. These limits are saved in the TEST data set using an ODS statement and may be exponentiated in an additional DATA step if findings on the original scale are desired. In this case, the lower and upper bounds are 0.7932 and 0.9425.

Within-subject variation for the test and reference formulations were 0.09851 and 0.1110, respectively. Variation associated with subject-by-formulation was negligible, with an estimated standard deviation of 1.24E-17.

No significant evidence of sequence ($p = 0.7706$) or period ($p = 0.0835$) effects was detected; however, the formulations were observed to be significantly different ($p = 0.0066$).

The FDA guidance (2001) states that "In the Random statement, Type=FA0(2) could possibly be replaced by Type=CSH. This guidance recommends that Type=UN not be used, as it could result in an invalid (i.e., non-negative definite) estimated covariance matrix."

There is however a similar issue with use of TYPE=FA0(2) or CSH. These are *constrained* structures in that their application does not allow all positive definite covariance structures to be estimable (Patterson and Jones, 2002). For example, in the CSH structure, the estimate for correlation, ρ, is constrained by PROC MIXED so that it lies in the interval $-1 \le \rho \le 1$. The FA0(2) structure similarly imposes constraints such that the estimate for the subject-by-formulation standard deviation is zero or greater.

The choice of FA0(2) is somewhat arbitrary and reflective in regulatory application of George Box's statement "All models are wrong, but some are useful." In applications, it has been shown that this procedure protects the Type I error rate (of key concern to regulators) when variance estimates of interest are constrained to be positive or null (Patterson and Jones, 2004). It has the additional benefit of providing variance-covariance estimates that are readily interpretable (i.e., non-negative); however, statisticians using these procedures should recognize that the variance-covariance estimates may be biased as a consequence of this choice. Note that restricted maximum likelihood modeling (REML) is also specified as a recommended option, and those desiring to apply a different method should first talk with the FDA. Although REML estimates do possess less bias than maximum likelihood estimates, they do not produce estimates which maximize the likelihood function. The FDA has thus chosen a model which is readily interpretable and potentially biased, but which protects their risk of making a false positive decision of bioequivalence.

Note that if the TYPE=UN option is specified in Program 8.2, removing this constraint, a confidence interval of $(-0.2220, -0.0710)$ is derived as subject-by-formulation variation is estimated to be negative.

8.3 Assessing Dose Linearity

As discussed in the introduction, when an oral tablet is taken it will be absorbed, distributed, metabolized and eliminated. Imagine that the body is made up of many different compartments. The drug amount that enters circulation is then both simultaneously distributed into and eliminated from these compartments. Typically, a drug is eliminated from the body through urine or feces. In order to facilitate this process, the liver often metabolizes the compound to a chemical form that is more readily eliminated. If all of the eliminations from one compartment to another are proportional to the concentration of the compound in the compartment, the drug has *dose linearity*. Dose linearity implies that the elimination rate from the body is proportional as well.

The following two conditions are therefore implied. First, the elimination rate is proportional to drug concentration with k denoting the proportion. This constant is often replaced with clearance. As can be seen, dose linearity implies that clearance is a constant. Second, with repeated dosing there exists a point in time in which there is an equilibrium between the amount of drug that enters the body and the amount that leaves. Once this equilibrium has been reached, the drug is said to have achieved *steady state*.

As can be seen, it would be impossible to really tell if a drug were truly dose linear; however, a drug that has dose linearity (or that at least approximates dose linearity) should have certain properties. One of these is a proportional increase in concentration in the blood or plasma as measured with AUC and $C_{\max}$, with increasing dose. This situation is often referred to as *dose proportionality*. A second property is that the clearance ought to be stationary with time. Both of these properties can be studied in human pharmacokinetic studies.

8.3.1 Assessing Dose Proportionality

Dose proportionality is a proportional increase in concentration as measured with AUC and $C_{\max}$, with increasing dose. Using AUC, this implies that

$$AUC = \alpha d,$$

where d represents the dose. There are many different models to examine dose proportionality. A discussion to the relative merits and demerits of these models can be found is Smith (2004). This section will focus on the following model

$$\ln AUC = \alpha + \beta \ln d.$$

This model is referred to as the *power model* on an ln-transformed AUC. The same model is used for $C_{\max}$ as discussed in the previous section. Notice that when we exponentiate both sides of this equation we have

$$AUC = \exp(\alpha)d^{\beta}.$$

Thus, when $\beta = 1$, the drug is dose proportional.

In assessing dose proportionality, we may first consider a hypothesis test of the following

$$H_0 : \beta = 1.$$

We may consider declaring dose proportionality if we fail to reject this hypothesis. Smith et al. (2000) argue that it is much more natural to think of dose proportionality as an equivalence problem, implying that the structure for testing dose proportionality should be

$$H_{01} : \quad \beta \leq 1 - t,$$
$$H_{02} : \quad \beta \geq 1 + t.$$

Unlike bioequivalence, the equivalence region currently has no set regulatory standard. Smith et al. (2000) considers the following:

1. Examine dose-normalized concentrations.

2. Define some high dose (h) and low dose (l), where $r = h/l$, for which we wish to examine dose proportionality.

3. Set the following structure for testing the equivalence of the dose-normalized concentrations:

$$H_{03}: \quad \mu_h - \mu_l \leq -\ln\theta,$$

$$H_{04}: \quad \mu_h - \mu_l \geq \ln\theta,$$

where μ_h and μ_l denote the dose-normalized mean of the high dose and low doses on the natural log scale, respectively.

Smith et al. (2000) show that this structure is equivalent to the structure of H_{01} and H_{02} with $t = \ln\theta/\ln r$. Notice that $t \to \infty$ when $r \to 1$ and $t \to 0$ when $r \to \infty$. Thus, there exists a set of r's such that both H_{01} and H_{02} will be rejected.

The largest of these r's becomes the largest ratio of doses for which dose proportionality can be declared. Smith et al. (2000) show that if H_{01} and H_{02} are tested at the α level and that if L and H define the lower and upper limit of a $(1-2\alpha)100\%$ confidence interval, this largest ratio ρ_1 is

$$\rho_1 = \theta^{(1/max(1-L,U-1))}.$$

Furthermore, Smith et al. (2000) describe three possible conclusions that can be drawn about dose proportionality:

1. Definitely dose proportional.

2. Definitely not dose proportional.

3. Inconclusive.

"Definitely not dose proportional" is concluded if either the null hypothesis H_{01} or the null hypothesis H_{02} cannot be rejected. Smith et al. (2000) show that sometimes there exists a smallest ratio, ρ_2, in which dose proportionality definitely does not exist and is given by

$$\rho_2 = \theta^{(1/\max(L-1,1-U))}.$$

It should be pointed out that this equation cannot be solved unless $L > 1$ or $U < 1$.

Data for examination of dose proportionality most often comes from first human, dose escalation type trials in which randomization-to-period is not done. Occasionally, a stand-alone study is performed to ensure that period effects do not confound inference.

EXAMPLE: First human-dose study

Consider the simulated data from a hypothetical first human dose trial (FHDAUC data set can be found on the book's companion Web site). Here 16 subjects received ascending doses of drug, and AUC(0-∞) [ng/mL/h] was measured after each administration.

Program 8.3 shows how to perform the dose proportionality analysis of the AUC data using a random-intercept model in PROC MIXED.

Program 8.3 Analysis of AUC data from the first human-dose study

```
data fhdauc;
    set fhdauc;
    lndose=log(dose);
    lnauc=log(auc);
    study=1;
proc mixed data=fhdauc method=reml;
    class subject;
    model lnauc=lndose/s ddfm=kenwardroger cl alpha=0.1;
    random subject;
    ods output solutionf=one;
data one;
    set one;
    if effect='lndose' then delete;
    rho1=(4/3)**(1/max(1-lower,upper-1));
    if lower < 1 and upper > 1 then rho2=.;
    else rho2=(4/3)**(1/max(lower-1,1-upper));
proc print data=one noobs;
    var rho1 rho2;
    run;
```

Output from Program 8.3

```
Covariance Parameter Estimates

 Cov Parm      Estimate

 subject       0.04175
 Residual      0.01861

         Solution for Fixed Effects

                        Standard
 Effect      Estimate     Error      DF    t Value    Pr > |t|

 Intercept   -0.2583     0.07518    36.9    -3.44      0.0015
 lndose       1.0395     0.01235    30.7    84.20     <.0001

 Effect      Alpha     Lower      Upper

 Intercept    0.1     -0.3851    -0.1314
 lndose       0.1      1.0186     1.0605

         Type 3 Tests of Fixed Effects

             Num     Den
 Effect       DF      DF     F Value    Pr > F

 lndose        1     30.7    7088.96    <.0001


      rho1          rho2

   116.630     5344869.90
```

Output 8.3 shows that β significantly differs from unity ($p < 0.0001$) with 90% confidence bounds of $(1.0186, 1.0605)$. Between-subject variation in ln-AUC was 0.0418, and within-subject variation was observed to be 0.0186.

Output 8.3 also extracts information from PROC MIXED in order to calculate ρ_1 and ρ_2. Notice that $\theta = 4/3$ in Program 8.3. The choice for θ is user-defined. In this case the resulting equivalence region is $(0.75, 1.33)$, as compared to the equivalence region for bioequivalence of $(0.80, 1.25)$. Dose proportionality is concluded as the 90% confidence interval for β falls within these limits.

In this first human-dose study, one derives $\rho_1 = 117$ and $\rho_2 = 5,340,000$. One would interpret this to mean that dose proportionality has been demonstrated up to a 117-fold range of doses. Said another way, if we knew the concentration (c) for some dose a, then a good prediction of the concentration for a dose $117a$ would be $117c$. The value of ρ_2 in this case is so large that it has little practical meaning. All and all, since it is unlikely in clinical practice to have available a 117-fold range of doses, this molecule demonstrates a significant degree of dose proportionality.

Note that, in this example, $\rho_1 = 117$ and the dose range studied is 200-fold. It is quite possible for ρ_1 to be greater than the dose range studied. This brings up the question of how we should interpret ρ_1 if this is the case. Assume that in this example ρ_1 turned out to be 500. There would obviously be danger in saying that a good prediction of the concentration for a dose $500a$ would be $500c$, since we are obviously extrapolating beyond available data. With this said, however, it is still useful to think of ρ_1 as a measure of dose proportionality. That is, a value of 500 indicates a greater degree of dose proportionality than a value of 300 would have.

There are certainly other covariance structures that could be chosen in Program 8.3. For instance, a random intercept and slope model can be generated with the following RANDOM statement:

```
random subject subject*lndose;
```

8.3.2   Assessing Time Stationarity of Clearance

In the beginning of this section, clearance was defined as the rate of elimination divided by the concentration. When an oral dose is administered, only a fraction of the total dose is absorbed by the body. It can be shown that for a single dose of a drug that

$$\text{Clearance} = \frac{Fd}{AUC(0 - \infty)},$$

where F is the fraction adsorbed, d is the dose and $AUC(0 - \infty)$ is what has been called AUC so far in this chapter. On the other hand, when a drug is at steady state, it can be shown that

$$\text{Clearance} = \frac{Fd}{AUC(0 - \tau)},$$

where τ is the dosing interval (24 hours if dosed once a day), d is the amount of drug delivered at each dosing interval, and $AUC(0 - \tau)$ is the area under the concentration when the drug is at steady state. The implication of these two formulas is that if $AUC(0 - \infty) = AUC(0 - \tau)$ for one dose, then clearance does not change with repeated dosing. If this is true for all doses, then clearance is said to be *stationary*. Dose linearity implies that clearance is a constant, which implies stationarity.

If the half-life of a drug is short enough, it may be possible to calculate $AUC(0 - \infty)$ for each individual in the first day of dosing of a multiple dose study and find AUC(0-τ) at the last day. Subjects must be at steady state and measures should be taken to ensure this.

The shorter the half-life, the quicker steady state can be achieved. Thus, since we need a short half-life to be able to do this, the drug would most likely be at steady state. Having a short enough half-life to be able to use day 1 data, however, is pretty unlikely.

Another possibility is doing the first human dose study in the same subjects as the multiple dose study. Usually, when this is done, these studies are combined into one protocol. Again, however, the opportunity to do this does not present itself very often. The reason these options are appealing is it allows us to use a subject as its own control. The most common case, however, is that the first human dose study and multiple dose study are done in separate groups of subjects. We can, however, still combine this information to examine stationarity of clearance.

EXAMPLE: Multiple-dose study

Consider the MDAUC data set that can be found on the book's companion Web site. The data set contains simulated data from a hypothetical multiple-dose study. This study tested the compound examined in the first human-dose trial described above. It is assumed that the multiple-dose study was performed in a different set of subjects from the first study. Here, 16 subjects were dosed with 20 to 200 mg of drug repeatedly for several days, and $AUC(0\text{-}\tau)$ [ng/mL/h] was measured over the dosing interval on the final day.

Program 8.4 examines stationarity of clearance in the multiple-dose study using PROC MIXED. Stationarity of clearance would be concluded if the estimated geometric means from the 20 mg first human dose trial and estimated geometric means from the 20 mg multiple dose trial are equivalent and the estimated geometric means from the 200 mg first human dose trial and estimated geometric means from the 200 mg multiple dose trial are equivalent.

The PROC MIXED code in Program 8.4 fits a separate power model for both studies. The ESTIMATE statements are then used to derive estimates for the single dose mean AUC at 20 mg, the multiple dose mean AUC at 20 mg, their ratio, the single dose mean at 200 mg, the multiple dose mean at 200 mg, and their ratio, respectively. These estimates are saved in the OUT data set and then exponentially transformed to the original scale in order to compute 90% confidence intervals for each scenario. Program 8.4 suppresses the standard PROC MIXED output using ODS LISTING statements. Note that the log-doses of 20 and 200 mg are denoted as 2.995732 and 5.298317, respectively.

Program 8.4 Analysis of AUC data from the multiple-dose study

```
data mdauc;
    set mdauc;
    lnauc=log(auc);
    lndose=log(dose);
    study=2;
* Merge the data from the first human-dose and multiple-dose studies;
data cl;
    set fhdauc mdauc;
ods listing close;
proc mixed data=cl method=reml;
    class subject study;
    model lnauc=lndose study lndose*study/ddfm=kenwardroger cl alpha=0.1;
    random subject*study;
    estimate 'Single dose AUC 20 mg' intercept 1 lndose 2.995732
        study 1 lndose*study 2.995732;
    estimate 'Multiple dose AUC 20 mg' intercept 1 lndose 2.995732
        study 1 lndose*study 0 2.995732;
    estimate 'Ratio single dose:multiple dose 20 mg' study 1 -1
        lndose*study 2.995732 -2.995732/cl alpha=0.1;
```

```
      estimate 'Single dose AUC 200 mg' intercept 1 lndose 5.298317
         study 1 lndose*study 5.298317;
      estimate 'Multiple dose AUC 200 mg' intercept 1 lndose 5.298317
         study 0 1 lndose*study 0 5.298317;
      estimate 'Ratio single dose:multiple dose 200 mg' study 1 -1
         lndose*study 5.298317 -5.298317/cl alpha=0.1;
      ods output estimates=out;
 data out;
      set out;
      gmean=exp(estimate);
      lbound=exp(lower);
      ubound=exp(upper);
      drop estimate stderr df tvalue alpha lower upper;
 proc print data=out noobs;
      ods listing;
      run;
```

Output from Program 8.4

Label	Probt	gmean	lbound	ubound
Single dose AUC 20 mg	<.0001	17.385	15.617	19.354
Multiple dose AUC 20 mg	<.0001	16.738	13.697	20.455
Ratio single dose:multiple dose 20 mg	0.7371	1.039	0.860	1.255
Single dose AUC 200 mg	<.0001	190.365	170.987	211.939
Multiple dose AUC 200 mg	<.0001	214.759	177.887	259.273
Ratio single dose:multiple dose 200 mg	0.2667	0.886	0.740	1.062

Output 8.4 lists the estimated geometric means and associated 90% confidence limits. The 90% confidence interval of the ratio of geometric means for 20 mg is $(0.860, 1.255)$ and the 90% confidence interval of the ratio of geometric mean for 200 mg is $(0.740, 1.062)$. Although neither confidence interval meets the strict bioequivalence $(0.8, 1.25)$ criteria, at this point of drug development, given the limited amount of data, stationarity of clearance would be assumed, until and unless compelling future data indicated otherwise.

Since dose proportionality can be concluded and dose stationarity seems highly tenable, this molecule seems to exhibit dose linearity.

8.4   Summary

In this chapter, pharmacokinetic terminology frequently encountered in clinical trial assessment of human pharmacokinetics was reviewed, and the collection, measurement, and statistical assessment of pharmacokinetic data obtained in Phase I and clinical pharmacology studies were described. REML models were used to examine the most common pharmacokinetic data of practical interest, and tests for assessing bioequivalence and dose linearity were developed and implemented using PROC MIXED.

References

Atkinson, A., Daniels, C., Dedrick, R., Grudzinskas, C., Markey, S. (2001). *Principles of Clinical Pharmacology*. San Diego: Academic Press.

Anderson, S., Hauck W.W. (1996). "The transitivity of bioequivalence testing: potential for drift." *International Journal of Clinical Pharmacology and Therapeutics*. 34, 369–374.

Barrett, J.S., Batra, V., Chow, A., Cook, J., Gould, A.L., Heller, A., Lo, MW., Patterson, S.D., Smith, B.P., Stritar, J.A., Vega, J.M., Zariffa, N. (2000). "PhRMA Perspective on Population and Individual Bioequivalence and Update to the PhRMA Perspective on Population and Individual Bioequivalence." *Journal of Clinical Pharmacology*. 40, 561–572.

Benet, L.Z. (1999). "Understanding Bioequivalence Testing." *Transplantation Proceedings*. 31, Supplement 1, 7S–9S.

Box, G.E., Cox, D.R. (1964). "An analysis of transformations." *Journal of the Royal Statistical Society, Series B*. 26, 211–243.

Chow, S.C., Liu, J. (2000). *Design and Analysis of Bioavailability and Bioequivalence Studies*, second edition. New York: Marcel Dekker.

Crow, E.L., Shimizu, K. (1988). *Lognormal distributions*. New York: Marcel Dekker.

FDA Guidance. (1992). *Statistical procedures for bioequivalence studies using a standard two treatment cross-over design*.

FDA Guidance. (February, 1999). *Guidance for Industry: Population Pharmacokinetics*.

FDA Guidance. (January, 2001). *Guidance for Industry: Statistical Approaches to Establishing Bioequivalence*.

FDA Guidance. (April, 2003). *Guidance for Industry: Exposure-Response Relationships—Study Design, Data Analysis, and Regulatory Applications*.

Hauck, W.W., Anderson S. (1992). "Types of bioequivalence and related statistical considerations." *International Journal of Clinical Pharmacology, Therapy, and Toxicology*. 30, 181–187.

Hauck, W.W., Hyslop, T., Anderson, S., Bois, F.Y., Tozer, T.N. (1995). "Statistical and regulatory considerations for multiple measures in bioequivalence testing." *Clinical Research and Regulatory Affairs*. 12, 249–265.

Hauschke, D., Steinijans, V.W., Diletti, E. (1990). "A distribution free procedure for the statistical analysis of bioequivalence studies." *International Journal of Clinical Pharmacology, Therapy, and Toxicology*. 28, 72–78.

Hinkelmann, K., Kempthorne, O. (1994). *Design and Analysis of Experiments*, Volume I: Introduction to Experimental Design. New York: Wiley.

Jones, B., Kenward M.G. (2003). *Design and Analysis of Cross-over Trials*, second edition. London: Chapman and Hall.

Julious, S., Debarnot, C. A-M. (1999). "Why are pharmacokinetic data summarized by arithmetic means?" *Journal of Biopharmaceutical Statistics*. 10, 55–71.

Kenward, M., Roger, J. (1997). "Small sample inference for fixed effects from restricted maximum likelihood." *Biometrics*. 33, 983–997.

Lacey, L.F., Keene, O.N., Bye, A. (1995). "Glaxo's exprerience of different absorption rate metrics of immediate release and extended release dosage forms." *Drug Information Journal*. 29, 821–840.

Lacey, L.F., Keene, O.N., Pritchard, J.F., Bye, A. (1997). "Common noncompartmental pharmacokinetic variables: are they normally or log-normally distributed?" *Journal of Biopharmaceutical Statistics*. 7, 171–178.

Machado, S., Miller, R., Hu, C. (1999). "A regulatory perspective on pharmacokinetic and pharmacodynamic modelling." *Statistical Methods in Medical Research*. 8, 217–45.

Milliken, G.A., Johnson, D.E. (1992). *Analysis of Messy Data*, Volume 1: Designed Experiments, New York: Chapman and Hall.

Patterson, H.D. (1950). "The construction of balanced designs for experiments involving sequences of treatments." *Biometrika*. 39, 32.

Patterson, H.D., Thompson, R. (1971). "Recovery of Inter-block information when block sizes are unequal." *Biometrika*. 58, 545–554.

Patterson, S.D. (2001). "A review of the development of biostatistical design and analysis techniques for assessing in vivo bioequivalence, Part 1." *Indian Journal of Pharmaceutical Sciences*. 63, 81–100.

Patterson, S.D., Jones, B. (2002). "Bioequivalence and the pharmaceutical industry." *Pharmaceutical Statistics*. 1, 83–95.

Patterson, S.D., Jones, B. (2004). "Simulation assessments of statistical aspects of bioequivalence in the pharmaceutical industry." *Pharmaceutical Statistics*. 3, 13–23.

Rowland, M., Tozer, T.N. (1980). *Clinical Pharmacokinetics Concepts and Applications.* Philadelphia: Lea and Febiger.

Schuirmann, D.J. (1987). "A comparison of the two one sided tests procedure and the power approach for assessing the equivalence of average bioavailability." *Journal of Pharmacokinetics and Biopharmaceutics.* 15, 657–680.

Senn, S. (2002) *Cross-over Trials in Clinical Research,* second edition. New York: John Wiley & Sons.

Sheiner L.B., Beal, S.L., Sambol, N.C. (1989). "Study Designs for Dose-Ranging." *Clinical Pharmacology and Therapeutics.* 46, 63–77.

Sheiner, L.B. (1997). "Learning versus confirming in clinical drug development." *Clinical Pharmacology and Therapeutics.* 61, 275–291.

Sheiner, L.B., Steimer, J-L. (2000). "Pharmacokinetic-pharmacodynamic modeling in drug development." *Annual Reviews of Pharmacology and Toxicology.* 40, 67–95.

Smith, B., Vandenhende, F.R., DeSante, K., Farid, N., Welch, P., Callaghan, J., Forgue, S. (2000). "Confidence interval criteria for assessment of dose proportionality." *Pharmaceutical Research.* 17, 1278–1283.

Smith, B.P. (2004). "Assessment of Dose Proportionality." Chapter 6, Part 2 of *Pharmacokinetics in Drug Development: Clinical Study Design and Analysis,* Volume 1. Bonate, P., Howard, D., editors. Arlington,VA: AAPS Press. 363–382.

Vonesh, E.F., Chinchilli, V.M. (1997). *Linear and Nonlinear models for the analysis of repeated measurements.* New York: Marcel Dekker.

Wellek, S. (2003) *Testing Statistical Hypotheses of Equivalence.* London: Chapman and Hall.

Westlake, W.J. (1986). "Bioavailability and bioequivalence of pharmaceutical formulations." *Biopharmaceutical Statistics for Drug Development.* K. Peace, editor. New York: Marcel Dekker. 329–352.

Yeh, K.C., Kwan, K.C. (1978). "A comparison of numerical integrating algorithms of trapezoidal, Legrange, and spline approximation." *Journal of Pharcokinetics and Biopharmaceutics.* 6, 79–81.

Zariffa, N.M-D., Patterson, S.D. (2001). "Population and Individual Bioequivalence: Lessons from Real Data and Simulation Studies." *Journal of Clinical Pharmacology.* 41, 811–822.

Chapter 9

Allocation in Randomized Clinical Trials

Olga Kuznetsova
Anastasia Ivanova

This chapter discusses the operational aspects of randomization designs used in the pharmaceutical industry. Most of this chapter is devoted to the widely used permuted block design and its variations. Allocation procedures that are balanced on baseline covariates, such as stratified randomization, and covariate-adaptive randomization, are also described. All methods are illustrated by examples that include SAS code to generate allocation sequences. Although all of the examples in this chapter refer to clinical trials, the described randomization methods can also be used in a non-clinical setting, i.e., in animal studies.

We refer the reader to Rosenberger and Lachin (2002) for a thorough coverage of other aspects of randomization such as randomization-based inference and covariate-adjusted analysis. A chapter in Senn's book (1997) offers excellent insights on randomization in clinical trials.

9.1 Introduction

Randomization, an allocation of subjects to treatment regimens using a random element, is an essential component of clinical trials. Randomization promotes comparability of the treatment groups with respect to known as well as unknown covariates and thus reduces the chance for bias in the evaluation of the treatment effect. It can also serve as a basis for a randomization approach to inference (Rosenberger and Lachin, 2002).

Several types of bias are considered in the context of randomization. *Selection bias* occurs in an unmasked (unblinded) trial when the investigator, either consciously or otherwise, uses knowledge of the upcoming treatment assignment to help decide whom to enroll (Blackwell and Hodges, 1957). *Observer bias* is the bias in the evaluation of responses

Olga Kuznetsova is Associate Director, Clinical Biostatistics, Merck, USA. Anastasia Ivanova is Associate Professor, Department of Biostatistics, University of North Carolina-Chapel Hill, USA.

to a treatment that can occur if some of the assignments are known to the investigator. *Imbalance* in a covariate strongly associated with the study outcome can also bias the results. Although it is possible to adjust for known covariates, imbalance in unobserved covariates can lead to *accidental bias* (Efron, 1971). Various allocation procedures differ in how susceptible they are to these types of bias, and that fact influences the choice of an allocation procedure for a clinical trial. Another consideration that needs to be taken into account when selecting an allocation procedure is the procedure's ability to achieve a targeted allocation ratio.

Complete randomization, in which each subject is allocated at random with the probability determined by the targeted allocation ratio, is totally unpredictable and does not lead to selection bias in a trial unmasked to the investigator. Its significant drawback, though, is the risk of undesirable imbalance in the number of subjects allocated to each arm. Such imbalance can negatively affect the power of treatment comparisons, especially when the total number of subjects in the trial is small. Also, if the subjects' prognostic factors exhibit a time trend during the course of the trial, a considerable imbalance in treatment assignments throughout the trial can lead to accidental bias. Restricted randomization implemented through a permuted block design (Rosenberger and Lachin, 2002) provides a good balance in the treatment assignments throughout the enrollment and is more commonly used. Other designs that maintain a good balance in treatment assignments at any time and have low potential for selection bias have recently been suggested (Berger et al., 2003).

After an appropriate randomization procedure is selected, the sequence of random assignments for the study subjects is generated and documented in the randomization (allocation) schedule.

It might be desirable to maintain balance in important prognostic baseline covariates. Balance is typically achieved through stratified randomization, in which a separate restricted allocation schedule is prepared for each combination of levels of the stratification factors. Alternatively, balance in baseline covariates can be maintained with a dynamic allocation, in which a new subject is assigned to the treatment group that results in the best balance (in some sense) across his or her set of covariates. A fixed allocation schedule cannot be prepared for dynamic procedures, as the sequence of treatment assignments depends on the covariates of the subjects entering the trial.

The randomization schedule or a sequence of treatment assignments for a dynamic allocation procedure can be easily generated with SAS software, e.g., using the PLAN procedure or random number generators. The flexibility of the PLAN procedure often allows the same schedule to be generated in many different ways; we will cover a variety of options in the examples throughout this chapter.

To save space, some SAS code has been shortened and some output is not shown. The complete SAS code and data sets used in this book are available on the book's companion Web site at `http://support.sas.com/publishing/bbu/companion_site/60622.html`.

9.2 Permuted Block Randomization

Permuted block randomization is commonly used in clinical trials to allocate subjects to the treatment arms in required ratios. It is well accepted in the clinical community, provides good balance in treatment assignments, and supports the needs of drug packaging and distribution.

9.2.1 Permuted Block Designs

The permuted block schedule consists of a sequence of blocks that contain the treatment assignments in desired ratios; the treatment assignments are randomly permuted within the blocks. Consider, for example, a study in which subjects are to be assigned to three treatments (Treatment 1, Treatment 2, and Treatment 3) using a $2:2:1$ ratio. The smallest permuted block that provides the $2:2:1$ treatment ratio is the block of size 5 and is a random permutation of

$$(1,1,2,2,3).$$

An example of a sequence of treatment assignments (with blocks bracketed to illustrate the technique) is

$$(2,1,1,3,2),(1,1,2,3,2),(3,2,1,2,1),\ldots.$$

A larger block size (a multiple of 5) can also be used. If the block size of 10 is chosen, the blocks will each be a random permutation of

$$(1,1,1,1,2,2,2,2,3,3).$$

Stratifying the randomization is easy with the permuted block schedule. Since any set of blocks constitutes a permuted block schedule on its own, each stratum can be assigned a set of blocks from one common schedule. This property is useful in multi-center trials where, as recommended by the ICH Guidance entitled "Guidance on Statistical Principles for Clinical Trials" (ICH E-9), the allocation is typically stratified by center. The stratification is implemented by assigning a set of blocks to each center.

The permuted block allocation provides a good balance in treatment assignments: when most of the blocks in the allocation schedule are filled, the treatment ratio of allocated subjects at the end of the study is close to the planned one. Unless there are many strata and thus, many unfilled blocks, the balance stays reasonably tight throughout the enrollment as well. This feature improves the efficiency of an interim analysis and helps mitigate the accidental bias if a time trend in treatment effect is present.

These properties made the permuted block randomization an industry standard for masked clinical trials. For unmasked trials, where selection bias is an issue, other approaches such as a permuted block design with a random block size or maximal procedure (Berger et al., 2003) might offer benefit.

9.2.2   Choice of Block Size in Permuted Block Designs

The choice of block size for a masked permuted block schedule requires some consideration. It is preferred to have a block size big enough to include at least two subjects on each treatment to mitigate the impact of rare instances when a subject's allocation may need to be unblinded due to a serious adverse event. Having at least two subjects on each treatment within a permuted block will not narrow the list of possible assignments for other subjects from the block that the unblinded subject belongs to. This provision would lessen the potential for selection bias (if some of the subjects on the block are yet to be enrolled) and for observer bias.

This approach will allow using the minimal block size when the smallest block with required ratio already includes at least two subjects on each treatment (e.g., a $3:2:2$ allocation ratio). When the smallest block includes only one subject on any of the treatments, e.g., a $2:2:1$ ratio, the preferred block size will be at least twice as large as the minimal block size (that is, at least 10).

The outlined block size selection strategy is intended to lessen the impact of unblinding and might not be advisable for a multi-center trial with small centers. For example, if the minimal block size is 5 and most of the centers are expected to enroll about six subjects, using a block size of 10 will lead to predominantly incomplete blocks and thus to suboptimal balance in treatment assignments. A compromise can often be reached by using one of the variations of permuted block designs (see Sections 9.3.2 and 9.3.4); however, if the possibility of unblinding is low, the most practical option might be to use the smallest block size. In reality, designs with a block size of 2, in which an unblinding of one subject leads to the unblinding of the other subject in the same block, are rarely used.

9.2.3   Generating Permuted Block Schedules

Permuted block schedules can be easily generated using PROC PLAN. PROC PLAN is a powerful procedure and often allows to generate the same randomization schedule using several different sets of options. We will present one of the options in the example below.

EXAMPLE: Three-arm clinical trial with a $2:2:1$ randomization ratio

Consider a study in which 120 subjects are allocated to three treatment arms (Treatment 1, Treatment 2, and Treatment 3) with a $2:2:1$ ratio using the permuted block randomization with a block size 5. As is typically the case, it was decided to prepare an allocation schedule for more than 120 subjects by generating a total of 60 permuted blocks.

Program 9.1 uses PROC PLAN to generate a permuted block schedule for the three-arm trial. The first option (SEED=56789) in the PROC PLAN statement specifies a random seed to make the schedule reproducible. If no seed is provided, the time of the day will be used as a default seed and each run of the program will result in a different schedule. The random seed used in the schedule should not be disclosed until the study is unblinded. The next statement

```
factors block=60 ordered treatment=5;
```

requests that the BLOCK variable be created in the output data set with values ordered from 1 to 60. For each value of BLOCK, five observations with five levels of the TREAT variable will be generated in a random order. Next,

```
output out=schedule treatment nvals=(1 1 2 2 3);
```

specifies the name of the output data set (SCHEDULE) as well as numeric values (1, 1, 2, 2, and 3) that will be assigned to the five levels of the TREATMENT variable. PROC PLAN is followed by a DATA step that assigns consecutive allocation numbers (AN variable) from 1 to 300 to all observations on the list. The treatment assignment is stored in the TREATMENT variable.

Program 9.1 Three-arm clinical trial with a $2:2:1$ randomization ratio

```
ods listing close;
proc plan seed=56789;
    factors block=60 ordered treatment=5;
    output out=schedule treatment nvals=(1 1 2 2 3);
data schedule;
    set schedule;
    an=_n_;
    label treatment='Treatment group'
          block='Block number'
          an='Allocation number';
proc print data=schedule noobs label;
    var an treatment block;
    ods listing;
    run;
```

Partial output from Program 9.1

```
Allocation     Treatment      Block
 number          group       number

    1              3            1
    2              2            1
    3              2            1
    4              1            1
    5              1            1
    6              1            2
    7              1            2
    8              2            2
    9              3            2
   10              2            2
```

Output 9.1 shows the first ten observations in the SCHEDULE data set. These observations represent the allocation of the first ten subjects (two permuted blocks).

To generate 30 permuted blocks of size 10, the PROC PLAN options in Program 9.1 need to be modified in the following way:

```
proc plan seed=56789;
   factors block=30 ordered treatment=10;
   output out=schedule treat nvals=(1 1 1 1 2 2 2 2 3 3);
```

The described approach to generating permuted block schedules relies on drawing permuted blocks with replacement from a set of all existing permuted blocks with specified contents. The resulting schedule does not necessarily use all existing permuted blocks of a given structure and the same block can appear on the schedule multiple times. This approach is simple to implement and, in most circumstances, adequately serves the needs of clinical researchers.

An alternative approach was described by Zelen (1974) for two-arm studies with a 1:1 allocation that uses the permuted blocks of size 4. Zelen proposed to list all six existing permuted blocks,

$$(1,1,2,2), (1,2,1,2), (1,2,2,1), (2,1,1,2), (2,1,2,1), (2,2,1,1),$$

and then randomly permute them to form a randomization sequence for the first 24 subjects. The procedure can be repeated as many times as needed to produce the randomization schedule for any number of subjects. With this approach, the sets of all existing permuted blocks are repeatedly used. This strategy can also be implemented with PROC PLAN. Section 9.3.4 describes the circumstances under which the Zelen approach can be beneficial.

9.3   Variations of Permuted Block Randomization

Although permuted block randomization provides a satisfactory allocation for the majority of clinical trials, some variations of it are also employed.

9.3.1   Permuted Block Design with a Variable Block Size

In a study unmasked to the investigator, permuted block design with block of variable size is sometimes employed (ICH E-9), with the intention of making it harder for the investigator to guess the next treatment assignment and thus lessen the potential for selection bias. This strategy and its limitations are discussed in Rosenberger and Lachin

(2002). They mention that, under a commonly used model for selection bias, variable block allocation "yields a substantial potential for selection bias in an unmasked study that is approximately equal to that associated with the average block size M". On the other hand, variable block allocation reduces the expected number of assignments that may be predicted with certainty compared to the sequence of permuted blocks of size M.

A permuted block randomization with variable block size can be generated in SAS using random number generators.

EXAMPLE: A 1 : 1 randomization schedule with a variable block size

Consider a clinical trial with 20 subjects who are to be allocated to two treatments in a 1:1 ratio using permuted blocks of size 2, 4, or 6. The size of each permuted block will be chosen at random with probabilities 0.25, 0.5, and 0.25, respectively. To provide a schedule for a minimum of 20 subjects, 10 permuted blocks need to be generated. The set of 10 blocks can include up to 60 subjects; however, only the first 20 treatment assignments will be used to allocate the study subjects.

Program 9.2 demonstrates how to generate a randomization schedule for the two-arm trial. First, the BLOCKS data set with 10 blocks of variable size is created. The size of each block is computed based on the HALF_SIZE variable which is defined as follows

```
half_size=rantbl(12345,0.25,0.5,0.25);
```

This statement invokes the RANTBL function (with the random seed 12345) that assigns values 1, 2, or 3 with probabilities 0.25, 0.5, and 0.25 to the HALF_SIZE variable. After the block size is determined (it is equal to two times HALF_SIZE since there are two treatment groups), HALF_SIZE observations are created in each treatment group within the block. After that, a random value uniformly distributed over (0,1) is added to the data set (RANDOM variable) to randomly permute the treatment assignments within each block. The RANDOM variable is generated using the RANUNI function with the seed set to 678. The last DATA step assigns consecutive allocation numbers (AN variable) to all observations on the list.

Program 9.2 A 1 : 1 randomization schedule with a variable block size

```
data blocks;
    do block=1 to 10;
        half_size=rantbl(12345,0.25,0.5,0.25);
        do j=1 to half_size;
            do treatment=1,2;
                random=ranuni(678);
                output;
            end;
        end;
    end;
proc sort data=blocks out=schedule;
    by block random;
data schedule;
    set schedule;
    an=_n_;
    label treatment='Treatment group'
          block='Block number'
          an='Allocation number';
proc print data=schedule noobs label;
    var an treatment block;
    run;
```

Output from Program 9.2

Allocation number	Treatment group	Block number
1	2	1
2	1	1
3	1	1
4	2	1
5	1	2
6	1	2
7	2	2
8	2	2
9	2	3
10	1	3
11	1	3
12	2	3
13	1	4
14	2	4
15	2	4
16	1	4
17	1	4
18	2	4
19	2	5
20	1	5
21	2	5
22	1	5
23	2	6
24	1	6
25	1	7
26	2	7
27	2	7
28	1	7
29	2	8
30	1	8
31	1	9
32	2	9
33	2	9
34	1	9
35	1	10
36	2	10

Output 9.2 displays the allocation schedule generated by Program 9.2. The schedule uses the following sequence of block sizes

$$(4, 4, 4, 6, 4, 2, 4, 2, 4, 2).$$

It should be noted that, although in the example above, the first 20 subjects happened to be split equally (10 and 10) between the two groups, an unbalanced allocation of $9 : 11$ might have also been an outcome. In general, when a sequence of permuted blocks with a random block size is used, the maximum imbalance in treatment assignments between the two groups is determined by the largest block size allowed. In studies of small size, the need to achieve an exact balance in treatment assignments might be a consideration when choosing an allocation procedure.

9.3.2 Constrained Block Randomization

Clinical trials are often designed to have unequal allocation to different treatment arms. This might be based on efficiency considerations, ethical considerations (placing more subjects on the treatment believed to be more efficient), the need to accumulate more experience with the experimental drug, or simply to meet regulatory exposure requirements.

In multi-arm trials, the number of subjects enrolled in each arm may reflect the roles played by the arms in the trial's objective. For example, consider a three-arm study with an experimental drug arm, an active control arm and a placebo arm. The primary objective of the trial is to compare the experimental drug to the active control and the placebo arm is included in the study to perform safety assessments. Power considerations may call for a size of 500 subjects in each of the two active treatment arms, and an additional 200 subjects to be enrolled in the placebo group.

The allocation ratio of $5:5:2$ will lead to a large block size (the smallest block size is 12), which, in turn, can lead to logistical problems described below. However, changing the treatment ratio to a similar, but more manageable ratio of $2:2:1$, may not be acceptable since it requires enrolling 50 additional placebo subjects.

When the allocation ratio calls for a large block size, permuted blocks can turn out to be quite unbalanced with some treatment groups gathered at the beginning of the block and other groups at the end. To illustrate, consider a three-arm clinical trial with a $5:5:2$ treatment ratio. With this ratio, the smallest block size is 12, and we can theoretically find the following unbalanced block

$$(1, 1, 1, 1, 1, 2, 2, 2, 2, 2, 3, 3).$$

Highly unbalanced blocks of this kind are a concern in multi-center trials stratified by center. If some of the centers enroll less than a full block of subjects, these centers might end up with an extreme imbalance in treatment assignments, or even without all treatment arms represented. That may lead to problems at the analysis stage and, if there are many small centers, the described phenomenon will have a negative impact on the overall balance of treatment assignments.

To avoid these problems when the block size is large, constrained randomization methods proposed by Youden (1964, 1972) can be used. Constrained block randomization does not use every existing permuted block with a given treatment ratio but only the better balanced blocks specified in advance.

Consider again the trial with a $5:5:2$ allocation to three treatments (Treatments 1, 2, and 3). Blocks of size 12 can be constructed in the following way: five permuted blocks $(1,2)$ or $(2,1)$ are lined up and then two Treatment 3 assignments are randomly inserted (one among the first six assignments and the other one among the last six assignments). An example of a block generated with the described algorithm is given below

$$(2, 1, 1, 2, \underline{3}, 1, 2, 2, \underline{3}, 1, 2, 1).$$

Another way to build a constrained randomization schedule with a block of size 12 is to generate a random permutation of

$$(1, 1, 1, 2, 2, 3)$$

followed by a random permutation of

$$(1, 1, 2, 2, 2, 3),$$

or vice versa, at random. This set of assignments is less restrictive compared to the allocation algorithm described above.

It is worth noting that other constrained block randomization algorithms have been proposed in the literature. For example, Berger et al. (2003) introduced a randomization procedure (known as the *maximal procedure*) which uses any sequence in a block for which the maximum imbalance does not exceed a prespecified number.

With constrained randomization, blocks left incomplete at the end of enrollment will exhibit a better balance compared to a regular permuted blocks schedule, which might be especially important if the trial uses interim analyses. Constrained randomization is also less susceptible to accidental bias caused by a time trend in a prognostic covariate.

Constrained randomization schedules can easily be generated using PROC PLAN by imposing a constraint on a full factorial design in which the number of factors is equal to the block size (Song and Kuznetsova, 2003).

9.3.3 Allocation for Clinical Trials with Treatment Group Splitting

A common example of constrained randomization—perhaps not recognized as such—arises when treatment groups are split for a washout period or safety extension.

EXAMPLE: Dose-ranging study with a placebo washout

Consider a multi-center dose-ranging study with six treatment arms in which each arm is to be split in a 2 : 1 ratio at the end of the main study period for a placebo washout. Two thirds of each treatment arm will continue on the base study treatment, while the remaining third will be switched to placebo. This study design results in 12 different treatment regimens defined by a combination of a treatment arm (one of 6) and whether a subject will be switched to placebo or stay on the base study treatment during the placebo washout period.

At the base study start, the subjects are randomized to one of the 12 regimens (Arm 1 without switch, Arm 1 with a switch, Arm 2 without switch, Arm 2 with a switch,..., Arm 6 without switch, Arm 6 with a switch) in a

$$2 : 1 : 2 : 1 : 2 : 1 : 2 : 1 : 2 : 1 : 2 : 1$$

ratio. The minimal block size (18) can be too large for the study if most centers are expected to enroll about 6 to 9 subjects. Therefore, a constrained randomization schedule will be employed in which each block of 18 is built of three sub-blocks of six consecutive allocations in the following way:

- Each sub-block of six contains all six base study arms.

- In addition, to evenly spread the allocations to placebo washout across the three sub-blocks, two of the six base study arms within each sub-block are assigned to placebo washout and each sub-block has a different pair of the base study treatments assigned to the placebo washout.

- The sub-blocks of six are distributed among the centers to achieve the balance across the six base study arms within the centers and the balance in 12 treatment regimens across the centers.

Program 9.3 implements the described constrained permuted block allocation algorithm and generates a schedule with ten blocks of size 18. First, PROC PLAN outputs ten blocks (BLOCK variable) of six arms (ARM variable assumes values 1 to 6 for each value of BLOCK) into the WASHOUT data set. It assigns to each of the 6 arms a value of 1, 2, or 3 (WASHOUT variable) by randomly permuting the set of values 1, 1, 2, 2, 3, and 3, among the six observations of each block:

Program 9.3 Dose-ranging trial with a placebo washout, Part 1

```
ods listing close;
proc plan seed=56789;
    factors block=10 ordered arm=6 ordered;
    treatments washout=6;
    output out=washout washout nvals=(1 1 2 2 3 3);
data washout;
    set washout;
    label block='Block number'
          arm='Base study arm'
          washout='Washout';
proc print data=washout noobs label;
    ods listing;
    run;
```

Partial output from Program 9.3

Block number	Base study arm	Washout
1	1	3
1	2	2
1	3	3
1	4	1
1	5	2
1	6	1
2	1	1
2	2	2
2	3	3
2	4	2
2	5	3
2	6	1
3	1	2
3	2	1
3	3	2
3	4	3
3	5	1
3	6	3
4	1	2
4	2	1
4	3	2
4	4	3
4	5	3
4	6	1

Output 9.3 lists the first four blocks in the WASHOUT data set. Within each block, The WASHOUT variable determines which of the six treatment arms will be switched to placebo in the first, second or third sub-block of six, respectively. In the first block of 18 (BLOCK=1), the first sub-block of six will have Arms 4 and 6 switched to placebo washout (WASHOUT=1), the second sub-block of six will have Arms 2 and 5 switched to placebo washout (WASHOUT=2) and the third sub-block of six will have Arms 1 and 3 switched to placebo washout (WASHOUT=3).

The next step is to build the constrained blocks of 18 based on the WASHOUT data set. To do that, for each BLOCK=1 to 10, for each ARM=1 to 6 within the block, three observations (with the value of variable SUBBLOCK equal to 1, 2, and 3) are output into the REGIMENS data set. The SWITCH variable is set to YES when the SUBBLOCK=WASHOUT (a sub-block where a respective arm switches to placebo) and NO otherwise. The treatment regimen is then defined as $(2 * ARM - 1)$ if the arm does not switch to placebo and as $2 * ARM$ if the arm switches to placebo. A random key (RANDOM) is added to each observation in order to randomly permute the treatment regimens within each sub-block of 6. As in Program 9.2, this variable is uniformly distributed over $(0, 1)$.

Program 9.4 Dose-ranging trial with a placebo washout, Part 2

```
data regimens;
    set washout;
    length switch $3.;
    do subblock=1 to 3;
        if washout=subblock then do;
            switch='Yes';
            treatment=arm*2;
        end;
        else do;
            switch='No';
            treatment=arm*2-1;
        end;
        random=ranuni(6789);
        output;
    end;
    label block='Block number'
        arm='Base study arm'
        washout='Washout'
        subblock='Sub-block'
        switch='Switch to placebo?'
        random='Random key'
        treatment='Treatment group';
proc print data=regimens noobs label;
    var block arm washout subblock switch random treatment;
    run;
```

Partial output from Program 9.4

Block number	Base study arm	Washout	Sub-block	Switch to placebo?	Random key	Treatment group
1	1	3	1	No	0.71089	1
1	1	3	2	No	0.93229	1
1	1	3	3	Yes	0.40721	2
1	2	2	1	No	0.05740	3
1	2	2	2	Yes	0.75990	4
1	2	2	3	No	0.42393	3
1	3	3	1	No	0.28561	5
1	3	3	2	No	0.65511	5
1	3	3	3	Yes	0.05164	6
1	4	1	1	Yes	0.92032	8

Output 9.4 lists the first ten observations in the REGIMENS data set.

To complete the randomization schedule, the REGIMENS data set needs to be sorted by block (BLOCK), subblock (SUBBLOCK) and randomly generated key (RANDOM) to have the treatment regimens randomly permuted within each sub-block of 6 (see Program 9.5).

Program 9.5 Dose-ranging trial with a placebo washout, Part 3

```
proc sort data=regimens out=schedule;
    by block subblock random;
data schedule;
    set schedule;
    an=_n_;
    label an='Allocation number';
proc print data=schedule noobs label;
    var an arm switch block subblock treatment;
    run;
```

Partial output from Program 9.5

Allocation number	Base study arm	Switch to placebo?	Block number	Sub-block	Treatment group
1	2	No	1	1	3
2	3	No	1	1	5
3	6	Yes	1	1	12
4	5	No	1	1	9
5	1	No	1	1	1
6	4	Yes	1	1	8
7	5	Yes	1	2	10
8	4	No	1	2	7
9	3	No	1	2	5
10	2	Yes	1	2	4
11	6	No	1	2	11
12	1	No	1	2	1
13	3	Yes	1	3	6
14	5	No	1	3	9
15	1	Yes	1	3	2
16	2	No	1	3	3
17	4	No	1	3	7
18	6	No	1	3	11
19	3	No	2	1	5
20	2	No	2	1	3

Output 9.5 lists the treatment assignments of the first 20 subjects in the dose-ranging trial. The first column is the subject's allocation number, the second column (Base study arm) indicates the arm to which the subject will be assigned during the base study, and the switch column shows whether or not the subject will be switched to placebo at the end of the base study. The column "Treatment group" contains the treatment regimen (one of 12) that the subject is allocated to at randomization. The other two columns help to explain the allocation algorithm implemented in Programs 9.3, 9.4, and 9.5.

9.3.4   Balanced-Across-Centers Allocation

A large number of incomplete blocks can seriously impair the balance of treatment assignments in clinical trials with stratified randomization (Hallstrom and Davis, 1988). As the last block in each stratum is often left incomplete at the end of enrollment, stratified assignment can result in as many incomplete blocks as there are strata. If the number of subjects in each stratum is small the imbalance in treatment assignments might be considerable. Typically, this is of concern in a study with numerous small centers. For example, in a 30-center study stratified by center, as many as 30 blocks might be incomplete. Additional stratification by gender may result in as many as $2 \times 30 = 60$ incomplete blocks.

To avoid having a large number of incomplete blocks, a central randomization algorithm can be employed to allocate the subjects. With central randomization, a subject is assigned the next available allocation number (and the respective treatment) in his or her stratum (for example, gender), regardless of what center this subject belongs to. This allocation scheme can be easily implemented with the Interactive Voice Response System (IVRS). However, if centers are small, the central randomization may result in some centers being very unbalanced in treatment assignments, up to a point of having only one treatment assigned to all subjects at the center. That may lead to bias if a center happens to deviate from the other centers in efficacy or safety assessments due to a training issue. Also, it will lead to a suboptimal drug re-supply pattern, where the drug kits at the center with an unbalanced allocation will have to be replenished more frequently than would have been necessary with a balanced allocation. Thus, even when an IVRS solution is available, there are merits to having the allocation balanced within the centers.

If no IVRS or similar technology is available to support a central randomization algorithm in a study, the randomization is stratified by center because of drug distribution, if nothing else. In a study with a large number of small centers it may lead to a large number of incomplete blocks.

The imbalance caused by incomplete blocks can be lessened by having the blocks balanced across the centers. Consider a study with 40 small centers and a $2:1:2$ ratio of allocation to Treatments 1, 2, and 3. The subject allocation is stratified by center to facilitate drug distribution. Stratification by gender is desired but the small size of the centers, which are expected to enroll six to nine subjects, makes it problematic. Almost all the permuted blocks of five will likely be left incomplete.

To improve the balance in treatment assignments within each stratum, the randomization schedule for each stratum will be prepared following a modified version of Zelen's (1974) approach described in Section 9.2.3. There are 30 different permutations of $(1, 1, 2, 3, 3)$. These blocks can be arranged in six balanced sets of five blocks so that the five permuted blocks within each set will have a $2:1:2$ allocation ratio across the row of subjects allocated first, second, and so on. Thus, when the five blocks within the first balanced set are distributed among Centers 1 through 5 (see Table 9.1), a balance is established across these five centers among the subjects allocated first, second, and so on, at their respective center. If the same number of subjects are allocated to all five centers, there will be a perfect balance in treatment assignments across the centers even if all five blocks are incomplete. Of course, in real life the number of subjects enrolled will vary across the centers but even in this case the described algorithm provides a better balance

Table 9.1 Permuted Blocks for the First Ten Centers in a Balanced-across-centers Allocation Scheme

Allocation order within each center	Center									
	1	2	3	4	5	6	7	8	9	10
1st subject	2	1	3	1	3	2	3	1	1	3
2nd subject	3	2	1	3	1	3	1	3	1	2
3rd subject	1	3	3	1	2	1	1	2	3	3
4th subject	3	1	1	2	3	1	3	3	2	1
5th subject	1	3	2	3	1	3	2	1	3	1

than an allocation schedule built from randomly selected permuted blocks. Further, the second balanced set of five blocks will be sent out to Centers 6 through 10; the procedure will be repeated until all the blocks are generated for all 40 study centers.

The described randomization schedule can be generated in SAS (Song and Kuznetsova, 2003). The algorithm is implemented in Program 9.6 using PROC PLAN. First, the six balanced sets of blocks are generated as six 5×5 Latin squares, where the value of TREATMENT=1 is mapped to 2, and the vector of 4 remaining values $(2, 3, 4, 5)$ is mapped to one of the six permutations of $(1, 1, 3, 3)$. This is accomplished by calling PROC PLAN with numeric values for treatment specified as (21133), (21313), (21331), (23113), (23131) and (23311). After that, all six balanced sets of five blocks each are put together, the order of the sets is randomly permuted, and the order of the blocks within each set is randomly permuted.

Program 9.6 Balanced-across-centers allocation

```
proc plan seed=6767;
    factors row=5 col=5 ordered/noprint;
    treatments treatment=5 cyclic 4;
    output out=sq1 treatment nvals=(2 1 1 3 3);
proc plan seed=7878;
    factors row=5 col=5 ordered/noprint;
    treatments treatment=5 cyclic 4;
    output out=sq2 treatment nvals=(2 1 3 1 3);
proc plan seed=8989;
    factors row=5 col=5 ordered/noprint;
    treatments treatment=5 cyclic 4;
    output out=sq3 treatment nvals=(2 1 3 3 1);
proc plan seed=9797;
    factors row=5 col=5 ordered/noprint;
    treatments treatment=5 cyclic 4;
    output out=sq4 treatment nvals=(2 3 1 1 3);
proc plan seed=8791;
    factors row=5 col=5 ordered/noprint;
    treatments treatment=5 cyclic 4;
    output out=sq5 treatment nvals=(2 3 1 3 1);
proc plan seed=6917;
    factors row=5 col=5 ordered/noprint;
    treatments treatment=5 cyclic 4;
    output out=sq6 treatment nvals=(2 3 3 1 1);
data sixsq;
    set sq1(in=in1) sq2(in=in2) sq3(in=in3) sq4(in=in4) sq5(in=in5) sq6(in=in6);
    retain r1-r6;
    if _n_=1 then do;
        r1=ranuni(4565);
        r2=ranuni(7271);
        r3=ranuni(8171);
        r4=ranuni(6171);
        r5=ranuni(7567);
        r6=ranuni(9231);
    end;
    random=r1*in1+r2*in2+r3*in3+r4*in4+r5*in5+r6*in6;
proc sort data=sixsq out=schedule;
    by random row;
data schedule;
    set schedule;
    by random row;
```

```
    an=_n_;
    center=ceil(_n_/5);
    label an='Allocation number'
          center='Center'
          treatment='Treatment group';
proc print data=schedule noobs label;
    var an center treatment;
    run;
```

Partial output from Program 9.6

Allocation number	Center	Treatment group
1	1	2
2	1	3
3	1	1
4	1	3
5	1	1
6	2	1
7	2	2
8	2	3
9	2	1
10	2	3
11	3	3
12	3	1
13	3	3
14	3	1
15	3	2
16	4	1
17	4	3
18	4	1
19	4	2
20	4	3

Output 9.6 displays the treatment assignments of the first 20 subjects. It is easy to see that the obtained assignments are identical to the assignments at Centers 1 through 4 in Table 9.1.

The benefits of balancing across the centers are most obvious when the block size is large, e.g., in the case of a $5:5:2$ allocation ratio. Balancing across centers can also be advantageous when a stratification of the schedule by another factor, e.g., gender, is contemplated in a study with small centers. Such a schedule will make the treatment groups better balanced in gender and will provide at least as good a balance in overall treatment assignments as a randomization not stratified by gender.

In the extreme case of a very small stratum, e.g., when only two to three subjects per center are expected to be enrolled in a study with a minimal block size of 8, the schedule built of incomplete blocks balanced across the centers can be considered for the stratum. More examples can be found in Cho et al. (2004).

A balanced-across-centers allocation schedule is a special case (a single-factor case) of the factorial stratification proposed by Sedransk (1973). In the multi-factor case, the factorial stratification provides a close balance across all first, second, third, and so on assignments in all strata. Factorial stratification is most beneficial if subjects are evenly distributed across the strata, which is often not the case.

9.4   Allocations Balanced on Baseline Covariates

When certain subject characteristics (baseline covariates) are known to affect the response, it is often important to have the treatment groups balanced with respect to these covariates:

- A balanced allocation helps minimize bias due to covariate imbalances unaccounted for in the analysis model.

- A balance with respect to important covariates improves the efficiency of the analysis (Senn, 1997). Note, however, that in large trials the gain in power resulting from the use of a balanced allocation is typically small compared to post-stratified analysis without balanced allocation (McEntegart, 2003). At the same time, as pointed out by McEntegart, even a large trial can suffer from a substantial loss of power due to an unbalanced allocation in the presence of many small strata. Also, an allocation balanced by baseline predictors can considerably improve the efficiency of interim and subgroup analyses.

- The results of a trial with treatment groups balanced with respect to major covariates are more convincing for the medical community. Serious imbalance in important covariates might raise concerns even if it is adjusted for in the analysis.

It is debatable to what extent the balance in prognostic factors should be pursued, when statistical analysis can account for any imbalances in covariates; see, for example, a heated discussion following Atkinson (1999). The decision to balance or not to balance in a particular study can be guided by an assessment of the probability to conclude with an imbalance that is either perceived as dangerously high in the clinical community or exceeds the boundaries within which the model assumptions can be trusted. More insights can be gained from McEntegart (2003).

It is recommended that the baseline predictors balanced upon be included in the analysis model (International Conference on Harmonization, 1998; Gail, 1988; Simon, 1979; Kalish and Begg, 1987; Senn, 1997; Scott et al., 2002). A failure to do so might result in an incorrect Type I error rate. When the randomization-based analysis is performed, it should follow the randomization procedure (Rosenberger and Lachin, 2002).

9.4.1   Stratified Permuted Block Randomization

The most common way to achieve balance in a given factor is stratified randomization which relies on creating a separate randomization schedule for each level of the factor. If there are several baseline covariates to balance upon, a separate restricted randomization schedule (most commonly, a permuted block schedule) is prepared for each stratification cell defined by a combination of factor levels. For example, in a study balanced by gender (male or female) and smoking status (smoker, ex-smoker or never-smoker), a separate schedule is generated for each of the six strata formed by a combination of the factor levels.

The stratified randomization approach has its limitations: only a small number of factors can be balanced upon. If the number of strata is large, some strata will have few or no subjects resulting in an inadequate balance across factor levels (Therneau, 1993; Rosenberger and Lachin, 2002; Hallstrom and Davis, 1988).

9.4.2   Covariate-Adaptive Allocation Procedures

When there are too many important prognostic factors for stratification to handle, one of the covariate-adaptive allocation procedures can be used to provide a balance in selected covariates. Such procedures are dynamic in nature—the treatment assignment of a subject depends on the subject's vector of covariates and thus is determined only when the subject arrives. It is conceptually different from the regular stratified randomization method that relies on a fixed allocation schedule prepared for each stratum prior to the study start.

Minimization, pioneered by Taves (1974) and expanded by Pocock and Simon (1975), is the most commonly used covariate-adaptive allocation procedure. Minimization produces a marginal balance in each individual factor but not in individual factor interaction cells. For example, when balancing on gender and smoking status, the balance in treatment assignments is achieved across factor levels (across males, females, smokers, ex-smokers, and never-smokers) but not within each cell, e.g., male smokers, as is the case with the stratified randomization. Because of that, minimization can simultaneously be balanced on a large number of factors even in a moderate size trial.

Minimization achieves the balance in treatment assignments across factor levels by choosing the allocation for the new subjects that would result in the least possible imbalance (in some sense) across the set of his or her baseline characteristics.

In what follows we will describe the minimization algorithm proposed by Taves with a variance imbalance function popularized by Freedman and White (1976). Due to its simplicity, this algorithm is frequently used in clinical trials (McEntegart, 2003) and is often described in the literature (Senn, 1997; Scott et al., 2002).

9.4.3   Taves Minimization Algorithm

Consider a parallel study in which subjects are to be allocated equally to two treatment groups (Treatment A and Treatment B) and the allocation needs to be balanced in gender (male or female) and smoking status (smoker, ex-smoker, or never-smoker).

When the minimization algorithm is described, it is convenient to assign scores of 1 and -1 to the treatment groups A and B, respectively. Suppose a male ex-smoker arrives for allocation when there are 20 allocated subjects in the trial. To select the treatment assignment for this subject, we count the number of males and the number of ex-smokers in each of the treatment arms. Assume that there are five males in the Treatment A group and seven males in the Treatment B group. The imbalance across males, defined as the difference in number of males allocated to A versus B, is $5 - 7 = -2$. Also, there are three ex-smokers in the Treatment A group and two ex-smokers in the Treatment B group. The imbalance across ex-smokers is $3 - 2 = 1$. The total imbalance, defined as the sum of imbalances across males and across ex-smokers, is $(-2) + 1 = -1$. The negative total imbalance indicates that, overall, Treatment A is underrepresented among males and among ex-smokers combined. Thus, to improve the balance, the male ex-smoker is allocated to Treatment A. When the group totals are equal, the subject is allocated to one of the treatment arms at random with equal probability.

Given the covariates and treatment allocations of the subjects already in the study, the treatment assignment of a new subject is fully determined by his or her set of covariates, except when a tie in group totals is encountered. This is why Scott et al. (2002) referred to the Taves minimization algorithm as "largely nonrandom"—that is, deterministic assignments occur more often than random ones. Nevertheless, simulations show that assignments at random occur often enough to provide a reasonably rich set of possible allocation sequences for a given sequence of covariates (Kuznetsova and Troxell, 2004). Thus, in a masked trial, a largely deterministic nature of the Taves minimization algorithm does not lead to selection bias. However, in a single-center unmasked trial, a large share of the treatment assignments will be predictable, providing a considerable opportunity for selection bias.

9.4.4   Pocock-Simon Minimization Algorithm

Pocock and Simon (1975) extended the Taves minimization algorithm to make treatment assignments less predictable. This is achieved by using an additional random element at each treatment assignment. A subject is allocated to the treatment that results in the least imbalance with probability $p < 1$ rather than $p = 1$. If p is close to 1 ($p = 0.9$ or $p = 0.95$ are often used) the Pocock-Simon procedure still has good balancing properties and somewhat less potential for selection bias in an unmasked trial. The ICH E-9 guidance recommends using a random element at each allocation.

To define the Pocock-Simon minimization algorithm, consider a clinical study with subjects allocated in a 1 : 1 ratio to Treatments A and B. The allocation scheme in this study needs to be balanced by gender (male or female) and the smoking status (smoker, ex-smoker, or never-smoker). As was explained above, the Pocock-Simon algorithm extends the Taves algorithm by introducing a random element to each allocation step. In this trial, a subject will be assigned to the treatment that results in a smaller imbalance with probability $p = 0.9$ and to the opposite treatment with probability $p = 0.1$. When the tie in total imbalances is encountered, the treatment will be assigned by a toss of a fair coin.

The Pocock-Simon procedure, which can be expanded to more than two treatment groups, allows the use of different measures of imbalance across a factor level. If some of the factors are considered more important than others, they can be included in combined imbalance with higher weight. If an interaction between the two factors is known to affect the response, the interaction should be included as a factor in the minimization algorithm (Pocock and Simon, 1975).

The minimization approach has been shown to provide a good marginal balance in a large number of factors simultaneously (Taves, 1974; Pocock and Simon, 1975; Therneau, 1993; Begg and Iglewicz, 1980; Birkett, 1985; Zielhuis et al., 1990; Weir and Lees, 2003). By McEntegart's (2003) estimate, minimization has been used in more than 1,000 trials, including several prestigious mega-trials.

9.4.5 Implementation of Minimization Algorithms in SAS

Below we describe the %ASSIGN macro that must be invoked each time a new subject is available for allocation. The set of the subject's covariates is specified through the macro parameters. The macro assigns a treatment to the new subject and also updates the ALL_ANS data set that stores the allocation numbers, covariates, and treatment assignments of all allocated study subjects. This data set will have one observation for each study subject and will include the following variables:

- The study subjects will be identified by their allocation numbers (AN variable), assigned to them in the order they were allocated.

- The treatment assignments of the study subjects will be stored in the TREATMENT variable. Treatments A and B will be coded by 1 and −1, respectively.

- The set of covariates of each subject will be described by five 0/1 variables, C1 to C5. The first three variables, C1, C2, and C3, are 0/1 indicators of the subject's level of the smoking status (smoker, ex-smoker, or never-smoker, respectively), while C4 and C5 are the indicators of the subject's gender (male or female, respectively). For example, a male ex-smoker will have the following set of variables: C1=0, C2=1, C3=0, C4=1, C5=0.

- Lastly, there will be five variables, M1 to M5, to store the marginal imbalances, that is, the differences in the number of subjects allocated to A versus B across smokers (M1), ex-smokers (M2), never-smoker (M3), males (M4), and females (M5), respectively, that result after the subject is allocated.

Before the first subject is allocated, the ALL_ANS data set must be initialized by setting all of the variables to 0 in the following step as shown in Program 9.7.

Program 9.7 Initialize the parameters in the ALL_ANS data set

```
data all_ans;
    label an='Allocation Number'
        c1='Smoker'
        c2='Ex-smoker'
        c3='Never-smoker'
        c4='Male'
        c5='Female'
        m1='Imbalance across Smokers'
        m2='Imbalance across Ex-smokers'
        m3='Imbalance across Never-smokers'
        m4='Imbalance across Males'
        m5='Imbalance across Females'
        treatment='Treatment';
    input an c1-c5 m1-m5 treatment;
    datalines;
    0 0 0 0 0 0 0 0 0 0 0 0 0
    ;
    run;
```

Each time a new subject arrives for allocation, the %ASSIGN macro is called to determine the treatment assignment of the new subject and update the ALL_ANS data set. The macro works in the following way:

- The macro reads the current marginal imbalances into a one-observation data set (ASSIGN data set). It creates indicator variables C1 to C5 that describe the covariates of the new subject and assigns them values of the macro parameters &C1 to &C5. The last macro parameter, &P, is the probability of assigning a new subject to Treatment B.

- The treatment assignment for the new subject is determined by the scalar product of the vectors (C1, C2, C3, C4, C5) and (M1, M2, M3, M4, M5), the so-called total imbalance (TOTIMB variable). If TOTIMB=0, the TREATMENT variable is set to 1 or −1 with probability 0.5. If TOTIMB is positive, the TREATMENT variable is set to −1 (Treatment B) with probability &P and 1 (Treatment A) with probability 1-&P. If TOTIMB is negative, TREATMENT=1 (Treatment A) with probability &P and value TREATMENT=−1 (Treatment B) with probability 1 to &P.

- After the new subject has been assigned to a treatment group, the marginal imbalances M1 to M5 are updated in the ASSIGN data set. This data set, which contains the allocation number, covariate indicators C1 to C5, treatment assignment for the new subject, and updated marginal imbalances, is appended to the ALL_ANS data set.

We need to go through these steps every time a subject arrives for allocation. The %ASSIGN macro is defined in Program 9.8.

Program 9.8 The %ASSIGN macro

```
%macro assign(c1,c2,c3,c4,c5,p);
data assign;
    set all_ans(keep=m1-m5 an) end=lastobs;
    if lastobs;
    c1=&c1; c2=&c2; c3=&c3; c4=&c4; c5=&c5;
    totimb=c1*m1+c2*m2+c3*m3+c4*m4+c5*m5;
    * Assign -1 or 1 with probability 0.5 if totimb=0;
    if totimb=0 then treatment=2*rantbl(0,0.5,0.5)-3;
```

```
         * Assign 1 with probability &p and -1 with probability 1-&p if totimb>0,
           otherwise assign -1 with probability &p and 1 with probability 1-&p;
         else treatment=-sign(totimb)*(2*rantbl(0,1-&p,&p)-3);
         an=an+1;
         m1=m1+c1*treatment;
         m2=m2+c2*treatment;
         m3=m3+c3*treatment;
         m4=m4+c4*treatment;
         m5=m5+c5*treatment;
         keep an c1-c5 m1-m5 treatment;
    data all_ans;
        set all_ans assign;
        if an>0;
        run;
    %mend assign;
```

Program 9.9 invokes the %ASSIGN macro to allocate the first three subjects in a study that involves a never-smoking male, never-smoking female, and a smoker male in the order of arrival. These subjects will be assigned to the treatment that produces better balance with probability 0.9 and thus &P=0.9.

Program 9.9 Pocock-Simon minimization algorithm with $p = 0.9$

```
* 1st subject: never-smoking male;
%assign(c1=0,c2=0,c3=1,c4=1,c5=0,p=0.9);
* 2nd subject: never-smoking female;
%assign(c1=0,c2=0,c3=1,c4=0,c5=1,p=0.9);
* 3rd subject: smoker male;
%assign(c1=1,c2=0,c3=0,c4=1,c5=0,p=0.9);
proc print data=all_ans noobs;
    var an treatment c1-c5 m1-m5;
    run;
```

Output from Program 9.9

an	treatment	c1	c2	c3	c4	c5	m1	m2	m3	m4	m5
1	-1	0	0	1	1	0	0	0	-1	-1	0
2	1	0	0	1	0	1	0	0	0	-1	1
3	1	1	0	0	1	0	1	0	0	0	1

Output 9.9 lists the allocation numbers, treatment assignments, and values of the C1 to C5 and M1 to M5 variables. The three subjects were assigned to Treatments B, A, and A, respectively.

To change the probability of assigning subjects to the treatment that produces better balance, we must change the &P macro parameter. For example, increasing the value of &P to 0.95 will result in a tighter balance with respect to baseline covariates.

To implement the Taves minimization algorithm, the &P macro parameter is set to 1.

There are other approaches to balancing an allocation on baseline covariates. Atkinson (1982) proposed an approach based on optimal design considerations that focuses on minimizing the variance of treatment contrasts in the presence of covariates rather than on balancing over the covariates to minimize the bias. A different approach (Miettinen, 1976) is based on stratifying by a single risk score that accounts for the effect of all known covariates.

The CPMP Points to Consider document on adjustment for baseline covariates (EAEMP CPMP, 2003) discourages the use of covariate-adaptive allocation procedures. The issues involved are discussed by Roes (2004). He shows that some of the arguments against dynamic allocation procedures (e.g., predictability of the assignments) apply to stratified randomization as well, and might be even more pronounced with stratified randomization. The utility of dynamic allocation procedures is well described by McEntergart (2003).

9.5 Summary

In this chapter we describe randomization procedures used by the pharmaceutical industry. We devote most attention to the popular permuted block design and its variations brought in by variable block size, constrained randomization, and randomization, balanced across the centers. For each of the described approaches, we provide SAS code to create a randomization schedule.

We also discuss covariate-adaptive allocation procedures. When there is a need to balance the treatment groups with respect to baseline covariates, the stratified permuted block randomization is typically used. However, if the number of baseline covariates to balance upon is large, stratified randomization might not be feasible. In this case, one of the dynamic covariate-adaptive allocation procedures can be used to achieve the required balance. We describe the popular minimization procedure—a largely deterministic Taves (1974) version and a version by Pocock and Simon (1975) where a random element is added at each allocation step. Both algorithms are implemented in a SAS macro. An example that illustrates the use of the macro to randomize the patients dynamically is provided.

The randomization designs not covered in this chapter include biased coin designs (Efron, 1971), the maximal procedure (Berger et al., 2003), and designs based on urn models (see the review in Wei and Lachin, 1988). The maximal procedure (Berger et al., 2003) was proposed as an alternative to a sequence of permuted blocks of small sizes. The maximal procedure maintains an imbalance no larger than a prespecified positive integer number b throughout the enrollment and provides less potential for selection bias than the set of randomized blocks of size $2b$. Although originally developed for trials with two treatments and a $1:1$ treatment ratio, the maximal procedure can be easily expanded for more than two treatments. Each of these designs achieves a certain trade-off between the treatment balance and the amount of randomization the design provides.

Randomization designs can be viewed as adaptive allocations that use previous treatment assignments to modify the probability of the next assignment to ensure good balance in treatment assignments and to provide some amount of randomization. A completely different class of randomization designs are response-adaptive allocation procedures. These designs use responses observed so far in the trial to change allocation away from perfectly balanced allocation based on a certain objective. We refer the reader to Hu and Ivanova (2004) for the most recent review of response adaptive designs.

Acknowledgments

We gratefully acknowledge advice and helpful interactions with Deborah Shapiro, Larry Gould, Meehyung Cho, Inna Perevozskaya, Chau Thach, Cindy Song, and John Troxell of Merck and Co., Inc., with whom the first author has collaborated on randomization issues over the years.

References

Atkinson, A.C. (1982). "Optimum biased coin designs for sequential clinical trials with prognostic factors." *Biometrics*. 69, 61–67.

Atkinson, A.C. (1999). "Optimum biased-coin designs for sequential treatment allocation with covariate information with Discussion (1999)." *Statistics in Medicine.* 18, 1741–1755.

Begg, C. B., Iglewicz, B. (1980). "A treatment allocation procedure for sequential clinical trials." *Biometrics.* 36, 81–90.

Berger, V.W., Ivanova, A., Knoll, M. (2003). "Minimizing Predictability while Retaining Balance through the Use of Less Restrictive Randomization Procedures." *Statistics in Medicine.* 22, 3017–3028.

Birkett, N. J. (1985). "Adaptive allocation in randomized controlled trials." *Controlled Clinical Trials.* 6, 146–155.

Blackwell, D., Hodges, J. Jr. (1957). "Design for the control of selection bias." *Annals of Mathematical Statistics.* 28, 449–460.

Cho, M., Kuznetsova, O., Perevozskaya, I., Thach, C. (2004). Allocation Procedures in Clinical Trials Globally and Locally. Merck Technical Report, 101.

Committee for Proprietary Medicinal Products. (2003). "Points to consider on adjustment for baseline covariates." The European Agency for Evaluation of Medicinal Products. CPMP/EWP/2863/99.

Efron, B. (1971). "Forcing a sequential experiment to be balanced." *Biometrika.* 58, 403–417.

Follmann, D., Proschan, M. (1994). "The effect of estimation and biasing strategies on selection bias in clinical trials with permuted blocks." *Journal of Statistical Planning and Inference.* 39, 1–17.

Freedman, L. S., White, S. J. (1976). "On the use of Pocock and Simon's method for balancing treatment numbers over prognostic factors in the controlled clinical trial." *Biometrics.* 32, 691–694.

Gail, M. N. (1988). "The effect of pooling across strata in perfectly balanced studies." *Biometrics.* 44, 151–163.

Hallstrom, A., Davis, K. (1988). "Imbalance in treatment assignments in stratified blocked randomization." *Controlled Clinical Trials.* 9, 375–382.

Hu, F., Ivanova, A. (2004). "Adaptive design." *Encyclopedia of Biopharmaceutical Statistics.* 2d ed. New York: Marcel Dekker.

International Conference on Harmonization (ICH) (1998). "Guidance on Statistical Principles for Clinical Trials." Federal Register. International Conference on Harmonization, E-9 Document. 63, 179, 49583–49598.

Kalish, L. A., Begg, C. B. (1987). "The impact of treatment allocation procedures on nominal significance levels and bias." *Controlled Clinical Trials.* 8, 121–135.

Kuznetsova, O.M., Troxell, J. K. (2004). "Maximum imbalance in treatment assignments in a clinical trial that uses the minimization allocation procedure." *2004 Proceedings of the American Statistical Association.* Biopharmaceutical Section.

McEntegart, D. J. (2003). "The pursuit of balance using stratified and dynamic randomization techniques: An overview." *Drug Information Journal.* 37, 293–308.

Miettinen, O. S. (1976). "Stratification by a multivariate confounder score." *American Journal of Epidemiology.* 104, 609–620.

Pocock, S. J., Simon, R. (1975). "Sequential treatment assignment with balancing for prognostic factors in the controlled clinical trials." *Biometrics.* 31, 103–115.

Roes, Kit C.B. (2004). "Regulatory Perspectives: Dynamic allocation as a balancing act." *Pharmaceutical Statistics.* 3, 187–191.

Rosenberger, W.F., Lachin, J.M. (2002). *Randomization in Clinical Trials: Theory and Practice.* New York: Wiley.

Scott, N. W., McPherson G.C., Ramsay, C.R., Campbell. M.K. (2002). "The method of minimization for allocation to clinical trials: a review." *Controlled Clinical Trials.* 23, 662–674.

Sedransk, N. (1973). "Allocation of sequentially available units to treatment groups." *Proceedings of the 39th International Statistical Institute.* Book 2, 393–400.

Senn, S. (1997). *Statistical Issues in Drug Development.* Chichester: Wiley.

Simon, R. (1979). "Restricted randomization designs in clinical trials." *Biometrics.* 35, 503–512.

Song, C., Kuznetsova, O.M. (2003). "Implementing Constrained or Balanced-Across-the-Centers Randomization with SAS v8 Procedure PLAN." *Proceedings of the PharmaSUG Annual Meeting*, 473–479.

Taves, D. R. (1974). "Minimization: A new method of assigning patients to treatment and control groups." *Clinical Pharmacology Therapeutics.* 15, 443–453.

Therneau, T. M. (1993). "How many stratification factors are 'too many' to use in a randomization plan?" *Controlled Clinical Trials.* 14, 98–108.

Wei, L.J. (1978). "An application of an urn model to the design of sequential controlled clinical trials." *Journal of the American Statistical Association.* 73, 559–563.

Wei, L.J., Lachin, J.M. (1988). "Properties of the urn randomization in clinical trials." *Controlled Clinical Trials.* 9, 345–364.

Weir, C. J., Lees, K. R. (2003). "Comparison of stratification and adaptive methods for treatment allocation in an acute stroke clinical trial." *Statistics in Medicine.* 22, 705–726.

Youden, W. J. (1964). "Inadmissible random assignments." *Technometrics.* 6, 103–104.

Youden, W. J. (1972). "Randomization and experimentation." *Technometrics.* 14, 13–22.

Zelen, M. (1974). "The randomization and stratification of patients to clinical trials." *Journal of Chronic Disease.* 27, 365–375.

Zielhuis, G. A. , Straatman, H., Hof-Grootenboer, A. E. van 'T, van Lier, H. J. J., Rach, G. H., van den Broek, P. (1990). "The choice of a balanced allocation method for a clinical trial in otitis media with effusion." *Statistics in Medicine.* 9, 237–246.

Chapter 10

Sample-Size Analysis for Traditional Hypothesis Testing: Concepts and Issues

Ralph G. O'Brien
John Castelloe

Sample-size analysis continues to be transformed by ever-improving strategies, methods, and software. Using these tools intelligently depends on what the investigators understand about statistical science and what they know and conjecture about the particular research questions driving the study planning. This chapter covers only the most common type of sample-size analysis—power analysis, i.e., studying the chance that a given hypothesis test will be "statistically significant," $p \leq \alpha$. We focus on the core concepts and issues that the collaborating statistician must master and that key investigators must understand.

Ralph O'Brien is Professor, Quantitative Health Sciences, Cleveland Clinic, USA (obrien.ralph@gmail.com). John Castelloe is Senior Research Scientist, SAS Institute, USA (John.Castelloe@sas.com).

We begin by reviewing p values and discuss how to conduct sample-size analyses that focus on the *classical* Type I and Type II error rates, α and β. Then we go further to consider two other error rates, the *crucial* Type I error rate, α^*, which is the chance that the null hypothesis is true even though $p \leq \alpha$, and the *crucial* Type II error rate, β^*, defined as the chance that the null hypothesis is false in some particular way even though $p > \alpha$. We argue that α^* and β^* are just as relevant (if not more so) than α and β. These issues are explored in depth through two examples stemming from a straightforward clinical trial.

10.1 Introduction

In their "Perspectives on Large-Scale Cardiovascular Clinical Trials for the New Millennium," Dr. Eric Topol and colleagues (1997) provide a fine preamble to our discussions:

> The calculation and justification of sample size is at the crux of the design of a trial. Ideally, clinical trials should have adequate power, $\approx 90\%$, to detect a clinically relevant difference between the experimental and control therapies. Unfortunately, the power of clinical trials is frequently influenced by budgetary concerns as well as pure biostatistical principles. Yet an underpowered trial is, by definition, unlikely to demonstrate a difference between the interventions assessed and may ultimately be considered of little or no clinical value. From an ethical standpoint, an underpowered trial may put patients needlessly at risk of a new therapy without being able to come to a clear conclusion.

In addition, it must be stressed that investigators do not plan studies in a vacuum. They design them based on their knowledge and thoughtful conjectures about the subject matter, on results from previous studies, and on sheer speculation. They may already be far along in answering a research question, or they may be only beginning. Richard Feynman, the 1965 Nobel Laureate in Physics and self-described "curious character," stated this somewhat poetically (1999, P. 146):

> Scientific knowledge is a body of statements of varying degrees of uncertainty,
>   some mostly unsure,
>   some nearly sure,
>   none absolutely certain.

This reflects what we will call The March of Science, which for clinical research is sketched in Figure 10.1.

As we step forward, our sample-size considerations need to reflect what we know. At any point, but especially at the beginning, the curious character inside us should be free to conduct observational, exploratory, or pilot studies because, as Feynman said, "something wonderful can come from them." Such studies are still "scientific" but they are for generating new and more specific hypotheses, not for testing them. Accordingly, little or no formal sample-size analyses may be called for. But to become "nearly sure" about our answers, we typically conduct convincing confirmatory studies under specific protocols. This often requires innovative and sophisticated statistical planning, which is usually heavily scrutinized by all concerned, especially by the reviewers. No protocol is ever perfect, but paraphrasing the New York Yankee catcher and populist sage, Yogi Berra, *Don't make the wrong mistake.*

Medical research is still dominated by traditional (frequentist) hypothesis testing and classical power analysis. Here, investigators and reviewers typically ask, "What is the chance (inferential power) that some given key p-value will be significant, i.e. less than some specified Type I error rate, α?" Thus, we cannot understand inferential power without knowing what p-values are and what they are not. Researchers rely on them to

Figure 10.1 March of science in clinical research

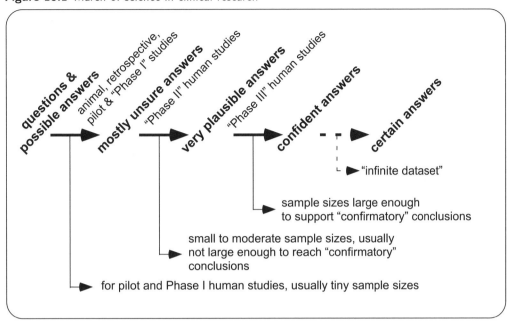

assess whether a given null hypothesis is true, but *p*-values are random variables, so they can mislead us into making Type I and II errors. The respective *classical* error rates are called α and $\beta = 1 - power$. All of this is reviewed in detail.

This chapter also considers other error rates that relate directly to two crucial questions that researchers should address. First, if a test turns out to be significant, what is the chance that its null hypothesis is actually true (Type I error)? A great many researchers think that this chance is at most α. They might say something like, "We will use $\alpha = 0.05$ as our level for statistical significance, so if we get a significant result, we will be more than 95% confident that the treatments are different with respect to this outcome." Researchers want to be able to make statements like this, but *this particular logic is wrong.* Likewise, if a test turns out to be non-significant, they might ask, "What is the chance that its null hypothesis is actually false (Type II error) to some particular degree?" Many researchers think this is the usual Type II error rate, β. It is not.

So, what is an appropriate way to do this? We describe something we call the *crucial* Type I error rate (here, α^*), which is the chance that the null hypothesis is true even after obtaining significance, $p \le \alpha$. Similarly, the crucial Type II error rate (β^*) is the chance that the null hypothesis is false in some particular way even though a $p > \alpha$ result has occurred. We argue that α^* and β^* are just as relevant (if not more so) than α and β. We demonstrate how crucial error rates can be guesstimated if investigators are willing to state and justify their current belief about the chance that the null hypothesis is indeed false. Importantly, for a given α level, greater inferential power reduces both crucial error rates.

All these concepts will be developed and illustrated by carrying out a sample-size analysis for a basic two-group trial to compare two treatments for children with severe malaria: usual care only versus giving an adjuvant drug known to reduce high levels of lactic acid. Two planned analyses will be covered. The first compares the groups with respect to a binary outcome, death within the first ten days. The second compares them on a continuous outcome, the ratio of two amino acids measured in plasma, using baseline values as covariates. The principles covered apply to any traditional statistical test being used to try to reject a null hypothesis, including analyses far more complex than those discussed here.

While obtaining an appropriate and justifiable sample size is important, going through the analytical process itself may be just as vital in that it forces the research team to work collaboratively with the statistician to delineate and critique the rationale undergirding the study and all the components of the research protocol. The investigators must specify tight research questions, the specific research design, the various measures, and an analysis plan. They must come to agree on and justify reasonable conjectures for what the "infinite dataset" may be for their study. In essence, they must imagine how the entire study will proceed before the first subject is recruited. The "group think" on this can be invaluable.

Our reader audience includes both collaborating statisticians and content investigators. While the examples given here involve clinical trials, the principles apply broadly across all of science. Therefore, we present almost no mathematical details.

The SAS procedures POWER and GLMPOWER are the primary computational engines, but we use only a small portion of their capabilities. Far more information can be found in the current SAS/STAT User's Guide.

To save space, some SAS code has been shortened and some output is not shown. The complete SAS code and data sets used in this book are available on the book's companion Web site at `http://support.sas.com/publishing/bbu/companion_site/60622.html`.

10.2   Research Question 1: Does "QCA" Decrease Mortality in Children with Severe Malaria?

According to a report released in 2003 by the World Health Organization, malaria remains one of the world's foremost health problems, killing at least one million people annually, mostly children under five years old in sub-Saharan Africa. Lactic acidosis (toxic levels of lactic acid in the blood) is a frequent complication in severe malaria and is an incremental statistical predictor ("independent risk factor") of death. Moreover, a plausible biological rationale supports the hypothesis that lactic acidosis is a contributing cause of death.

Dr. Peter Stacpoole of the University of Florida has spent decades investigating the safety and efficacy of dichloroacetate (DCA) for treating lactic acidosis in genetic and acquired diseases. In 1997–99, he collaborated with Dr. Sanjeev Krishna of the University of London to lead a team that conducted a small, randomized, double-blind, controlled trial of quinine-only versus quinine+DCA in treating lactic acidosis in Ghanaian children with severe malaria (Agbenyega et al., 2003). They concluded that a single infusion of DCA was well-tolerated, did not appear to interfere with quinine and, as hypothesized, reduced blood lactate levels. The sample size of $N = 62 + 62$ was much too small to support comparing mortality rates. The authors concluded that a large prospective study was warranted.

From now on the story is fictionalized. Suppose "quadchloroacetate" (QCA) has the same molecular structure as DCA at the active biological site, and has now been shown in large animal and human studies to be clinically equivalent to DCA in quickly reducing abnormally high blood lactate levels. However, QCA is less expensive to produce (about US$1/dose) and has a longer shelf-life, especially in tropical climates.

"Dr. Sol Capote" heads the malaria research group at "Children's Health International (CHI)," and he and his colleagues are now designing a large clinical trial to be coordinated from "Jamkatnia" in West Africa. Dr. Capote is an experienced investigator, so he knows that substantial thought, effort, and experience must go into developing the sample-size analysis and the rest of the statistical considerations.

The CHI study will use a randomized, double-blind design to compare usual care only (UCO) versus usual care plus a single dose of QCA. After reviewing all previous human studies of both DCA and QCA, the CHI team is convinced that a single dose of QCA is very likely to be safe. Accordingly, after consulting with Jamkatnian health officials and a bioethicist, they decide that two-thirds of the subjects should get QCA.

10.3 p-Values, α, β and Power

The primary efficacy analysis will yield a p-value that compares the mortality rates of control versus QCA. Smaller p-values indicate greater statistical separation between the two samples, but *how that p-value is determined is an issue that is not critical to understanding the essential concepts in sample-size analysis.* In this case, that p-value may come from one of the many methods to compare two independent proportions, including the likelihood ratio chi-square test, as used here, or it may come from a logistic or hazard modeling that includes co-predictors. Regardless of what test is used to get the p-value, if p is small enough ("significant") and the QCA mortality rates are better, Dr. Capote will report that the study supported the hypothesis that QCA reduces mortality in children with severe malaria complicated with lactic acidosis. If the p-value is not small enough ("not significant"), then he will report that the data provided insufficient evidence to support the hypothesis.

10.3.1   Null and Non-Null Distributions of p-Values; Type I and Type II Errors

Dr. Capote's quest here is to answer the following question: Does QCA decrease mortality in children with severe malaria? While Mother Nature knows the correct answer, only if we were able to gather an infinitely large, perfectly clean dataset could we figure this out ourselves. Rather, we must design a study or, usually, a series of studies, that will yield sample datasets that give us a solid chance of inferring what Mother Nature knows. Unfortunately, Lady Luck builds randomness into those sample datasets, and thus even the best studies can deliver misleading answers.

Please study the top distribution in Figure 10.2. Here, there is no difference between the two groups' mortality in the infinite dataset, so *regardless of the sample size*, all values for $0 < p < 1$ are equally likely. Accordingly, there is a 5% chance that $p \leq 0.05$, or a $100\alpha\%$ chance that $p \leq \alpha$ (in practice, these percentages are rarely exact because the data are discrete or they fail to perfectly meet the test's underlying mathematical assumptions). If there is no true effect, but $p \leq \alpha$ indicates otherwise, then this result triggers a Type I error, which is why α is called the Type I error rate. α should be chosen after some thought; it should not be automatically set at 0.05.

What if QCA has some true effect, good or bad? Then the non-null (non-central) distribution of the p-value will be skewed toward 0.0, as in the middle and bottom plots of Figure 10.2. The middle one comes from presuming (1) true mortality rate of 0.28 for UCO and 0.21 for QCA, which is a 25% reduction in mortality; (2) 700 patients randomized to UCO versus 1400 to QCA, and (3) the p-value arises from testing whether the two mortality proportions differ (non-directional) using the likelihood ratio chi-square statistic. The bottom plot conforms to presuming that QCA cuts mortality by 33%.

Inferential power is the chance that $p \leq \alpha$ when the null hypothesis is false, which is why α could be called the "null power." If there is some true effect, but $p > \alpha$, then a Type II error is triggered. Consider the middle plot, which is based on a 25% reduction in mortality. Using the common Type I error rate, $\alpha = 0.05$, the power is 0.68, so the Type II error rate is $\beta = 0.32$. By tolerating a higher α-level, we can increase power (decrease β). Here, using $\alpha = 0.20$. the power is 0.87, so $\beta = 0.13$. If QCA is more effective (bottom plot, 33% reduction in mortality), then the power rises to 0.91 with $\alpha = 0.05$ and 0.98 with $\alpha = 0.20$. Again, we will never know the true power, because Mother Nature will never tell us the true mortality rates in the two groups, and Lady Luck will always add some natural randomness into our outcome data.

10.3.2   Balancing Type I and II Error Rates

Recall that Topol et al. (1997) advocated that the power should be around 90%, which puts the Type II error rate around 10%. We generally agree, but stress that there should

Figure 10.2 Distribution of the *p*-value under the null hypothesis and two non-null scenarios

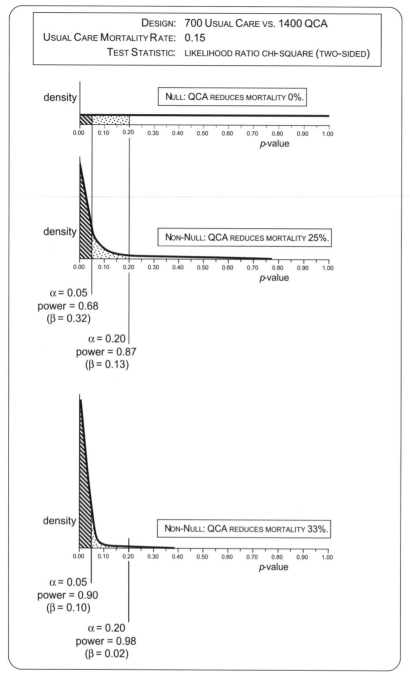

be no standard power threshold that is accepted blindly as being satisfactory across all situations. Why do so many people routinely perform power analyses using $\alpha = 0.05$ and 80% power ($\beta = 0.20$)? Rarely do they give it any thought.

Consider the middle plot in Figure 10.2. We could achieve a much better Type II error rate of 13% if we are willing to accept a substantially greater Type I error rate of 20%. Investigators should seek to obtain α versus β values that are in line with the consequences of making a Type I error versus a Type II error. In some cases, making a Type II error may be far more costly than making a Type I error. In particular, in the early stages of the March of Science, making a Type I error may only extend the research to another stage.

This is undesirable, of course, as are all mistaken inferences in science, but making a Type II error may be far more problematic, because it may halt a line of research that would ultimately be successful. So it might be justified to use $\alpha = 0.20$ (maybe more) in order to reduce β as much as possible. Using such high α values is not standard, so investigators adopting this philosophy must be convincing in their argument.

10.4 A Classical Power Analysis

Dr. Capote and his team plan their trial as follows.

Study Design

This trial follows the small ($N = 62 + 62$) DCA trial reported by Agbenyega et al. (2003). It will be double-blind, but instead of a 1:1 allocation, the team would like to consider giving one patient usual care only for every two patients that get QCA, where the QCA is given in a single infusion of 50 mg/kg.

Subjects

Study patients will be less than 13 years old with severe malaria complicated by lactic acidosis. "Untreatable" cases (nearly certain to die) will be excluded. These terms will require operational definitions and the CHI team will formulate the other inclusion/exclusion criteria and state them clearly in the protocol. They think it is feasible to study up to 2100 subjects in a single malaria season using centers in Jamkatnia alone. If needed, they can add more centers in neighboring "Gabrieland" and increase the total size to 2700. Drop-outs should not be a problem in this study, but all studies must consider this and enlarge recruitment plans accordingly.

Primary Efficacy Outcome Measure

Death before Day 10 after beginning therapy. Almost all subjects who survive to Day 10 will have fully recovered. Time to death (i.e., survival analysis) is not a consideration.

Primary Analysis

To keep this story and example relatively simple, we will limit our attention to the basic relative risk that associates treatment group (UCO vs. QCA) with death (no or yes). For example, if 10% died under QCA and 18% died under UCO, then the estimated relative risk would be $0.10/0.18 = 0.55$ in favor of QCA. p-values will be based on the likelihood ratio chi-square statistic for association in a 2×2 contingency table. The group's biostatistician, "Dr. Phynd Gooden," knows that the test of the treatment comparison could be made with greater power through the use of a logistic regression model that includes baseline measurements such as a severity score or lactate levels, etc. (as was done in Holloway et al., 1995). In addition, this study will be completed in a single malaria season, so performing interim analyses is not feasible. These issues are beyond the scope of this chapter.

10.4.1 Scenario for the Infinite Dataset

A prospective sample-size analysis requires the investigators to characterize the hypothetical infinite dataset for their study. Too often, sample-size analysis reports fail to explain the rationale undergirding the conjectures. If we explain little or nothing, reviewers will question the depth of our thinking and planning, and thus the scientific integrity of our proposal. We must be as thorough as possible and not apologize for having to make some sound guesstimates. All experienced reviewers have had to do this themselves.

Dr. Stacpoole's $N = 62 + 62$ human study (Agbenyega et al., 2003) had eight deaths in each group. This yields 95% confidence intervals of [5.7%, 23.9%] for the quinine-only mortality rate (using the EXACT statement's BINOMIAL option in the FREQ procedure) and [0.40, 2.50] for the DCA relative risk (using the asymptotic RELRISK option in the OUTPUT statement in PROC FREQ). These wide intervals are of little help in specifying the scenario. However, CHI public health statistics and epidemiologic studies in the literature indicate that about 19% of these patients die within ten days using quinine only. This figure will likely be lower for a clinical trial, because untreatable cases are being excluded and the general level of care could be much better than is typical. Finally, the Holloway et al. (1995) rat study obtained a DCA relative risk of 0.67 [95% CI: 0.44, 1.02], and the odds ratios adjusting for baseline covariates were somewhat more impressive, e.g., OR= 0.46 (one-sided $p = 0.021$).

Given this information, the research team conjectures that the mortality rate is 12-15% for usual care. They agree that if QCA is effective, then it is reasonable to conjecture that it will cut mortality 25-33% (relative risk of 0.67-0.75).

Topol et al. (1999) wrote about needing sufficient power to detect a clinically relevant difference between the experimental and control therapies." Some authors speak of designing studies to detect the smallest effect that is clinically relevant. How do we define such things? Everyone would agree that mortality reductions of 25-33% are clinically relevant. What about 15%? Even a 5% reduction in mortality would be considered very clinically relevant in a disease that kills so many people annually, especially because a single infusion of QCA is relatively inexpensive. Should the CHI team feel they must power this study to detect a 5% reduction in mortality? As we shall see, this is infeasible. It is usually best to ask: "What do we actually know at this point? What do we think is possible? What scenarios are supportable? Will the reviewers agree with us?"

10.4.2   What Allocation Ratio? One-Sided or Two-Sided Test?

Dr. Gooden is aware of the fact that the likelihood ratio chi-square test for two independent proportions can be more powerful when the sample sizes are unbalanced. Her first task is to assess how the planned 1:2 (UCO: QCA) allocation ratio affects the power. As shown in Program 10.1, this is relatively easy to do in PROC POWER. Its syntax is literal enough that we will not explain it, but note particularly the GROUPWEIGHTS statement.

Table 10.1 displays results obtained using Program 10.1 and some simple further computations. For this conjecture of 15% mortality versus $0.67 \times 15\%$ mortality, the most efficient of these four designs is the 1:1 allocation ratio. It has a power of 0.930 or $\beta = 0.070$ with $N_{\text{total}} = 2100$ ($\alpha = 0.05$), and to get a 0.90 power requires $N_{\text{total}} = 1870$. Compared to the 1:1 design, the 1:2 design has a 36% larger Type II error rate ("relative Type II risk ratio") at $N_{\text{total}} = 2100$ and requires 2064 subjects to achieve a 0.90 power. Thus, the 1:2 design has a relative efficiency of $1870/2064 = 0.91$ and requires about 10% more subjects to achieve 0.90 power (relative inefficiency: $2064/1870 = 1.10$). The relative inefficiencies for the 2:3 and 1:3 designs are 1.03 and 1.29, respectively.

Table 10.1 Effect of the Allocation Ratio, $N_{UCO} : N_{QCA}$, on Power, β, and Sample Size for Two-Sided $\alpha = 0.05$ Assuming 15% Mortality with Usual Care and a Relative Risk of 0.67 in Favor of QCA

		Allocation ratio ($N_{UCO} : N_{QCA}$)			
		1:1	2:3	1:2	1:3
	Power	0.930	0.923	0.905	0.855
$N_{\text{total}} = 2100$	β	0.070	0.077	0.095	0.145
	Relative Type II risk ratio	1.00	1.10	1.36	2.07
	N_{total}	1870	1925	2064	2420
Power = 0.90	Relative efficiency	1.00	0.97	0.91	0.77
	Relative inefficiency	1.00	1.03	1.10	1.29

Program 10.1 Compare allocation weights

```
* Powers at Ntotal=2100;
proc power;
    TwoSampleFreq
    GroupWeights =  (1 1) (2 3) (1 2) (1 3)  /* UCO:QCA */
    RefProportion = .15  /* Usual Care Only (UCO) mortality rate*/
    RelativeRisk =  .67   /* QCA mortality vs. UCO mortality */
    alpha = .05
    sides = 1 2
    Ntotal = 2100
    test = lrchi
    power = .;

* Ntotal values for power = 0.90;
proc power;
    TwoSampleFreq
    GroupWeights =  (1 1) (2 3) (1 2) (1 3)/* UCO:QCA */
    RefProportion = .15  /* Usual Care Only (UCO) mortality rate*/
    RelativeRisk =  .67   /* QCA mortality vs. UCO mortality */
    alpha = .05
    sides = 1 2
    Ntotal = .
    test = lrchi
    power = .90;
    run;
```

Note that Dr. Gooden uses SIDES=1 2 in Program 10.1 to consider both one-sided and two-sided tests. Investigators and reviewers too often dogmatically call for two-sided tests only because they believe using one-sided tests is not trustworthy. But being good scientists, Dr. Capote's team members think carefully about this issue. Some argue that the scientific question is simply whether QCA is efficacious versus whether it is not efficacious, where "not efficacious" means that QCA has no effect on mortality or it increases mortality. This conforms to the one-sided test. For the design, scenario, and analysis being considered here, the one-sided test requires 1683 subjects versus 2064 for the two-sided test, giving the two-sided test a relative inefficiency of 1.23. At $N = 2100$, the Type II error rate for the one-sided test is $\beta = 0.052$, which is 45% less than the two-sided rate of $\beta = 0.095$. On the other hand, other members argue that it is important to assess whether QCA increases mortality. If it does, then the effective Type II error rate for the one-sided test is 1.00. This logic causes many to never view one-sided tests favorably under any circumstances. After considering these issues with Dr. Gooden, Dr. Capote decides to take the traditional approach and use a two-sided test.

For some endpoints, such as for rare adverse events or in trials involving rare diseases, the argument in favor of performing one-sided tests is often compelling. Suppose there is some fear that a potential new treatment for arthritis relief could increase the risk of gastrointestinal bleeding in some pre-specified at-risk subpopulation, say, raising this from an incidence rate in the first 30 days from 8% to 24%, a relative risk of 3.0. A balanced two-arm trial with $N = 450 + 450$ subjects may be well-powered for testing efficacy (arthritis relief), but suppose the at-risk group is only 20% of the population being sampled, so that only about $N = 90 + 90$ will be available for this planned sub-group analysis. Using $\alpha = 0.05$, the likelihood ratio test for comparing two independent proportions will provide 0.847 power for the two-sided test and 0.910 power for the one-sided test. Thus, using a one-sided test cuts the Type II error rate from 0.153 to 0.090, a 41% reduction. Stated differently, using a two-sided test increases β by 70%. However, if this research aim is concerned only with detecting an increase in GI bleeding, why not use the statistical hypothesis—the one-sided version—that conforms to that aim? If using the two-sided test increases the Type II error rate by 70%, why is that more trustworthy?

For completeness, and because it takes so little time to do, Dr. Gooden also uses PROC POWER to find the approximate optimal allocation ratio. After iterating the group weights, she settles on using Program 10.2 to show that while the theoretical optimal is approximately 0.485:0.515, the balanced (0.500:0.500) design has almost the same efficiency.

Program 10.2 Find optimal allocation weights

```
proc power;
    TwoSampleFreq
    GroupWeights =        /* UCO : QCA */
    (.50 .50) (.49 .51) (.485  .515) (.48 .52) (.45 .55) (.33 .66)
    RefProportion = .15  /* Usual Care Only (UCO) mortality rate*/
    RelativeRisk =  .67   /* QCA mortality vs. UCO mortality */
    alpha = .05
    sides = 2
    Ntotal = .
    test =  LRchi /* likelihood ratio chi-square */
    power = .90
    nfractional;
    run;
```

Output from Program 10.2

Index	Weight1	Weight2	Fractional N Total	Actual Power	Ceiling N Total
1	0.500	0.500	1868.510571	0.900	1869
2	0.490	0.510	1867.133078	0.900	1868
3	0.485	0.515	1867.002923	0.900	1868
4	0.480	0.520	1867.245653	0.900	1868
5	0.450	0.550	1876.616633	0.900	1877
6	0.330	0.660	2061.667869	0.900	2062

Should the study use the less efficient 1:2 design? After substantial debate within his team, Dr. Capote decides that the non-statistical attributes of the 1:2 design give it more practical power than the 1:1 design. First, nobody has safety concerns about giving a single dose of QCA. Second, Jamkatnian health officials and parents will prefer hearing that two out of three subjects will be treated with something that could be life-saving for some. Third, the extra cost associated with a 10% increase in the sample size is not prohibitive. Given that this study's set-up costs are high and the costs associated with data analysis and reporting are unaffected by the sample size, the total cost will increase by only about 3%.

10.4.3 Obtaining and Tabling the Powers

The stage is now set to carry out and report the power analysis. Please examine Program 10.3 together with Output 10.3, which contains the essential part of the results. In SAS 9.1, PROC POWER provides plain graphical displays of the results (not shown here), but lacks corresponding table displays. As this chapter was going to press, a general-purpose SAS macro, %Powtable, was being developed to help meet this need; see the book's website. Here, Dr. Gooden uses the ODS OUTPUT command and the TABULATE procedure to create a basic table.

Program 10.3 Power analysis for comparing mortality rates

```
options ls=80 nocenter FORMCHAR="|----|+|---+=|-/\<>*";

proc power;
    ODS output output=MortalityPowers;
    TwoSampleFreq
    GroupWeights =  (1 2) /* 1 UCO : 2 QCA*/
    RefProportion = .12 .15 /* UCO mortality rate */
    RelativeRisk = .75 .67  /* QCA rate vs UCO rate*/
    alpha = .01 .05 .10
    sides = 2
    Ntotal = 2100 2700
    test = LRchi  /* likelihood ratio chi-square */
    power = .;
    plot vary (panel by RefProportion RelativeRisk);

/* Avoid powers of 1.00 in table */
data MortalityPowers;
    set MortalityPowers;
    if power>0.999 then power999=0.999;
      else power999=power;

proc tabulate data=MortalityPowers format=4.3 order=data;
    format Alpha 4.3;
    class RefProportion RelativeRisk alpha NTotal;
    var Power999;
    table
    RefProportion="Usual Care Mortality"
        * RelativeRisk="QCA Relative Risk",
    alpha="Alpha"
        * Ntotal="Total N"
        * Power999=""*mean=" "/rtspace=28;
    run;
```

Output from Program 10.3

```
-------------------------------------------------------------
|                        |                 Alpha             |
|                        |-----------------------------------|
|                        |  .010   |  .050   |  .100   |
|                        |---------+---------+---------|
|                        | Total N | Total N | Total N |
|                        |---------+---------+---------|
|                        |2100|2700|2100|2700|2100|2700|
|------------------------+----+----+----+----+----+----|
|Usual Care  |QCA Relative|    |    |    |    |    |    |
|Mortality   |Risk        |    |    |    |    |    |    |
|------------+------------|    |    |    |    |    |    |
|0.12        |0.75        |.329|.437|.569|.677|.687|.780|
|            |------------+----+----+----+----+----+----|
|            |0.67        |.622|.757|.823|.905|.893|.948|
|------------+------------+----+----+----+----+----+----|
|0.15        |0.75        |.438|.566|.677|.783|.781|.864|
|            |------------+----+----+----+----+----+----|
|            |0.67        |.757|.872|.905|.960|.948|.981|
-------------------------------------------------------------
```

Figure 10.3 Plots for the mortality analysis showing how changing the reference proportion or the relative risk rate affects power.

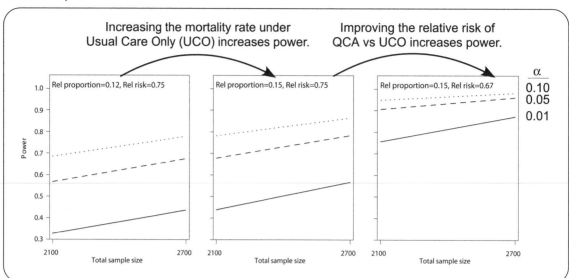

Figure 10.3 juxtaposes three plots that were produced by using the same ODS output data set, MortalityPowers, with a SAS/GRAPH program not given here, but which is available at this book's companion Website. This shows concretely how power increases for larger UCO mortality rates or better (smaller) relative risks for QCA versus UCO.

Colleagues and reviewers should have little trouble understanding and interpreting the powers displayed as per Output 10.3. If the goal is to have 0.90 power using $\alpha = 0.05$, then $N = 2100$ will suffice only under the most optimistic scenario considered: that is, if the usual care mortality rate is 15% and QCA reduces that risk 33%. $N = 2700$ seems to be required to ensure adequate power over most of the conjecture space.

Tables like this are valuable for teaching central concepts in traditional hypothesis testing. We can see with concrete numbers how power is affected by various factors. While we can set N and α, Mother Nature sets the mortality rate for usual care and the relative risk associated with QCA efficacy.

Let us return to the phrase from Topol et al. (1997) that called for clinical trials to have adequate power "to detect a clinically relevant difference between the experimental and control therapies." With respect to our malaria study, most people would agree that even a true 5% reduction in mortality is clinically relevant and it would also be economically justifiable to provide QCA treatment given its low cost and probable safety. But could we detect such a small effect in this study? If QCA reduces mortality from 15% to $0.95 \times 15\% = 14.25\%$, then the proposed design with $N = 900 + 1800$ has only 0.08 power (two-sided $\alpha = 0.05$). In fact, under this 1:2 design and scenario, it will require almost 104,700 patients to provide 0.90 power. This exemplifies why confirmatory trials (Phase III) are usually designed to detect plausible outcome differences that are considerably larger than "clinically relevant." The plausibility of a given scenario is based on biological principles and from data gathered in previous relevant studies of all kinds and qualities. By ordinary human nature, investigators and statisticians tend to be overly optimistic in guesstimating what Mother Nature might have set forth, and this causes our studies to be underpowered. This problem is particularly relevant when new therapies are tested against existing therapies that might be quite effective already. It is often the case that potentially small but important improvements in therapies can be reliably assessed only in very large trials. Biostatisticians are unwelcome and even sometimes disdained when they bring this news, but they did not make the Fundamental Laws of Chance—they are only charged with policing and adjudicating them.

10.5 Beyond α and β: Crucial Type I and Type II Error Rates

Are α and β (or power= $1 - \beta$) the only good ways to quantify the risk of making Type I and Type II errors? While they may be the classical rates to consider and report, they fail to directly address two fundamental questions:

- If the trial yields traditional statistical significance ($p \leq \alpha$), what is the chance this will be an incorrect inference?
- If the trial does not yield traditional statistical significance ($p > \alpha$), what is the chance this will be an incorrect inference?

To answer these in some reasonable way, we need to go beyond using just α and β.

10.5.1 A Little Quiz: Which Study Provides the Strongest Evidence?

Table 10.2 summarizes outcomes from three possible QCA trials. Which study has the strongest evidence that QCA is effective? Studies #1 and #2 have $N = 150 + 300$ subjects, whereas #3 has $N = 700 + 1400$ subjects. Studies #1 and #3 have identical 0.79 estimates of relative risk, but with $p = 0.36$, Study #1 does not adequately support QCA efficacy. Choosing between Studies #2 and #3 is harder. They have the same p-value, so many people would argue that they have the same inferential support. If so, then #2 is the strongest result, because its relative risk of 0.57 is substantially lower than the relative risk of 0.79 found in Study #3. However, Study #3 has nearly five times the sample size, so it has greater power. How should that affect our assessment?

Table 10.2 Which Study Has the Strongest Evidence That QCA is Effective?

	Deaths/N		Mortality		Relative risk		LR test
Study	UCO	QCA	UCO	QCA	RR	[95% CI]	p-value
#1	21/150	33/300	14.0%	11.0%	0.79	$[0.47, 1.31]$	0.36
#2	21/150	24/300	14.0%	8.0%	0.57	$[0.33, 0.992]$	0.05
#3	98/700	154/1400	14.0%	11.0%	0.79	$[0.62, 0.995]$	0.05

10.5.2 To Answer the Quiz: Compare the Studies' Crucial Error Rates

Suppose that Mother Nature has set the true usual care mortality rate at 0.15 and the QCA relative risk at 0.67, the most powerful scenario we considered above. We have already seen (Figure 10.3, Output 10.3) that with $N = 700 + 1400$ subjects and using $\alpha = 0.05$ (two-sided), the power is 90%. With 150 subjects getting usual care and 300 getting QCA, the power is only about 33%.

Now, in addition, suppose that Dr. Capote and his team are quite optimistic that QCA is effective. This does not mean they have lost their ordinary scientific skepticism and already believe that QCA is effective. Consider another Feynman-ism (1999, P. 200):

> The thing that's unusual about good scientists is that they're not so sure of themselves as others usually are. They can live with steady doubt, think "maybe it's so" and act on that, all the time knowing it's only "maybe."

Dr. Capote's team understands that even for the most promising experimental treatments, the clear majority fail to work when tested extensively. In fact, Lee and Zelen (2000) estimated that among 87 trials completed and reported out by the Eastern Cooperative Oncology Group at Harvard from 1980-1995, only about 30% seem to have been testing therapies that had some clinical efficacy.

Let us suppose that Dr. Capote's team conducted 1000 independent trials looking for significant treatment effects, but Mother Nature had set things up so that 700 effects were actually null. What would we expect to happen if Dr. Capote ran all 1000 trials at average powers of 33%? 90%? Table 10.3 presents some straightforward computations that illustrate what we call the *crucial* Type I and Type II error rates. With 700 null tests, we would expect to get 35 (5%) Type I errors (false positives). From the 300 non-null hypotheses tested with 33% power, we would expect to get 99 true positives. Thus, each "significant" test ($p \leq 0.05$) has an $\alpha^* = 35/134 = 0.26$ chance of being misleading. Note how much larger this is than $\alpha = 0.05$. Some people (including authors of successful statistics books) confuse α and α^*, and hence they also misinterpret what p-values are. A p-value of 0.032 does *not* imply that there is a 0.032 chance that the null hypothesis is true.

Table 10.3 Expected Results for 1000 Tests Run at $\alpha = 0.05$. Note: The true hypothesis is null in 700 tests. For the 300 non-null tests, the average power is 33% or 90%.

	Result of hypothesis test	
	$p \leq 0.05$ ("significant")	$p > 0.05$ ("not significant")
33% average power		
700 true null	5% of 700 = 35	95% of 700 = 665
300 true non-null	33% of 300 = 99	67% of 300 = 201
	Crucial Type I error rate: $\alpha^* = 35/134 = \mathbf{0.26}$	Crucial Type II error rate: $\beta^* = 201/866 = \mathbf{0.23}$
90% average power		
700 true null	5% of 700 = 35	95% of 700 = 665
300 true non-null	90% of 300 = 270	10% of 300 = 30
	Crucial Type I error rate: $\alpha^* = 35/305 = \mathbf{0.11}$	Crucial Type II error rate: $\beta^* = 30/695 = \mathbf{0.04}$

The crucial Type II error rate, β^*, is defined similarly. With 33% power, we would expect to get 201 Type II errors (false negatives) to go with 665 true negatives; thus $\beta^* = 210/866 = 0.23$. Note that this is not equal to $\beta = 0.67$.

10.5.3   Greater Power Reduces Both Types of Crucial Error Rates

A key point illustrated in Table 10.3 is that *greater power reduces both types of crucial error rates*. In other words, statistical inferences are generally more trustworthy when the underlying power is greater. Let us return to Table 10.2. Again, which study has the strongest evidence that QCA is effective? Even under our most powerful scenario, a $p \leq 0.05$ result has a 0.26 chance of being misleading when using $N = 150 + 300$, as per Study #2. This falls to 0.11 using $N = 700 + 1400$ (Study #3). Both studies may have yielded $p = 0.05$, but they do not provide the same level of support for inferring that QCA is effective. Study #3 provides the strongest evidence that QCA has *some* degree of efficacy. This concept is poorly understood throughout all of science.

10.5.4   The March of Science and Sample-Size Analysis

Consistent with Lee and Zelen (2000), we think that investigators designing clinical trials are well served by considering α^* and β^*. (Note that Lee and Zelen's definition is reversed from ours in that our α^* and β^* correspond to their β^* and α^*, respectively.) Ioannidis (2005b) used the same logic in arguing "why most published research findings are false." Wacholder et al. (2004) described the same methodology to more carefully infer whether a genetic variant is really associated with a disease. Their "false positive report probability" is identical to α^*. Also, readers familiar with accuracy statistics for medical tests will see that $1 - \alpha$ and β are isomorphic to the specificity and sensitivity of the diagnostic method and $1 - \alpha^*$ and $1 - \beta^*$ are isomorphic with the positive and negative predictive values.

Formally, let γ be the probability that the null hypothesis is false. We like to think of γ as measuring where the state of knowledge currently is in terms of confirming the non-null hypothesis; in short, its location along its March of Science (Figure 10.1). Thus, for novel research hypotheses, γ will be nearer to 0. For mature hypotheses that are ready for solid confirmation with say, a large Phase III trial, γ will be markedly greater than 0. We might regard $\gamma = 0.5$ as scientific equipoise, saying that the hypothesis is halfway along its path to absolute confirmation in that we consider the null and non-null hypothesis as equally viable. Lee and Zelen's (2000) calculations put γ around 0.3 for Phase III trials coordinated in the Eastern Cooperative Oncology Group.

Given γ, α and some β set by some particular design, sample size, and non-null scenario, we can apply Bayes' Theorem to get

$$\alpha^* = \text{Prob}[H_0 \text{ true} \mid p \le \alpha] = \frac{\alpha(1 - \gamma)}{\alpha(1 - \gamma) + (1 - \beta)\gamma}$$

and

$$\beta^* = \text{Prob}[H_0 \text{ false} \mid p > \alpha] = \frac{\beta\gamma}{\beta\gamma + (1 - \alpha)(1 - \gamma)}.$$

To be precise, "H_0 false" really means "H_0 false, as conjectured in some specific manner." For the example illustrated first in Table 10.3, we have $\gamma = 0.30$, $\alpha = 0.05$ and $\beta = 0.67$, thus

$$\alpha^* = \frac{(0.05)(1 - 0.30)}{(0.05)(1 - 0.30) + (1 - 0.67)(0.30)} = 0.261$$

and

$$\beta^* = \frac{(0.67)(0.30)}{(0.67)(0.30) + (1 - 0.05)(1 - 0.30)} = 0.232.$$

In Bayesian terminology, $\gamma = 0.3$ is the prior probability that QCA is effective, and $1 - \alpha^* = 0.739$ is the posterior probability given that $p \le \alpha$. However, nothing here involves Bayesian *data analysis* methods, which have much to offer in clinical research, but are not germane to this chapter. Some people are bothered by the subjectivity involved in specifying prior probabilities like γ, but we counter by pointing out that there are many other subjectivities involved in sample-size analysis for study planning, especially the conjectures made in defining the infinite dataset. Indeed, we find that most investigators are comfortable specifying γ, at least with a range of values, and that computing various α^* and β^* values of interest gives them much better insights into the true inferential strength of their proposed (frequentist) analyses.

10.6   Research Question 1, Continued: Crucial Error Rates for Mortality Analysis

In developing the statistical considerations for the QCA/malaria trial, Dr. Gooden understands the value in assessing its crucial Type I and Type II error rates, and she presses her CHI colleagues to complete the exercise faithfully. As mentioned before, they are optimistic that QCA is effective, but to compute crucial error rates, they must now quantify that optimism by setting γ. Initial discussions place γ near 0.75, but the 0.30 value reported by Lee and Zelen (2000) tempers their thinking substantially. They settle on $\gamma = 0.50$. Dr. Gooden will also use $\gamma = 0.30$.

Program 10.4 Compute crucial error rates for mortality endpoint

```
options ls=80 nocenter FORMCHAR="|----|+|---+=|-/\<>*";

proc power;
  ODS output output=MortalityPowers;
  TwoSampleFreq
    GroupWeights = (1 2)     /* 1 UCO for every 2 QCA  */
    RefProportion = .12 .15  /* UCO mortality rate     */
    RelativeRisk = .75 .67   /* QCA rate vs UCO rate   */
    alpha = .01 .05
    sides = 2
    Ntotal = 2700
    test = LRchi  /* likelihood ratio chi-square */
    power = .;
  plot key=OnCurves;
  run;

* Call %CrucialRates macro, given in Appendix B of this chapter;
%CrucialRates(   PriorPNullFalse = .30 .50,
                 Powers = MortalityPowers,
                 CrucialErrRates = MortalityCrucRates   )

proc tabulate data=MortalityCrucRates format=4.3 order=data;
title3 "Crucial Error Rates for QCA Malaria Trial";
  format alpha 4.3;
  class RefProportion RelativeRisk alpha gamma TypeError NTotal;
  var CrucialRate;
  table
    Ntotal="Total N: ",
    RefProportion="Usual Care Mortality"
      * RelativeRisk="QCA Relative Risk",
    alpha="Alpha"
      * gamma="PriorP[Null False]"
      * TypeError="Crucial Error Rate"
      * CrucialRate=""*mean=" "
  / rtspace = 26;
  run;
```

Program 10.4 gives the code to handle this. First, a more focused version of Program 10.3 computes the powers. Second, a macro called %CrucialRates (given in Appendix B and available on the book's web site) converts the PROC POWER results into crucial Type I and Type II error rates. Finally, PROC TABULATE organizes these crucial rates effectively.

Output 10.4 shows that the most optimistic case considered here presumes that the mortality rate is 0.15 under usual care, and it takes QCA to have a prior probability of $\gamma = 0.50$ of being effective, where "effective" is a QCA relative risk of 0.67. If so, then by using $\alpha = 0.05$, the crucial Type I and Type II error rates are $\alpha^* = 0.050$ and $\beta^* = 0.040$, respectively, which seem very good. However, for $\alpha = 0.01$, β^* rises to 0.115. Now consider the most pessimistic case. If $\gamma = 0.30$ and the non-null scenario has a mortality rate of 0.12 under usual care and a QCA relative risk is 0.75, then using $\alpha = 0.05$ gives $\alpha^* = 0.147$ and $\beta^* = 0.127$. The team from Children's Health International decides that they can tolerate these values and, thus, planning continues around $N = 2700$.

After going through this process, Dr. Capote remarks that if all clinical trial protocols were vetted in this manner, a great many of them would show crucial Type I and Type II error rates that would severely temper any inferences that can be made. This is true.

Output from Program 10.4

Total N: 2700

Usual Care Mortality	QCA Relative Risk	Alpha							
		.010				.050			
		PriorP[Null False]				PriorP[Null False]			
		0.3		0.5		0.3		0.5	
		Crucial Error Rate		Crucial Error Rate		Crucial Error Rate		Crucial Error Rate	
		Type I	Type II	Type I	Type II	Type I	Type II	Type I	Type II
0.12	0.75	.051	.196	.022	.362	.147	.127	.069	.254
	0.67	.030	.095	.013	.197	.114	.041	.052	.091
0.15	0.75	.040	.158	.017	.305	.130	.089	.060	.186
	0.67	.026	.053	.011	.115	.108	.018	.050	.040

10.7 Research Question 2: Does "QCA" Affect the "Elysemine : Elysemate" Ratios (EER)?

This section expands Dr. Capote's planning to consider a test that compares the UCO and QCA arms with respect to a continuous outcome, adjusted for baseline covariates. PROC GLMPOWER is used to perform the calculations.

10.7.1 Rationale Behind the Research Question

Now the team turns to investigating potential adverse effects.

A descriptive analysis being completed in Jamkatnia has compared 34 children who have severe malaria with 42 healthy children on some 27 measures related to metabolic functioning, including two amino acids, "elysemine" and "elysemate" (both fictitious). Elysemine is synthesized by the body from elysemate, which is abundant in food grains and meat. Phagocytes (a type of white blood cell) need elysemine to fight infection. Low plasma elysemine levels have been shown to be an incremental risk factor for death in critically ill adults and children, especially in very premature infants. Thus, a suppressed elysemine:elysemate ratio (EER) seems to be associated with a weakened immune system. In addition, plasma elysemine concentrations fall, and plasma elysemate concentrations rise, in response to extended periods of physical exertion, such as marathon running. Of course, typical marathon runners have no problem rapidly converting elysemate to elysemine and their EERs rebound within two hours.

This Jamkatnian study is of keen interest because the children with malaria had a median EER of 2.00 (inter-95% range: 1.10-3.04) compared to 2.27 (inter-95% range: 1.50-3.28) for the healthy children ($p = 0.01$, two-tailed median test). The researchers now rationalize that children with severe malaria may show reduced EERs, because the parasite attacks red blood cells and this reduces blood oxygen levels. Given that so many measures were analyzed in an exploratory manner, this $p = 0.01$ result is supportive, but not confirmatory. Nevertheless, it stirs great attention.

Related to this was a study of seven healthy adult human volunteers who were given a single standard dose of QCA and monitored intensively for 24 hours in a General Clinical Research Center. The data are summarized in Table 10.4. By four hours post infusion, their EERs fell by a geometric average of 14.9% ($p = 0.012$; 95% CI: 4.9-23.8% reduction via one-sample, two-sided t test comparing log(EER) values measured pre and post). In that the EER may already be suppressed in these diseased children, any further reduction caused by QCA would be considered harmful. On the other hand, EERs could rebound (rise) more quickly as QCA reduces lactic acid levels and thus helps restore metabolic normality. Accordingly, now the research question is "Does QCA increase or decrease elysemine:elysemate ratios in children with severe malaria complicated by lactic acidosis?"

Table 10.4 Elysemine and Elysemate Levels from Seven Healthy Adults Given QCA

Subject	Baseline			4 Hours After QCA			EER4/EER0
	E'mine0	E'mate0	EER0	E'mine4	E'mate4	EER4	
1	288	143	2.01	260	167	1.56	0.77
2	357	163	2.19	302	135	2.24	1.02
3	285	122	2.34	246	129	1.91	0.82
4	349	143	2.44	317	157	2.02	0.83
5	332	127	2.61	285	152	1.88	0.72
6	329	119	2.76	294	114	2.58	0.93
7	389	114	3.41	365	118	3.09	0.91
Geometric mean	331	132	2.51	293	138	2.13	0.85
Upper 95% limit	367	149	2.94	331	158	2.63	0.95
Lower 95% limit	298	117	2.14	260	120	1.73	0.76

Review of Study Design and Subjects

To reiterate, this double-blinded trial will randomize 900 subjects to receive usual care only (UCO) and 1800 to receive a single infusion of 50 mg/kg QCA. Study patients will be less than 13 years old diagnosed with severe malaria complicated by lactic acidosis.

Continuous Outcome Measure and Baseline Covariates

Our focus here is on the elysemine:elysemate ratio measured four hours post-infusion (EER4). The three primary covariates being considered are the baseline (five minutes prior to QCA infusion) measures of log EER0, plasma lactate level, and log parasitemia, the percentage of red blood cells infected.

It should be mentioned that elysemine and elysemate assays are expensive to conduct, costing about US$60 for each time, and thus costing US$120 for each subject.

Planned Analysis

Ratio measurements like EER are usually best handled after being log transformed; for ease of understanding we shall use $\log_2(\text{EER4})$, so that a 1.0 log discrepancy between two values equates to having one value twice that of the other.

Scenarios for the Infinite Datasets

Figure 10.4 Scenario for EER4 distributions of the Usual Care Only and QCA arms. Note: The medians, as well as the geometric means, are 2.0 and 1.8, and the common inter-95% relative spread is $3.16/1.26 = 2.85/1.14 = 2.5$.

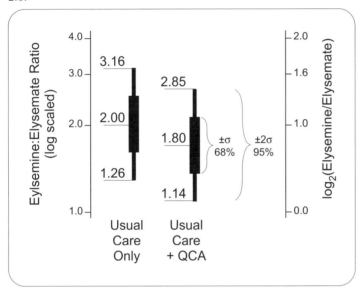

Based on the Jamkatnian study reviewed above, the investigators speculate that the median EER4 for the Usual Care Only arm is 2.0. See Figure 10.4. Two scenarios for the QCA arm are considered, a 10% decrease in EER4 (2.0 versus 1.8; as per Figure 10.4) and a 15% decrease (2.0 versus 1.7). Assuming that $\log_2(\text{EER4})$ has a Normal distribution, EER4 medians of 2.0 versus 1.8 (or 1.7) become $\log_2(\text{EER4})$ means of 1.00 versus 0.85 (or 0.77).

Making conjectures for the spread is a knotty problem, and the values chosen have critical influence on the sample-size analysis. Dr. Gooden usually takes a pragmatic approach based on the fact that, for a Normal distribution, the inter-95% range spans about four standard deviations. Thus, when the outcome variable is Normal, it is sufficient to estimate or guesstimate the range of the middle 95% of the infinite dataset for a group and divide by four to set the scenario for the standard deviation.

Here, Dr. Gooden takes log(EER4) to be Normal, i.e. EER4 is logNormal, so the process is a bit more complex. Let $\text{EER4}_{0.025}$ and $\text{EER4}_{0.975}$ be the 2.5% and 97.5% quantiles of a distribution of EER4 values. With respect to $\log_2(\text{EER4})$ values, the approximate standard deviation is

$$\sigma = \frac{\log_2(\text{EER4}_{0.975}) - \log_2(\text{EER4}_{0.025})}{4} = \frac{\log_2(\text{EER4}_{0.975}/\text{EER}_{0.025})}{4}.$$

Define $\text{RS95} = \text{EER4}_{0.975}/\text{EER4}_{0.025}$ to be the inter-95% relative spread of EER4. For the Jamkatnian study, these were $3.04/1.10 = 2.76$ and $3.28/1.50 = 2.18$. To be conservative, Dr. Capote sets RS95 to be either 2.5 (as per Figure 10.4) or 3.0. Both arms are assumed to have the same relative spread. These give values for σ of $\log_2(2.5)/4 = 0.33$ and $\log_2(3.0)/4 = 0.40$.

Now Dr. Gooden needs to have the team decide how strongly the three baseline covariates are correlated to $\log_2(\text{EER4})$. Technically, this correlation is the partial multiple correlation, R, of $X_1 = \log_2(\text{EER0})$, $X_2 = $ plasma lactate, and $X_3 = \log_2(\text{parasitemia})$ with $Y = \log_2(\text{EER4})$, controlling for treatment group, but this terminology is not likely to be well understood by the CHI team. Is there any existing data on this? Not for children infected with malaria. So, Gooden asks Dr. Capote's group to imagine that some baseline index is computed by taking a linear combination of the three covariates $(b_1 X_1 + b_2 X_2 + b_3 X_3)$ in such a way that this index is maximally correlated with $\log_2(\text{EER4})$ within the two treatment groups. Dr. Gooden needs to know what R might be in the infinite dataset, but she does not simply ask them this directly, because few investigators have good understandings about what a given correlation value, say $\rho = 0.30$, conveys. Instead, she shows them a version of Figure 10.5 *that has the values of the correlations covered from view.*

The strongest correlation is most likely to be between $\log_2(\text{EER0})$ and $\log_2(\text{EER4})$. The team agrees and suspects that this is at least $\rho = 0.20$, even if the malaria and the treatments have a substantial impact on the metabolic pathways affecting EER. Using plasma lactate and parasitemia to also predict $\log_2(\text{EER4})$ can only increase R. Looking at the scatterplots in Figure 10.5, the team agrees that R is, conservatively, between 0.20 and 0.50.

Figure 10.5 Scatterplots showing eight degrees of correlation. The order of presentation is unsystematic to aid in eliciting more careful conjectures.

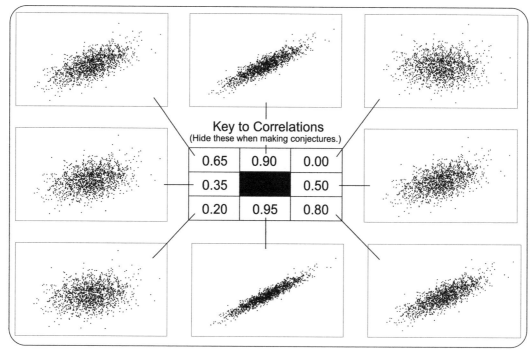

Finally, Dr. Capote wants a minimal risk of committing a Type I or Type II error for this question, so he would like to keep both α and β levels below 0.05. We will investigate the crucial error rates, α^* and β^*, later.

Classical Power Analysis

In order for SAS to compute the powers for this problem, two programming steps are necessary. First, Program 10.5 creates an "exemplary" dataset that conforms to the conjectured infinite dataset.

Program 10.5 Build and print an exemplary dataset

```
data EER;
  group = "UCO";
    CellWgt = 1;
    meanlog2EER_a = log2(2.0);
    meanlog2EER_b = log2(2.0);
    output;
  group = "QCA";
    CellWgt = 2;
    meanlog2EER_a = log2(1.8);
    meanlog2EER_b = log2(1.7);
    output;
  run;

proc print data=EER;
  run;
```

The PROC PRINT output shows that there are only two exemplary cases in the dataset, one to specify the UCO group and the other to specify the QCA group.

Output from Program 10.5

Obs	group	CellWgt	meanlog2EER_a	meanlog2EER_b
1	UCO	1	1.00000	1.00000
2	QCA	2	0.84800	0.76553

Secondly, Program 10.6 analyzes the exemplary dataset using PROC GLMPOWER.

Program 10.6 Use PROC GLMPOWER to see range of N_{total} values

```
proc GLMpower data=EER;
  ODS output output=EER_Ntotals;
  class group;
  model meanlog2EER_a meanlog2EER_b = group;
  weight CellWgt;
  power
    StdDev = 0.33 0.40   /* log2(2.5)/4 and log2(3.0)/4  */
    Ncovariates = 3
    CorrXY = .2 .35 .50
    alpha = .01 .05
    power = 0.95 0.99
    Ntotal = .;
  run;
```

Lastly, Program 10.7 summarizes the N_{total} values in a basic, but effective manner (Output 10.7). Again, more sophisticated methods are possible.

Program 10.7 Table the N_{total} values

```
* Augment GLMPOWER output to facilitate tabling ;
data EER_Ntotals; set EER_Ntotals;
  if dependent = "meanlog2EER_a" then EEratio = "2.0 vs 1.8";
  if dependent = "meanlog2EER_b" then EEratio = "2.0 vs 1.7";;
  if UnadjStdDev = 0.33 then RelSpread95 = 2.5;
  if UnadjStdDev = 0.40 then RelSpread95 = 3.0;
  run;

proc tabulate data=EER_Ntotals format=5.0 order=data;
  format Alpha 4.3 RelSpread95 3.1;
  class EEratio alpha RelSpread95 CorrXY NominalPower;
  var Ntotal;
  table
    EEratio="EE Ratios: "
      * alpha="Alpha"
      * NominalPower="Power",
    RelSpread95="95% Relative Spread"
      * CorrXY="Partial R for Covariates"
      * Ntotal=""*mean=" "
    /rtspace=35;
  run;
```

Output from Program 10.7

```
-------------------------------------------------------------------------
|                                 |            95% Relative Spread        |
|                                 |---------------------------------------|
|                                 |        2.5       |        3.0         |
|                                 |------------------+------------------- |
|                                 |  Partial R for   |   Partial R for    |
|                                 |   Covariates     |    Covariates      |
|                                 |------------------+------------------- |
|                                 |0.20 |0.35 |0.50 |0.20 |0.35 |0.50 |
|--------------------------------+-----+-----+-----+-----+-----+-----|
|EE Ratios:|Alpha    |Power      |     |     |     |     |     |     |
|----------+---------+-----------|     |     |     |     |     |     |
|2.0 vs 1.8|.010     |0.95       | 369 | 336 | 288 | 537 | 492 | 420 |
|          |         |-----------+-----+-----+-----+-----+-----+-----|
|          |         |0.99       | 495 | 453 | 387 | 723 | 663 | 567 |
|          |---------+-----------+-----+-----+-----+-----+-----+-----|
|          |.050     |0.95       | 267 | 246 | 210 | 393 | 360 | 306 |
|          |         |-----------+-----+-----+-----+-----+-----+-----|
|          |         |0.99       | 378 | 345 | 297 | 552 | 507 | 432 |
|----------+---------+-----------+-----+-----+-----+-----+-----+-----|
|2.0 vs 1.7|.010     |0.95       | 156 | 144 | 123 | 228 | 210 | 180 |
|          |         |-----------+-----+-----+-----+-----+-----+-----|
|          |         |0.99       | 210 | 192 | 165 | 306 | 282 | 240 |
|          |---------+-----------+-----+-----+-----+-----+-----+-----|
|          |.050     |0.95       | 114 | 105 |  90 | 168 | 153 | 132 |
|          |         |-----------+-----+-----+-----+-----+-----+-----|
|          |         |0.99       | 162 | 147 | 126 | 234 | 216 | 183 |
-------------------------------------------------------------------------
```

Upon scanning the results in Output 10.7, Drs. Capote and Gooden decide that $N_{\text{total}} = 100 + 200$ may be minimally sufficient, and Gooden focuses on this by using Program 10.8.

Program 10.8 Compute and table powers at $N_{\text{total}} = 300$ for EER4 outcome

```
proc GLMpower data=EER;
  ODS output output=EER_powers;
  class group;
  model meanlog2EER_a meanlog2EER_b = group;
  weight CellWgt;
  power
    StdDev = 0.33 0.40  /* log2(2.5)/4 and log2(3.0)/4 */
    Ncovariates = 3
    CorrXY = .2 .35 .5
    alpha = .01 .05
    Ntotal = 300
    power = .;
  run;

* Augment GLMPOWER output to facilitate tabling ;
data EER_powers; set EER_powers;
  if dependent = "meanlog2EER_a" then EEratio = "2.0 vs 1.8";
  if dependent = "meanlog2EER_b" then EEratio = "2.0 vs 1.7";;
  if UnadjStdDev = 0.33 then RelSpread95 = 2.5;
  if UnadjStdDev = 0.40 then RelSpread95 = 3.0;
  if power > .999 then power999 = .999;
    else power999 = power;
  run;

proc tabulate data=EER_powers format=4.3 order=data;
  format Alpha 4.3 RelSpread95 3.1;
  class EEratio alpha RelSpread95 CorrXY Ntotal;
    var power999;
    table
      Ntotal="Total Sample Size: ",
      EEratio="EE Ratios: "
        * alpha="Alpha",
      RelSpread95="95% Relative Spread"
        * CorrXY="Partial R for Covariates"
        * power999=""*mean=" "
  /rtspace=35;
  run;
```

Output 10.8 shows that only in the most pessimistic scenario does the power drop a little below 0.90 using $N_{\text{total}} = 300$ and $\alpha = 0.05$, and the mid-range scenarios even have substantial power at $\alpha = 0.01$. Furthermore, with $N_{\text{total}} = 300$, the assay costs associated with this aim will run about $300 \times \text{US\$}120 = \text{US\$}36000$, which is deemed practical. The CHI team still wants to assess the crucial Type I and Type II error rates.

Output from Program 10.8

```
Total Sample Size: 300
-----------------------------------------------------------------
|                            |      95% Relative Spread          |
|                            |-----------------------------------|
|                            |      2.5       |      3.0         |
|                            |-----------------+-----------------|
|                            |Partial R for   |Partial R for    |
|                            | Covariates     |  Covariates     |
|                            |-----------------+-----------------|
|                            |0.20|0.35|0.50|0.20|0.35|0.50|
|----------------------------+----+----+----+----+----+----|
|EE Ratios:     |Alpha       |    |    |    |    |    |    |
|---------------+------------|    |    |    |    |    |    |
|2.0 vs 1.8     |.010        |.893|.922|.959|.717|.764|.838|
|               |------------+----+----+----+----+----+----|
|               |.050        |.969|.979|.991|.884|.910|.946|
|---------------+------------+----+----+----+----+----+----|
|2.0 vs 1.7     |.010        |.999|.999|.999|.989|.994|.998|
|               |------------+----+----+----+----+----+----|
|               |.050        |.999|.999|.999|.998|.999|.999|
-----------------------------------------------------------------
```

10.7.2   Crucial Type I and Type II Error Rates

Based on the current state of knowledge reviewed above, Dr. Capote's team believes that while this hypothesis is important to investigate seriously, there is only a 20-30% chance that QCA affects EER. Accordingly, Dr. Gooden uses Program 10.9 to convert the results given in Output 10.8 to the crucial error rates.

Program 10.9 Compute and table crucial error rates for EER4 outcome

```
%CrucialRates (   PriorPNullFalse= .20 .30,
                  Powers = EER_powers,
                  CrucialErrRates = EERCrucRates )

proc tabulate data=EERCrucRates format=4.3 order=data;
title3 "Crucial Error Rates for EER Outcome";
  format Alpha 4.3 RelSpread95 3.1;
  class TypeError gamma EEratio alpha RelSpread95 CorrXY Ntotal;
  var CrucialRate;
  table
    Ntotal="Total N: ",
    EEratio="EE Ratios: "
      * RelSpread95="95% Relative Spread"
      * CorrXY="Partial R for Covariates",
    alpha="Alpha"
      * gamma="PriorP[Null False]"
      * TypeError="Crucial Error Rate"
      * CrucialRate=""*mean=" "
    / rtspace = 32;
    run;
```

Output from Program 10.9

```
Total N: 300
```

			.010 PriorP[Null False] 0.2 Crucial Error Rate Type I	.010 0.2 Type II	.010 0.3 Type I	.010 0.3 Type II	.050 0.2 Type I	.050 0.2 Type II	.050 0.3 Type I	.050 0.3 Type II
EE Ratios:	95% Relative Spread	Partial R for Covariates								
2.0 vs 1.8	2.5	0.20	.043	.026	.025	.044	.171	.008	.107	.014
		0.35	.042	.019	.025	.033	.170	.005	.106	.009
		0.50	.040	.010	.024	.017	.168	.002	.105	.004
	3.0	0.20	.053	.067	.032	.109	.184	.030	.117	.050
		0.35	.050	.056	.030	.093	.180	.023	.114	.039
		0.50	.046	.039	.027	.065	.174	.014	.110	.024
2.0 vs 1.7	2.5	0.20	.038	.000	.023	.000	.167	.000	.104	.000
		0.35	.038	.000	.023	.000	.167	.000	.104	.000
		0.50	.038	.000	.023	.000	.167	.000	.104	.000
	3.0	0.20	.039	.003	.023	.005	.167	.000	.105	.001
		0.35	.039	.002	.023	.003	.167	.000	.105	.000
		0.50	.039	.000	.023	.001	.167	.000	.104	.000

Dr. Capote likes what he sees here using $\alpha = 0.01$, because almost all the α^* and β^* values are less than 0.05. The CHI team decides to use $\alpha = 0.01$ and $N_{\text{total}} = 100 + 200$ subjects for the EER component of this trial.

10.7.3   Using Baseline Covariates in Randomized Studies

What are the consequences of failing to use helpful *baseline* covariates when comparing adjusted group means in randomized designs? What are the consequences of using worthless baseline covariates—those that have no value whatsoever in predicting the outcome (Y)? Researchers face this question because each additional covariate requires another parameter to be estimated, and this decreases by 1 the degrees of freedom for error for the F test of the group differences.

The question is easily addressed, and the answer surprises many. The power values displayed in Table 10.5 were obtained by modifying the PROC GLMPOWER code in Program 10.8. Here, we limit our focus to the case with EER medians of 2.0 versus 1.8, a 95% relative spread of 2.5, $N_{\text{total}} = 300$, and $\alpha = 0.01$. On the other hand, we consider several more values for R (SAS Code: `CorrXY = 0 .20 .35 .50 .70`) and three possible values for the number of covariates (SAS Code: `Ncovariates = 0 3 50`).

Table 10.5 Powers Obtained Using or Not Using Baseline Covariates in Randomized Studies

Number of covariates used	Multiple partial correlation (R)				
	0.00	0.20	0.35	0.50	0.70
0	.878	.878	.878	.878	.878
3	.878	.893	.922	.959	.996
50	.877	.892	.921	.959	.996

The point here is obvious. In a *randomized design*, there is virtually no cost associated with using worthless *baseline* covariates, because they are uncorrelated with the group assignment. The only cost is that the nominal null F distributions change, but in this case, the 0.01 critical values for F(1, 298) and F(1, 248) are 6.72 and 6.74, respectively, which are virtually equal. On the other hand, there is a high cost to be paid by not using baseline covariates that have some value in predicting the outcome. This concept holds for both continuous and categorical outcomes.

10.8   Crucial Error Rates When the Null Hypothesis Is Likely to Be True

Suppose "Dr. Art Ary" is planning a small trial to obtain some sound human data on a novel biologic called nissenex, which could reduce percent atheroma volume in patients with atherosclerosis. Even Dr. Ary is skeptical about nissenex, however, giving it a 2% chance of being truly effective: $\gamma = 0.02$. Using a reasonable characterization of the infinite dataset presuming nissenex is really efficacious, the power for the key hypothesis test is judged to be 0.83 at $\alpha = 0.05$ and $N = 120$. Accordingly, the crucial error rates are $\alpha^* = 0.75$ and $\beta^* = 0.004$. Thus, three out of four significant tests will be misleading.

Does this high α^* value imply that the study should not be run? No. If this trial yields $p \leq 0.05$, it would push this line of research forward to a $1 - 0.75 = 0.25$ chance that nissenex is effective, a major shift from the prior $\gamma = 0.02$. If $p > 0.05$, then there is a $1 - 0.004 = 0.996$ chance that nissenex has null or near-null efficacy, perhaps solidifying Dr. Ary's initial skepticism. Thus, either outcome will help Dr. Ary decide whether to continue with further studies. He also considers using $\alpha = 0.20$, which gives 0.95 power and makes $\alpha^* = 0.91$ and $\beta^* = 0.001$.

$1 - \alpha^* = 0.09$ is considerably weaker than the 0.25 computed for $\alpha = 0.05$, and there is little practical difference in β^* values (0.004 versus 0.001). Thus, Dr. Ary will use $\alpha = 0.05$, but he understands that, given his current prior skepticism regarding the efficacy of nissenex in treating atherosclerosis, not even a $p \leq 0.05$ outcome will sway him to prematurely publicly tout nissenex as effective. It will, of course, encourage him and his sponsors to design and carry out a more confirmatory study. This is prudent scientific practice. If everyone followed this practice, the scientific literature would not be cluttered with "significant" findings that fail to replicate in further, larger studies and meta-analyses, provided that any such work takes place (Ioannidis, 2005a, b).

10.9   Table of Crucial Error Rates

Table 10.6 shows how α^* and β^* depend on γ, α, and β. Type I errors are more frequent early in the March of Science (low γ), whereas Type II errors are more frequent later in the March. Reducing either α or β reduces both α^* and β^*. Note also that when $\gamma = 0.50$ and $\alpha = \beta$, then $\alpha = \alpha^* = \beta = \beta^*$.

Table 10.6 Crucial Type I and Type II Error Rates as a Function of γ, α and Power

$\gamma : \Pr[H_0 \text{ false}]$	Power	β	$\alpha^*: \Pr[H_0 \text{ true} \mid p \le \alpha]$				$\beta^*: \Pr[H_0 \text{ false} \mid p > \alpha]$			
			α: 0.01	0.05	0.10	0.20	α: 0.01	0.05	0.10	0.20
0.05	0.30	0.70	.388	.760	.864	.927	.036	.037	.039	.044
	0.50	0.50	.275	.655	.792	.884	.026	.027	.028	.032
	0.70	0.30	.213	.576	.731	.844	.016	.016	.017	.019
	0.80	0.20	.192	.543	.704	.826	.011	.011	.012	.013
	0.90	0.10	.174	.514	.679	.809	.005	.006	.006	.007
	0.95	0.05	.167	.500	.667	.800	.003	.003	.003	.003
0.30	0.30	0.70	.072	.280	.438	.609	.233	.240	.250	.273
	0.50	0.50	.045	.189	.318	.483	.178	.184	.192	.211
	0.70	0.30	.032	.143	.250	.400	.115	.119	.125	.138
	0.80	0.20	.028	.127	.226	.368	.080	.083	.087	.097
	0.90	0.10	.025	.115	.206	.341	.041	.043	.045	.051
	0.95	0.05	.024	.109	.197	.329	.021	.022	.023	.026
0.50	0.30	0.70	.032	.143	.250	.400	.414	.424	.438	.467
	0.50	0.50	.020	.091	.167	.286	.336	.345	.357	.385
	0.70	0.30	.014	.067	.125	.222	.233	.240	.250	.273
	0.80	0.20	.012	.059	.111	.200	.168	.174	.182	.200
	0.90	0.10	.011	.053	.100	.182	.092	.095	.100	.111
	0.95	0.05	.010	.050	.095	.174	.048	.050	.053	.059
0.70	0.30	0.70	.014	.067	.125	.222	.623	.632	.645	.671
	0.50	0.50	.008	.041	.079	.146	.541	.551	.565	.593
	0.70	0.30	.006	.030	.058	.109	.414	.424	.438	.467
	0.80	0.20	.005	.026	.051	.097	.320	.329	.341	.368
	0.90	0.10	.005	.023	.045	.087	.191	.197	.206	.226
	0.95	0.05	.004	.022	.043	.083	.105	.109	.115	.127

10.10   Summary

In writing a single chapter on sample-size analysis, one strives for either breadth or depth. We opted to cover two examples in depth, and thus we failed to even mention any of the vast array of other tools now available to help investigators carefully assess and justify their choices for sample sizes across the statistical landscape. What have we not discussed? The list of topics and references is too long to begin and would soon be outdated anyway.

What readers need to understand is that if they have a sample-size analysis issue, there may be good methodological articles and strategies that address it. If no such help can be found, then Monte Carlo simulation can provide results that are entirely satisfactory. In fact, some excellent statisticians now use simulation for all such problems, even for traditional ones that have sound "mathematical" solutions that are widely used.

We hope the two examples given here provide a rich context to fashion good strategies to address other problems that may be encountered. Though the methods may vary widely, the core concepts and issues do not.

Acknowledgments

The authors thank Drs. Walter Ambrosius, Alex Dmitrienko, and Jean Lightner Norum for skillfully reviewing a draft of this chapter. Comments from other readers are always welcome.

References

Agbenyega, T., Planche, T., Bedu-Addo, G., Ansong, D., Owusu-Ofori, A., Bhattaram, V.A., Nagaraja, N.V., Shroads, A.L., Henderson, G.N., Hutson, A.D., Derendorf, H., Krishna, S., Stacpoole, P.W. (2003). "Population kinetics, efficacy, and safety of Dichloroacetate for lactic acidosis due to severe malaria in children." *Journal of Clinical Pharmacology.* 43, 386–396.

Agresti, A. (1980). "Generalized odds ratios for ordinal data." *Biometrics.* 36, 59–67.

Feynman, R.P. (1999). *The Pleasure of Finding Things Out.* Cambridge, MA: Perseus Books.

Holloway, P.A., Knox, K., Bajaj, N., Chapman, D., White, N.J., O'Brien, R., Stacpoole, P.W., Krishna, S. (1995). "Plasmodium Berghei infection: Dichloroacetate improves survival in rats with lactic acidosis." *Experimental Parasitology.* 80, 624–632.

Ioannidis, J.P.A. (2005a). "Contradicted and Initially Stronger Effects in Highly Cited Clinical Research." *JAMA.* 294, 218–228.

Ioannidis, J.P.A. (2005b). "Why Most Published Research Findings Are False." *PLoS Medicine.* 2, 696–701.

Johnson, N.L., Kotz, S., Balakrishnan, N. (1994). *Continuous Univariate Distributions*, Vol. 1 (2nd Ed.). New York: John Wiley.

Lee, S.J., Zelen, M. (2000). "Clinical trials and sample size considerations: Another perspective." *Statistical Science.* 15, 95–100.

SAS Institute Inc. (2006). *SAS/STAT User's Guide.* Cary, NC: SAS Institute Inc.

Stacpoole, P.W., Nagaraja, N.V., Hutson, A.D. (2003). "Efficacy of dichloroacetate as a lactate-lowering drug." *Journal of Clinical Pharmacology.* 43, 683–691.

Topol, E.J., Califf, R.M., Van de Werf, F., Simoons, M., Hampton, J., Lee, K.L., White, H., Simes, J., Armstrong, P.W. (1997). "Perspectives on large-scale cardiovascular clinical trials for the new millennium." *Circulation.* 95, 1072–1082.

Wacholder, S., Chanock, S., Garcia-Closas, M., El ghormli, L., Rothman, N. (2004). "Assessing the probability that a positive report is false: An approach for molecular epidemiological studies." *Journal of the National Cancer Institute.* 96, 434–442.

Appendix A   Guidelines for "Statistical Considerations" Sections

A well-developed statistical considerations section persuades reviewers that solid skill and effort have gone into framing the research questions, planning the study, and forming an appropriate team. The writing should be crafted for the clinical researcher who is a good "para-statistician," as well as for the professional biostatistician. The "Statistical Considerations" section should be mostly self-contained and thus may reiterate information found elsewhere in the proposal.

A.1   Components

Design.   Summarize the study design. It may be helpful to use appropriate terms such as *randomized, double blind, cross-over, controlled, comparative, case-control, prospective, retrospective, longitudinal,* and *cohort.*

Research questions.   Strictly speaking, not all studies are driven by testable hypotheses, but all studies have research questions that should be delineated in your Specific Aims section. Summarize the outcome measures and describe how you expect them to be related to the components of the study design and other predictor variables. Restate/translate your primary research questions into specific estimates of effects and their confidence intervals, and/or into statistical hypotheses or other methods. Similar descriptions regarding secondary questions are valuable, too.

Statistical analysis plan.   Specify what statistical methods will be used to analyze the primary and secondary outcome measures. Cite statistical references for non-routine methods. (Example: The two groups will be compared on KMOB830430 and its metabolites using estimates and 95% confidence limits for the generalized odds ratio (Agresti, 1980), which is directly related to the common Wilcoxon rank-sum test.) These sections often state what statistical software package and version will be used, but this usually provides little or no information about what actually will be done.

Randomization (if appropriate).   Specify how the randomization will be done, especially if it involves blocking or stratification to control for possible confounding factors.

Sample-size analyses.   State the proposed sample size and discuss its feasibility. Estimate the key inferential powers, or other measures of statistical sensitivity/precision, such as the expected widths of key confidence intervals. Strive to make your sample-size analyses congruent with the statistical methods proposed previously, and discuss any incongruencies. State how you arrived at the conjectures for all the unknowns that underlie the sample-size analysis, citing specific articles and/or summarizing analyses of "preliminary" data or analyses presented in unpublished works. If a sample-size analysis was not performed, state this categorically and explain why. For example, the proposal may be only a small pilot study.

Data management.   Summarize the schema for collecting, checking, entering, and managing the data. What database software will be used? How will the database be tapped to build smaller analysis datasets? Note how you will meet modern standards for data security.

Technical support.   Who will perform the necessary database and statistical work? If such people are less experienced, who will supervise the work?

Appendix B   SAS Macro Code to Automate the Programming

In the interest of simplicity, the SAS code provided above avoided all macro programming, except for using the %CrucialRates macro. However, analysts can profit greatly by making elementary use of the SAS Macro Language. Below is the full program that was used in developing the EER example. Note how the parameters that shape the problem are specified only once at the beginning. Due to rounding, the results obtained with this code differ slightly from those given above.

```
options ls=80 nocenter;
/*******************************************************************\
  Program Name: EER_SSAnalysis060722.sas
          Date: 22 July 2006

  Investigator: "Sol Capote, MD; CHI Malaria Research Group"
  Statistician: "Phynd Gooden, PhD" (Actually, Ralph O'Brien)
```

```
              Purpose: Sample-size analysis for comparing usual care only vs.
                       QCA on elysemine:elysemate ratios at 4 hours (EER4).
                       Assumes data will be logNormal in distribution with same
                       relative range in the two groups.
\******************************************************************************/


*options symbolgen mlogic mprint;              * for macro debugging;
*options FORMCHAR="|----|+|---+=|-/\<>*";      * for ordinary text tables;

title1 "Usual Care Only (UCO) vs. Usual Care + QCA (QCA)";
title2 "Difference in 4-hour elysemine-elysemate ratio (EER4),adjusted";
title3 "for three baseline covariates: EER0, plasma lactate, parasitemia";


/******************************************************************\
This program is structured so that all the defining values are set
through %let macro statements at the start of the code.
\******************************************************************/


/*******************************\
   BEGIN TECHNICAL SPECIFICATIONS
/*******************************/


* Set label for Y;
* --------------;
%let Ylabel = EE Ratio;


* Set variable labels for the two groups ;
* -------------------------------------;
%let GrpLabel_1 = UCO;
%let GrpLabel_2 = DCA;


/*********************************************************************************\
Each distribution is logNormal with different medians, but same relative
spread (defined below). This is the same as saying that the distributions
have different means but the same coefficients of variation.
\*********************************************************************************/


* Supply guesstimates for medians ;
* ------------------------------ ;
%let Ymedian0 = 2.0;     * median for control arm, only one scenario ;
%let Ymedian1A = 1.8;    * median for experimental arm, scenario A ;
%let Ymedian1B = 1.7;    * median for experimental arm, scenario B ;


* Supply guesstimates for the 95% relative spread, defined below;
* ----------------------------------------------------------- ;
%let YRelSpread95_1 = 2.5;  * YRelSpread95, scenario 1 ;
%let YRelSpread95_2 = 3.0;  * YRelSpread95, scenario 2 ;


*Set NCovariates and supply guesstimates for PrtlCorr(XXX,logY);
* ---------------------------------------------------------;
%let NCovariates = 0 3 50;  * number of covariates ;
%let PrtlCorr_XXXlogY = .2 .35 .50 ; * Multiple partial correlation (R) ;
                                     * between covariates ("XXX") and   ;
                                     * logY, within treatment groups.   ;


* Supply prior probabilities that null is false;
```

```
* --------------------------------------------;
%let PriorPNullFalse = .20 .30;


/*****************************\
   END TECHNICAL SPECIFICATIONS
\*****************************/




/******************************************************************\
=====
Notes
=====
```

1. Each distribution is logNormal with different medians, but same relative spread (defined below). This is the same as saying that the distributions have different means but the same coefficients of variation. Under logNormality, medians are also geometric means.

2. Let Y025, Y500 and Y975 be the 2.5%, 50%, and 97.5% quantiles for Y, i.e., Y500 is the median of Y and Y025 and Y975 are the limits of the mid-95% range for Y.

3. Define YRelSpread95 = Y975/Y025 to be the inter-95% relative spread. These relative spreads are taken to be equal in control and experimental groups.

4. Log(Y025), log(Y500), and log(Y925) are the 2.5% quantile, the median, and the 97.5% quantiles for logY.

   If Y ~ logNormal, then log(Y) ~ Normal, so

   $$\text{mean\_logY} = \text{median\_logY} = \log(Y500).$$

   Let SD_logY be the standard deviation of logY. Then, log(Y025) and log(Y925) are each 1.96*SD_log2Y units from mean_logY, so

   $$\text{SD\_logY} = [\log(Y025) - \log(Y025)]/(2*1.96),$$

   where 1.96 is the 97.5% quantile (Z975) of the standard Normal, Z ~ N(0,1).

   Taking 1.96 to "equal" 2, we have,

   $$\text{SD\_logY} = [\log(Y025) - \log(Y025)]/4,$$

   With YRelSpread95 = Y975/Y025, we get,

   $$\text{SD\_logY} = \log(\text{YRelSpread95})/4.$$

5. It some cases it may be more convenient to use another relative spread, say YRelSpread90 or YRelSpread50. Using Z900 = 1.65 and Z750 = 0.67, we have

   $$\text{SD\_logY} = \log(\text{YRelSpread90})/(2*1.65)$$

   and

```
        SD_logY = log(YRelSpread50)/(2*0.67).
```

Whereas [log(Y750) - log(Y250)] is the interquartile range for logY
YRelSpread50 could be called the interquartile relative range for Y.

6. One can show that the coefficient of variation is

```
        CoefVar_Y = sqrt(exp(SD_logY**2 - 1)).
```

See page 213 of Johnson, Kotz, Balakrishnan (1994), Continuous
Univariate Distributions, Vol. I.

7. All logs are taken at base 2, but this choice is irrelevant for
sample-size analysis.

```
\*********************************************************************/


/********\
 Main code
\********/

%let SD_log2Y_1 = %sysevalf(%sysfunc(log2(&YRelSpread95_1))/4);
%let SD_log2Y_2 = %sysevalf(%sysfunc(log2(&YRelSpread95_2))/4);

data exemplary;
  group = "&GrpLabel_1";
    CellWgt = 1;
    mean_log2Y_A = log2(&Ymedian0);
    mean_log2Y_B = log2(&Ymedian0);
    output;
  group = "&GrpLabel_2";
    CellWgt = 2;
    mean_log2Y_A = log2(&Ymedian1A);
    mean_log2Y_B = log2(&Ymedian1B);
    output;
run;

proc print data=exemplary;
run;


proc GLMpower data=exemplary;
  ODS output output=Ntotals;
  class group;
  model mean_log2Y_A mean_log2Y_B = group;
  weight CellWgt;
  power
    StdDev = &SD_log2Y_1 &SD_log2Y_2
    NCovariates = &NCovariates
    CorrXY = &PrtlCorr_XXXlogY
    alpha = .01 .05
    power = 0.95 0.99
    Ntotal = .;
run;
```

```
data Ntotals;
    set Ntotals;
    if dependent = "mean_log2Y_A"
      then comparison = "&Ymedian0 vs &Ymedian1A";
    if dependent = "mean_log2Y_B"
      then comparison = "&Ymedian0 vs &Ymedian1B";
    if UnadjStdDev = &SD_log2Y_1
      then YRelSpread95 = &YRelSpread95_1;
    if UnadjStdDev = &SD_log2Y_2
      then YRelSpread95 = &YRelSpread95_2;
run;


proc tabulate data=Ntotals format=5.0 order=data;
  format Alpha 4.3 YRelSpread95 3.1;
  class comparison alpha YRelSpread95 CorrXY
    NominalPower Ncovariates;
  var Ntotal;
  table
    Ncovariates="Number of covariates; ",
    comparison="&Ylabel: "
      * alpha="Alpha"
      * NominalPower="Power",
    YRelSpread95="95% Relative Spread"
      * CorrXY="Partial R for Covariates"
      * Ntotal=""*mean=" "
    /rtspace=35;
run;


proc GLMpower data=exemplary;
  ODS output output=powers;
  class group;
  model mean_log2Y_A mean_log2Y_B = group;
  weight CellWgt;
  power
    StdDev = &SD_log2Y_1 &SD_log2Y_2
    Ncovariates = &NCovariates
    CorrXY = &PrtlCorr_XXXlogY
    alpha = .01 .05
    Ntotal = 300
    power = .;
run;


data powers;
    set powers;
    if dependent = "mean_log2Y_A"
      then comparison = "&Ymedian0 vs &Ymedian1A";
    if dependent = "mean_log2Y_B"
      then comparison = "&Ymedian0 vs &Ymedian1B";
    if UnadjStdDev = &SD_log2Y_1
      then YRelSpread95 = &YRelSpread95_1;
    if UnadjStdDev = &SD_log2Y_2
      then YRelSpread95 = &YRelSpread95_2;
    if power > .999
```

```
      then power999 = .999;
      else power999 = power;
run;


proc tabulate data=powers format=4.3 order=data;
  format Alpha 4.3 YRelSpread95 3.1;
  class comparison alpha YRelSpread95 CorrXY
    Ntotal Ncovariates;
  var power999;
  table
    Ntotal="Total Sample Size: "
      * Ncovariates="Number of covariates; ",
    comparison="&Ylabel.: "
      * alpha="Alpha",
      YRelSpread95="95% Relative Spread"
      * CorrXY="Partial R for Covariates"
      * power999=""*mean=" "
    /rtspace=35;
run;


%macro CrucialRates (PriorPNullFalse = ,
                     Powers = powers,
                     CrucialErrRates = CrucialErrRates
                     );
   /*******************************************************************\
   Converts Alphas and Powers to Crucial Error Rates
   -------------------------------------------------
     <> PriorPNullFalse= value1 value2 ... value10
         This is gamma = PriorP[Ho false].
     <> Powers= InputDSName
         "InputDSName" corresponds to ODS output statement in PROC POWER
           or PROC GLMPOWER, such as
             proc power;
               ODS output output=MoralityPowers;
     <> CrucialErrRates= OutputDSName
         "OutputDSName" is SAS dataset name; default: "CrucialErrRates"
   \*******************************************************************/

   data &CrucialErrRates;
     set &Powers;
     array PrNullFalseV{10} _temporary_ (&PriorPNullFalse);
     beta = 1 - power;
     iPNF = 1;
     do until (PrNullFalseV{iPNF} = .);
         gamma = PrNullFalseV{iPNF};
         /* Compute Crucial Type I error rate */
         TypeError = "Type  I";
         CrucialRate
           = alpha*(1 - gamma)/(alpha*(1 - gamma) + (1 - beta)*gamma);
         output;
         /* Compute Crucial Type II error rate */
         TypeError = "Type II";
         CrucialRate
           = beta*gamma/(beta*gamma + (1 - alpha)*(1 - gamma));
         output;
```

```
              iPNF + 1;
          end;
      run;
%mend; *CrucialRates;

%CrucialRates (  PriorPNullFalse = &PriorPNullFalse,
                 Powers=powers,
                 CrucialErrRates = CrucRates );

proc tabulate data=CrucRates format=4.3 order=data;  *&UniversalText;
title3 "Crucial Error Rates for QCA Malaria Trial";
   format Alpha 4.3 YRelSpread95 3.1;
   class TypeError gamma comparison alpha YRelSpread95 CorrXY
     Ntotal Ncovariates;
   var CrucialRate;
   table
     Ntotal="Total Sample Size: "
       * Ncovariates="Number of covariates; ",
     comparison="&Ylabel.: "
       * YRelSpread95="95% Relative Spread"
       * CorrXY="Partial R for Covariates",
     alpha="Alpha"
       * gamma="PriorP[Null False]"
       * TypeError="Crucial Error Rate"
       * CrucialRate=""*mean=" "
     / rtspace = 32;
run;
```

Chapter 11

Design and Analysis of Dose-Ranging Clinical Studies

Alex Dmitrienko
Kathleen Fritsch
Jason Hsu
Stephen Ruberg

Dose-ranging clinical studies play a very important role in early evaluation of safety and efficacy profiles of experimental drugs. The chapter reviews popular statistical methods used in dose-response analysis, including trend tests, dose-response modeling, and multiple tests for identifying safe and effective doses. The testing and estimation procedures discussed in this chapter are illustrated using examples from dose-ranging clinical trials.

11.1 Introduction

Dose-ranging studies are conducted at early stages of drug development (both pre-clinical and clinical stages) to evaluate safety and efficacy of experimental drugs. A large number of publications deal with the analysis of dose-ranging studies designed to test several doses (or dose regimens) of an experimental drug versus a control. This area of research, commonly referred to as *dose-response analysis*, covers the relationship between dose and clinical response.

Alex Dmitrienko is Principal Research Scientist, Global Statistical Sciences, Eli Lilly and Company, USA. Kathleen Fritsch is Mathematical Statistician, Division of Biometrics III, Food and Drug Administration, USA. (The views presented in this chapter are those of the author and do not necessarily represent the views of the FDA. Kathleen Fritsch's contribution to this chapter was completed prior to her employment with the FDA). Jason Hsu is Professor, Department of Statistics, The Ohio State University, USA. Stephen Ruberg is Director, Global Medical Information Sciences, Eli Lilly and Company, USA.

There are several distinct research topics within dose-response analysis (Ruberg, 1995a):

1. Assessment of the dose-related trend in the response variable.
2. Estimation of the shape of the dose-response function.
3. Determination of optimal doses (including identification of the *minimum effective dose*, which is of increasing concern as safety issues arise with marketed products).

The three objectives of dose-response analysis are obviously intertwined; however, it is generally prudent to examine them one at a time.

First, before making a decision to invest more research dollars in an experimental drug, drug developers need to ensure that a drug effect, manifested by a positive dose-response relationship, is present. A positive dose-response relationship, defined as a non-decreasing shape in the response variable from the lowest dose to the highest dose, plays a key role in early drug evaluation. As pointed out by Kodell and Chen (1991), "evidence of a dose-response relationship is taken to be a more compelling finding than evidence of a positive effect that does not appear to be dose-related".

Once it has been demonstrated that an overall dose-related trend is present, one needs to characterize the dose-response function and determine efficacious and safe doses. To understand the importance of this step, note that a significant overall trend is rarely accompanied by significant treatment differences at all dose levels. In most cases, significant treatment differences are observed only at some doses included in a dose-ranging study and it is critical to identify the range of doses over which a positive dose-response is truly present. This information is used to determine a *therapeutic window* (a contiguous interval within which every dose is safe and effective) and will ultimately guide the selection of doses for registration studies.

Dose-response testing and modeling have received much attention in the clinical trial literature. For a general overview of issues arising in dose-response studies and a discussion of relevant statistical methods, see Ruberg (1995a, 1995b), Chuang-Stein and Agresti (1997) and Phillips (1997, 1998). Also, Westfall et al. (1999, Section 8.5) provide an overview of dose-finding methods in clinical trials with a large number of SAS examples.

This chapter summarizes popular statistical methods used in dose-response analysis, e.g., trend tests, dose-response models, and dose-finding strategies based on multiple tests. The statistical procedures introduced in this chapter are illustrated using examples from dose-ranging clinical trials.

11.1.1 Clinical Trial Examples

To illustrate the dose-response tests and models that will be introduced in this chapter, we will use the following three clinical trials examples. The first one is a cross-over trial representative of trials conducted at early stages of clinical development. The other two trials use a parallel group design and serve as examples of dose-ranging Phase II or Phase III trials.

11.1.2 Clinical Trial in Diabetes Patients (Cross-Over Design)

A Phase I study in patients with Type II diabetes was conducted to test four doses of an experimental drug versus placebo using a cross-over design with two periods. Patients enrolled into the study were assigned to one of four groups (six patients to each group). Each group of patients received 24-hour placebo and experimental drug infusions on two occasions. The primary pharmacodynamic objective of the study was to examine effects of the selected doses on fasting serum glucose. Specifically, the primary pharmacodynamic endpoint was defined as the fasting serum glucose concentration at the end of a 24-hour infusion.

Figure 11.1 Fasting serum glucose level in the diabetes trial example

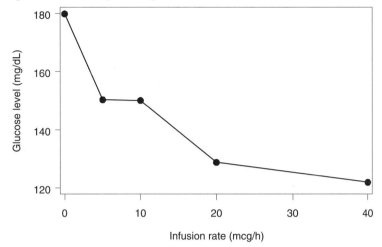

Figure 11.1 displays the dose-response relationship observed in the diabetes trial. The mean fasting serum glucose level decreases as the infusion rate increases in a monotone fashion. The mean glucose levels are based on least square means computed from a mixed model with a fixed group effect and a random subject effect.

The data set with fasting serum glucose measurements (DIABETES data set) can be found on the book's companion Web site.

To save space, some SAS code has been shortened and some output is not shown. The complete SAS code and data set used in this book are available on the book's companion Web site at `http://support.sas.com/publishing/bbu/companion_site/60622.html`.

11.1.3   Clinical Trial in Patients with Asthma (Parallel Group Design)

A Phase II trial in 108 patients with mild to moderate asthma was conducted to compare three doses of an experimental drug to placebo. The drug was administered daily (50, 250, and 1000 mg/day). The efficacy of the experimental drug was studied using spirometry measurements. The primary efficacy endpoint was the forced expiratory volume in one second (FEV1).

Figure 11.2 displays the results of the trial. It shows the relationship between the mean FEV1 improvement (estimated using least square means) and daily dose in the asthma trial. The observed dose-response relationship is not monotone—this is an example of the

Figure 11.2 Mean improvement in FEV1 in the asthma trial example

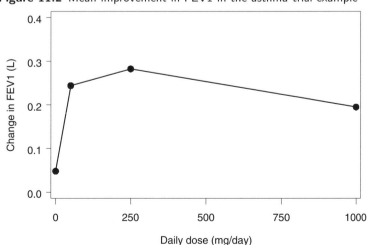

so called *umbrella-shaped* relationship. The mean FEV1 improvement achieves its maximum at the medium dose (250 mg/day) and then drops to a lower value. The treatment effect at the high dose is even less than that at the low dose. Note, however, that the non-monotone trend depicted in Figure 11.2 may be due to sampling variability, and the underlying dose-response relationship may still be positive. A statistical trend test is required to draw conclusions about the true dose-response curve.

The data set with FEV1 measurements (ASTHMA data set) can be found on the book's companion Web site.

11.1.4 Clinical Trial in Patients with Hypertension (Parallel Group Design)

Consider a Phase II study in 68 patients with hypertension. The patients were randomly assigned to receive one of three doses of an antihypertensive drug (10, 20 or 40 mg/day) or placebo. The primary objective of the study was to examine the effect of the selected doses on diastolic blood pressure (DBP).

Figure 11.3 depicts the relationship between the mean reduction in diastolic blood pressure (based on least square means) and daily dose of the antihypertensive therapy. The overall dose-response trend is positive. The treatment effect seems to achieve a plateau at the 20 mg/day dose.

Figure 11.3 Mean reduction in diastolic blood pressure in the hypertension trial example

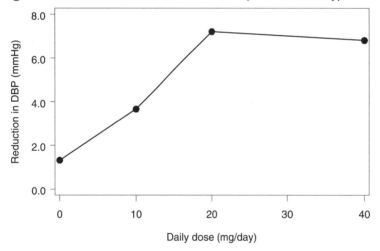

The data set with diastolic blood pressure measurements (HYPERTENSION data set) can be found on the book's companion Web site.

11.1.5 Overview

As was stated above, dose-response analysis plays an important role in both pre-clinical and clinical evaluation of compounds. Although some of the methods discussed below can be used in toxicological studies, this chapter focuses on a clinical setting which includes pharmacological studies and randomized clinical trials. See Chapter 5 for a brief overview of dose-response approaches used in toxicological studies. Also, this chapter deals mainly with continuous endpoints (contrast and some other tests can also be used with categorical variables). Chuang-Stein and Agresti (1997) give a general discussion of approaches to dose-response studies with categorical endpoints.

Section 11.2 outlines general principles that need to be considered in designing dose-ranging trials. The section describes issues arising in the design of dose-ranging studies and discusses the choice of research hypotheses in trials with negative and positive controls.

Section 11.3 discusses statistical procedures for addressing the first objective of dose-response analysis, i.e., procedures for testing dose-related trends. The section reviews both parametric and non-parametric methods. We also briefly discuss sample size calculations in dose-ranging trials based on general contrast tests (see Section 11.3.7).

Section 11.4 deals with the second objective—it briefly introduces regression-based approaches to examining the shape of dose-response functions, including methods based on linear and sigmoid models. Lastly, Section 11.5 considers the third objective of dose-response analysis and reviews approaches to finding the optimal dose based on multiple pairwise and treatment-control comparisons. This section introduces multiplicity adjustment procedures that are derived using the principles of closed and partition testing. It is worth noting that, in general, dose-response analyses are performed in Phase II trials and thus adjustments for multiplicity are not always considered. A strict control of the Type I error rate is mandated only in registration trials.

11.2 Design Considerations

11.2.1 Types of Dose-Ranging Studies

To define dose-ranging studies, we may consider the entire class of studies that involve multiple doses of an active compound. This broad class includes dose-escalation or dose-titration studies (i.e., the dose could go up or down for the same trial subject), parallel group studies (subjects are randomized to distinct dose groups) or cross-over studies involving sequences of various dose administrations. When considering the design of a dose-ranging study, we must first consider patient safety, especially in the early stages of drug development. Regardless of how much in vitro or animal testing has been completed, human subjects must be exposed to a new chemical entity on a gradual basis both in terms of the amount of the drug to be administered as well as the duration of exposure.

11.2.2 Dose-Escalation Studies

The typical dose-ranging trials that are done initially in humans are dose-escalation trials in which there is a single administration of a very small dose of the drug to a small number of subjects—usually four to ten volunteers. Doses are escalated, depending on the safety and pharmacokinetic responses of the patients, in an effort to explore the boundary of tolerability to the drug. Such designs are often replicated using repeated administration of a fixed dose of the drug for one- or two-week intervals, again to gradually increase exposure and establish tolerable doses of the new compound. It is worth noting that the assignment of subjects to dose groups in dose-escalation trials is non-random. Subjects are randomized within each dose group but not across the groups as in parallel group designs.

We can consider having the same patients progress through the entire dose escalation or having independent groups of volunteers for each step of the way. The choice depends on the nature of the responses and the objectives of the trial:

- If the response is one that persists, independent groups will be required to avoid prolonged washout periods.
- If the objective is to collect safety information on as many subjects as safely possible, independent groups will be preferred. Of course, using the same patients throughout the entire escalation of doses allows more direct comparisons of responses across doses and thereby more precise estimates of parameters of interest.
- Lastly, in dose-escalation studies, we need to be prepared to deal with incomplete data caused by dropouts, especially when the dropout is due to intolerance.

11.2.3 Cross-Over Design

In the early stages of drug development, but after tolerability has been established as described above, larger studies with greater sophistication can be initiated. An example of a cross-over design that embodies elements of randomization and dose escalation is shown in Table 11.1. A very useful aspect of the design is that, in the first period, two-thirds of the patients receive the lower dose of the drug. As the periods of the study progress, more patients are exposed to the higher dose, and all patients are exposed to the lower dose before the final period.

Table 11.1 A Cross-Over Study with Three Periods

Sequence	Period 1	Period 2	Period 3
1	Placebo	Lower dose	Higher dose
2	Lower dose	Placebo	Higher dose
3	Lower dose	Higher dose	Placebo

This design is usually appealing to clinicians because of the safety of patients inherent in the dose escalation embedded in the cross-over periods. In addition, the use of placebo in each period helps maintain the double-blind nature of the trial, which is usually desirable, and allows for the separation of treatment and period effects. All of the usual considerations need to be made when considering a cross-over versus parallel design (e.g., within- and between- subject variability, stability of the disease state).

11.2.4 Parallel Group Design

The most common and simple design is the controlled, randomized, parallel dose response study. In this study design, patients are randomly allocated to one of several active dose groups or control, most often placebo. While a placebo control leads to the most clear interpretation of results, active controls can also be used. The parallel design is most popular in Phase II development when larger studies are done in order to explore safety and effectiveness of a new drug. Since the only difference between treatment groups is the dose of the drug, the design leads to more straightforward analysis and interpretation as will be described in subsequent sections.

11.2.5 Factorial Design

In a growing number of situations, researchers want to study the effect of more than one aspect of the dose of a new compound on a disease state. Several examples come to mind: the amount of the dose and the frequency of dosing (i.e., once a day or twice a day) or the size of a bolus injection and the subsequent infusion rate of a compound. Furthermore, in some disease states combination therapy is the norm, or at least a common phenomenon, such as chemotherapy and antihypertensive therapy. It may be of interest to study multiple doses of each drug in combination with each other. Obviously, such explorations of dose response lead to factorial studies. Statisticians have long recognized the value and economy of conducting factorial experiments, but their use in clinical trials is less common. This may be due to the complexity (e.g., packaging and blinding clinical medications) of many clinical trials. Such designs may be most relevant if one is interested in finding the optimal dosing regimen for a drug or combination of drugs.

11.2.6 Choice of a Control and Research Hypothesis

Negative and positive controls play a key role in dose-ranging clinical trials. As pointed out in many publications, a significant dose-response trend in the absence of a control group cannot serve as evidence of a drug effect. In studies with a negative control, a significant dose-response relationship can be observed even if the lowest dose is less efficacious than the control and the highest dose is generally comparable to the control.

Most commonly, dose-ranging clinical trials are designed to investigate the effect of multiple doses of an experimental drug compared to a negative control (placebo). Placebo is a "dummy" treatment that appears as identical as possible to the test treatment but does not contain the test drug. As stated in the (amended) Helsinki Declaration and ICH E10 guidance document, the use of a placebo may be justified in studies where no proven therapeutic method exists. Using a placebo as the control may also be justified if the disease does not cause serious harm (e.g., seasonal allergy) or when treatment may be optional (e.g., treatment for pain or depression). However, if the disease causes mortality, the inclusion of a placebo-controlled group may not be ethical. Even if the disease does not cause mortality but causes irreversible morbidity, that is, the disease invariably progresses unless treated (e.g., Type II diabetes), the inclusion of a placebo group may be unethical as well.

11.2.7   Superiority and Non-Inferiority Testing

In comparing an experimental drug with a negative control, we hope to prove that the new drug is *superior* to the negative control by a clinically meaningful amount. If μ_{negative} is the mean response of the negative control, and μ_{drug} is the mean of the experimental drug, then the drug is considered efficacious if

$$\mu_{\text{drug}} > \mu_{\text{negative}} + \delta_{\text{sup}},$$

where δ_{sup} is a prespecified non-negative quantity representing a clinically important treatment difference for establishing superiority.

An active control is a drug which has been proven to be efficacious, typically already approved and in use. In comparing a new (experimental) drug against an active control, clinical researchers may hope to prove that the new drug is superior to the active control. If μ_{active} is the mean of the active control, then the new drug is considered efficacious if

$$\mu_{\text{new}} > \mu_{\text{active}} + \delta_{\text{sup}},$$

where δ_{sup} again represents a clinically important treatment difference.

This is done primarily in the European Union. The rationale for demanding proven superiority before approving a drug is that, by limiting the number of drugs and thus allocating each a large share of market, it might be possible to better negotiate a cost-effective managed care plan.

Alternatively, clinical researchers may hope to demonstrate that the new drug is *non-inferior* to the active control, i.e., show that

$$\mu_{\text{new}} > \mu_{\text{active}} - \delta_{\text{non-inf}},$$

where $\delta_{\text{non-inf}}$ is the so-called *non-inferiority margin* that defines a clinically meaningful difference in non-inferiority trials.

This approach is often taken in the United States. In such cases, $\delta_{\text{non-inf}}$ may be a fraction (e.g., 50%) of the presumably known improvement the active control provides over the negative control. For example, in the U.S., a new drug for some mild digestive disorder might be approved if it is shown to maintain at least half the treatment effect of another drug on the market. The rationale for approving such a drug is to let the consumers make their own decisions, based on cost, efficacy, and side effect considerations.

In dose-ranging trial with a non-inferiority objective, in addition to the new treatment and the active control groups, it is desirable to include a negative control group as well. This is done to ensure *assay sensitivity*. It is best to proceed to compare the new drug with the active control only after establishing that the non-inferiority trial is sensitive enough to detect the known difference between the active and negative controls. Otherwise, clinical researchers may fail to detect a difference between the new treatment and active control (and conclude that the new treatment is non-inferior to the active control) simply due to lack of power. Obviously, if there is a statistically significant difference between the new treatment and the active control, the difference stands whether the active control is found to be different from the placebo or not.

11.2.8   One-Sided and Two-Sided Testing

Of particular relevance when designing a study to assess a dose-response relationship is the matter of one-sided versus two-sided testing. This issue is debated passionately by some, but it is clear that in the vast majority of situations, the researcher has a known interest in the direction of the desired response. This would naturally imply a one-sided alternative with a suitable sample size to carry out the appropriate statistical test. The size of the test also needs to be considered carefully in the design of the trial.

11.2.9   Control of Type I and Type II Error Rates

When taking a hypothesis testing approach, we must consider the importance of Type I and Type II errors. If the response of interest is an efficacy response, then the Type I error is of greatest interest (i.e., we do not want to conclude a drug is effective when it truly is not). However, if the response of interest is a safety variable (e.g., QTc prolongation), the Type II error plays a more important role (i.e., we do not want to conclude there is no effect on safety when in fact there is). Often, this is under-appreciated and in the case of a safety study it may be perfectly acceptable to have a larger than usual Type I error rate such as 0.10 or 0.15 to shrink the Type II error for a limited or fixed sample size that may be needed for a practical clinical trial.

11.3   Detection of Dose-Response Trends

It is commonly agreed that the first step in analyzing dose-response trends and more complex problems related to dose-finding is the assessment of the overall effect of an experimental drug. This step is essentially a screening test to determine whether or not the drug effect exists at all.

Mathematically, the goal of the overall drug effect assessment is to test the null hypothesis of no treatment effect. For example, in the diabetes trial example, clinical researchers were interested in testing whether or not there was any drug-related improvement in the response variable as the infusion rate increased. The corresponding null hypothesis of no treatment effect is given by

$$H_0 : \mu_P = \mu_{D1} = \mu_{D2} = \mu_{D3} = \mu_{D4},$$

where μ_P is the mean glucose level in the placebo period and μ_{D1}, μ_{D2}, μ_{D3}, μ_{D4} are the mean glucose levels in the four dosing periods. The null hypothesis of no treatment effect is typically tested against the ordered alternative. Under the ordered alternative, the response is assumed to increase monotonically with dose:

$$H_A : \mu_P \leq \mu_{D1} \leq \mu_{D2} \leq \mu_{D3} \leq \mu_{D4} \text{ and } \mu_P < \mu_{D4}.$$

In this section we will discuss statistical methods for assessing dose-related trends in the response variable. We will begin with simple contrast tests, discuss powerful isotonic methods (Bartholomew and Williams tests) and introduce a non-parametric approach to testing dose-response trends.

11.3.1   Contrast Tests

The simplest way of testing the null hypothesis of no treatment effect is based on the F test associated with a simple ANOVA model. However, the F test is not especially powerful in a dose-response setting because it does not take advantage of the dose order information. The ANOVA-based approach to examining the overall drug effect can be greatly improved if we select a contrast that mimics the most likely shape of the dose-response curve and carries out a trend test associated with this contrast.

To define contrast tests, consider a general dose-ranging study with m doses of a drug tested versus a placebo. Let θ_0 be the true value of the response variable in the placebo group and $\theta_1, \ldots, \theta_m$ denote the true values of the response variable in the m dose groups. The θ parameters can represent mean values of the response variable when it is continuous or incidence rates when the response variable is binary. Further, let $\widehat{\theta}_0, \ldots, \widehat{\theta}_m$ be the sample estimates of the θ parameters in the placebo and dose groups.

A contrast is defined by $m + 1$ contrast coefficients (one for each treatment group) denoted by

$$c_{0m}, \ldots, c_{mm}.$$

The coefficients are chosen in such a way that they add up to 0, i.e.,

$$\sum_{i=0}^{m} c_{im} = 0.$$

Contrasts are often defined with integer coefficients since the coefficients are standardized when the test statistic is computed.

Once a contrast has been selected, the corresponding t statistic is computed by dividing the weighted sum of the sample estimates $\widehat{\theta}_0, \ldots, \widehat{\theta}_m$ by its standard error:

$$t = \frac{\sum_{i=0}^{m} c_{im}\widehat{\theta}_i}{\text{SE}\left(\sum_{i=0}^{m} c_{im}\widehat{\theta}_i\right)}.$$

In parallel group studies with an equal number of patients in each group (say, n), this statistic follows a t distribution with $(m+1)(n-1)$ degrees of freedom.

Popular contrast tests based on linear, modified linear, and maximin contrasts are defined below.

Linear Contrast

The linear contrast is constructed by assigning integer scores (from 0 to m) to placebo and m ordered doses and then centering them around 0 (Tukey et al., 1985; Rom et al., 1994). In other words, linear contrast coefficients are given by:

$$c_{im} = i - m/2.$$

Linear contrast coefficients for dose-ranging studies with 2, 3, and 4 dose groups are displayed in Table 11.2.

As was noted above, each contrast roughly mimics the dose-response shape for which it is most powerful. The linear contrast test is sensitive to a variety of positive dose-response shapes, including a linear dose-response relationship. If the test is significant, this does not necessarily imply that the true dose-response relationship is linear.

Secondly, the linear contrast introduced above was originally proposed for dose-response studies with equally spaced doses and equal sample sizes across the treatment groups. However, the same coefficients are frequently used even when these assumptions are violated. When the doses are not equally spaced, the linear contrast essentially becomes an ordinal contrast that ignores the actual dose levels and replaces them with ordinal scores.

Maximin and Modified Linear Contrasts

The maximin contrast was derived by Abelson and Tukey (1963) and possesses an interesting optimal property. The contrast maximizes the test's power against the worst-case configuration of the dose effects under the ordered alternative H_A:

$$H_A : \theta_0 \leq \ldots \leq \theta_m \text{ and } \theta_0 < \theta_m.$$

The maximin contrast is defined as follows:

$$c_{im} = \sqrt{i - \frac{i^2}{m+1}} - \sqrt{i + 1 - \frac{(i+1)^2}{m+1}}.$$

Along with the optimal maximin contrast, Abelson and Tukey (1963) also proposed a simple modification of the linear contrast that performs almost as well as the maximin contrast. The modified linear contrast (termed the *linear-2-4 contrast* by Abelson and Tukey) is similar to the linear contrast. The only difference is that the end coefficients are quadrupled and the coefficients next to the end coefficients are doubled. Unlike the maximin coefficients, the modified linear coefficients are easy to compute and remember.

To understand the difference between the linear and maximin tests, note that, unlike the linear tests, the maximin and modified linear tests assign large weights to the treatment means in the placebo and highest dosing groups. As a result, these tests tend to ignore the drug effect in the intermediate dosing groups. Although the maximin and modified linear tests are generally more powerful than the simple linear test, they may perform poorly in trials with non-monotonic dose-response curves.

Maximin and modified linear coefficients for dose-ranging studies with 2, 3, and 4 dose groups are shown in Table 11.2.

Table 11.2 Linear, Modified Linear, and Maximin Contrasts in Dose-Ranging Studies with 2, 3, and 4 Dose Groups

	Contrast coefficients				
Contrast	Placebo	Dose 1	Dose 2	Dose 3	Dose 4
2 dose groups					
Linear	−1	0	1		
Modified linear	−4	0	4		
Maximin	−0.816	0	0.816		
3 dose groups					
Linear	−3	−1	1	3	
Modified linear	−12	−2	2	12	
Maximin	−0.866	−0.134	0.134	0.866	
4 dose groups					
Linear	−2	−1	0	1	2
Modified linear	−8	−2	0	2	8
Maximin	−0.894	−0.201	0	0.201	0.894

Note: The linear and modified linear coefficients have been standardized.

11.3.2 Contrast Tests in the Diabetes Trial Example

Program 11.1 analyzes changes in fasting serum glucose levels between the placebo and four dosing periods in the diabetes trial example. The raw data (DIABETES data set) used in the program can be found on the book's companion Web site. To account for the cross-over design, the program uses the MIXED procedure to fit a simple mixed model with a fixed group effect and a random subject effect. The drug effect is assessed using the overall F test as well as the three contrast tests introduced in this section. The contrast coefficients included in the three CONTRAST statements are taken from Table 11.2. Lastly, the program uses the Output Delivery System (ODS) to select relevant sections of the PROC MIXED output.

Program 11.1 F and contrast tests in the diabetes trial example

```
proc mixed data=diabetes;
    ods select tests3 contrasts;
    class subject rate;
    model glucose=rate/ddfm=satterth;
    repeated/type=un subject=subject;
    contrast "Linear" rate -2 -1 0 1 2;
    contrast "Modified linear" rate -8 -2 0 2 8;
    contrast "Maximin" rate -0.894 -0.201 0 0.201 0.894;
    run;
```

Output from Program 11.1

```
         Type 3 Tests of Fixed Effects

             Num    Den
Effect        DF     DF    F Value    Pr > F

rate           4   20.2      90.49    <.0001


                  Contrasts

             Num    Den
Label         DF     DF    F Value    Pr > F

Linear         1   21.3     151.14    <.0001
Modified linear 1   21.6     163.63    <.0001
Maximin        1   21.6     163.56    <.0001
```

Output 11.1 lists the test statistics and associated p-values produced by the overall F test and the linear, modified linear and maximin tests. The test statistics are very large and offer strong evidence against the null hypothesis of no drug effect.

11.3.3 Contrast Tests in the Asthma Trial Example

A similar SAS program can be used to analyze dose-response trends in a parallel group setting, e.g., in the asthma trial introduced in Section 11.1. The only change that needs to be made in Program 11.1 is deleting the REPEATED statement in PROC MIXED. Program 11.2 carries out the overall F test and three contrast tests to analyze mean changes in FEV1 collected in the asthma trial example. The contrast coefficients for this four-arm trial come from Table 11.2. The raw data (ASTHMA data set) used in the program can be found on the book's companion Web site.

Program 11.2 *F*- and contrast tests in the asthma trial example

```
proc mixed data=asthma;
    ods select tests3 contrasts;
    class dose;
    model change=dose;
    contrast "Linear" dose -3 -1 1 3;
    contrast "Modified linear" dose -12 -2 2 12;
    contrast "Maximin" dose -0.866 -0.134 0.134 0.866;
    run;
```

Output from Program 11.2

```
            Type 3 Tests of Fixed Effects

              Num      Den
Effect         DF       DF     F Value    Pr > F

dose            3      104       1.70     0.1707


                    Contrasts

              Num      Den
Label          DF       DF     F Value    Pr > F

Linear          1      104       1.86     0.1758
Modified linear 1      104       1.86     0.1760
Maximin         1      104       1.85     0.1764
```

Output 11.2 shows the test statistics and p-values produced by the overall F test and the three contrast tests introduced earlier in this section. The four test statistics indicate that the non-monotone (umbrella-shaped) dose-response relationship depicted in Figure 11.2 is actually consistent with the null hypothesis of no drug effect. Note that, in the presence of an umbrella-shaped dose-response curve, the contrast test statistics are comparable in magnitude and are similar to the test statistic of the overall F test.

11.3.4 Isotonic Tests

An alternative approach to testing dose-response trends is based on *isotonic* methods[1]. Unlike contrast tests, isotonic tests rely heavily on the assumption of a monotone dose-response relationship. This assumption is reasonable in most dose-ranging studies because higher doses generally produce stronger treatment effects compared to lower doses.

Two most popular isotonic tests were proposed by Bartholomew (1961) and Williams (1971, 1972). Both of them are based on maximum likelihood estimates of the treatment means under the monotonicity constraint and, as a result, are computationally intensive. Due to the computational complexity of the Bartholomew test, this section will focus on the Williams test for balanced dose-ranging studies with continuous endpoints. It is worth noting the Williams approach can be extended to test for trends in the binary case (Williams, 1988) or in a non-parametric setting (Shirley, 1977).

The Williams trend test is based on a comparison between the highest dose and placebo groups under the monotonicity constraint. To be precise, the Williams test statistic W_m is a two-sample t statistic in which the sample mean in the highest dose group is replaced

---

[1]The word *isotonic* is used in medical literature to describe muscular contractions. In this context, *isotonic* refers to a monotonically increasing dose-response function.

with the maximum likelihood estimate under the ordered alternative H_A:

$$H_A : \mu_0 \leq \ldots \leq \mu_m.$$

The W_m statistic is given by

$$W_m = \frac{\widehat{\mu}_m - \bar{X}_0}{s\sqrt{2/n}},$$

where $\bar{X}_0$ is the regular sample mean of the response variable in the placebo group, $\widehat{\mu}_m$ is the maximum likelihood estimate of the average response in the highest dose group under the ordered alternative, s is the pooled sample standard deviation, and n is the sample size per group. In the balanced case, the maximum likelihood estimate $\widehat{\mu}_m$ is defined as follows

$$\widehat{\mu}_m = \max\left[\frac{\sum_{i=1}^{m} \bar{X}_i}{m}, \frac{\sum_{i=2}^{m} \bar{X}_i}{m-1}, \ldots, \bar{X}_m\right].$$

Although the Williams statistic is similar to the two-sample t statistic, it no longer follows a t distribution. To find the critical values of the null distribution of W_m, we need to use a rather complicated algorithm given in Williams (1971). This algorithm is implemented in the PROBMC function. Using this function, we have written a macro for carrying out the Williams test in dose-ranging trials (the %WilliamsTest macro can be found on the book's companion Web site).

The %WilliamsTest macro has three arguments described below:

- DATASET is the data set to be analyzed.
- GROUP is the name of the group variable in the data set.
- VAR is the name of the response variable in the data set.

The macro assumes that the input data set is sorted by the group variable and the first group is the placebo group. Note also that the Williams test was developed for parallel group designs and thus the %WilliamsTest macro will not work with cross-over trials.

To illustrate the use of the Williams test, we will apply it to test for a monotonic dose-related relationship in the asthma trial (see Program 11.3).

Program 11.3 Williams test in the asthma trial example

```
%WilliamsTest(dataset=asthma,group=group,var=change);
```

Output from Program 11.3

```
Williams statistic P-value

        1.7339  0.0538
```

Output 11.3 displays the Williams statistic and associated p-value. The p-value is marginally significant ($p = 0.0538$) and thus the true dose-response relationship in the asthma trial is unlikely to be positive. Also, it is instructive to compare the p-value produced by the Williams test to those produced by popular contrast tests (Output 11.2). The Williams p-value is much smaller than the p-values listed in Output 11.2 which indicates that in this example the Williams test is more sensitive to the dose-response trend than the linear, modified linear, and maximin contrast tests.

11.3.5 Jonckheere Test

So far we have discussed dose-response tests for normally or nearly normally distributed endpoints. The assumption of normality may not be met (or may be difficult to justify) in smaller proof-of-concepts trials in which case clinical researchers need to consider a non-parametric test for dose-related trends.

A popular non-parametric trend test for dose-ranging studies was proposed by Jonckheere (1954). A similar test was also described by Terpstra (1952) and, for this reason, the Jonckheere test is sometimes referred to as the Jonckheere-Terpstra test. Another non-parametric trend test was recently proposed by Neuhäuser, Liu, and Hothorn (1998). This test is based on a simple modification of Jonckheere's approach and is more powerful than the original Jonckheere test. However, the Neuhäuser-Liu-Hothorn test is not yet available in SAS.

The Jonckheere test is based on counting the number of times a measurement from one treatment group is smaller (or larger) than a measurement from another treatment group and comparing the obtained counts to the counts that would have been observed under the null hypothesis of no drug effect. To define the Jonckheere approach, let X_{ij} be the measurement from the jth patient in the ith treatment group, $i = 0, \ldots, m$. The magnitude of response in two treatment groups, say, kth and lth groups, can be assessed in a non-parametric fashion using the Mann-Whitney statistic U_{kl}. This statistic is equal to the number of times $X_{kj} < X_{lj'}$ plus one-half the number of times $X_{kj} = X_{lj'}$. The Jonckheere statistic, defined for all pairwise comparisons as a sum of the Mann-Whitney statistic, combines the information across the $m + 1$ groups.

$$T = \sum_{k=0}^{m} \sum_{l=k+1}^{m} U_{kl}.$$

Once the T statistic has been computed, statistical inferences can be performed using a normal approximation or exact methods. First, we can compute a standardized test statistics which is asymptotically normally distributed and then find a p-value from this normal distribution. Alternatively, a p-value can be computed from the exact distribution of the T statistic.

The Jonckheere test is implemented in the FREQ procedure, which supports both the asymptotic and exact versions of this test. The Jonckheere test (in its asymptotic form) is requested by adding the JT option to the TABLES statement. To illustrate, Program 11.4 carries out the Jonckheere test to examine the dose-response relationship in the asthma trial.

Program 11.4 Jonckheere test in the asthma trial example

```
proc freq data=asthma;
    tables dose*change/noprint jt;
    run;
```

Output from Program 11.4

```
Statistics for Table of dose by change

  Jonckheere-Terpstra Test

Statistic               2450.5000
Z                          1.4498
One-sided Pr >  Z          0.0736
Two-sided Pr > |Z|         0.1471


Effective Sample Size = 108
Frequency Missing = 4
```

Output 11.4 shows that the standardized Jonckheere statistic for testing the dose-related trend in FEV1 changes is 1.4498. The associated two-sided p-value is fairly large ($p = 0.1471$), suggesting that the dose-response relationship is not positive.

It is worth noting that PROC FREQ also supports the exact Jonckheere test. To request an exact p-value, we need to use the EXACT statement as shown below

```
proc freq data=asthma;
   tables dose*change/noprint jt;
   exact jt;
   run;
```

It is important to remember that exact calculations may take a very long time (even in SAS 9.1) and, in some cases, even cause SAS to run out of memory.

11.3.6   Comparison of Trend Tests

Several authors, including Shirley (1985), Hothorn (1997) and Phillips (1997, 1998), studied the power of popular trend tests in a dose-ranging setting, including contrast tests (linear and maximin), Williams, Bartholomew and Jonckheere tests.

It is commonly agreed that contrast tests can be attractive in dose-ranging trials because they gain power by pooling information across several dose groups. This power advantage is most pronounced in cases when the shape of the dose-response curve matches the pattern of the contrast coefficients. The contrast tests introduced in this section perform best when the response increases with increasing dose. However, when a non-monotonic (e.g., umbrella-shaped) dose-response curve is encountered, contrast tests become less sensitive and may fail to detect a dose-response relationship. The same is also true for the Jonckheere test.

The isotonic tests (Williams and Bartholomew tests) tend to be more robust than contrast tests and perform well in dose-ranging trials with monotonic and non-monotonic dose-response curves (as shown in Output 11.3). To summarize his conclusions about the performance of various trend tests, Phillips (1998) stated that

> specialized tests such as Williams' and Bartholomew's tests which are tailored to the special ordering of dose groups are powerful against a variety of different patterns of results, and therefore, should be used to establish whether or not there is any drug effect.

11.3.7   Sample Size Calculations Based on Linear Contrast Tests

In this subsection, we will briefly discuss a problem that plays a key role in the design of dose-ranging trials—calculation of the trial's size. Consider a clinical trial with a parallel group design in which m doses of an experimental drug are tested against a placebo. The same number of patients, n, will be enrolled in each arm of the trial. The total sample size, $n(m+1)$, is chosen to ensure that, under a prespecified alternative hypothesis, an appropriate trend test will have $1 - \beta$ power to detect a positive dose-response relationship at a significance level α. Here β is the Type II error probability.

Sample size calculations for isotonic tests (Williams and Bartholomew tests) tend to be rather complicated, and we will focus on general contrast tests introduced in Section 11.3.1. As before, θ_0 will be the true value of the response variable in the placebo group, and $\theta_1, \ldots, \theta_m$ will denote the true values of the response variable in the m dose groups (the response can be a continuous variable or proportion). Suppose that clinical researchers are interested in rejecting the null hypothesis of no drug effect in favor of

$$H_A : \theta_0 = \theta_0^*, \; \theta_1 = \theta_1^*, \ldots, \theta_m = \theta_m^*,$$

where $\theta_0^*, \theta_1^*, \ldots, \theta_m^*$ reflect the assumed magnitude of the response in the placebo and m dose groups. These values play the role of the so-called *smallest clinically meaningful difference* which is specified when sample size calculations are performed in two-arm clinical trials.

The presence of a positive dose-response relationship will be tested using a general contrast test with the contrast coefficients $c_{0m}, c_{1m}, \ldots, c_{mm}$. Assume that the standard deviation of $\widehat{\theta}_i$ is $\sigma/\sqrt{n}$. It is easy to show that, under H_A, the test statistic t, defined in Section 11.3.1, is approximately normally distributed with mean

$$\frac{\sqrt{n} \sum_{i=0}^{m} c_{im} \theta_i^*}{\sigma \sqrt{\sum_{i=0}^{m} c_{im}^2}}$$

and standard deviation 1. Therefore, using simple algebra, the sample size in each arm is given by

$$n = \frac{\sigma^2 (z_{\alpha/2} + z_\beta)^2 \sum_{i=0}^{m} c_{im}^2}{\left(\sum_{i=0}^{m} c_{im} \theta_i^* \right)^2}.$$

EXAMPLE: Clinical trial in patients with hypercholesterolemia

A clinical trial will be conducted to study the efficacy and safety profiles of three doses of a cholesterol-lowering drug compared to placebo. The primary trial endpoint is the reduction in LDL cholesterol after a 12-week treatment. Clinical researchers hypothesize that the underlying dose-response relationship will be linear, i.e.,

$$H_A : \theta_0 = 0 \text{ mg/dL}, \; \theta_1 = 7 \text{ mg/dL}, \; \theta_2 = 14 \text{ mg/dL}, \; \theta_3 = 21 \text{ mg/dL},$$

where θ_0 is the placebo effect and θ_1, θ_2, and θ_3 represent the drug effect in the low, medium, and high dose groups, respectively. The standard deviation of LDL cholesterol changes is expected to vary between 20 and 25 mg/dL. The dose-ranging trial needs to be powered at 90% ($\beta = 0.1$) with a two-sided significance level of 0.05 ($\alpha/2 = 0.025$).

Program 11.5 computes the sample size in this dose-ranging trial using the linear contrast test with $c_{03} = -3$, $c_{13} = -1$, $c_{23} = 1$ and $c_{33} = 3$ (see Table 11.2).

Program 11.5 Sample size calculations in the hypercholesterolemia dose-ranging trial

```
data sample_size;
    * Mean effects in each arm;
    theta0=0; theta1=7; theta2=14; theta3=21;
    * Linear contrast test coefficients;
    c0=-3; c1=-1; c2=1; c3=3;
    alpha=0.025; * One-sided significance level;
    beta=0.1;
    z_alpha=probit(1-alpha);
    z_beta=probit(1-beta);
    do sigma=20 to 25;
        sum1=theta0*c0+theta1*c1+theta2*c2+theta3*c3;
        sum2=c0*c0+c1*c1+c2*c2+c3*c3;
        n=ceil(sum2*(sigma*(z_alpha+z_beta)/sum1)**2);
        output;
    end;
proc print data=sample_size noobs;
    var sigma n;
    run;
```

Output from Program 11.5

sigma	n
20	18
21	19
22	21
23	23
24	25
25	27

Output 11.5 lists the patient numbers for the selected values of σ. It is important to remember that these numbers define the size of the analysis population. In order to compute the number of patients that will be enrolled in each treatment group, we need to make assumptions about the dropout rate. With a 10% drop-out rate, the sample size in the first scenario ($\sigma = 20$) will be $18/0.9 = 20$ patients per arm.

11.4   Regression Modeling

In this section we will introduce statistical methods for addressing the second objective of dose-ranging studies, namely, characterization of the dose-response curve. While in the previous section we were concerned with hypothesis testing, this section will focus on modeling dose-response functions and estimation of their parameters. We will consider approaches based on linear and sigmoid (or E_{max}) models.

It is important to note that the modeling approaches discussed below can be used to study effects of either actual dose or pharmacokinetic exposure parameters (e.g., area under the time-plasma concentration curve or maximum plasma concentration) on the response variable of interest. When the drug's distribution follows simple pharmacokinetic models, e.g., a one-compartmental model with a first-order transfer rate, the exposure parameters are proportional to the dose (this is known as *pharmacokinetic dose-proportionality*). In this case, the two approaches yield similar results. However, in the presence of complex pharmacokinetics, dose-proportionality may not hold and exposure-response models are more informative than dose-response models.

11.4.1   Linear Models

Here we will consider a simple approach to modeling the dose- or exposure-response relationship in dose-ranging trials. This approach relies on linear models. Linear models often serve as a reasonable first approximation to an unknown dose-response function and can be justified in a variety of dose-ranging trials unless the underlying biological processes exhibit ceiling or plateau effects. If a linear model is deemed appropriate, it is easy to fit using PROC REG (parallel group designs) or PROC MIXED (parallel group or cross-over designs).

As an illustration, we will fit a linear model to approximate the relationship between the drug exposure (measured by the maximum plasma concentration of the study drug, $C_{\max}$) and mean fasting glucose level in the diabetes trial. Program 11.6 fits a linear repeated-measures model to the glucose and $C_{\max}$ measurements in the Diabetes data set using PROC MIXED. A repeated-measures model specified by the REPEATED statement is used here to account for the cross-over nature of this trial. The program also computes predicted fasting glucose levels and a confidence band for the true exposure-response function. The predicted values and confidence intervals are requested using the OUTPREDM option in the MODEL statement.

Program 11.6 Linear exposure-response model in the diabetes trial

```
proc mixed data=diabetes;
    class subject;
    model glucose=cmax/outpredm=predict;
    repeated/type=un subject=subject;
proc sort data=predict;
    by cmax;
axis1 minor=none value=(color=black)
    label=(color=black angle=90 'Glucose level (mg/dL)')
    order=(100 to 200 by 20);
axis2 minor=none value=(color=black) label=(color=black 'Cmax (ng/mL)');
symbol1 value=none i=join color=black line=20;
symbol2 value=none i=join color=black line=1;
symbol3 value=none i=join color=black line=20;
symbol4 value=dot i=none color=black;
proc gplot data=predict;
    plot (lower pred upper glucose)*cmax/overlay frame haxis=axis2 vaxis=axis1;
    run;
    quit;
```

Output from Program 11.6

| | | Standard | | | |
Effect	Estimate	Error	DF	t Value	Pr > |t|
Intercept	175.78	2.1279	23	82.61	<.0001
cmax	-0.03312	0.003048	23	-10.87	<.0001

Solution for Fixed Effects

Output 11.6 lists estimates of the intercept (INTERCEPT) and slope (CMAX) of the linear exposure-response model produced by PROC MIXED. The intercept represents the baseline (placebo) effect and the slope indicates the rate at which the mean glucose concentration decreases with increasing exposure.

Figure 11.4 depicts the predicted fasting glucose concentration as a function of $C_{\max}$ along with a series of 95% confidence intervals that form a confidence band. It is clear from

Figure 11.4 Fasting glucose levels predicted by a linear model in the diabetes trial

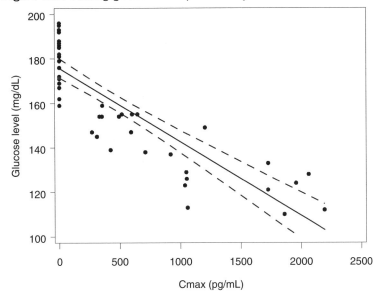

Figure 11.4 that the linear model provides a fairly poor fit to the fasting glucose data. Most of the dots representing individual glucose measurements lie below the fitted line and outside the confidence band. The pattern of individual measurements suggests that the true exposure-response relationship is likely to be curvilinear.

11.4.2 Sigmoid Models

We saw in the previous subsection that linear models may not always provide an adequate approximation to exposure-response functions. Non-linear models perform better in practice and are generally more appealing from a clinical perspective. One way to improve the performance of simple linear models is by applying a non-linear transformation of the independent variable, e.g., fitting a linear model with a log-transformed AUC or $C_{\max}$. Another popular non-linear modeling approach relies on the sigmoid model (also referred to as the $E_{\max}$ model). This model describes the true treatment effect E as a sigmoid function of drug exposure d:

$$E = E_0 + \frac{E_{\max}d^\gamma}{d^\gamma + d_{50}^\gamma}.$$

In this model,

- E_0 is the baseline effect at the zero exposure level.
- $E_{\max}$ represents the maximum possible effect of the experimental drug on the response variable relative to baseline.
- d_{50} is the exposure level at half-maximal effect (sometimes called the *median effective dose*). It is easy to verify that the treatment effect E is equal to $E_0 + E_{\max}/2$ when $d = d_{50}$.
- γ determines the steepness of the sigmoid function. The γ parameter, known as the *Hill coefficient* in the pharmacodynamic literature, typically ranges between 1 and 3.

In cross-over trials, the introduced sigmoid model needs to be modified to account for repeated measurements on the same subject. This is achieved by including a random subject term in the model. Program 11.7 uses a sigmoid model with a random subject term to approximate the exposure-response curve in the diabetes trial:

$$E = E_0 - \frac{(E_{\max} + \eta)d}{d + d_{50}},$$

where η is a random subject term and the γ parameter is set to 1 to simplify non-linear modeling in this small data set. Note that E is a decreasing function of d since the experimental drug is expected to lower the glucose level. To fit this complex non-linear model, the program relies on the NLMIXED procedure, a powerful procedure that supports a broad class of non-linear models with random effects.

In Program 11.7, the response variable (GLUCOSE) is modeled as a non-linear function of a fixed exposure effect (CMAX) and a random subject effect (SUB). The subject effects are assumed to be normally distributed. The SUBJECT option in the RANDOM statement identifies groups of related observations made on the same individual.

The initial values of the baseline and maximum effects, E_0 and $E_{\max}$, as well as the median effective dose, d_{50}, were taken from the dose-response function depicted in Figure 11.1. The initial values of both the within-subject (INTRASUBJECT) and between-subject (INTERSUBJECT) variances were 1.

Program 11.7 Sigmoid exposure-response model in the diabetes trial

```
proc nlmixed data=diabetes;
    ods select ParameterEstimates;
    parms e0=180 emax=60 d50=900 intrasigma=1 intersigma=1;
    prediction=e0-(sub+emax)*cmax/(cmax+d50);
    model glucose~normal(prediction,intersigma);
    random sub~normal(0,intrasigma) subject=subject;
    predict e0-emax*cmax/(cmax+d50) out=predict;
proc sort data=predict;
    by cmax;
axis1 minor=none value=(color=black)
    label=(color=black angle=90 'Glucose level (mg/dL)')
    order=(100 to 200 by 20);
axis2 minor=none value=(color=black) label=(color=black 'Cmax (ng/mL)');
symbol1 value=none i=join color=black line=20;
symbol2 value=none i=join color=black line=1;
symbol3 value=none i=join color=black line=20;
symbol4 value=dot i=none color=black;
proc gplot data=predict;
    plot (lower pred upper glucose)*cmax/
        overlay frame haxis=axis2 vaxis=axis1;
    run;
    quit;
```

Output from Program 11.7

| | | Standard | | | | |
Parameter	Estimate	Error	DF	t Value	Pr > |t|	Alpha
e0	179.36	2.0873	47	85.93	<.0001	0.05
emax	87.1425	16.3395	47	5.33	<.0001	0.05
d50	899.75	396.02	47	2.27	0.0277	0.05
intrasigma	3.0613	148.03	47	0.02	0.9836	0.05
intersigma	101.12	24.8511	47	4.07	0.0002	0.05

Parameter Estimates

```
               Parameter Estimates

Parameter      Lower      Upper     Gradient

e0            175.16     183.56     0.000378
emax          54.2717    120.01    -0.00006
d50           103.05     1696.45    0.001309
intrasigma   -294.73     300.85     0.003211
intersigma    51.1297    151.12    -0.00003
```

Output 11.6 shows estimates of the parameters of the sigmoid model. The estimated E_0, E_{max}, and d_{50} parameters are generally close to their initial values. According to the fitted sigmoid model, the experimental drug is theoretically capable of reducing the mean glucose level to 92 mg/dL ($= 179 - 87$). The half-maximal effect is achieved at the median effective dose (D50) which is equal to 900 mg/dL. The within-subject variance (INTRASUBJECT) is considerably smaller than the between-subject variance (INTERSUBJECT).

Figure 11.5 Fasting glucose levels (mg/dL) predicted by a sigmoid model in the diabetes trial

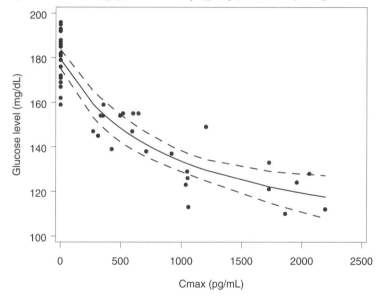

Figure 11.5 displays the fasting glucose concentration predicted by the sigmoid model with a 95% confidence band. The fitted dose-response curve has a concave shape and fits the data well. There is no obvious pattern in the residuals and many observations fall within the 95% confidence band around the fitted curve.

The fitted model can be used for an informal determination of doses or drug exposures to be investigated in Phase II trials (Ruberg, 1995b). Suppose, for example, that the objective of the diabetes trial was to reduce the glucose concentration to a "normal" level defined as 126 mg/dL. We can see from Figure 11.5 that the lower confidence limit is below this clinically important threshold when the maximum plasma concentration is greater than 1200 pg/mL. Secondly, the mean glucose concentration falls below the 126 mg/dL threshold around $C_{max} = 1700$ pg/mL. Using this information, we could define the *minimally effective exposure level* as a value between 1200 and 1700 pg/mL and *effective exposure levels* as values exceeding 1700 pg/mL.

11.5   Dose-Finding Procedures

As was indicated in Section 11.1, once a positive dose-response has been established, clinical researchers are interested in determining the optimal dose range for their experimental drug known as the *therapeutic window*. As an example, the usual adult dosage range for prescription ibuprofen is 200 to 800 mg three or four times per day, not to exceed 3200 mg total daily. In other words, ibuprofen's therapeutic window extends from 600 mg to 3200 mg per day. The therapeutic window is more conservative for over-the-counter (OTC) versions. The usual adult dosage range for OTC ibuprofen is 200 to 400 mg every four to six hours, not to exceed 1200 mg total daily.

The therapeutic window is defined by the minimum effective dose (MED) and maximum tolerated dose (MTD). As the name implies, the MED is chosen to ensure the efficacy of both it and all dose levels higher than it (up to the upper limit of the therapeutic window). Likewise, as the maximum dose, an MTD is chosen to ensure the safety of both it and all dose levels lower than it. In the drug approval process, safety (including the determination of the MTD) is frequently assessed through simple comparisons of adverse events across treatment arms. The Center for Drug Evaluation and Research at the FDA is currently studying how to provide a more detailed statistical assessment of the safety profile of a drug by taking into account the time course or other important factors.

In this section, we will concentrate on the statistical determination of the MED in dose-ranging studies with a placebo control, given that a safe range of doses has been tentatively determined. For more information on simultaneous tests for identifying the MED and MTD, see Bauer, Brannath, and Posch (2001) and Tamhane and Logan (2002). Further, Bauer et al. (1998) describe testing strategies for dose-ranging studies with both negative and positive controls.

11.5.1   MED Estimation under the Monotonicity Assumption

The problem of estimating the MED is often stated as a problem of stepwise multiple testing. Clinical researchers begin with the highest dose or the dose corresponding to the largest treatment difference and proceed in a stepwise fashion until they encounter a non-significant treatment difference. The immediately preceeding dose is defined as the minimum effective dose (MED).

To simplify the statement of the multiple testing problem arising in MED estimation, it is common to make the following two assumptions known as monotonicity constraints:

- All doses are no worse than placebo.
- If Dose k is not efficacious, the lower doses (Doses 1 through $k-1$) are not efficacious either.

These assumptions are reasonable in clinical trials with a positive dose-response relationship. However, they are not met when the true dose-response function is umbrella-shaped and stepwise tests relying on the monotonicity assumptions break down in the presence of non-monotone dose-response curves.

Under the monotonicity constraints, the MED estimation problem is easily expressed in terms of sequential testing of m null hypotheses associated with the m dose levels. To define the null hypotheses, assume that the drug's effect is expressed as a positive shift. Further, the observations, y_{ij}, are assumed to be normally distributed and follow an ANOVA model

$$y_{ij} = \mu_i + \varepsilon_{ij},$$

where $\mu_0, \ldots, \mu_m$ are the true treatment means in the placebo and m dose groups and ε_{ij}'s are residuals. Given this model, the null hypotheses used in MED estimation are displayed in Table 11.3. The clinically important difference is denoted by δ. When the clinically important difference is assumed to be 0, the kth null hypothesis, H_{0k}^M, simplifies to

$$\mu_0 = \mu_1 = \ldots = \mu_k.$$

The first null hypothesis in Table 11.3, H_{01}^M, states that Dose 1 is ineffective, and therefore its rejection implies that Dose 1 is the MED. Likewise, if H_{01}^M is retained but H_{02}^M is rejected, Dose 2 is declared the MED, etc.

Table 11.3 Null Hypotheses Used in MED Estimation under the Monotonicity Constraint

Null hypothesis	Interpretation
$H_{01}^M : \mu_0 \leq \mu_1 < \mu_0 + \delta$	Dose 1 is ineffective
$H_{02}^M : \mu_0 \leq \mu_1, \mu_2 < \mu_0 + \delta$	Doses 1 and 2 are ineffective
$\vdots$	$\vdots$
$H_{0k}^M : \mu_0 \leq \mu_1, \ldots, \mu_k < \mu_0 + \delta$	Doses $1, \ldots, k$ are ineffective
$\vdots$	$\vdots$
$H_{0m}^M : \mu_0 \leq \mu_1, \ldots, \mu_m < \mu_0 + \delta$	All doses are ineffective

The stepwise testing approach goes back to early work by Tukey, Ciminera, and Heyse (1985), Mukerjee, Robertson, and Wright (1987), Ruberg (1989) and others who proposed multiple-contrast methods for examining dose-related trends. This section focuses on stepwise contrast tests considered by Tamhane, Hochberg, and Dunnett (1996) and Dunnett and Tamhane (1998). We can also construct stepwise testing procedures for estimating the MED using other tests, e.g., isotonic tests described in Section 11.3. For more information about MED estimation procedures based on the Williams test, see Dunnett and Tamhane (1998) and Westfall et al. (1999, Section 8.5.3).

11.5.2   Multiple-Contrast Tests

The null hypotheses displayed in Table 11.3 will be tested using *multiple-contrast* tests. It is important to understand the difference between multiple-contrast tests introduced here and the tests discussed in Section 11.3 (known as *single-contrast* tests). As their name implies, single-contrast tests rely on a single contrast and are used mainly for studying the overall drug effect. Each of multiple-contrast tests discussed in this subsection actually relies on a family of contrasts

$$(c_{0m}(1), \ldots, c_{mm}(1)), \quad (c_{0m}(2), \ldots, c_{mm}(2)), \quad \ldots, \quad (c_{0m}(m), \ldots, c_{mm}(m)).$$

These contrasts are applied in a stepwise manner to test $H_{01}^M, \ldots, H_{0m}^M$. The k null hypothesis, H_{0k}^M, is tested using the following t statistic

$$t_k = \frac{\sum_{i=0}^{m} c_{im}(k)\widehat{\mu}_i - \delta}{\text{SE}\left(\sum_{i=0}^{m} c_{im}(k)\widehat{\mu}_i\right)},$$

where $\widehat{\mu}_0, \ldots, \widehat{\mu}_m$ are the sample treatment means in the placebo and m dose groups. Assuming n patients in each treatment group, each of the t statistics follows a t distribution with $\nu = (m+1)(n-1)$ degrees of freedom.

It is important to note that the contrasts can be constructed on any scale when the clinically significant difference (δ) is 0. The contrast coefficients are automatically standardized when the test statistic is computed. However, when $\delta > 0$, the contrasts need to be on the correct scale because the standardization occurs only after δ is subtracted. To ensure the correct scale is used, the contrast coefficients are defined in such a way that $c_{0m}(k)$ is negative, $c_{km}(k)$ is positive and the positive coefficients add up to 1.

Popular multiple-contrast tests used in dose-finding are defined below.

Pairwise Contrasts

The easiest way to compare multiple dose groups to placebo is to consider all possible dose-placebo comparisons. The associated test is based on a family of pairwise contrasts. The following coefficients are used when the Dose k is compared to placebo:

$$c_{0m}(k) = -1, \quad c_{km}(k) = 1$$

and all other coefficients are equal to 0.

Helmert Contrasts

Unlike the pairwise contrast, the Helmert test combines information across several dose groups. Specifically, when comparing Dose k to placebo, this test assumes that the lower doses (Doses 1 through $k-1$) are not effective and pools them with placebo. The kth Helmert contrast is defined as follows:

$$c_{0m}(k) = \ldots = c_{(k-1)m}(k) = -\frac{1}{k}, \quad c_{km}(k) = 1.$$

The other coefficients are equal to 0. Helmert contrasts are most powerful when the lower doses are similar to placebo.

Reverse Helmert Contrasts

The reverse Helmert contrast test is conceptually similar to the regular Helmert test. The reverse Helmert test combines information across doses by assuming that the lower doses are as effective as Dose k. Thus, the kth reverse Helmert contrast are given by

$$c_{0m}(k) = -1, \quad c_{1m}(k) = \cdots = c_{km}(k) = \frac{1}{k}.$$

The remaining coefficients are equal to 0. Reverse Helmert contrasts are most powerful when the treatment effect quickly plateaus and the larger doses are similar to the highest dose.

Linear Contrasts

The linear contrast test assigns weights to the individual doses that increase in a linear fashion. Dose k is compared to placebo using the following contrast:

$$c_{0m}(k) = -\frac{k}{2l}, \quad c_{1m}(k) = \frac{1}{l}\left(1 - \frac{k}{2}\right), \quad \ldots, \quad c_{km}(k) = \frac{k}{2l},$$

where

$$l = \frac{k}{4}\left(\frac{k}{2} + 1\right) \text{ if } k \text{ is even}, \quad l = \frac{1}{2}\left(\left[\frac{k}{2}\right] + 1\right)^2 \text{ if } k \text{ is odd}.$$

and $[k/2]$ is the largest integer in $k/2$. As before, the other coefficients are equal to 0. Linear contrasts are most powerful when the MED is near the middle of the range of doses tested in a trial.

Put simply, the Helmert, reverse Helmert and linear contrasts each correspond to different shapes of the dose-response curve for which they are most powerful. Figure 11.6 displays the treatment effect configurations that lead to the maximum expected value of the t statistic for each type of contrast.

Figure 11.6 Treatment effect configurations leading to the maximum expected value of the t statistic under the monotonicity constraint

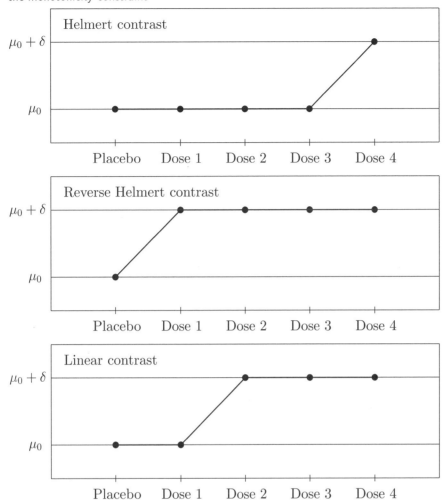

11.5.3 Closed Testing Procedures Based on Multiple Contrasts

In this section, we will consider the MED estimation problem from a multiple testing perspective and focus on MED estimation procedures that control the overall Type I error rate. As was pointed out in the introduction, control of the Type I error probability is required in registration trials and is less common in Phase II trials.

Two general stepwise procedures for testing $H_{01}^M, \ldots, H_{0m}^M$ can be constructed based on the principle of closed testing proposed by Marcus, Peritz, and Gabriel (1976). The principle has provided a foundation for numerous multiple tests and has found a large number of applications in multiplicity problems arising in clinical trials. For example, Kodell and Chen (1991) and Rom, Costello, and Connell (1994) applied the closed testing principle to construct multiple tests for dose-ranging studies.

Very briefly, the closed testing principle is based on a hierarchical representation of a multiplicity problem. In general, we need to consider all possible intersections of the null hypotheses of interest (known as a *closed family of hypotheses*) and test each intersection at the same significance level. After that, the results need to be combined to make inferences about the original null hypotheses. In this case, the family of null hypotheses defined in Table 11.3 is already a closed family and therefore closed testing procedures have a simple sequentially rejective form.

Two closed procedures for estimating the MED are introduced below. The first procedure begins with the most significant dose-placebo comparison and then works downward. For this reason, this procedure is known as the *step-down procedure*. Note that the idea behind the step-down approach is that the order in which the dose-placebo tests are examined is driven by the data. By contrast, the *fixed-sequence procedure* requires that the testing sequence be specified prior to data analysis.

11.5.4   Step-Down MED Estimation Procedure

Consider the m contrast test statistics, $t_1, \ldots, t_m$, and order them from the most significant to the least significant. The ordered t statistics will be denoted by

$$t_{(1)} \geq \ldots \geq t_{(m)}.$$

The step-down MED estimation procedure is defined below:

- Consider the most significant t statistic, $t_{(1)}$, and compare it to the prespecified critical value (e.g., two-sided 0.05 critical value) derived from the null distribution of $t_1, \ldots, t_m$ with $\nu = (m+1)(n-1)$ degrees of freedom and correlation matrix ρ. This critical value is denoted by c_1. If $t_{(1)}$ exceeds c_1, the corresponding null hypothesis is rejected and we proceed to the second most significant t statistic.

- The next statistic, $t_{(2)}$, is compared to c_2 which is computed from the null distribution of $t_1, \ldots, t_{m-1}$ with the same number of degrees of freedom and correlation matrix as above (c_2 is less than c_1). If $t_{(2)} > c_2$, reject the corresponding null hypothesis and examine $t_{(3)}$, etc.

- The step-down procedure terminates as soon as it encounters a null hypothesis which cannot be rejected.

Let l be the index of the lowest dose which is significantly different from placebo. When testing stops, the monotonicity assumption implies that the null hypotheses $H_{0l}^M, \ldots, H_{0m}^M$ should be rejected and therefore Dose l is declared the MED.

11.5.5   Fixed-Sequence MED Estimation Procedure

This procedure relies on the assumption that the order in which the null hypotheses are tested is predetermined. Let $t_{[1]}, \ldots, t_{[m]}$ denote the *a priori* ordered t contrast statistics (any ordering can be used as long as it is prespecified). The fixed-sequence procedure is defined as follows:

- Compare $t_{[1]}$ to the prespecified critical value of the t distribution with ν degrees of freedom (denoted by c). If $t_{[1]}$ is greater than c, the corresponding null hypothesis is rejected and the next test statistic is examined.

- The next test statistic, $t_{[2]}$, is also compared to c and the fixed-sequence procedure proceeds in this manner until it fails to reject a null hypothesis.

The dose corresponding to the last rejected null hypothesis is declared the MED.

The step-down and fixed-sequence procedures protect the Type I error rate with respect to the entire family of null hypotheses shown in Table 11.3, given the monotonicity constraints. To be more precise, the two procedures control the probability of erroneously rejecting any true null hypothesis in the family regardless of which and how many other null hypotheses are true. This is known as the control of the *familywise error rate* in the *strong sense*.

11.5.6   Pairwise Multiple-Contrast Test in the Hypertension Trial

To illustrate the use of stepwise tests in MED estimation, we will begin with the pairwise contrast test. Unlike other contrast tests, the pairwise test is easy to implement in practice because the associated correlation matrix has a very simple structure. In the balanced case, the correlation between any two pairwise t statistics is 0.5 and, because of this property, we can carry out the pairwise contrast test by a stepwise application of the well-known Dunnett test.

In the following program, we will apply the step-down and fixed-sequence versions of the Dunnett test to find the MED in the hypertension trial. First, Program 11.8 assesses the overall drug effect in the hypertension trial using the three contrast tests defined in Section 11.3.

Program 11.8 Contrast tests in the hypertension trial

```
proc mixed data=hypertension;
   ods select contrasts;
   class dose;
   model change=dose;
   contrast "Linear" dose -3 -1 1 3;
   contrast "Modified linear" dose -12 -2 2 12;
   contrast "Maximin" dose -0.866 -0.134 0.134 0.866;
   run;
```

Output from Program 11.8

	Contrasts			
Label	Num DF	Den DF	F Value	Pr > F
Linear	1	64	8.61	0.0046
Modified linear	1	64	7.70	0.0072
Maximin	1	64	7.62	0.0075

Output 11.8 shows the test statistics and p-values of the linear, modified linear, and maximin tests. All three p-values are very small which indicates that the response improves in a highly significant manner with increasing dose in the hypertension trial.

Now that a positive dose-response relationship has been established, we are ready for a dose-finding exercise. First, we will use the step-down version of the Dunnett test and then apply the fixed-sequence Dunnett test.

11.5.7   Step-Down Dunnett Test in the Hypertension Trial

Program 11.9 carries out the step-down Dunnett test to determine the MED in the hypertension trial with the clinically important difference $\delta = 0$. The program uses PROC MIXED to compute the t statistics associated with the three dose-placebo comparisons. The statistics are ordered from the most significant to the least significant and compared to successively lower critical values. Specifically, the critical values of the step-down procedure are computed from the Dunnett distribution for the case of three, two, and one dose-placebo comparisons. These critical values are found using the PROBMC function. Note that the PROBMC function assumes a balanced case. Therefore, we computed the average number of patients per group in the hypertension trial (17) and plugged the number into the well-known formula for calculating degrees of freedom ($4(17 - 1) = 64$).

Program 11.9 Step-down Dunnett test in the hypertension trial

```
ods listing close;
* Compute t statistics for three dose-placebo comparisons;
proc mixed data=hypertension;
    class dose;
    model change=dose;
    lsmeans dose/pdiff adjust=dunnett;
    ods output diffs=dunnett;
* Order t statistics;
proc sort data=dunnett;
    by descending tvalue;
* Compute critical values (based on Dunnett distribution);
data critical;
    set dunnett nobs=m;
    format c 5.2;
    order=_n_;
    c=probmc("DUNNETT2",.,0.95,64,m-_n_+1);
    label c='Critical value';
proc print data=critical noobs label;
    var order dose tvalue c;
    ods listing;
    run;
```

Output from Program 11.9

order	dose	t Value	Critical value
1	20	2.80	2.41
2	40	2.54	2.26
3	10	1.10	2.00

Output 11.9 shows the t statistics computed in the hypertension trial and the associated critical values of the step-down Dunnett test. The dose groups are ordered by their t statistic and the ORDER variable shows the order in which the doses will be compared to placebo. The following algorithm is used to determine significance of the t statistics listed in Output 11.9.

- The most significant test statistic ($t_{(1)} = 2.80$) arises when we compare the medium dose to placebo. This statistic is greater than the critical value ($c_1 = 2.41$) and therefore we proceed to the next dose-placebo comparison.
- The test statistic for the high dose vs. placebo comparison ($t_{(2)} = 2.54$) also exceeds the corresponding critical value ($c_2 = 2.26$). Due to this significant result, we will now compare the low dose versus placebo.
- Examining the last dose-placebo comparison, we see that the test statistic ($t_{(3)} = 1.10$) is less than $c_3 = 2.00$.

Since the last dose-placebo comparison did not yield a significant result, the medium dose (20 mg/day) is declared the MED. The multiple comparisons summarized in Output 11.9 enable clinical researchers to provide the following characterization of the dose-response function in the hypertension trial. First, as was shown in Output 11.8, the overall dose-response trend in diastolic blood pressure change is positive and highly significant. Further, we have concluded from Output11.9 that the 20 mg/day dose is the MED and thus the positive dose-response trend in the hypertension trial is due to statistically significant treatment differences at the 20 mg/day and 40 mg/day dose levels.

It is important to remember that the step-down testing approach relies heavily on the monotonicity assumption. When this assumption is not met, the approach may lead to results that look counterintuitive. Suppose, for example, that we conclude significance for the low and medium doses but not for the high dose. In this case, the low dose is declared the MED even though we did not actually reject the null hypothesis of no drug effect at the high dose.

11.5.8   Fixed-Sequence Pairwise Test in the Hypertension Trial

Program 11.10 takes a different approach to the dose-finding problem in the hypertension trial. It identifies the MED using the fixed-sequence version of the Dunnett test. To understand the difference between the two approaches, recall that the step-down procedure (see Output 11.9) compares ordered t statistics to successively lower Dunnett critical values whereas the fixed-sequence procedure compares the same t statistics to a constant critical value derived from a t distribution (it is the two-sided 95th percentile of the t distribution with 64 degrees of freedom). Additionally, the fixed-sequence procedure requires that the three tests for individual dose-placebo comparisons be ordered before the dose-finding exercise begins. We will assume a natural testing sequence here: high dose vs. placebo, medium dose vs. placebo, and low dose vs. placebo.

Program 11.10 Fixed-sequence pairwise test in the hypertension trial

```
ods listing close;
* Compute t statistics for three dose-placebo comparisons;
proc mixed data=hypertension;
    class dose;
    model change=dose;
    lsmeans dose/pdiff adjust=dunnett;
    ods output diffs=dunnett;
* Prespecified testing sequence;
data dunnett;
    set dunnett;
    if dose=10 then order=3;
    if dose=20 then order=2;
    if dose=40 then order=1;
* Order t statistics;
proc sort data=dunnett;
    by order;
* Compute critical values (based on t distribution);
data critical;
    set dunnett;
    format c 5.2;
    c=tinv(0.975,64);
    label c='Critical value';
proc print data=critical noobs label;
    var order dose tvalue c;
    ods listing;
    run;
```

Output from Program 11.10

order	dose	t Value	Critical value
1	40	2.54	2.00
2	20	2.80	2.00
3	10	1.10	2.00

Output 11.10 lists *t* statistics for the three dose-placebo comparisons and critical values of the fixed-sequence procedure. Note that the testing sequence is no longer data-driven—the doses are ordered according to the prespecified rule (the ORDER variable indicates the order in which the doses will be compared to placebo). Also, the critical values in Output 11.10 are uniformly smaller than those shown in Output 11.9 with the exception of the low dose vs. placebo test. This means that the fixed-sequence dose-finding procedure is likely to find more significant differences than the step-down procedure, provided that the prespecified testing sequence is consistent with the true dose-response curve.

Proceeding in a stepwise fashion, we see that the test statistics associated with the two highest doses (2.54 and 2.80) are greater than the critical value. However, the low dose is not significantly different from placebo because its *t* statistic is too small. Given this configuration of *t* values, we conclude that the medium dose (20 mg/day dose) is the MED. When we compare this to Output 11.9, it is easy to see that the two MED estimation procedures resulted in the same minimally effective dose. However, the conclusions are not guaranteed to be the same, especially when the dose-response curve is not perfectly monotone.

11.5.9 Other Multiple-Contrast Tests in the Hypertension Trial

The dose-finding procedures described so far relied on pairwise comparisons. Intuitively, these procedures are likely to be inferior (in terms of power) to procedures that pool information across dose levels. For example, consider procedures based on a stepwise application of the Helmert or linear contrasts. In general, the problem of computing critical values for the joint distribution of *t* statistics in stepwise procedures presents a serious computational challenge, especially in the case of unequal sample sizes. This is due to the complex structure of associated correlation matrices. Although exact numerical methods have been discussed in the literature (see, for example, Genz and Bretz, 1999), simulation-based approaches are generally more flexible and attractive in practice. Here we will consider a simulation-based solution proposed by Westfall (1997).

Westfall (1997) proposed a Monte Carlo method that can be used in dose-finding procedures based on non-pairwise contrasts. Instead of computing critical values for *t* statistics as was done in Program 11.9, the Monte Carlo algorithm generates adjusted *p*-values. The adjusted *p*-values are then compared to a prespecified significance level (e.g., two-sided 0.05 level) to test the null hypotheses listed in Table 11.3. The algorithm is implemented in the %SimTests macro described in Westfall et al. (1999, Section 8.6).[2]

Program 11.11 calls the %SimTests macro to carry out the reverse Helmert multiple-contrast test in the hypertension trial. Reverse Helmert contrasts were chosen because they perform well in clinical trials in which the treatment effect reaches a plateau and other multiple-contrast tests can be used if a different dose-relationship is expected.

As shown in the program, to invoke the %SimTests macro, we need to define several parameters, including the family of reverse Helmert contrasts, the treatment means in the hypertension trial, and their covariance matrix. The parameters are specified in the %Estimates and %Contrasts macros using SAS/IML syntax. Once these parameters have been defined, the macro approximates adjusted two-sided *p*-values for the three dose-placebo tests using 100,000 simulations.

[2]As this book was going to press, the methods in the %SimTests and %SimIntervals macros were being completely hard-coded in the GLIMMIX procedure in SAS/STAT software. The methodology is described in Westfall and Tobias (2006).

Program 11.11 Reverse Helmert multiple-contrast test in the hypertension trial

```
ods listing close;
* Compute treatment means and covariance matrix;
proc mixed data=hypertension;
    class dose;
    model change=dose;
    lsmeans dose/cov;
    ods output lsmeans=lsmeans;
    run;
%macro Estimates;
    use lsmeans;
    * Treatment means;
    read all var {estimate} into estpar;
    * Covariance matrix;
    read all var {cov1 cov2 cov3 cov4} into cov;
    * Degrees of freedom;
    read point 1 var {df} into df;
%mend;
%macro Contrasts;
    * Reverse Helmert contrasts;
    c={1 -1 0 0, 1 -0.5 -0.5 0, 1 -0.3333 -0.3333 -0.3334};
    c=c`;
    * Labels;
    clab={"Dose 1 vs Placebo", "Dose 2 vs Placebo", "Dose 3 vs Placebo"};
%mend;
* Compute two-sided p-values using 100,000 simulations;
%SimTests(nsamp=100000,seed=4533,type=LOGICAL,side=B);
proc print data=SimTestOut noobs label;
    var contrast adjp seadjp;
    format adjp seadjp 6.4;
    ods listing;
    run;
```

Output from Program 11.11

| | | SEAdj |
Contrast	AdjP	P
Dose 1 vs Placebo	0.2753	0.0000
Dose 2 vs Placebo	0.0425	0.0002
Dose 3 vs Placebo	0.0200	0.0002

Output 11.11 lists Monte Carlo approximations to the adjusted p-values for the three dose-placebo comparisons as well as associated standard errors. The standard errors are quite small which indicates that the approximations are reliable. Comparing the computed p-values to the 0.05 threshold, we see that the highest two doses are significantly different from placebo, whereas the low dose not separate from placebo. As a consequence, the medium dose (20 mg/day dose) is declared the MED. This conclusion is consistent with the conclusions we reached when applying the pairwise multiple-contrast test.

11.5.10   Limitations of Tests Based on the Monotonicity Assumption

It is important to remember that the dose-finding procedures described in the previous subsection control the Type I error rate only if the true dose-response relationship is monotone. When the assumption of a monotonically increasing dose-response relationship is not met, the use of multiple-contrast tests can result in a severely inflated probability of false-positive outcomes (Bauer, 1997; Bretz, Hothorn and Hsu, 2003). This phenomenon will be illustrated below.

Non-monotone dose-response shapes, known as the *U-shaped* or *umbrella-shaped* curves, are characterized by a lower response at higher doses (see, for example, Figure 11.1). When the dose-response function is umbrella-shaped and the true treatment difference at the highest dose is not clinically meaningful (i.e., $\mu_m - \mu_0 < \delta$), the highest dose should not be declared effective. However, if the remaining portion of the dose-response function is monotone, some reasonable tests that control the Type I error rate under the monotonicity constraint may finally declare the highest dose effective with a very high probability.

11.5.11   Hypothetical Dose-Ranging Trial Example

To illustrate, consider a hypothetical trial in which five doses of an experimental drug are compared to placebo. Table 11.4 shows the true treatment means μ_0, μ_1, μ_2, μ_3, μ_4, and μ_5 as well as true standard deviation σ (the standard deviation is chosen in such a way that $\sigma/\sqrt{n} = 1$). Assume that the clinically important treatment difference δ is 0.

Table 11.4 Hypothetical Dose-Ranging Trial

	Placebo	Group 1	Group 2	Group 3	Group 4	Group 5
n	5	5	5	5	5	5
Mean	0	0	0	2.5	5	0
SD	2.236	2.236	2.236	2.236	2.236	2.236

It can be shown that the sequential test based on pairwise contrasts (i.e., stepwise Dunnett test) controls the familywise error rate under any configuration of true treatment means. However, when other contrast tests are carried out, the familywise error rate becomes dependent on the true dose-response shape and can become considerably inflated when an umbrella-shaped dose-response function (similar to the one shown in Table 11.4) is encountered.

Consider, for example, the linear contrast test. If linear contrasts are used for detecting the MED in this hypothetical trial, the probability of declaring Dose 5 efficacious will be 0.651 (using a t test with 24 error degrees of freedom). To see why the linear contrast test does not control the Type I error rate, recall that, in testing H_{05}^M defined in Table 11.3, positive weights are assigned to the treatment differences $\widehat{\mu}_4 - \widehat{\mu}_1$ and $\widehat{\mu}_3 - \widehat{\mu}_2$. The expected value of the numerator of the t-statistic for testing H_{05} is given by

$$\left(\frac{5}{9}(0) + \frac{3}{9}(5) + \frac{1}{9}(2.5) \right) - \left(\frac{1}{9}(0) + \frac{3}{9}(0) + \frac{5}{9}(0) \right) = 1.94.$$

As a result, the test statistic has a non-central t distribution with a positive non-centrality parameter and thus the linear contrast test will reject the null hypothesis H_{05}^M more often than it should.

The reverse Helmert contrast test, like the linear contrast test, can also have too high an error rate when the true dose-response curve is umbrella-shaped. In the hypothetical trial example, the probability of declaring the highest dose efficacious will be 0.377 (again, based on a t test with 24 degrees of freedom). This error rate is clearly much higher than 0.05. Combining the signal from the third and fourth doses into the estimate $\widehat{\mu}_5$ causes the reverse Helmert contrast test to reject the hypothesis too often.

The regular Helmert contrast test does control the Type I error rate in studies with umbrella-shaped dose-response curves. However, it can have an inflated Type I error rate under other configurations. In particular, the Type I error probability can be inflated if the response at lower doses is worse than the response observed in the placebo group. Figure 11.7 shows the general shapes of treatment effect configurations that can lead to inflated Type I error rates for each type of contrast.

Figure 11.7 Treatment effect configurations that can lead to inflated Type I error rates

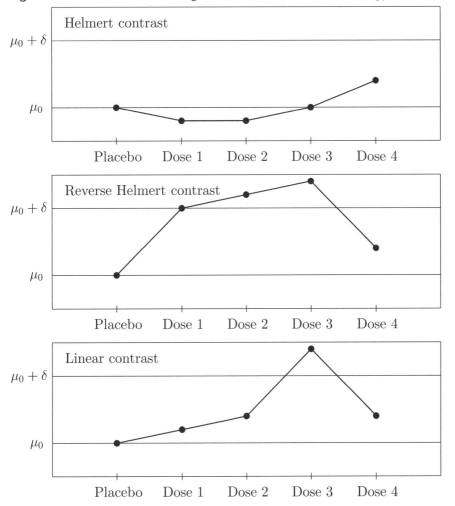

The choice of which contrast test to use in dose-finding problems is thus clear:

- If a monotonically increasing dose-response relationship can be assumed, clinical researchers need to choose contrasts according to which dose-response shape in Figure 11.6 is expected.

- If a monotone dose-response relationship cannot be assumed, we must choose pairwise contrasts to protect the overall Type I error probability.

11.5.12   MED Estimation in the General Case

In this section we will discuss a dose-finding approach that does not rely on the assumption of a monotone dose-response relationship.

The closed testing principle is widely used in a dose-ranging setting to generate stepwise tests for determining optimal doses. Here we will focus on applications of another powerful principle known as the *partitioning principle* to dose finding. The advantage of the partitioning principle is two-fold. First, as shown by Finner and Strassburger (2002), the partitioning principle can generate multiple tests at least as powerful as closed tests.

Secondly, the partitioning principle makes drawing statistical inferences relatively clear-cut: the statistical inference given is simply what is consistent with the null hypotheses which have not been rejected (Stefansson, Kim, and Hsu, 1988; Hsu and Berger, 1999).

The partitioning principle is based on partitioning the entire parameter space into disjoint null hypotheses which correspond to useful scientific hypotheses to be proved. Since these null hypotheses are disjoint, exactly one of them is true. Therefore, testing each null hypothesis at a prespecified α level, e.g., $\alpha = 0.05$, controls the familywise error rate in the strong sense without any multiplicity adjustments.

To demonstrate how the partitioning principle can be used in dose finding, consider the dose-ranging study introduced earlier in this section. This study is conducted to test m doses of a drug versus placebo. Assume that positive changes in the response variable correspond to a beneficial effect. Table 11.5 shows a set of partitioning hypotheses that can be used in finding the MED.

Table 11.5 Partitioning Hypotheses for Finding the MED

Null hypothesis	Interpretation
$H_{00}^P : \mu_0 + \delta < \mu_1, \ldots, \mu_m$	All doses are effective
$H_{01}^P : \mu_1 \leq \mu_0 + \delta < \mu_2, \ldots, \mu_m$	Doses $2, \ldots, m$ are effective but Dose 1 is ineffective
$\vdots$	$\vdots$
$H_{0k}^P : \mu_k \leq \mu_0 + \delta < \mu_{k+1}, \ldots, \mu_m$	Doses $k+1, \ldots, m$ are effective but Dose k is ineffective
$\vdots$	$\vdots$
$H_{0m}^P : \mu_m \leq \mu_0 + \delta$	Dose m is ineffective

It is easy to check that the null hypotheses displayed in Table 11.5 are mutually exclusive and only one of them can be true. For example, if H_{0m}^P is true (Dose m is ineffective), H_{0k}^P (Doses $k+1, \ldots, m$ are effective but Dose k is ineffective) must be false. Because of this interesting property, testing each null hypothesis at a prespecified α level controls the familywise error rate in the strong sense at the α level. Note that the last hypothesis, H_{00}^P, does not really need to be tested to determine which doses are effective; however, it is useful toward pivoting the tests to obtain a confidence set. Secondly, the union of the hypotheses is the entire parameter space and thus inferences resulting from testing these hypotheses are valid without any prior assumption on the shape of the dose-response curve.

As the null hypotheses listed in Table 11.5 partition the entire parameter space, drawing useful inferences from testing $H_{01}^P, \ldots, H_{0m}^P$ is relatively straightforward. We merely state the inference which is logically consistent with the null hypotheses that have not been rejected. Consider, for example, the hypertension trial in which four doses were tested against placebo ($m = 4$). The union of H_{04}^P (Dose 4 is ineffective) and H_{03}^P (Dose 4 is effective but Dose 3 is ineffective) is "either Dose 3 or Dose 4 is ineffective". Thus, the rejection of H_{03}^P and H_{04}^P implies that both Dose 3 and Dose 4 are effective.

In general, if $H_{0k}^P, \ldots, H_{0m}^P$ are rejected, we conclude that Doses $k, \ldots, m$ are all efficacious because the remaining null hypotheses in Table 11.5, namely, $H_{01}^P, \ldots, H_{0(k-1)}^P$ are consistent with this inference. As a consequence, Dose k is declared the MED.

11.5.13 Shortcut Testing Procedure

The arguments presented in the previous section lead to a useful shortcut procedure for determining the MED in dose-ranging trials. The shortcut procedure is defined in Table 11.6.

The dose-finding procedure presented in Table 11.6 is equivalent to the fixed-sequence procedure based on pairwise comparisons of individual doses to placebo (see Section 11.5.6). The SAS code for implementing this procedure can be found in Program 11.10.

Table 11.6 Shortcut Testing Procedure for Finding the MED (All null hypotheses are tested at a prespecified α level such as a two-sided 0.05 level)

Step	Null hypothesis	Decision rule
1	$H_{0m}^S : \mu_m \leq \mu_0 + \delta$	If retained, there are no effective doses. If rejected, infer $\mu_m > \mu_0 + \delta$ and go to Step 2.
2	$H_{0(m-1)}^S : \mu_{m-1} \leq \mu_0 + \delta$	If retained, Dose m is the MED. If rejected, infer $\mu_{m-1} > \mu_0 + \delta$ and go to Step 3.
$\vdots$	$\vdots$	$\vdots$
k	$H_{0k}^S : \mu_k \leq \mu_0 + \delta$	If retained, Dose $k+1$ is the MED. If rejected, infer $\mu_k > \mu_0 + \delta$ and go to Step $k+1$.
$\vdots$	$\vdots$	$\vdots$
m	$H_{01}^S : \mu_1 \leq \mu_0 + \delta$	If retained, Dose 2 is the MED. If rejected, infer $\mu_1 > \mu_0 + \delta$ and Dose 1 is the MED.

It is important to remember that the parameter space should be partitioned in such a way that dose levels expected to show a greater response are tested first and the set of dose levels inferred to be efficacious is contiguous. For example, suppose that the dose-response function in a study with four active doses is expected to be umbrella-shaped. In this case we might partition the parameter space so that dose levels are tested in the following sequence: Dose 3, Dose 4, Dose 2, and Dose 1. However, it is critical to correctly specify the treatment sequence. It is shown in Bretz, Hothorn, and Hsu (2003) that the shortcut procedure that begins at the highest dose suffers a significant loss of power when the dose-response function is not monotone.

11.5.14   Simultaneous Confidence Intervals

Another important advantage of using the partitioning approach is that, unlike closed testing procedures, partitioning procedures are easily inverted to derive simultaneous confidence intervals for the true treatment means or proportions. For example, Hsu and Berger (1999) demonstrated how partitioning-based simultaneous confidence sets can be constructed for multiple comparisons of dose groups to a common control.

To define the Hsu-Berger procedure for computing stepwise confidence intervals, consider the ANOVA model introduced in the beginning of Section 11.5 and let

$$l_k = \bar{y}_k - \bar{y}_0 - t_{\alpha,\nu} s \sqrt{2/n}$$

be the lower limit of the naive one-sided confidence interval for $\mu_k - \mu_0$. Here s is the pooled sample standard deviation and $t_{\alpha,\nu}$ is the upper 100αth percentile of the t distribution with $\nu = 2(n-1)$ degrees of freedom, and n is the number of patients per group. The naive limits are not adequate in the problem of multiple dose-placebo comparisons because their simultaneous coverage probability is less than its nominal value, $100(1-\alpha)\%$. To achieve the nominal coverage probability, the confidence limits $l_1, \ldots, l_m$ need to be adjusted downward. The adjusted lower limit for the true treatment difference $\mu_k - \mu_0$ will be denoted by l_k^*.

Assume that the doses will be compared with placebo beginning with Dose m, i.e., Dose m vs. placebo, Dose $m-1$ vs. placebo, etc. As before, any other testing sequence can be used as long as it is not data-driven. A family of one-sided stepwise confidence intervals with a $100(1-\alpha)\%$ coverage probability is defined in Table 11.7.

Program 11.12 uses the Hsu-Berger method to compute one-sided confidence intervals with a simultaneous 95% coverage probability for the treatment means in the hypertension

Table 11.7 Simultaneous Confidence Intervals in Dose-Ranging Trials

Step	Adjusted confidence limits
1	If $l_m > \delta$, let $l_m^* = \delta$ and go to Step 2. Otherwise, let $l_m^* = l_m$ and stop.
2	If $l_{m-1} > \delta$, let $l_{m-1}^* = \delta$ and go to Step 3. Otherwise, let $l_{m-1}^* = l_{m-1}$ and stop.
$\vdots$	$\vdots$
k	If $l_{m-k+1} > \delta$, let $l_{m-k+1}^* = \delta$ and go to Step $k+1$. Otherwise, let $l_{m-k+1}^* = l_{m-k+1}$ and stop.
$\vdots$	$\vdots$
$m+1$	If all doses are efficacious, let $l_k^* = \min(l_1, \ldots, l_m)$, $k = 1, \ldots, m$.

Note: If Dose k is not efficacious, the adjusted confidence limits $l_1^*, \ldots, l_{k-1}^*$ are not defined.

study[3]. As in Program 11.10, we will consider a testing sequence that begins at the high dose (high dose vs. placebo, medium dose vs. placebo, and low dose vs. placebo). Also, the clinically important reduction in diastolic blood pressure will be set to 0 ($\delta = 0$).

Program 11.12 Simultaneous confidence intervals in the hypertension trial

```
ods listing close;
* Compute mean squared error;
proc glm data=hypertension;
    class dose;
    model change=dose;
    ods output FitStatistics=mse(keep=rootmse);
* Compute n and treatment mean in placebo group;
proc means data=hypertension;
    where dose=0;
    var change;
    output out=placebo(keep=n1 mean1) n=n1 mean=mean1;
* Compute n and treatment mean in dose groups;
proc means data=hypertension;
    where dose>0;
    class dose;
    var change;
    output out=dose(where=(dose^=.)) n=n2 mean=mean2;
* Prespecified testing sequence;
data dose;
    set dose;
    if _n_=1 then set mse;
    if _n_=1 then set placebo;
    if dose=10 then order=3;
    if dose=20 then order=2;
    if dose=40 then order=1;
* Order doses;
proc sort data=dose;
    by order;
```

---

[3]Program 11.12 is based on SAS code published in Dmitrienko et al. (2005, Section 2.4).

```
data dose;
    set dose;
    retain reject 1;
    format lower adjlower 5.2;
    delta=0;
    lower=mean2-mean1-tinv(0.95,n1+n2-2)*sqrt(1/n1+1/n2)*rootmse;
    if reject=0 then adjlower=.;
    if reject=1 and lower>delta then adjlower=delta;
    if reject=1 and lower<=delta then do; adjlower=lower; reject=0; end;
    label lower='Naive 95% lower limit'
          adjlower='Adjusted 95% lower limit';
proc print data=dose noobs label;
    var order dose lower adjlower;
    ods listing;
    run;
```

Output from Program 11.12

order	dose	Naive 95% lower limit	Adjusted 95% lower limit
1	40	1.82	0.00
2	20	2.33	0.00
3	10	-1.26	-1.26

Output 11.12 displays the naive and adjusted confidence limits for the mean reduction in diastolic blood pressure in the three dose groups compared to placebo as well as the order in which the doses are compared to placebo (ORDER variable). As was explained above, adjusted limits that result in a confidence region with a 95% coverage probability are computed in a stepwise manner. The lower limits of the naive one-sided confidence interval for the high and medium doses are greater than the clinically important difference $\delta = 0$ and thus the corresponding adjusted limits are set to 0. The mean treatment difference between the low dose and placebo is not significant (the naive confidence interval contains 0) and therefore the adjusted limit is equal to the naive limit. If the hypertension trial included more inefficient dose groups, the adjusted confidence limits for those doses would remain undefined.

11.6   Summary

This chapter reviews most important issues arising in the analysis of dose-ranging trials. The following topics are discussed in the chapter.

- **Assessment of the dose-related trend.** Dose-response analysis begins with an overall assessment of dose-related trends in the response variable. The chapter reviews contrast-based, isotonic (Bartholomew and Williams test), and non-parametric (Jonckheere test) approaches to testing dose-response trends.

- **Estimation of the shape of the dose-response function.** The next step after the assessment of the overall drug effect is the characterization of the underlying dose-response relationship. This is achieved by modeling dose-response functions and estimation of their parameters The chapter introduces two dose-response models (linear and sigmoid) that are widely used in dose-ranging trials.

- **Determination of the optimal dose.** The last step in dose-response analysis is the determination of the *therapeutic window* defined by the minimum effective and maximum tolerated doses. Two popular approaches to the estimation of the minimum effective dose (MED) are considered in the chapter. The first one, based on the *principle of closed testing*, relies on a sequential application of contrast tests to determine the smallest dose that is significantly different from placebo. The closed testing procedures control the overall Type I error rate only if the underlying dose-response function is monotone. An alternative approach which relies on the *partitioning principle* can be safely used even when the assumption of monotonicity is not met.

Statistical methods introduced in this chapter are illustrated using examples from cross-over and parallel group clinical trials.

References

Abelson, R.B., Tukey, J.W. (1963). "Efficient utilization of non-numerical information in quantitative analysis: General theory and the case of simple order." *The Annals of Mathematical Statistics.* 34, 1347–1369.

Bartholomew, D.J. (1961). "Ordered tests in the analysis of variance." *Biometrika.* 48, 325–332.

Bauer, P. (1997). "A note on multiple testing procedures in dose finding." *Biometrics.* 53, 1125–1128.

Bauer, P., Brannath, W., Posch, M. (2001). "Multiple testing for identifying effective and safe treatments." *Biometrical Journal.* 43, 605–616.

Bauer, P., Röhmel, J., Maurer, W., Hothorn, L. (1998). "Testing strategies in multi-dose experiments including active control." *Statistics in Medicine.* 17, 2133–2146.

Bretz, F., Hothorn, L.A., Hsu, J.C. (2003). "Identifying effective and/or safe doses by stepwise confidence intervals for ratios." *Statistics in Medicine.* 22, 847–858.

Chuang-Stein, C., Agresti, A. (1997). "A review of tests for detecting a monotone dose-response relationship with ordinal response data." *Statistics in Medicine.* 16, 2599–2618.

Dmitrienko, A., Molenberghs, G., Chuang-Stein, C., Offen, W. (2005). *Analysis of Clinical Trials Using SAS: A Practical Guide.* Cary, NC: SAS Institute Inc.

Dunnett, C.W., Tamhane, A.C. (1998). "Some new multiple-test procedures for dose finding." *Journal of Biopharmaceutical Statistics.* 8, 353–366.

Finner, H., Strassburger, K. (2002). "The partitioning principle: a powerful tool in multiple decision theory." *Annals of Statistics.* 30, 1194–1213.

Genz, A., Bretz, F. (1999). "Numerical computation of the multivariate t probabilities with application to power calculation of multiple contrasts." *Journal of Statistical Computation and Simulation.* 63, 361–378.

Hothorn, L.A. (1997). "Modifications of the closure principle for analyzing toxicological studies." *Drug Information Journal.* 30, 403–412.

Hsu, J.C., Berger, R.L. (1999). "Stepwise confidence intervals without multiplicity adjustment for dose-response and toxicity studies." *Journal of the American Statistical Association.* 94, 468–482.

Jonckheere, A.R. (1954). "A distribution-free K sample test against ordered alternatives." *Biometrika.* 41, 133–145.

Kodell, R.L., Chen, J.J. (1991). "Characterization of dose-response relationships inferred by statistically significant trend tests." *Biometrics.* 47, 139–146.

Marcus, R., Peritz, E., Gabriel, K.R. (1976). "On closed testing procedure with special reference to ordered analysis of variance." *Biometrika.* 63, 655–660.

Mukerjee, H., Robertson, T., Wright, F.T. (1987). "Comparison of several treatments with a control using multiple contrasts." *Journal of the American Statistical Association.* 82, 902–910.

Neuhäuser, M., Liu, P.Y., Hothorn, L.A. (1998). "Nonparametric tests for trend: Jonckheere's test, a modification and a maximum test." *Biometrical Journal.* 40, 899–909.

Phillips, A. (1997). "Design and analysis of dose response studies: reality versus regulatory requirements." *Drug Information Journal.* 31, 737–744.

Phillips, A. (1998). "A review of the performance of tests used to establish whether there is a drug effect in dose-response studies." *Drug Information Journal*. 32, 683–692.

Rom, D.M., Costello, R.J., Connell, L.T. (1994). "On closed test procedures for dose-response analysis." *Statistics in Medicine*. 13, 1583–1596.

Ruberg, S.J. (1989). "Contrasts for identifying the minimum effective dose." *Journal of the American Statistical Association*. 84, 816–822.

Ruberg, S.J. (1995a). "Dose response studies. Some design considerations." *Journal of Biopharmaceutical Statistics*. 5, 1–14.

Ruberg, S.J. (1995b). "Dose response studies. Analysis and interpretation." *Journal of Biopharmaceutical Statistics*. 5, 15–42.

Shirley, E.A. (1977). "A nonparametric equivalent of Williams' test for contrasting increasing dose levels of a treatment." *Biometrics*. 33, 386–389.

Shirley, E.A. (1985). "The value of specialized tests in studies where ordered group means are expected." *Statistics in Medicine*. 4, 489–496.

Stefansson, G., Kim, W.C., Hsu, J.C. (1988). "On confidence sets in multiple comparisons." *Statistical Decision Theory and Related Topics IV*. Edited by S.S. Gupta and J.O. Berger. New York: Academic Press. 89–102.

Stewart, W.H., Ruberg, S.J. (2000). "Detecting dose response with contrasts." *Statistics in Medicine*. 19, 913–921.

Tamhane, A.C., Hochberg, Y., Dunnett, C.W. (1996). "Multiple test procedures for dose finding." *Biometrics*. 52, 21–37.

Tamhane, A.C., Logan, B.R. (2002). "Multiple test procedures for identifying the minimum effective and maximum safe doses of a drug." *Journal of the American Statistical Association*. 97, 293–301.

Terpstra, T.J. (1952). "The asymptotic normality and consistency of Kendall's test against trend, when ties are present in one ranking." *Indigationes Mathematicae*. 14, 327–333.

Tukey, J.W., Ciminera, J.L., Heyse, J.F. (1985). "Testing the statistical certainty of a response to increasing doses of a drug." *Biometrics*. 41, 295–301.

Westfall, P.H. (1997). "Multiple testing of general contrasts using logical constraint and correlations." *Journal of the American Statistical Association*. 92, 299–306.

Westfall, P.H., Tobias, R.D. (2006). "Multiple testing of general contrasts: truncated closure and the extended Shaffer-Royen method." *Journal of the American Statistical Association*. In press.

Westfall, P.H., Tobias, R.D., Rom, D., Wolfinger, R.D., Hochberg, Y. (1999). *Multiple Comparisons and Multiple Tests Using the SAS System*. Cary, NC: SAS Institute Inc.

Williams, D.A. (1971). "A test for differences between treatment means when several dose levels are compared with a zero dose control." *Biometrics*. 27, 103–117.

Williams, D.A. (1972). "The comparison of several dose levels with a zero dose control." *Biometrics*. 28, 519–531.

Williams, D.A. (1988). "Tests for differences between several small proportions." *Applied Statistics*. 37, 421–434.

Chapter 12

Analysis of Incomplete Data

Geert Molenberghs
Caroline Beunckens
Herbert Thijs
Ivy Jansen
Geert Verbeke
Michael Kenward
Kristel Van Steen

Relying on Rubin's standard missing-data taxonomy, and using simple algebraic derivations, this chapter argues that some methods that are commonly used to handle incomplete longitudinal data are based on poor principles and are unnecessarily restrictive. We define *longitudinal clinical trial data* as complete case analyses and methods based on *last observation carried forward* (LOCF), for which the *missing completely at random* (MCAR) assumption is required.

Because flexible software is available that can analyze longitudinal sequences of unequal length, this chapter proposes a shift to a likelihood-based ignorable analysis that is carried out using SAS software.

Geert Molenberghs is Professor, Center for Statistics, Universiteit Hasselt, Belgium. Caroline Beunckens is Research Assistant, Center for Statistics, Universiteit Hasselt, Belgium. Herbert Thijs is Postdoctoral Fellow, Center for Statistics, Universiteit Hasselt, Belgium. Ivy Jansen is Postdoctoral Fellow, Center for Statistics, Universiteit Hasselt, Belgium. Geert Verbeke is Professor, Biostatistical Center, Katholieke Universiteit Leuven, Belgium. Michael Kenward is Professor, Medical Statistics Unit, London School of Hygiene and Tropical Medicine, United Kingdom. Kristel Van Steen is Professor, Universitiet Gent, Belgium.

12.1   Introduction

In a longitudinal clinical trial, each unit is measured on several occasions. It is not unusual in practice for some sequences of measurements to terminate early for reasons outside the control of the investigator. Any unit so affected is called a *dropout*. It might therefore be necessary to accommodate the dropout in the modeling process.

When referring to the missing-value (or non-response) process we will use the terminology of Little and Rubin (1987, Chapter 6). A non-response process is said to be *missing completely at random* (MCAR) if the missingness is independent of both unobserved and observed data. A process is said to be *missing at random* (MAR) if, conditional on the observed data, the missingness is independent of the unobserved measurements. A process that is neither MCAR nor MAR is termed *non-random* (MNAR). In the context of likelihood inference, and when the parameters describing the measurement process are functionally independent of the parameters describing the missingness process, MCAR and MAR are *ignorable*, while a non-random process is non-ignorable.

Many methods are formulated as selection models (Little and Rubin, 1987) as opposed to pattern-mixture models (PMM) (Little 1993, 1994a). A selection model factors the joint distribution of the measurement and response mechanisms into the marginal measurement distribution and the response distribution, conditional on the measurements. This is intuitively appealing since the marginal measurement distribution would be of interest also with complete data. Little and Rubin's taxonomy is most easily developed in the selection setting. Parameterizing and making inference about the effect of treatment and its evolution over time is straightforward in the selection model context.

12.1.1   Incomplete Data in Clinical Trials

In the specific case of a clinical trial setting, standard methodology used to analyze longitudinal data subject to non-response is mostly based on such methods as *last observation carried forward* (LOCF), *complete case analysis* (CC), or simple forms of imputation. This is often done without questioning the possible influence of these assumptions on the final results, even though several authors have written about this topic. A relatively early account is given in Heyting, Tolboom, and Essers (1992). Mallinckrodt et al. (2003a,b) and Lavori, Dawson, and Shera (1995) propose direct-likelihood and multiple-imputation methods, respectively, to deal with incomplete longitudinal data. Siddiqui and Ali (1998) compare direct-likelihood and LOCF methods.

It is unfortunate that there is such a strong emphasis on methods like LOCF and CC, since they are based on extremely strong assumptions. In particular, even the strong MCAR assumption does not suffice to guarantee that an LOCF analysis is valid. On the other hand, under MAR, valid inference can be obtained through a likelihood-based analysis, without the need for modeling the dropout process. As a consequence, we can simply use, for example, linear or generalized linear mixed models (Verbeke and Molenberghs, 2000), without additional complication or effort. We will argue that such an analysis not only enjoys much wider validity than the simple methods but in addition is simple to conduct, *without additional data manipulation* using such tools as, for example, the SAS MIXED or NLMIXED procedure. Thus, clinical trial practice should shift away from the *ad hoc* methods and focus on likelihood-based, ignorable analyses instead. As will be argued further, the cost involved in having to specify a model will arguably be mild to moderate in realistic clinical trial settings. Thus, we promote the use of direct-likelihood ignorable methods and argue against the use of the LOCF and CC approaches.

At the same time, we cannot avoid a reflection on the status of MNAR approaches. In realistic settings, the reasons for dropout are varied, and it is therefore difficult to fully justify on a priori grounds the assumption of MAR. For example, the rate of and the reasons for dropout varied considerably across eleven clinical trials of similar design, of the same drug in the same indication. In one study, completion rates were 80% for drug and placebo. In another study, two-thirds of the patients who were taking a drug completed the study, while only one-third did so on placebo. In yet another study, 70% finished on placebo but only 60% on drug. Reasons for dropout also varied, even within the drug arm. For example, at low doses, more patients on drug dropped out due to lack of efficacy, whereas at higher doses dropout that was due to adverse events was more common. At first sight, this calls for a further shift towards MNAR models. However, some careful considerations have to be made, the most important one of which is that no modeling approach, whether either MAR or MNAR, can recover the lack of information that occurs due to incompleteness of the data.

First, under MAR, a standard analysis would follow, if it would be possible to be entirely sure of the MAR nature of the mechanism. However, it is only rarely the case that such an assumption is known to hold (Murray and Findlay, 1988). Nevertheless, ignorable analyses may provide reasonably stable results, even when the assumption of MAR is violated, in the sense that such analyses constrain the behavior of the unseen data to be similar to that of the observed data (Mallinckrodt et al., 2001a,b). A discussion of this phenomenon in the survey context has been given in Rubin, Stern, and Vehovar (1995). These authors argue that, in well-conducted experiments (some surveys and many confirmatory clinical trials), the assumption of MAR is often to be regarded as a realistic one. Second, and very important for confirmatory trials, an MAR analysis can be specified a priori without additional work relative to a situation with complete data. Third, while MNAR models are more general and explicitly incorporate the dropout mechanism, the inferences they produce are typically highly dependent on the untestable and often implicit assumptions built in regarding the distribution of the unobserved measurements given the observed ones. The quality of the fit to the observed data need not reflect at all the appropriateness of the implied structure governing the unobserved data. This point is irrespective of the MNAR route taken, whether a parametric model of the type of Diggle and Kenward (1994) is chosen, or a semiparametric approach such as in Robins, Rotnitzky, and Scharfstein (1998). Hence in any incomplete-data setting there cannot be anything that could be termed a definitive analysis. Based on these considerations, we recommend, for primary analysis purposes, the use of ignorable likelihood-based methods. In many examples, however, the reasons for dropout will be many and varied. It is therefore difficult to justify on a priori grounds the MAR assumption. Arguably, in the presence of MNAR missingness, a wholly satisfactory analysis of the data is not feasible.

In fact, modeling in this context often rests on strong (untestable) assumptions and relatively little evidence from the data themselves. Glynn, Laird, and Rubin (1986) indicated that this is typical for selection models. It is somewhat less the case for pattern-mixture models (Little, 1993; 1994a; Hogan and Laird, 1997), although caution should be used (Thijs, Molenberghs, and Verbeke, 2000). This awareness and the resulting skepticism about fitting MNAR models initiated the search for methods to investigate the results with respect to model assumptions and for methods that allow us to assess influences in the parameters describing the measurement process, as well as the parameters describing the non-random part of the dropout mechanism. Several authors have suggested various types of sensitivity analyses to address this issue (Molenberghs, Kenward, and Goetghebeur, 2001; Scharfstein, Rotnitzky, and Robins, 1999; Van Steen et al., 2001; Verbeke et al., 2001). Verbeke et al. (2001) and Thijs, Molenberghs, and Verbeke (2000) developed a local influence-based approach for the detection of subjects that strongly influence the conclusions. These authors focused on the Diggle and Kenward (1994) model for continuous outcomes. Van Steen et al. (2001) adapted these ideas to the model of Molenberghs, Kenward and Lesaffre (1997), for monotone repeated ordinal data. Jansen et al. (2003) focused on the model family proposed by Baker, Rosenberger, and DerSimonian (1992, henceforth referred to as BRD).

Thus, to explore the impact of deviations from the MAR assumption on the conclusions, we should ideally conduct a sensitivity analysis, within which MNAR models can play a major role, together with, for example, pattern-mixture models (Verbeke and Molenberghs, 2000, Chapters 18–20).

12.1.2 Outline

Three case studies used throughout this chapter, the exercise bike data and the mastitis in dairy cattle data, are introduced in Section 12.2. The general data setting is introduced in Section 12.3, as well as a formal framework for incomplete longitudinal data. A brief overview on the problems associated with simple methods is presented in Section 12.4. Next, methods valid under the MAR assumption are described in Sections 12.5 and 12.6, for continuous and categorical outcomes respectively. MNAR modeling is covered in Section 12.7, while Section 12.8 covers the aspects on sensitivity analysis.

To save space, some SAS code has been shortened and some output is not shown. The complete SAS code and data sets used in this book are available on the book's companion Web site at `http://support.sas.com/publishing/bbu/companion_site/60622.html`.

12.2 Case Studies

We will present three case studies which will be used throughout this chapter.

EXAMPLE: Exercise trial

In this heart failure study, the primary efficacy endpoint is based upon the ability to do physical exercise. This ability is measured in the number of seconds a subject is able to ride an exercise bike. The data collected in the study are included in the EXERCISE data set that can be found on the book's companion Web site. There are 25 subjects assigned to placebo (GROUP=0) and 25 to treatment (GROUP=1). The treatment consisted of the administration of ACE inhibitors. Four measurements were taken at monthly intervals (TIME variable). The Y variable represents the outcome scores transformed to normality.

All 50 subjects are observed at the first occasion, whereas there are 44, 41, and 38 subjects seen at the second, the third, and the fourth visits, respectively. Individual and mean response profiles per treatment arm are shown in Figures 12.1 and 12.2, respectively. The percentage of patients remaining in the study after each visit is tabulated in Table 12.1 per treatment arm.

Table 12.1 Percentage of Patients Remaining in the Study per Treatment Arm in the Exercise Trial

Visit	Placebo	Treatment
1	100%	100%
2	88%	88%
3	84%	80%
4	80%	72%

EXAMPLE: Mastitis data

This example, concerning the occurrence of the infectious disease mastitis in dairy cows, was introduced in Diggle and Kenward (1994) and reanalyzed in Kenward (1998). The MASTITIS data set, provided on the book's companion Web site, contains the milk yields (in thousands of liters) of 107 dairy cows from a single herd in two consecutive years: $Y_{ij}(i = 1, \ldots, 107; j = 1, 2)$. In the first year, all animals were supposedly free of mastitis; in

Figure 12.1 Individual response profiles in the placebo (left panel) and treatment (right panel) groups in the exercise trial

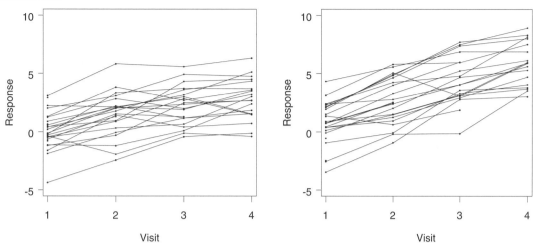

Figure 12.2 Mean response profiles in the placebo (solid curve) and treatment (dashed curve) groups in the exercise trial

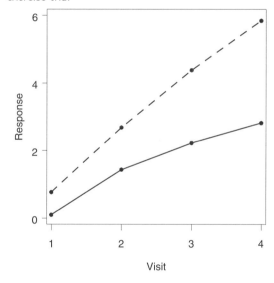

the second year, 27 became infected. Mastitis typically reduces milk yield, and the question of scientific interest is whether the probability of occurrence of mastitis is related to the yield that would have been observed had mastitis not occurred. A graphical representation of the complete data is given in Figure 12.3.

EXAMPLE: Depression trial

The DEPRESSION data set available on the book's companion Web site comes from a clinical trial including 342 patients with post-baseline data. The Hamilton Depression Rating Scale (HAMD17) is used to measure the depression status of the patients. For each patient, BASVAL represents the baseline HAMD17 assessment, Y is the HAMD17 score at Visits 4 through 8 and CHANGE is the change from baseline in the HAMD17 score at Visits 4 through 8. The YBIN variable is a binary outcome derived from the Y variable, i.e., Y=1 if the HAMD17 score is larger than 7, and 0 otherwise. Two treatment groups

Figure 12.3 Scatter plots of the Year 2 milk yield versus the Year 1 milk yield (left panel) the change in milk yield versus the Year 1 milk yield (right panel) in the mastitis example. Black dots represent Cows #4 and #5.

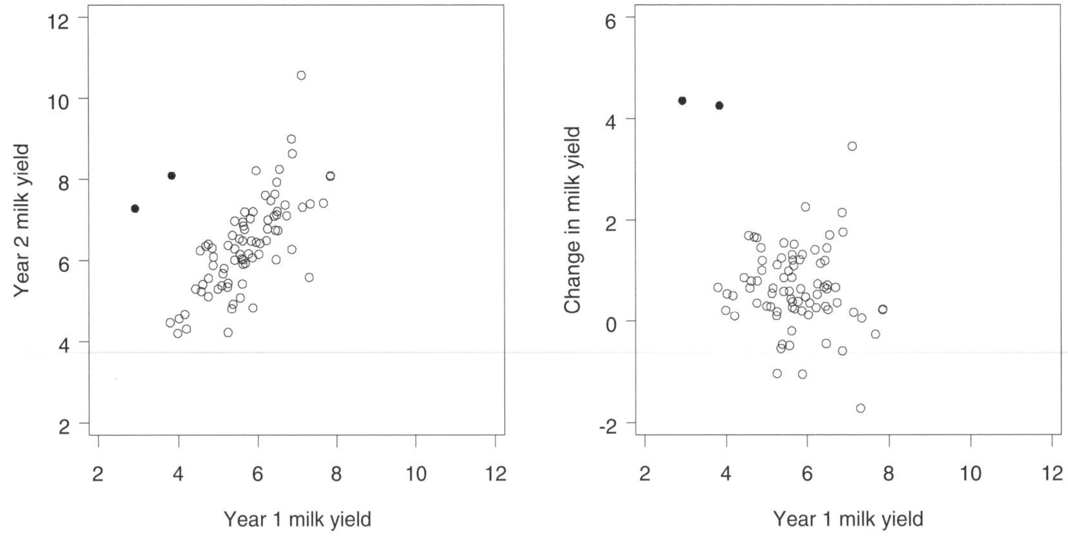

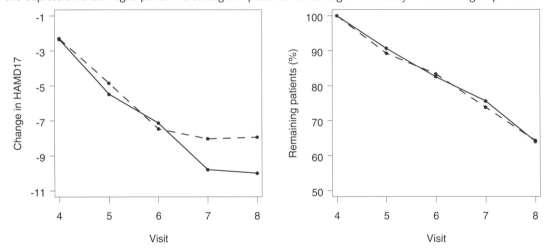

(primary dose of experimental drug, TRT=1, and placebo, TRT=4) are included in the data set.

Mean profiles of the HAMD17 changes and dropout rates in the two treatment groups are shown in Figure 12.4.

Figure 12.4 Left panel: Mean response profiles in Treatment groups 1 (solid curve) and 4 (dashed curve) in the depression trial. Right panel: Percentage of patients remaining in the study Treatment groups 1 and 4.

12.3 Data Setting and Modeling Framework

Assume that for subject $i = 1, \ldots, N$ in the study a sequence of responses Y_{ij} is designed to be measured at occasions $j = 1, \ldots, n$. The outcomes are grouped into a vector $Y_i = (Y_{i1}, \ldots, Y_{in})'$. In addition, define a dropout indicator D_i for the occasion at which dropout occurs and make the convention that $D_i = n + 1$ for a complete sequence. It is often necessary to split the vector Y_i into observed (Y_i^o) and missing (Y_i^m) components respectively. Note that dropout is a particular case of monotone missingness. In order to have a monotone pattern of missingness, there has to exist a permutation of the measurement components such that a measurement earlier in the permuted sequence is observed for at least those subjects that are observed at later measurements. For this definition to be meaningful, we need to have a balanced design in the sense of a common set of measurement occasions. Other patterns are called *nonmonotone* or *intermittent missingness*. When intermittent missingness occurs, it is best to use a vector of binary indicators $R_i = (R_{i1}, \dots, R_{in})'$ rather than the dropout indicator D_i.

In principle, we would like to consider the density of the full data $f(y_i, d_i | \theta, \psi)$, where the parameter vectors θ and ψ describe the measurement and missingness processes, respectively. Covariates are assumed to be measured but, for notational simplicity, they are suppressed from notation.

The taxonomy, constructed by Rubin (1976), further developed in Little and Rubin (1987), and informally sketched in Section 12.1, is based on the factorization

$$f(y_i, d_i | \theta, \psi) = f(y_i | \theta) f(d_i | y_i, \psi), \tag{12.3.1}$$

where the first factor is the marginal density of the measurement process and the second one is the density of the missingness process, conditional on the outcomes. Factorization (12.3.1) forms the basis of *selection modeling* as the second factor corresponds to the (self-)selection of individuals into *observed* and *missing* groups. An alternative taxonomy can be built based on so-called *pattern-mixture models* (Little, 1993; Little, 1994a). These are based on the factorization

$$f(y_i, d_i | \theta, \psi) = f(y_i | d_i, \theta) f(d_i | \psi). \tag{12.3.2}$$

Indeed, (12.3.2) can be seen as a mixture of different populations, characterized by the observed pattern of missingness.

Rubin (1976) and Little and Rubin (1987) have shown that, under MAR and mild regularity conditions (parameters θ and ψ are functionally independent), likelihood-based and Bayesian inferences are valid when the missing data mechanism is ignored (see also Verbeke and Molenberghs, 2000; Molenberghs and Verbeke, 2005). Practically speaking, the likelihood of interest is then based upon the factor $f(y_i^o | \theta)$. This is called *ignorability*.

The practical implication is that a software module with likelihood estimation facilities and with the ability to handle incompletely observed subjects manipulates the correct likelihood, providing valid parameter estimates and likelihood ratio values. Dmitrienko et al. (2005, Chapter 5) provide detailed guidelines on how to implement such analyses. They also issue a number of cautionary remarks. An important one is that the flexibility and ease of MAR (and hence ignorable) analyses do not rule out the option of an MNAR mechanism to operate. These authors also focused primarily on continuous outcomes. In this chapter, both discrete outcomes, as well as modeling approaches under MNAR, are of primary interest.

12.4   Simple Methods and MCAR

We will briefly review a number of relatively simple methods that have been and are still in extensive use. They have been discussed in some detail in Dmitrienko et al. (2005, Chapter 5). We will focus on complete case analysis (CC) and last observation carried forward (LOCF). The latter is a single or simple imputation method, which shares a certain number of pitfalls with the other methods. Multiple imputation, on the other hand (Section 12.5.2) is valid under MAR and therefore is discussed in Section 12.5.

Complete Case Analysis

A *complete case* (CC) analysis considers only those cases for which all n_i measurements were recorded. The method is simple since it restores balance, in the sense that a *rectangular data matrix* is obtained. However, the drawbacks surpass the advantages. Apart from considerable information loss, leading to inefficient estimates and tests with less than optimal power, often severe bias is to be expected. The method is, therefore, not recommended. See Dmitrienko et al. (2005, Chapter 5) for details.

Last Observation Carried Forward

Alternatively, balance can be restored by substituting the last obtained measurement for the missing ones. This technique is termed *last observation carried forward* (LOCF) or *last value carried forward*. The practical advantages are the same as with CC, but the issues are manifold. The technique has been discussed in detail in Dmitrienko et al. (2005, Chapter 5) and insightful illustrations of the issues are provided in Molenberghs et al. (2004). An important issue is that filled-in values are treated as actual data. Further, Molenberghs et al. (2004) have illustrated that the bias resulting from this method can be both conservative and liberal, contradicting common belief that the appeal of the method lies in it being conservative for the assessment of treatment effect in superiority trials.

Moreover, since direct likelihood is perfectly feasible in the sense that it is valid under MAR and easy to implement in standard software (Section 12.5), there is generally very little reason to apply LOCF. See also Mallinckrodt et al. (2003ab).

Available Case Methods

Available case methods (Little and Rubin, 1987) use as much of the data as possible. In a multivariate normal setting, means and variances are estimated based on the information available for individual outcomes, while covariances are estimated using pairs of outcomes. While this is a simple solution in this setting, the method is difficult to extend to other settings, such as continuous data to which regression models or linear mixed models are applied, or categorical data. Moreover, the method is still valid only under MCAR. The reason is that the method uses moments only; in our setting, these are the first and second moments of the multivariate normal. As such, it is a frequentist method. In the following section, we will show that using the available information in a likelihood framework extends validity to MAR.

12.5 MAR Methods

In this section, we give an overview of the most commonly used methods, which are valid under the MAR assumption. See Dmitrienko et al. (2005, Chapter 5) for details.

12.5.1 Direct Likelihood Analyses

As stated earlier, likelihood-based inference is valid whenever the mechanism is MAR, provided that the technical condition holds that the parameters describing the nonresponse mechanism are distinct from the measurement model parameters (Little and Rubin, 1987). The log-likelihood then partitions into two functionally independent components, one describing the measurement model, the other one the missingness model. This implies that likelihood-based software with facilities to handle incomplete records provides valid inferences. Such a likelihood-based ignorable analysis is also termed *likelihood-based MAR analysis*, or, as we will call it further on, a *direct likelihood analysis*.

Turning to SAS software, this implies that likelihood-based longitudinal analyses, conducted by means of the MIXED, NLMIXED, and GLIMMIX procedures, are valid, given that the MAR assumption holds or is considered a reasonable approximation. This is useful, even when inferences are conducted at the end of the planned measurement sequence. An important example is the assessment of treatment effect at the end of a clinical trial. A way to practically implement such an analysis is by specifying a sufficiently general treatment by time-mean profile model, supplemented with an unstructured variance-covariance structure. Appropriate use of the CONTRAST or ESTIMATE statement then leads to the required test. Technically, the score equations take the expected value of the incomplete measurements, given the observed ones, into account (Beunckens, Molenberghs, and Kenward, 2005). This implies that all information on a subject is used to assess treatment effect. If the treatment assignment is used as randomized, the method is fully consistent with the intention-to-treat principle.

12.5.2 Multiple Imputation

Apart from direct likelihood, other methods that are valid under MAR include multiple imputation (Rubin 1978, 1987, Rubin and Schenker 1986) (discussed here) and the Expectation-Maximization algorithm.

The key idea of the multiple imputation (MI) procedure is to replace each missing value with a set of M plausible values. Precisely, such values are drawn from the conditional distribution of the unobserved values, given the observed ones. The imputed data sets are then analyzed using standard complete data methods and software. Finally, the M obtained inferences thus obtained must be combined into a single one by means of the method proposed by Rubin (1978). Ample detail can be found in Dmitrienko et al. (2005, Chapter 5).

With the availability and ease of direct likelihood, it remains to be discussed when to use multiple imputation. First, the method can be used to conduct checks on direct likelihood. Second, MI is really useful when there are incomplete covariates, along with missing outcomes. Third, when several users want to conduct a variety of analyses on the same incomplete set of data, it is sensible to provide all of them with the same multiply imputed sets of data. Finally, multiple imputation can be used within the context of sensitivity analysis.

The SAS MI procedure is a multiple imputation procedure that creates multiply imputed data sets for incomplete p-dimensional multivariate data. It uses methods that incorporate appropriate variability across the M imputations. Once the M complete data sets are analyzed by using standard procedures, the MIANALYZE procedure can be used to generate valid statistical inferences about these parameters by combining results for the M complete data sets. See also Dmitrienko et al. (2005, Chapter 5).

12.5.3 The EM Algorithm

The Expectation-Maximization (EM) algorithm has been described in detail in Dmitrienko et al. (2005, Chapter 5). It is an alternative to direct likelihood and MI, in the sense that, in its basic form, it is valid under the same conditions. Dempster, Laird, and Rubin (1977) provided a very general description of the algorithm, showing its use in broad classes of missing data, latent variable, latent classes, random effects, and other data augmentation settings.

Within each iteration, there are two steps. In the E step, the expectation of the complete data log-likelihood is calculated. In exponential family settings, this is particularly easy since it reduces to the calculation of complete-data sufficient statistics. In the M step, the so-obtained function is maximized. Little and Rubin (1987), Schafer (1997), and McLachlan and Krishnan (1997) provide detailed descriptions and applications of the EM algorithm.

While the algorithm is elegant, the basic version does not provide precision estimates and a number of proposals have been made over the years, summarized in McLachlan and

Krishnan (1997), to rectify this situation. A number of applications of the EM algorithm are possible with the SAS MI procedure (Dmitrienko et al. 2005, Chapter 5).

12.6 Categorical Data

For non-Gaussian outcomes, there is no single, broadly applicable counterpart to the multivariate normal model, within which the linear mixed model is positioned. Therefore, it is important to carefully distinguish between three model families: marginal models, random-effects models, and conditional models. A comprehensive introduction of these and comparison between them has been provided in Dmitrienko et al. (2005, Chapter 5). Here, we provide a brief overview of the marginal family, with focus on generalized estimating equations (GEE), and put some emphasis on the random-effects family. In particular, we emphasize the generalized linear mixed model (GLMM).

12.6.1 Marginal Models

Marginal models describe the outcomes within an outcome sequence Y_i, conditional on covariates, but neither on other outcomes nor on unobserved (latent) structures. While full likelihood approaches exist (Molenberghs and Verbeke 2005), they are usually demanding in computational terms, explaining the popularity of generalized estimating equations (GEE), on which we will focus here. In their basic form, they are valid under MCAR, which explains why weighted GEE (WGEE) have been devised to allow for extension to the MAR framework. Whereas GEE can be fitted using the GENMOD procedure, there also is a linearization-based version, which can be fitted with the %GLIMMIX macro or procedure. A number of additional extensions of, and modifications to, GEE exist. They are not considered here. See Molenberghs and Verbeke (2005) for details.

Generalized Estimating Equations

Generalized estimating equations (Liang and Zeger, 1986) are useful when scientific interest focuses on the first moments of the outcome vector. Examples include time evolutions in the response probability, treatment effect, their interaction, and the effect of (baseline) covariates on these probabilities. GEE allows the researcher to use a "fix up" for the correlations in the second moments, and to ignore the higher order moments, while still obtaining valid inferences, at reasonable efficiency.

The GEE methodology is based on solving the equations

$$S(\beta) = \sum_{i=1}^{N} \frac{\partial \mu_i}{\partial \beta'} (A_i^{1/2} R_i A_i^{1/2})^{-1} (y_i - \mu_i) \;=\; 0, \tag{12.6.3}$$

in which the marginal covariance matrix V_i has been decomposed in the form $A_i^{1/2} R_i A_i^{1/2}$, with A_i the matrix with the marginal variances on the main diagonal and zeros elsewhere, and with $R_i = R_i(\alpha)$ the marginal correlation matrix, often referred to as the *working* correlation matrix. Usually, the marginal covariance matrix $V_i = A_i^{1/2} R_i A_i^{1/2}$ contains a vector $\boldsymbol{\alpha}$ of unknown parameters which is replaced for practical purposes by a consistent estimate.

Assuming that the marginal mean μ_i has been correctly specified as $h(\mu_i) = X_i\beta$, it can be shown that, under mild regularity conditions, the estimator $\widehat{\beta}$ obtained from solving (12.6.3) is asymptotically normally distributed with mean β and with covariance matrix

$$I_0^{-1} I_1 I_0^{-1}, \tag{12.6.4}$$

where

$$I_0 = \left(\sum_{i=1}^{N} \frac{\partial \mu_i}{\partial \beta}' V_i^{-1} \frac{\partial \mu_i}{\partial \beta'} \right), \quad I_1 = \left(\sum_{i=1}^{N} \frac{\partial \mu_i}{\partial \beta}' V_i^{-1} \mathrm{Var}(y_i) V_i^{-1} \frac{\partial \mu_i}{\partial \beta'} \right).$$

In practice, $\mathrm{Var}(y_i)$ in (12.6.4) is replaced by $(y_i - \mu_i)(y_i - \mu_i)'$, which is unbiased on the sole condition that the mean was again correctly specified.

Note that valid inferences can now be obtained for the mean structure, only assuming that the model assumptions with respect to the first-order moments are correct.

Liang and Zeger (1986) proposed moment-based estimates for the working correlation. To this end, first define deviations:

$$e_{ij} = \frac{y_{ij} - \mu_{ij}}{\sqrt{v(\mu_{ij})}}$$

and decompose the variance slightly more generally as above in the following way:

$$V_i = \phi A_i^{1/2} R_i A_i^{1/2},$$

where ϕ is an overdispersion parameter.

Weighted Generalized Estimating Equations

As stated before, GEE is valid under MCAR. To accommodate MAR missingness, Robins, Rotnitzky, and Zhao (1995) proposed a class of weighted estimating equations. The idea of WGEE is to weight each subject's measurements in the GEEs by the inverse probability that a subject drops out at that particular measurement occasion. This can be calculated as

$$\nu_{ij} = P(D_i = j) = \begin{cases} P(D_i = j | D_i \geq j) & j = 2, \\ P(D_i = j | D_i \geq j) \displaystyle\prod_{k=2}^{j-1} [1 - P(D_i = k | D_i \geq k)] & j = 3, \ldots, n_i, \\ \displaystyle\prod_{k=2}^{n_i} [1 - P(D_i = k | D_i \geq k)] & j = n_i + 1. \end{cases}$$

In the weighted GEE approach, which is proposed to reduce possible bias of $\widehat{\beta}$, the score equations to be solved when taking into account the correlation structure are:

$$S(\beta) = \sum_{i=1}^{N} \frac{1}{\nu_i} \frac{\partial \mu_i}{\partial \beta'} (A_i^{1/2} R_i A_i^{1/2})^{-1} (y_i - \mu_i) = 0$$

or

$$S(\beta) = \sum_{i=1}^{N} \sum_{d=2}^{n+1} \frac{I(D_i = d)}{\nu_{id}} \frac{\partial \mu_i(d)}{\partial \beta'} (A_i^{1/2} R_i A_i^{1/2})^{-1}(d)(y_i(d) - \mu_i(d)) = 0,$$

where $y_i(d)$ and $\mu_i(d)$ are the first $d - 1$ elements of y_i and μ_i respectively. We define $\frac{\partial \mu_i}{\partial \beta'}(d)$ and $(A_i^{1/2} R_i A_i^{1/2})^{-1}(d)$ analogously, in line with the definition of Robins, Rotnitzky, and Zhao (1995).

A Method Based on Linearization

Both versions of GEE studied so far can be seen as deriving from the score equations of corresponding likelihood methods. In a sense, GEE result from considering only a subvector of the full vector of scores, corresponding to either the first moments only (the outcomes themselves), or to the first and second moments (outcomes and cross-products thereof). On the other hand, they can be seen as an extension of the quasi-likelihood principles, where appropriate modifications are made to the scores to be sufficiently flexible and "work" at the same time. A classical modification is the inclusion of an overdispersion parameter, while in GEE also (nuisance) correlation parameters are introduced.

An alternative approach consists of linearizing the outcome, in the sense of Nelder and Wedderburn (1972), to construct a working variate, to which then weighted least squares is applied. In other words, iteratively reweighted least squares (IRLS) can be used (McCullagh and Nelder, 1989). Within each step, the approximation produces all elements typically encountered in a multivariate normal model, and hence corresponding software tools can be used. In case our models would contain random effects as well, the core of the IRLS could be approached using linear mixed models tools.

Write the outcome vector in a classical (multivariate) generalized linear models fashion:

$$y_i = \mu_i + \varepsilon_i$$

where, as usual, $\mu_i = E(y_i)$ is the systematic component and ε_i is the random component, typically following a multinomial distribution. We assume that $\text{Var}(y_i) = \text{Var}(\varepsilon_i) = \Sigma_i$. The model is further specified by assuming

$$\eta_i = g(\mu_i), \quad \eta_i = X_i\beta,$$

where η_i is the usual set of linear predictors, $g(\cdot)$ is a vector link function, typically made up of logit components, X_i is a design matrix and β are the regression parameters.

Estimation proceeds by iteratively solving

$$\sum_{i=1}^{N} X_i' W_i X_i \beta = \sum_{i=1}^{N} W_i y_i^*, \tag{12.6.5}$$

where a working variate y_i^* has been defined, following from a first-order Taylor series expansion of η_i around μ_i:

$$y_i^* = \widehat{\eta}_i + (y_i - \widehat{\mu}_i)\widetilde{F}_i^{-1}, \quad \widetilde{F}_i = \left.\frac{\partial \mu_i}{\partial \eta_i}\right|_{\beta=\widetilde{\beta}}. \tag{12.6.6}$$

The weights in (12.6.5) are specified as

$$W_i = F_i' \widetilde{\Sigma}_i^{-1} F_i. \tag{12.6.7}$$

In these equations $\widetilde{\beta}$ and $\widetilde{\Sigma}$ are evaluated at the current iteration step. Note that in the specific case of an identity link, $\eta_i = \mu_i$, $F_i = I_{n_i}$, and $y_i = y_i^*$, whence a standard multivariate regression follows.

The linearization based method can be implemented using the %GLIMMIX macro and PROC GLIMMIX, by ensuring no random effects are included. Empirically corrected standard errors can be obtained by including the EMPIRICAL option.

12.6.2 Random-Effects Models

Models with subject-specific parameters are differentiated from population-averaged models by the inclusion of parameters which are specific to the cluster. Unlike the correlated Gaussian outcomes, the parameters of the random effects and population-averaged models for correlated binary data describe different types of effects of the covariates on the response probabilities (Neuhaus, 1992).

The choice between population-averaged and random-effects strategies should heavily depend on the scientific goals. Population-averaged models evaluate the overall risk as a function of covariates. With a subject-specific approach, the response rates are modeled as a function of covariates and parameters, specific to a subject. In such models, interpretation of fixed-effects parameters is conditional on a constant level of the random-effects parameter. Population-averaged comparisons, on the other hand, make no use of within-cluster comparisons for cluster-varying covariates and are therefore not useful to assess within-subject effects (Neuhaus, Kalbfleisch and Hauck, 1991).

Whereas the linear mixed model is unequivocally the most popular choice in the case of normally distributed response variables, there are more options in the case of non-normal outcomes. Stiratelli, Laird, and Ware (1984) assume the parameter vector to be normally distributed. This idea has been carried further in the work on so-called *generalized linear mixed models* (Breslow and Clayton, 1993) which is closely related to linear and non-linear mixed models. Alternatively, Skellam (1948) introduced the beta-binomial model, in which the response probability of any response of a particular subject comes from a beta distribution. Hence, this model can also be viewed as a random-effects model. We will consider generalized linear mixed models.

Generalized Linear Mixed Models

Perhaps the most commonly encountered subject-specific (or random-effects) model is the generalized linear mixed model. A general framework for mixed-effects models can be expressed as follows.

As before, Y_{ij}, is the jth outcome measured for cluster (subject) i, $i = 1, \ldots, N$, $j = 1, \ldots, n_i$ and Y_i is the n_i-dimensional vector of all measurements available for cluster i. It is assumed that, conditional on q-dimensional random effects b_i, assumed to be drawn independently from the $N(0, D)$, the outcomes Y_{ij} are independent with densities of the form

$$f_i(y_{ij}|b_i, \beta, \phi) = \exp\left\{\phi^{-1}[y_{ij}\theta_{ij} - \psi(\theta_{ij})] + c(y_{ij}, \phi)\right\},$$

with $\eta(\mu_{ij}) = \eta(E(Y_{ij}|b_i)) = x_{ij}'\beta + z_{ij}'b_i$ for a known link function $\eta(\cdot)$, with x_{ij} and z_{ij} p-dimensional and q-dimensional vectors of known covariate values, with β a p-dimensional vector of unknown fixed regression coefficients, with ϕ a scale parameter, and with θ_{ij} the natural (or canonical) parameter. Further, let $f(b_i|G)$ be the density of the $N(0, G)$ distribution for the random effects b_i.

Due to the above independence assumption, this model is often referred to as a *conditional independence* model. This assumption is the basis of the implementation in the NLMIXED procedure. Just as in the linear mixed model case, the model can be extended with residual correlation, in addition to the one induced by the random effects. Such an extended model has been implemented in the SAS GLIMMIX procedure, and its predecessor the %GLIMMIX macro. This advantage is counterbalanced by bias induced by the optimization routine employed by GLIMMIX. It is important to realize that GLIMMIX can be used without random effects as well, thus effectively producing a marginal model, with estimates and standard errors similar to those obtained with GEE.

Note: The GLIMMIX procedure is experimental in SAS 9.1. There are advantages to using both PROC GLIMMIX and the %GLIMMIX macro, as we shall see.

In general, unless a fully Bayesian approach is followed, inference is based on the marginal model for Y_i which is obtained from integrating out the random effects. The likelihood contribution of subject i then becomes

$$f_i(y_i|\beta, G, \phi) = \int \prod_{j=1}^{n_i} f_{ij}(y_{ij}|b_i, \beta, \phi) \, f(b_i|G) \, db_i$$

from which the likelihood for β, D, and ϕ is derived as

$$L(\beta, G, \phi) = \prod_{i=1}^{N} f_i(y_i|\beta, G, \phi) = \prod_{i=1}^{N} \int \prod_{j=1}^{n_i} f_{ij}(y_{ij}|b_i, \beta, \phi) \ f(b_i|G) \ db_i. \tag{12.6.8}$$

The key problem in maximizing the obtained likelihood is the presence of N integrals over the q-dimensional random effects. In some special cases, these integrals can be worked out analytically. However, since no analytic expressions are available for these integrals, numerical approximations are needed. Here, we will focus on the most frequently used methods to do so. In general, the numerical approximations can be subdivided in those that are based on the approximation of the integrand, those based on an approximation of the data, and those that are based on the approximation of the integral itself. An extensive overview of the currently available approximations can be found in Tuerlinckx et al. (2004), Pinheiro and Bates (2000), and Skrondal and Rabe-Hesketh (2004). Finally, in order to simplify notation, it will be assumed that natural link functions are used, but straightforward extensions can be applied.

When integrands are approximated, the goal is to obtain a tractable integral such that closed-form expressions can be obtained, making the numerical maximization of the approximated likelihood feasible. Several methods have been proposed, but basically all come down to Laplace-type approximations of the function to be integrated (Tierney and Kadane, 1986).

A second class of approaches is based on a decomposition of the data into the mean and an appropriate error term, with a Taylor series expansion of the mean which is a non-linear function of the linear predictor. All methods in this class differ in the order of the Taylor approximation and/or the point around which the approximation is expanded. More specifically, consider the decomposition

$$Y_{ij} = \mu_{ij} + \varepsilon_{ij} \ = \ h(x_{ij}\beta + z_{ij}b_i) + \varepsilon_{ij} \tag{12.6.9}$$

in which $h(\cdot)$ equals the inverse link function $\eta^{-1}(\cdot)$, and where the error terms have the appropriate distribution with variance equal to $\mathrm{Var}(Y_{ij}|b_i) = \phi v(\mu_{ij})$ for $v(\cdot)$ the usual variance function in the exponential family. Note that, with the natural link function,

$$v(\mu_{ij}) = \frac{\partial h}{\partial \eta}(x_{ij}\beta + z_{ij}b_i).$$

Several approximations of the mean μ_{ij} in (12.6.9) can be considered. One possibility is to consider a linear Taylor expansion of (12.6.9) around current estimates $\widehat{\beta}$ and $\widehat{b_i}$ of the fixed effects and random effects, respectively. This will result in the expression

$$Y_i^* \equiv \widehat{W}_i^{-1}(Y_i - \widehat{\mu}_i) + X_i\widehat{\beta} + Z_i\widehat{b_i} \ \approx \ X_i\beta + Z_ib_i + \varepsilon_i^*, \tag{12.6.10}$$

with $\widehat{W}_i$ equal to the diagonal matrix with diagonal entries equal to $v(\widehat{\mu_{ij}})$, and for ε_i^* equal to $\widehat{W}_i^{-1}\varepsilon_i$, which still has mean zero. Note that (12.6.10) can be viewed as a linear mixed model for the pseudo data Y_i^*, with fixed effects β, random effects b_i, and error terms ε_i^*.

This immediately yields an algorithm for fitting the original generalized linear mixed model. Given starting values for the parameters β, G, and ϕ in the marginal likelihood, empirical Bayes estimates are calculated for b_i, and pseudo data Y_i^* are computed. Then, the approximate linear mixed model (12.6.10) is fitted, yielding updated estimates for β, G, and ϕ. These are then used to update the pseudo data and this whole scheme is iterated until convergence is reached.

The resulting estimates are called *penalized quasi-likelihood* estimates (PQL) in the literature (e.g., Molenberghs and Verbeke 2005), or *pseudo-quasi-likelihood* in the GLIMMIX documentation because they can be obtained from optimizing a quasi-likelihood function which involves only first- and second-order conditional moments, augmented with a penalty term on the random effects. The pseudo-likelihood terminology derives from the fact that the estimates are obtained by (restricted) maximum likelihood of the pseudo-response or working variable.

An alternative approximation is very similar to the PQL method, but is based on a linear Taylor expansion of the mean mu_{ij} in (12.6.9) around the current estimates $\widehat{\beta}$ for the fixed effects and around $b_i = 0$ for the random effects. The resulting estimates are called *marginal quasi-likelihood* estimates (MQL). See Breslow and Clayton (1993) and Wolfinger and O'Connell (1993) for details. Since the linearizations in the PQL and the MQL methods lead to linear mixed models, the implementation of these procedures is often based on feeding updated pseudo data into software for the fitting of linear mixed models. However, it should be emphasized that the results from these fittings, which are often reported intermediately, should be interpreted with great care. For example, reported (log)likelihood values correspond to the assumed normal model for the pseudo data and should not be confused with (log-)likelihood for the generalized linear mixed model for the actual data at hand. Further, fitting of linear mixed models can be based on maximum likelihood (ML) as well as restricted maximum likelihood (REML) estimation. Hence, within the PQL and MQL frameworks, both methods can be used for the fitting of the linear model to the pseudo data, yielding (slightly) different results. Finally, the quasi-likelihood methods discussed here are very similar to the method of linearization discussed in Section 12.6.1 for fitting generalized estimating equations (GEE). The difference is that here, the correlation between repeated measurements is modeled through the inclusion of random effects, conditional on which repeated measures are assumed independent, while, in the GEE approach, this association is modeled through a marginal working correlation matrix.

Note that, when there are no random effects, both this method and GEE reduce to a marginal model, the difference being in the way the correlation parameters are estimated. In both cases, it is possible to allow for misspecification of the association structure by resorting to empirically corrected standard errors. When this is done, the methods are valid under MCAR. In case we would have confidence in the specified correlation structure, purely model-based inference can be conducted, and hence the methods are valid when missing data are MAR.

A third method of numerical approximation is based on the approximation of the integral itself. Especially in cases where the above two approximation methods fail, this numerical integration proves to be very useful. Of course, a wide toolkit of numerical integration tools, available from the optimization literature, can be applied. Several of those have been implemented in various software tools for generalized linear mixed models. A general class of quadrature rules selects a set of abscissas and constructs a weighted sum of function evaluations over those. In the particular context of random-effects models, so-called *adaptive* quadrature rules can be used (Pinheiro and Bates 1995, 2000), where the numerical integration is centered around the EB estimates of the random effects, and the number of quadrature points is then selected in terms of the desired accuracy.

To illustrate the main ideas, we consider Gaussian and adaptive Gaussian quadrature, designed for the approximation of integrals of the form $\int f(z)\phi(z)dz$, for an known function $f(z)$ and for $\phi(z)$ the density of the (multivariate) standard normal distribution. We will first standardize the random effects such that they get the identity covariance matrix. Let δ_i be equal to $\delta_i = G^{-1/2}b_i$. We then have that δ_i is normally distributed with mean 0 and covariance I. The linear predictor then becomes $\theta_{ij} = x_{ij}'\beta + z_{ij}'G^{1/2}\delta_i$, so the variance components in G have been moved to the linear predictor. The likelihood contribution for subject i then equals

$$f_i(y_i|\beta, G, \phi) = \int \prod_{j=1}^{n_i} f_{ij}(y_{ij}|b_i, \beta, \phi)\, f(b_i|G)\, db_i. \tag{12.6.11}$$

Obviously, (12.6.11) is of the form $\int f(z)\phi(z)dz$ as required to apply (adaptive) Gaussian quadrature.

In Gaussian quadrature, $\int f(z)\phi(z)dz$ is approximated by the weighted sum

$$\int f(z)\phi(z)dz \approx \sum_{q=1}^{Q} w_q f(z_q).$$

Q is the order of the approximation. The higher Q, the more accurate the approximation will be. Further, the so-called nodes (or quadrature points) z_q are solutions to the Qth order Hermite polynomial, while the w_q are well-chosen weights. The nodes z_q and weights w_q are reported in tables. Alternatively, an algorithm is available for calculating all z_q and w_q for any value Q (Press et al., 1992).

Figure 12.5 Graphical illustration of Gaussian (left panel) and adaptive Gaussian (right panel) quadrature. The triangles indicate the position of the quadrature points, while the rectangles indicate the contribution of each point to the integral.

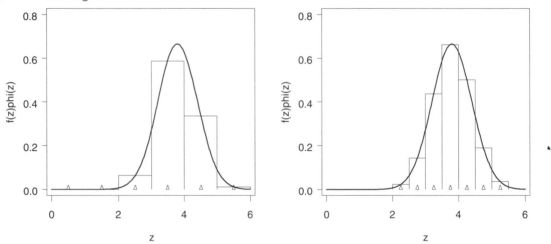

In the case of univariate integration, the approximation consists of subdividing the integration region in intervals, and approximating the surface under the integrand by the sum of surfaces of the so-obtained approximating rectangles. An example is given in the left panel of Figure 12.5, for the case of $Q = 10$ quadrature points. A similar interpretation is possible for the approximation of multivariate integrals. Note that the figure immediately highlights one of the main disadvantages of (non-adaptive) Gaussian quadrature, i.e., the fact that the quadrature points z_q are chosen based on $\phi(z)$, independent of the function $f(z)$ in the integrand. Depending on the support of $f(z)$, the z_q will or will not lie in the region of interest. Indeed, the quadrature points are selected to perform well in case $f(z)\phi(z)$ approximately behaves like $\phi(z)$, i.e., like a standard normal density function. This will be the case, for example, if $f(z)$ is a polynomial of a sufficiently low order. In our applications, however, the function $f(z)$ will take the form of a density from the exponential family, hence an exponential function. It may then be helpful to rescale and shift the quadrature points such that more quadrature points lie in the region of interest. This is shown in the right panel of Figure 12.5, and is called *adaptive* Gaussian quadrature.

In general, the higher the order Q, the better the approximation will be of the N integrals in the likelihood. Typically, adaptive Gaussian quadrature needs (many) fewer quadrature points than classical Gaussian quadrature. On the other hand, adaptive Gaussian quadrature requires for each unit the numerical maximization of a function of the form $\ln(f(z)\phi(z))$ for the calculation of $\hat{z}$. This implies that adaptive Gaussian quadrature is much more time consuming.

Since fitting of generalized linear mixed models is based on maximum likelihood principles, inferences for the parameters are readily obtained from classical maximum likelihood theory.

12.6.3   Marginal versus Random-Effects Models

Note that there is an important difference with respect to the interpretation of the fixed effects β. Under the classical linear mixed model (Verbeke and Molenberghs, 2000), we have that $E(Y_i)$ equals $X_i\beta$, such that the fixed effects have a subject-specific as well as a population-averaged interpretation. Under non-linear mixed models, however, this does no longer hold in general. The fixed effects now only reflect the conditional effect of covariates, and the marginal effect is not easily obtained anymore as $E(Y_i)$ is given by

$$E(Y_i) = \int y_i \int f_i(y_i|b_i)g(b_i)db_idy_i.$$

However, in a biopharmaceutical context, we are often primarily interested in hypothesis testing, and the random-effects framework can be used to this effect.

Note that both WGEE and GLMM are valid under MAR, with the extra condition that the model for weights in WGEE has been specified correctly. Nevertheless, the parameter estimates between both are rather different. This is due to the fact that GEE and WGEE parameters have a marginal interpretation, describing average longitudinal profiles, whereas GLMM parameters describe a longitudinal profile, conditional upon the value of the random effects. Let us now provide an overview of the differences between marginal and random-effects models for non-Gaussian outcomes (a detailed discussion can be found in Molenberghs and Verbeke 2004). The interpretation of the parameters in both types of model (marginal or random-effects) is completely different.

Figure 12.6 Representation of model families and corresponding inference (M stands for marginal and RE stands for random-effects). A parameter between quotes indicates that marginal functions but no direct marginal parameters are obtained, since they result from integrating out the random effects from the fitted hierarchical model.

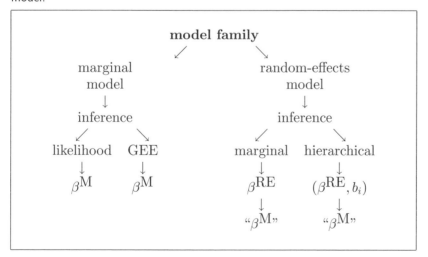

A schematic display is given in Figure 12.6. Depending on the model family (marginal or random-effects), we are led to either marginal or hierarchical inference. It is important to realize that in the general case the parameter β^M resulting from a marginal model is different from the parameter β^{RE} even when the latter is estimated using marginal inference. Some of the confusion surrounding this issue may result from the equality of these parameters in the very special linear mixed model case. When a random-effects model is considered, the marginal mean profile can be derived, but it will generally not produce a simple parametric form. In Figure 12.6 this is indicated by putting the corresponding parameter between quotation marks.

12.6.4   Analysis of Depression Trial Data

Let us now analyze the clinical depression trial, introduced in Section 12.2. The binary outcome of interest is the YBIN variable (it is equal to 1 if the HAMD17 score is larger than 7, and 0 otherwise). The primary null hypothesis has been tested using both GEE and WGEE (with the GENMOD procedure), as well as GLMM with adaptive and non-adaptive Gaussian quadrature (with PROC NLMIXED) in Dmitrienko et al. (2005, Chapter 5). Now, we will test the hypothesis using GEE based on linearization, as well as GLMM based on the penalized quasi-likelihood method. Both methods can be fitted using the GLIMMIX macro. Beginning with SAS 9.1, there is an experimental GLIMMIX procedure, which can be used as well. We include the fixed categorical effects of treatment, visit, and treatment-by-visit interaction, as well as the continuous, fixed covariates of baseline score and baseline score-by-visit interaction. A random intercept will be included when considering the random-effect models.

%GLIMMIX Macro for the Marginal Model

We will consider the marginal model:

$$Y_{ij} \sim \text{Bernoulli}(\mu_{ij}), \quad \text{logit}\left(\frac{\mu_{ij}}{1 - \mu_{ij}}\right) = X_i\beta, \quad (12.6.12)$$

where X_i is the design matrix for subject i, containing all covariates and an intercept.

Program 12.1 fits Model (12.6.12) with exchangeable working assumptions using the %GLIMMIX macro. The macro is based on fitting the iterative procedure, outlined in Section 12.6.1. The Generalized Linear Models shell linearizes the outcome and computes the weights, as in (12.6.6) and (12.6.7). Specific statements that govern this procedure are the ERROR and LINK statements. We chose the binomial error structure. The logit link is the default for this option, but we have still chosen to specify it, for clarity. The procedure is based on the MIXED procedure, used to solve iteratively reweighted least squares equations (12.6.5). Virtually all statements that are available in the MIXED procedure can be used. They are passed on, in string form, to the macro via the STMTS option. Note that TYPE=CS, referring to compound symmetry, must be used here. For unstructured working assumptions, we use TYPE=UN, for AR(1) this would be TYPE=AR(1), and for independence assumptions, TYPE=SIMPLE needs to be used. One set of assumptions, those corresponding to the PROC MIXED statement, are passed on via a separate statement, i.e., the PROCOPT string. Note that we have inserted the EMPIRICAL option, to ensure the empirically corrected standard errors are produced. Omitting the EMPIRICAL option produces the model-based standard errors. We have a choice between the updating methods that are available in the MIXED procedure.

To receive the output that would be produced by the MIXED procedure, we can use OPTIONS=MIXPRINTLAST within the STMTS option. To study the PROC MIXED output at each iteration, we add the MIXPRINTALL option. Arguably, the latter is primarily useful for debugging purposes.

Program 12.1 %GLIMMIX macro for the linearization-based method in the depression trial example

```
%glimmix(data=depression, procopt=%str(method=ml empirical),
    stmts=%str(
        class patient visit trt;
        model ybin = visit trt*visit basval basval*visit / solution;
        repeated visit / subject=patient type=cs rcorr;
        ),
    error=binomial,
    link=logit);
```

The typical GLIMMIX output consists of tables copied from the MIXED output, as well as some additional information. Typical output includes bookkeeping information such as model information, dimensions, and number of observations. Since we included the RCORR option in the REPEATED statement, the fitted correlation matrix of the measurements is given, which is to be interpreted as the working correlation matrix. In our case, this is a 5×5 correlation matrix with off-diagonal elements equal to 0.3317, the exchangeable working correlation. Output 12.1, the Covariance Parameter Estimates portion of the output, must be interpreted with caution, though, for reasons that we will explain.

Output from Program 12.1. Covariance Parameter Estimates

```
 Covariance Parameter Estimates

Cov Parm      Subject     Estimate

CS            PATIENT      0.3140
Residual                   0.6326
```

The working correlation is obtained from the usual compound-symmetry equation:

$$\frac{0.3140}{0.3140 + 0.6326} = 0.3317.$$

In the following output, the residual value is copied as the extra-dispersion parameter:

Output from Program 12.1. GLIMMIX Model Statistics

```
        GLIMMIX Model Statistics

Description                     Value

Deviance                      746.6479
Scaled Deviance              1180.3240
Pearson Chi-Square            673.9938
Scaled Pearson Chi-Square    1065.4703
Extra-Dispersion Scale          0.6326
```

It is best not to use this portion of output, since it is not appropriately adapted to the combination of generalized linear model and the repeated measures nature of the data. In case we had used independence working assumptions, there would have been a single covariance parameter only, which could then be considered the overdispersion parameter. Arguably, there is little basis to do so, and it is unlikely to see almost no overdispersion with independence and strong underdispersion with exchangeability. It might make more sense to consider the total variance in the exchangeable case, i.e., $0.3140 + 0.6326$, the overdispersion parameter. Similarly, it would be better to ignore the fit statistics portion of output, copied from the MIXED procedure.

The most relevant portion of the output is the Solution for Fixed Effects table, with its associated F tests.

Output from Program 12.1. Solution and Type 3 Tests for Fixed Effects

```
Solution for Fixed Effects

            Visit                     Standard
Effect      Number  TRT   Estimate    Error     DF    t Value  Pr > |t|

Intercept                 -1.2326     0.7853    168   -1.57    0.1184
VISIT       4              0.4457      1.2147    517    0.37    0.7138
...
VISIT*TRT   8     1       -0.6942      0.3810    517   -1.82    0.0690
...
BASVAL*VISIT 7            0.02439     0.05170    517    0.47    0.6373
BASVAL*VISIT 8            0           .          .      .       .


Type 3 Tests of Fixed Effects

            Num   Den
Effect      DF    DF    F Value   Pr > F

VISIT       4     517   0.39      0.8128
VISIT*TRT   5     517   1.22      0.3000
BASVAL      1     168   22.17     <.0001
BASVAL*VISIT 4    517   1.79      0.1291
```

These estimates, with both model-based and empirically corrected standard errors, are given in the linearization column of Table 12.2, next to the estimates for GEE and WGEE. Clearly, the estimates under exchangeability for the linearization-based method are very similar to those for GEE. Since we did not include the treatment as a main effect, the output immediately produces the treatment effects at the different time points. For instance, the estimate of the effect of treatment at the endpoint is -0.6942, with a corresponding p value equal to 0.0690. Comparing this with the result of GEE (the p-value of treatment effect at the last visit is 0.0633), it is clear that both are very similar. However, when a WGEE analysis is performed, the result of this contrast changes a lot (p-value is 0.0325).

PROC GLIMMIX for the Marginal Model

We can also use PROC GLIMMIX to fit models of a marginal type, subject-specific type, as well as models with subject-specific effects and residual association in addition to that. Program 12.2 relies on PROC GLIMMIX to fit a model identical to the model fitted by Program 12.1. Even though Program 12.2 is rather different, at first sight, from Program 12.1, the correspondence is almost immediate. Again, users of PROC MIXED for linear mixed models will recognize that the code here is very similar to that used in PROC MIXED. The reason is that, internally, PROC GLIMMIX calls PROC MIXED each time a linear mixed model needs to be fitted to newly updated pseudo data.

Table 12.2 Results of Marginal Models in the Depression Trial Example
Note: Parameter estimates (model-based standard errors; empirically corrected standard errors) for GEE, WGEE and the linearization based method are shown. Main visit effects and interaction terms of baseline and visit are not shown.

	GEE		WGEE		Linearization	
intercept	-1.22	$(0.77;0.79)$	-0.70	$(0.64;0.86)$	-1.23	$(0.75;0.79)$
visit 4	0.43	$(1.05;1.22)$	-0.08	$(0.88;1.85)$	0.45	$(1.05;1.22)$
visit 5	-0.48	$(0.91;1.23)$	-0.13	$(0.70;1.55)$	-0.47	$(0.92;1.23)$
visit 6	0.06	$(0.86;1.03)$	0.19	$(0.72;1.14)$	0.05	$(0.86;1.03)$
visit 7	-0.25	$(0.89;0.91)$	-0.28	$(0.82;0.88)$	-0.25	$(0.89;0.91)$
trt $\times$ visit 4	-0.24	$(0.54;0.57)$	-1.57	$(0.41;0.99)$	-0.22	$(0.53;0.56)$
trt $\times$ visit 5	-0.09	$(0.39;0.40)$	-0.67	$(0.21;0.65)$	-0.08	$(0.38;0.40)$
trt $\times$ visit 6	0.17	$(0.34;0.35)$	0.62	$(0.23;0.56)$	0.18	$(0.34;0.35)$
trt $\times$ visit 7	-0.43	$(0.36;0.35)$	0.57	$(0.30;0.37)$	-0.42	$(0.35;0.35)$
trt $\times$ visit 8	-0.71	$(0.38;0.38)$	-0.84	$(0.32;0.39)$	-0.69	$(0.37;0.38)$
baseline	0.08	$(0.04;0.04)$	0.07	$(0.03;0.05)$	0.08	$(0.04;0.04)$
baseline $\times$ visit 4	0.12	$(0.07;0.07)$	0.24	$(0.06;0.13)$	0.12	$(0.06;0.07)$
baseline $\times$ visit 5	0.09	$(0.05;0.07)$	0.07	$(0.04;0.08)$	0.09	$(0.05;0.07)$
baseline $\times$ visit 6	0.01	$(0.05;0.05)$	0.01	$(0.04;0.06)$	0.01	$(0.05;0.88)$
baseline $\times$ visit 7	0.02	$(0.05;0.05)$	0.03	$(0.05;0.05)$	0.02	$(0.05;0.05)$

Program 12.2 PROC GLIMMIX for the linearization-based method in the depression trial example

```
proc glimmix data=depression method=rspl empirical;
    class patient visit trt;
    model ybin (event='1')=visit trt*visit basval basval*visit/dist=binary solution;
    random _residual_/subject=patient type=cs;
    run;
```

The most important option in PROC GLIMMIX is METHOD. In Program 12.2, METHOD=RSPL specifies the PQL method based on REML for linear mixed models. Note that, strictly speaking, the "penalized" part of PQL is absent, since there are no random effects and hence no random-effects penalty. In this case, it is more straightforward to think of restricted pseudo-likelihood (RPL). An overview of the other available options is given in Table 12.3.

Table 12.3 Available Options for Specification of the Estimation Method in PROC GLIMMIX

GLIMMIX option	Quasi-likelihood type PQL/MQL	Inference pseudo-data ML/REML
METHOD=RSPL	PQL	REML
METHOD=MSPL	PQL	ML
METHOD=RMPL	MQL	REML
METHOD=MMPL	MQL	ML

The CLASS statement in PROC GLIMMIX specifies which variables should be considered as factors. Such classification variables can be either character or numeric. Internally, each of these factors will correspond to a set of dummy variables.

The MODEL statement names the response variable and all fixed effects, which determine the X_i matrices. By default, an intercept is added. The EVENT option (EVENT='1') specifies that the probability to be modeled is $P(Y_{ij} = 1)$ (the probability of a severe infection) rather than $P(Y_{ij} = 0)$. The SOLUTION option is used to request the printing of the estimates for all the fixed effects in the model, together with standard errors, t-statistics, corresponding p-values, and confidence intervals. The DIST option is used to specify the conditional distribution of the data, given the random effects. Various distributions are available, including the normal, Bernoulli, Binomial, and Poisson distribution. In Program 12.2, the Bernoulli distribution is specified (DIST=BINARY). The default link function is the natural link (in this example, it is the logit link).

The REPEATED statement of the %GLIMMIX macro corresponds to the RANDOM=_residual_ statement in PROC GLIMMIX. SAS refers to this as the R-side of the random statement. It is useful to think of it as the variance-covariance matrix of the outcome vector Y_i, of which the variances follow from the mean-variance link, but the correlation structure needs to be specified (as in GEE). When random effects are present, this structure refers to the residual correlation, in addition to the correlation induced by the random effects. Changing the TYPE option in the RANDOM statement to SIMPLE, CS or UN, respectively, combined with either omission or inclusion of the EMPIRICAL option in PROC GLIMMIX produces exactly the same results as with the %GLIMMIX macro; i.e., the results reported in the third panel of Table 12.2.

The output form PROC GLIMMIX in Program 12.2, although structured differently from the %GLIMMIX macro output, is largely equivalent. The fact that the empirically corrected standard errors are produced is acknowledged:

Output of Program 12.2. Part of the Model Information

```
                 The GLIMMIX Procedure

                    Model Information

Fixed Effects SE Adjustment     Sandwich - Classical
```

The fact that a marginal model (i.e., a model without random effects) is used is acknowledged through reference to the so-called R-side covariance parameters:

Output of Program 12.2. Part of the Dimensions

```
               Dimensions

R-side Cov. Parameters          2
```

PROC GLIMMIX in Program 12.2 took 11 iterations to converge. The final covariance parameters are equivalent to the ones produced by the %GLIMMIX macro:

Output of Program 12.2. Covariance Parameter Estimates

```
      Covariance Parameter Estimates

                                 Standard
Cov Parm    Subject    Estimate    Error

CS          PATIENT     0.3196    0.05209
Residual                0.6469    0.03955
```

PROC GLIMMIX also produces the fixed-effects parameters (again equivalent, up to numerical accuracy, to their %GLIMMIX macro counterparts) together with associated F tests.

Output from Program 12.2. Solution and Type 3 Tests for Fixed Effects

```
Solutions for Fixed Effects

                Visit                    Standard
Effect          Number  TRT   Estimate    Error     DF   t Value  Pr > |t|

Intercept                     -1.2331    0.7852    168    -1.57    0.1182
VISIT           4              0.4463    1.2146    517     0.37    0.7134
...
VISIT*TRT       8     1        -0.6939   0.3810    517    -1.82    0.0691
...
BASVAL*VISIT    7              0.02440   0.05170   517     0.47    0.6372
BASVAL*VISIT    8              0         .         .       .       .


Type III Tests of Fixed Effects

                Num    Den
Effect          DF     DF    F Value    Pr > F

VISIT           4      517     0.39     0.8129
VISIT*TRT       5      517     1.22     0.3003
BASVAL          1      168    22.18     <.0001
BASVAL*VISIT    4      517     1.79     0.1293
```

Again as long as PROC GLIMMIX is experimental (as in SAS 9.1), it is a good idea to use the %GLIMMIX macro as well. Also, having the macro as a backup may increase a program's chances of reaching convergence.

PROC GLIMMIX for the Random-Effects Model

PROC GLIMMIX supports the marginal and penalized quasi-likelihood methods. We will fit the following random-effects model, using the PQL method: assume that, conditional on subject-specific random intercepts, b_i, Y_{ij} is Bernoulli distributed with mean π_{ij}, modeled as

$$\text{logit}(\pi_{ij}) = X_i\beta + Z_i b_i, \tag{12.6.13}$$

in which X_i is the design matrix with covariates described above, β is the vector of fixed effects parameters, Z_i is a design matrix for the random effects (in this case a row of ones), and b_i is the random intercept assumed to be normally distributed with mean 0 and variance d.

The procedure has many more statements and options than those presented here, but we restrict our example to the basic statements needed to fit a generalized linear mixed model.

Program 12.3 fits this random-effects model using PQL based on REML estimation for the linear mixed models for the pseudo data.

Program 12.3 PROC GLIMMIX for the random-effects model using PQL maximization in the depression trial example

```
proc glimmix data=depression method=rspl;
    class patient visit trt;
    model ybin (event='1')=visit trt*visit basval basval*visit/dist=binary solution;
    random intercept/subject=patient;
    run;
```

Most of the statements used in Program 12.3 were described above, except for the RANDOM statement. The RANDOM statement defines the random effects in the model, i.e., the Z_i matrices containing the covariates with subject-specific regression coefficients. Note that when random intercepts are required (as in this example), this requirement should be specified explicitly, which is in contrast to the MODEL statement where an intercept is included by default. The SUBJECT option identifies the subjects in the DEPRESSION data set. The variable in the SUBJECT option can be continuous or categorical (specified in the CLASS statement); however, when it is continuous, PROC GLIMMIX considers a record to be from a new subject whenever the value of this variable is different from the previous record.

Suppose that random slopes for the time trend were to be included as well. This can be achieved by replacing the RANDOM statement in Program 12.3:

```
random intercept time/subject=patient type=un;
run;
```

Here TYPE=UN specifies that the random-effects covariance matrix G is a general unstructured 2×2 matrix. Special structures are available, e.g., equal variance for the intercepts and slopes, or independent intercepts and slopes.

The output of Program 12.3 includes information about the fitted model and the estimation procedure. The Residual PL estimation technique refers to PQL with REML (restricted or residual maximum likelihood) for the fitting of the linear models for the pseudo data:

Output of Program 12.3. Model Information

```
           The GLIMMIX Procedure

        Model Information

Data Set                     WORK.DEPRESSION
Response Variable            ybin
Response Distribution        Binary
Link Function                Logit
Variance Function            Default
Variance Matrix Blocked By   PATIENT
Estimation Technique         Residual PL
Degrees of Freedom Method    Containment
```

The Fit Statistics portion of the output gives minus twice the residual log-pseudo-likelihood value evaluated in the final solution, together with the information criteria of Akaike (AIC) and Schwarz (BIC), as well as a finite-sample corrected version of AIC (AICC). When REML estimation is used for the fitting of the linear mixed models for the pseudo data, an objective function is maximized which is called residual log-likelihood function, while, strictly speaking, the function is not a log-likelihood, and should not be used as a log-likelihood. We refer to Verbeke and Molenberghs (2000, Chapters 5 and 6) for a more detailed discussion with examples. Further, information criteria are statistics that are sometimes used to compare non-nested models which cannot be compared based on a formal testing procedure. The main idea behind information criteria is to compare models based on their maximized (residual) log-likelihood value (or equivalently minimized minus twice the log-likelihood value), while at the same time penalizing for including an excessive number of parameters. They should by no means be interpreted as formal statistical tests of significance. In specific examples, different information criteria can even lead to different model selections. An example of this is given in Section 6.4 of Verbeke and Molenberghs

(2000) in the context of linear mixed models. More details about the use of information criteria can be found in Akaike (1974), Schwarz (1978), and Burnham and Anderson (1998).

Output of Program 12.3. Fit Statistics and Covariance Parameter Estimates

```
               Fit Statistics

-2 Res Log Pseudo-Likelihood        3488.65
Pseudo-AIC  (smaller is better)     3490.65
Pseudo-AICC (smaller is better)     3490.65
Pseudo-BIC  (smaller is better)     3493.78
Pseudo-CAIC (smaller is better)     3494.78
Pseudo-HQIC (smaller is better)     3491.92
Pearson Chi-Square                   400.28
Pearson Chi-Square / DF                0.58


          Covariance Parameter Estimates

                                   Standard
Cov Parm     Subject    Estimate     Error


Intercept    PATIENT      2.5318     0.5343
```

Next, Output 12.3 presents two tables containing estimates and inferences for the fixed effects in the model. The reported inferences immediately result from the linear mixed model fitted to the pseudo data in the last step of the iterative estimation procedure. From the output, it is immediately clear the treatment effect at the last visit is clearly insignificant (p-value is 0.1286). Comparing this with the p-value obtained after fitting GLMM with adaptive Gaussian quadrature ($p = 0.0954$), we see that both yield an insignificant effect.

Output of Program 12.3. Solutions and Type 3 Tests of Fixed Effects

```
Solutions for Fixed Effects

            Visit                      Standard
Effect      Number  TRT   Estimate      Error    DF   t Value  Pr > |t|

Intercept                  -1.7036     1.0607    167   -1.61    0.1101
...
VISIT*TRT    8      1      -0.8426     0.5537    518   -1.52    0.1286
...
BASVAL*VISIT 7              0.02743    0.06897   518    0.40    0.6911
BASVAL*VISIT 8             0              .        .      .        .

Type III Tests of Fixed Effects

              Num    Den
Effect        DF     DF    F Value   Pr > F

VISIT          4     518     0.27    0.8959
VISIT*TRT      5     518     0.85    0.5141
BASVAL         1     518    18.12    <.0001
BASVAL*VISIT   4     518     1.20    0.3115
```

The Covariance Parameter Estimates portion of the output lists estimates and associated standard errors for the variance components in the model, i.e., for the elements in the random-effects covariance matrix G. In our example, this is the random-intercepts variance g.

Table 12.4 compares the parameter estimates and standard errors with the results obtained by fitting GLMM with adaptive Gaussian quadrature, using PROC NLMIXED. Obviously, there are some differences between these two methods for certain parameters. Since both approaches are likelihood-based, validity under the MAR assumption should be fulfilled. For PROC NLMIXED, the integral is approximated using adaptive Gaussian quadrature, which is quite an accurate method, if a sufficient number of quadrature points are chosen. However, PROC GLIMMIX (or %GLIMMIX macro) is based on penalized quasi-likelihood, and thus the mean is approximated by a Taylor expansion. This approximation can be relatively poor for random-effects models and hence the results thereof should be treated with caution.

Table 12.4 Results of Random-Effects Models in the Depression Trial Example
Note: Parameter estimates (standard errors) for GLMM with adaptive Gaussian quadrature (AGQ) and penalized-quasi-likelihood methods (PQL).

	PQL		AGQ	
intercept	−1.70	(1.06)	−2.31	(1.34)
visit 4	0.66	(1.48)	0.64	(1.75)
visit 5	−0.44	(1.29)	−0.78	(1.51)
visit 6	0.17	(1.22)	0.19	(1.41)
visit 7	−0.23	(1.25)	−0.27	(1.43)
treatment × visit 4	−0.29	(0.66)	−0.54	(0.82)
treatment × visit 5	−0.10	(0.53)	−0.20	(0.68)
treatment × visit 6	0.33	(0.49)	0.41	(0.64)
treatment × visit 7	−0.47	(0.52)	−0.68	(0.67)
treatment × visit 8	−0.84	(0.55)	−1.20	(0.72)
baseline	0.10	(0.06)	0.15	(0.07)
baseline × visit 4	0.14	(0.09)	0.21	(0.11)
baseline × visit 5	0.12	(0.07)	0.17	(0.09)
baseline × visit 6	0.007	(0.07)	0.01	(0.08)
baseline × visit 7	0.03	(0.07)	0.03	(0.08)

GLIMMIX Macro for the Random-Effects Model

Although still experimental in SAS 9.1, PROC GLIMMIX can be viewed as a formal procedure that has grown out of the %GLIMMIX macro, applied earlier for fitting generalized estimating equations (GEE) based on linearization. In GEE, the association between repeated measures is modeled through a marginal working correlation matrix. Now, this correlation is modeled via the inclusion of random effects, conditional on which repeated measures are assumed to be independent. This similarity implies that the same macro can be used for fitting generalized linear mixed models as well. Without going into much detail, we present in Program 12.4 the SAS code needed to repeat the previous analysis with the %GLIMMIX macro.

Program 12.4 %GLIMMIX macro for the random-effects model using PQL maximization in the depression trial example

```
%glimmix(
   data=depression,
   stmts=%str(
      class class patient visit trt;
      model ybin = visit trt*visit basval basval*visit/solution;
      random intercept/subject=patient;
      parms (4) (1)/hold=2;
      ),
   error=binomial);
```

The statements that appear in the STMTS statement are directly fed into the PROC MIXED calls needed for fitting the linear mixed models to the pseudo data. Note that by default the %GLIMMIX macro includes a residual overdispersion parameter. If the corresponding generalized linear mixed model does not contain such a parameter, it should explicitly be kept equal to one. This is done using the HOLD option in the PARMS statement.

Since PROC MIXED uses REML estimation by default, Program 12.4 requests PQL estimation based on REML fitting for the pseudo data. If ML fitting is required, this can be specified by adding the line

```
procopt=%str(method=ml),
```

to the GLIMMIX call in Program 12.4. If MQL is required, rather than the default PQL, this can be specified by adding the following line

```
options=MQL,
```

Without discussing the output from the %GLIMMIX macro in much detail, we here present some output tables, which can be compared with the output from PROC GLIMMIX (Program 12.3).

Partial Output of Program 12.4

```
 Covariance Parameter Estimates

Cov Parm       Subject    Estimate

Intercept      PATIENT     2.5318
Residual                   1.0000

           Fit Statistics

-2 Res Log Likelihood        3488.6
AIC (smaller is better)      3490.6
AICC (smaller is better)     3490.7
BIC (smaller is better)      3493.8
```

```
┌─────────────────────────────────────────────────────────────────────────┐
│ Solution for Fixed Effects                                                │
│                                                                           │
│                Visit                    Standard                          │
│ Effect         Number   TRT   Estimate    Error     DF   t Value  Pr > |t||
│                                                                           │
│ Intercept                     -1.7036    1.0607     167   -1.61    0.1101 │
│ ...                                                                       │
│ VISIT*TRT       8        1     -0.8426    0.5537     518   -1.52    0.1286 │
│ ...                                                                       │
│ BASVAL*VISIT    7              0.02743   0.06897     518    0.40    0.6911 │
│ BASVAL*VISIT    8                    0         .       .       .        . │
│                                                                           │
│ Type 3 Tests of Fixed Effects                                             │
│                                                                           │
│                 Num      Den                                              │
│ Effect          DF       DF    F Value   Pr > F                           │
│                                                                           │
│ VISIT            4       518     0.27     0.8959                          │
│ VISIT*TRT        5       518     0.85     0.5141                          │
│ BASVAL           1       518    18.12     <.0001                          │
│ BASVAL*VISIT     4       518     1.20     0.3115                          │
└─────────────────────────────────────────────────────────────────────────┘
```

Output 12.4 shows that, indeed, the residual overdispersion parameter was set to one. The obtained results are identical to the results produced by PROC GLIMMIX in Program 12.3.

12.7   MNAR Modeling

Even though the assumption of likelihood ignorability encompasses the MAR and not only the more stringent and often implausible MCAR mechanisms, it is difficult to exclude the option of a more general MNAR mechanism. One solution for continuous outcomes is to fit an MNAR model as proposed by Diggle and Kenward (1994). In the discrete case, Molenberghs, Kenward, and Lesaffre (1997) considered a global odds ratio (Dale) model.

However, as pointed out in the introduction and by several authors (Diggle and Kenward, 1994; Verbeke and Molenberghs 2000, Chapter 18), we must be extremely careful with interpreting evidence for or against MNAR in a selection model context, especially in large studies with a lot of power. We will return to these issues in Section 12.8.

12.7.1   Diggle-Kenward Model

To be consistent with notation introduced in Section 12.3, we assume a vector of outcomes Y_i is designed to be measured. If dropout occurs, Y_i is only partially observed. We denote the occasion at which dropout occurs by $D_i > 1$, and Y_i is split into the $(D_i - 1)$-dimensional observed component Y_i^o and the $(n_i - D_i + 1)$-dimensional missing component Y_i^m. In case of no dropout, we let $D_i = n_i + 1$, and Y_i equals Y_i^o. The likelihood contribution of the ith subject, based on the observed data (y_i^o, d_i), is proportional to the marginal density function

$$f(y_i^o, d_i | \theta, \psi) = \int f(y_i, d_i | \theta, \psi) \, dy_i^m = \int f(y_i | \theta) f(d_i | y_i, \psi) \, dy_i^m, \qquad (12.7.14)$$

in which a marginal model for Y_i is combined with a model for the dropout process, conditional on the response, and where θ and ψ are vectors of unknown parameters in the measurement model and dropout model, respectively.

Let $h_{ij} = (y_{i1}, \ldots, y_{i,j-1})$ denote the observed history of subject i up to time $t_{i,j-1}$. The Diggle-Kenward model for the dropout process allows the conditional probability for dropout at occasion j, given that the subject was still observed at the previous occasion, to depend on the history h_{ij} and the possibly unobserved current outcome y_{ij}, but not on future outcomes y_{ik}, $k > j$. These conditional probabilities $P(D_i = j | D_i \geq j, h_{ij}, y_{ij}, \psi)$ can now be used to calculate the probability of dropout at each occasion:

$$
P(D_i = j | y_i, \psi) = P(D_i = j | h_{ij}, y_{ij}, \psi)
$$

$$
= \begin{cases}
P(D_i = j | D_i \geq j, h_{ij}, y_{ij}, \psi) & j = 2, \\[2mm]
P(D_i = j | D_i \geq j, h_{ij}, y_{ij}, \psi) \times \prod_{k=2}^{j-1} [1 - P(D_i = k | D_i \geq k, h_{ik}, y_{ik}, \psi)] & j = 3, \dots, n_i, \\[2mm]
\prod_{k=2}^{n_i} [1 - P(D_i = k | D_i \geq k, h_{ik}, y_{ik}, \psi)] & j = n_i + 1.
\end{cases}
$$

Diggle and Kenward (1994) combine a multivariate normal model for the measurement process with a logistic regression model for the dropout process. More specifically, the measurement model assumes that the vector Y_i of repeated measurements for the ith subject satisfies the linear regression model $Y_i \sim N(X_i \beta, V_i)$, $i = 1, \dots, N$. The matrix V_i can be left unstructured or assumed to be of a specific form, e.g., resulting from a linear mixed model, a factor-analytic structure, or spatial covariance structure (Verbeke and Molenberghs, 2000).

In the particular case that a linear mixed model is assumed, we write (Verbeke and Molenberghs, 2000)

$$
Y_i = X_i \beta + Z_i b_i + \varepsilon_i, \tag{12.7.15}
$$

where Y_i is the n dimensional response vector for subject i, $1 \leq i \leq N$, N is the number of subjects, X_i, and Z_i are $(n \times p)$ and $(n \times q)$ known design matrices, β is the p dimensional vector containing the fixed effects, $b_i \sim N(0, G)$ is the q dimensional vector containing the random effects. The residual components $\varepsilon_i \sim N(0, \Sigma_i)$.

The logistic dropout model can, for example, take the form

$$
\text{logit} [P(D_i = j \mid D_i \geq j, h_{ij}, y_{ij}, \psi)] = \psi_0 + \psi_1 y_{i,j-1} + \psi_2 y_{ij}. \tag{12.7.16}
$$

More general models can easily be constructed by including the complete history $h_{ij} = (y_{i1}, \dots, y_{i;j-1})$, as well as external covariates, in the above conditional dropout model. Note also that, strictly speaking, we could allow dropout at a specific occasion to be related to all future responses as well. However, this is rather counter-intuitive in many cases. Moreover, including future outcomes seriously complicates the calculations since computation of the likelihood (12.7.14) then requires evaluation of a possibly high-dimensional integral. Note also that special cases of model (12.7.16) are obtained from setting $\psi_2 = 0$ or $\psi_1 = \psi_2 = 0$, respectively. In the first case, dropout is no longer allowed to depend on the current measurement, implying MAR. In the second case, dropout is independent of the outcome, which corresponds to MCAR.

Diggle and Kenward (1994) obtained parameter and precision estimates by maximum likelihood. The likelihood involves marginalization over the unobserved outcomes Y_i^m. Practically, this involves relatively tedious and computationally demanding forms of numerical integration. This, combined with likelihood surfaces tending to be rather flat, makes the model difficult to use. These issues are related to the problems to be discussed next.

Apart from the technical difficulties encountered during parameter estimation, there are further important issues surrounding MNAR based models. Even when the measurement model (e.g., the multivariate normal model) would beyond any doubt be the choice of preference for describing the measurement process *should the data be complete*, then the analysis of the actually observed, incomplete version is, in addition, subject to further untestable modeling assumptions.

When missingness is MAR, the problems are less complex, since it has been shown that, in a likelihood or Bayesian framework, it is sufficient to analyze the observed data, without explicitly modeling the dropout process (Rubin 1976, Verbeke and Molenberghs 2000). However, the very assumption of MAR is itself untestable. Therefore, ignoring MNAR models is as little an option as blindly shifting to one particular MNAR model. A sensible compromise between considering a single MNAR model on the one hand or excluding such models from consideration on the other hand, is to study the nature of such sensitivities and, building on this knowledge, formulate ways for conducting sensitivity analyses. Indeed, a strong conclusion, arising from most sensitivity analysis work, is that MNAR models have to be approached cautiously. This was made clear by several discussants to the original paper by Diggle and Kenward (1994), in particular by Laird, Little, and Rubin. An implication is that, for example, formal tests for the null hypothesis of MAR versus the alternative of MNAR should be approached with the utmost caution, a topic studied in detail by Jansen et al. (2005). These topics are taken up further in Section 12.8.

12.7.2   Implementation of Selection Models in SAS

We have developed a series of SAS programs to implement a special case of the Diggle-Kenward approach. For the measurement process, we considered a more specific case of model (12.7.15), with various fixed effects, a random intercept, and allowing Gaussian serial correlation. This means the covariance matrix V_i becomes

$$V_i = dJ_n + \sigma^2 I_n + \tau^2 H_i, \tag{12.7.17}$$

where J_n is an $n \times n$ matrix with all its elements equal to 1, I_n is the $n \times n$ identity matrix, and H_i is determined through the autocorrelation function $\rho^{u_{jk}}$, with u_{jk} the Euclidean distance between t_{ij} and t_{ik}, i.e.,

$$H_i = \begin{pmatrix} 1 & \rho^{u_{12}} & \cdots & \rho^{u_{1n}} \\ \rho^{u_{12}} & 1 & \cdots & \rho^{u_{2n}} \\ \vdots & \vdots & \ddots & \vdots \\ \rho^{u_{1n}} & \rho^{u_{2n}} & \cdots & 1 \end{pmatrix}.$$

The proposed SAS implementation can easily be adapted for another form of model (12.7.15), by just changing this V_i matrix.

Program 12.7, available on the book's companion Web site, plays the central role in fitting the Diggle-Kenward model. The following arguments need to be provided to the program:

- X and Z matrices that contain all X_i and Z_i design matrices ($i = 1, \ldots, N$).
- Y vector of all Y_i response vectors ($i = 1, \ldots, N$).
- INITIAL is a vector of initial values for the parameters in the model.
- NSUB and NTIME are the number of subjects N and the number of time points n respectively.

Within Program 12.7, the INTEGR module calculates the integral over the missing data, and the LOGLIK module evaluates the log-likelihood function $L(\theta, \psi)$. Finally, the log-likelihood function is maximized. Diggle and Kenward (1994) used the simplex algorithm (Nelder and Mead, 1965) for this purpose and Program 12.7 relies on the Newton-Raphson ridge optimization method (NLPNRR module of SAS/IML) since it combines stability and speed. However, in other analyses, it may be necessary to try several other optimization methods available in SAS/IML.

The following NLPNRR call is used in the program:

```
call nlpnrr(rc,xr,"loglik",initial,opt,con);
```

Here, LOGLIK is the module of the function we want to maximize. The initial values to start the optimization method are included in the INITIAL vector. The OPT argument indicates an options vector that specifies details of the optimization process: OPT[1]=1 request maximization, and the amount of output is controlled by OPT[2]. A constraint matrix is specified in CON, defining lower and upper bounds for the parameters in the first two rows ($d > 0$, $\tau^2 > 0$ and $\sigma^2 > 0$). Finally, all optimization methods return the following results: the scalar return code, RC, and a row vector, XR. The return code indicates the reason for the termination of the optimization process. A positive return code indicates successful termination, whereas a negative one indicates unsuccessful termination. That is, the result in the XR vector is unreliable. The XR vector contains the optimal point when the return code is positive.

Next, the program also calls the NLPFDD module, which is a subroutine that approximates derivatives by finite differences method,

```
call nlpfdd(maxlik,grad,hessian,"loglik",est);
```

Here, again LOGLIK is the module of the log-likelihood function. The vector that defines the point at which the functions and derivatives should be computed is EST. This module computes the function values MAXLIK (which is in this case the maximum likelihood, since EST is the maximum likelihood estimate) the gradient vector GRAD, and the Hessian matrix HESSIAN, which is needed to calculate the information matrix, and thus the standard errors STDE.

Analysis of Exercise Study Data

The program for fitting the Diggle-Kenward model is now applied to the analysis of the exercise study data. We will fit the model under the three different missingness mechanisms, MCAR, MAR, and MNAR. Further, we will also expand the logistic regression for the dropout model by allowing it to depend on covariates.

To obtain initial values for the parameters of the measurement model, Program 12.5 fits a linear mixed model to the exercise data using PROC MIXED. We assume a linear trend within each treatment group, which implies that each profile can be described with two parameters (intercept and slope). The error matrix is chosen to be of the form (12.7.17). For subject $i = 1, \ldots, 50$ on time point $j = 1, \ldots, 4$, the model can be expressed as

$$Y_{ij} = \beta_0 + \beta_1 (\text{group}_i - 1) + \beta_2\, t_j\, (\text{group}_i - 1) + \beta_3\, t_j\, \text{group}_i + \varepsilon_{ij},$$

where $\varepsilon_i \sim N(0, V_i)$ and $V_i = dJ_4 + \sigma^2 I_4 + \tau^2 H_i$, with

$$H_i = \begin{pmatrix} 1 & \rho & \rho^2 & \rho^3 \\ \rho & 1 & \rho & \rho^2 \\ \rho^2 & \rho & 1 & \rho \\ \rho^3 & \rho^2 & \rho & 1 \end{pmatrix}.$$

The intercept for the placebo group is $\beta_0 + \beta_1$, and the intercept for the treatment group is β_0. The slopes are β_2 and β_3, respectively.

Program 12.5 Linear mixed model in the exercise study example

```
proc mixed data=exercise method=ml;
    class group id;
    model y=group time*group/s;
    repeated/type=ar(1) local subject=id;
    random intercept/type=un subject=id;
    run;
```

The parameter estimates from the fitted linear mixed model (Program 12.5) are shown in Table 12.5 and will later be used as initial values in Program 12.7.

Table 12.5 Parameter Estimates of the Linear Mixed Model Used as Initial Values in Program 12.7

Parameter	Estimate	Initial value
β_0	−0.8233	−0.82
β_1	0.1605	0.16
β_2	0.9227	0.92
β_3	1.6451	1.65
d	2.0811	2.08
τ^2	0.7912	0.79
ρ	0.4639	0.46
σ^2	0.2311	0.23

Further, initial values for the parameters of the dropout model are also needed. As mentioned before, we will fit three models in turn, under the MCAR ($\psi_1 = \psi_2 = 0$), MAR ($\psi_2 = 0$) and MNAR mechanisms, respectively. The initial values for these parameters are given in Table 12.6.

Table 12.6 Initial Values for the Parameters of the Dropout Model

	Dropout Mechanism			
Parameter	MCAR	MAR	MNAR	MNAR + Covariate
ψ_0	1	$\widehat{\psi}_{0,MCAR}$	$\widehat{\psi}_{0,MAR}$	$\widehat{\psi}_{0,MNAR}$
ψ_1		1	$\widehat{\psi}_{1,MAR}$	$\widehat{\psi}_{1,MNAR}$
ψ_2			1	$\widehat{\psi}_{2,MNAR}$
γ				1

The parameters that will later be passed to Program 12.7 (X and Z matrices, Y and INITIAL vectors, and NSUB and NTIME parameters) are created using PROC IML. Program 12.6 illustrates the process of creating these variables when the missingness mechanism is MCAR.

Program 12.6 Creating matrices, vectors and numbers necessary for Program 12.7

```
proc iml;
    use exercise;
    read all var {id group time y} into data;
    id=data[,1];
    group1=data[,2];
    group0=j(nrow(data),1,1)-group1;
    time=data[,3];
    timegroup0=time#group0;
    timegroup1=time#group1;
```

```
   intercept=j(nrow(data),1,1);
   create x var {intercept group0 timegroup0 timegroup1}; append;
   y=data[,4];
   create y var {y}; append;
   z=j(nrow(y),1,1);
   create z var {z}; append;
   beta=-0.82//0.16//0.92//1.65;
   D=2.08;
   tau2=0.79;
   rho=0.46;
   sigma2=0.23;
   psi=1;
   initial=beta//D//tau2//rho//sigma2//psi;
   create initial var {initial}; append;
   nsub=50;
   create nsub var {nsub}; append;
   ntime=4;
   create ntime var {ntime}; append;
   quit;
```

Program 12.7 fits the Diggle-Kenward model under the MCAR assumption. To save space, the complete SAS code is provided on the book's companion Web site.

Program 12.7 Diggle-Kenward model under the MCAR assumption in the exercise study example

```
proc iml;
   use x; read all into x;
   use y; read all into y;
   use z; read all into z;
   use nsub; read all into nsub;
   use ntime; read all into ntime;
   use initial; read all into initial;
   g=j(nsub,1,0);
   start integr(yd) global(psi,ecurr,vcurr,lastobs);
   ...
   finish integr;

   start loglik(parameters) global(lastobs,vcurr,ecurr,x,z,y,nsub,ntime,nrun,psi);
   ...
   finish loglik;
   opt=j(1,11,0);
   opt[1]=1;
   opt[2]=5;
   con={. . . . 0 0 -1 0 .,
        . . . . . . 1 . .};
   call nlpnrr(rc,est,"loglik",initial,opt,con);
   call nlpfdd(maxlik,grad,hessian,"loglik",est);
   inf=-hessian;
   covar=inv(inf);
   var=vecdiag(covar);
   stde=sqrt(var);
   create result var {est stde}; append;
   quit;
```

To fit the model under the mechanisms MAR and MNAR, we need to change only a few lines in Program 12.7. For the model under MAR, we replace

```
psi[1]=parameters[9];
```

with

```
psi[1:2]=parameters[9:10];
```

while, under the MNAR assumption, it is replaced by

```
psi[1:3]=parameters[9:11];
```

Further, under the MAR and MNAR assumptions, we add one or two columns of dots, respectively, to the constraints matrix (CON). Parameter estimates resulting from the model fitted under the three missingness mechanisms, together with the estimates of the ignorable analysis using PROC MIXED, are listed in Table 12.7.

Table 12.7 Results under Ignorable, MCAR, MAR, and MNAR Assumptions with and without Covariate in the Dropout Model (Exercise Study Example)

	Dropout Mechanism									
	Ignorable		MCAR		MAR		MNAR		MNAR + Cov.	
Par.	Est.	(s.e.)	Est.	(s.e.)	Est.	(s.e.)	Est.	(s.e.)	Est.	(s.e.)
β_0	−0.82	(0.39)	−0.82	(0.40)	−0.82	(0.40)	−0.83	(0.40)	−0.82	(0.40)
β_1	0.16	(0.56)	0.16	(0.56)	0.16	(0.56)	0.17	(0.56)	0.16	(0.56)
β_2	0.92	(0.10)	0.92	(0.10)	0.92	(0.10)	0.93	(0.10)	0.92	(0.10)
β_3	1.65	(0.10)	1.65	(0.10)	1.65	(0.10)	1.66	(0.11)	1.64	(0.11)
d	2.08	(0.90)	2.08	(0.91)	2.08	(0.90)	2.09	(0.85)	2.07	(0.96)
τ^2	0.79	(0.54)	0.79	(0.55)	0.79	(0.55)	0.80	(0.70)	0.79	(0.45)
ρ	0.46	(1.10)	0.46	(1.13)	0.46	(1.12)	0.44	(1.12)	0.49	(1.13)
σ^2	0.23	(1.08)	0.23	(1.11)	0.23	(1.11)	0.21	(1.24)	0.25	(1.02)
ψ_0			−2.33	(0.30)	−2.17	(0.36)	−2.42	(0.88)	−1.60	(1.14)
ψ_1					−0.10	(0.14)	−0.24	(0.43)	−0.018	(0.47)
ψ_2							0.16	(0.47)	−0.13	(0.54)
γ									−0.66	(0.79)
-2ℓ			641.77		641.23		641.11		640.44	

Analysis of Exercise Study Data (Extended Model)

Next, we extend the Diggle-Kenward model by allowing the dropout process to depend on a covariate (GROUP). Thus, instead of (12.7.16), we use the following model

$$\text{logit}\left[P(D_i = j \mid D_i \geq j, h_{ij}, y_{ij}, \psi)\right] = \psi_0 + \psi_1 y_{i,j-1} + \psi_2 y_{ij} + \gamma \, \text{group}_i. \qquad (12.7.18)$$

Program 12.7 can easily be adapted to this case. First, in the INTEGR and LOGLIK modules, we add GROUP and GROUPI as global variables. Further, in the LOGLIK module, the γ parameter is specified as well:

```
group=parameters[12];
```

and where the information on a particular patient is selected, we add:

```
groupi = xi[1,2];
```

Next, in the INTEGR module, we replace

```
g=exp(psi[1]+psi[2]*lastobs+psi[3]*yd);
```

with

```
g=exp(psi[1]+psi[2]*lastobs+psi[3]*yd+group*groupi);
```

and in the LOGLIK module,

```
 g = exp(psi[1]+yobs[j-1]*psi[2]+yobs[j]*psi[3]);
```

is replaced by

```
 g = exp(psi[1]+yobs[j-1]*psi[2]+yobs[j]*psi[3]+group*groupi);
```

Finally, as in the MNAR program, we again add a column of dots to the constraints matrix (CON). The initial values used are listed in Table 12.6. The results produced by the program are displayed in Table 12.7. The results of the measurement model should be the same under the ignorable, MCAR, and MAR assumptions. As we can see from Table 12.7, this is more or less the case, except for some of the variance components, due to slight numerical variation. Adding the covariate to the dropout model results in a deviance change of 0.67, which means the covariate is not significant (*p*-value is 0.413). A likelihood-ratio test for the MAR versus MNAR assumption ($\psi_2 = 0$ or not) will not be fully trustworthy (Jansen et al., 2005). Note that, under the MNAR assumption, the estimates for ψ_1 and ψ_2 are more or less equal, but with different signs.

12.8   Sensitivity Analysis

Sensitivity to model assumptions has been reported for about two decades (Verbeke and Molenberghs, 2000; Molenberghs and Verbeke, 2005). In an attempt to formulate an answer to these concerns, a number of authors have proposed strategies to study sensitivity.

Broadly, we could define a sensitivity analysis as one in which several statistical models are considered simultaneously and/or where a statistical model is further scrutinized using specialized tools (such as diagnostic measures). This rather loose and very general definition encompasses a wide variety of useful approaches. The simplest procedure is to fit either a selected number of (MNAR) models which are all deemed plausible or to fit one model in which a preferred (primary) analysis is supplemented with a number of variations. The extent to which conclusions (inferences) are stable across such ranges provides an indication about the belief that can be put into them. Variations to a basic model can be constructed in different ways. The most obvious strategy is to consider various dependencies of the missing data process on the outcomes and/or on covariates. Alternatively, the distributional assumptions of the models can be changed.

Several publications have proposed the use of global and local influence tools (Verbeke et al., 2001; Verbeke and Molenberghs, 2000; Molenberghs and Verbeke, 2005). An important question is this: To what exactly are the sources causing an MNAR model to provide evidence for MNAR against MAR? There is evidence enough for us to believe that a multitude of outlying aspects is responsible for an apparent MNAR mechanism (Jansen et al., 2005). But it is not necessarily believable that the (outlying) nature of the missingness mechanism in one or a few subjects is responsible for an apparent MNAR mechanism. The consequence of this is that local influence should be applied and interpreted with due caution.

Further, within the selection model framework, Baker, Rosenberger, and DerSimonian (1992) proposed a model for multivariate and longitudinal binary data, subject to nonmonotone missingness. Jansen et al. (2003) extended this model to allow for (possibly continuous) covariates, and developed a local influence strategy.

Next, classical inference procedures account for the imprecision resulting from the stochastic component of the model. Less attention is devoted to the uncertainty arising from (unplanned) incompleteness in the data, even though the majority of clinical studies suffer from incomplete follow-up. Molenberghs et al. (2001) acknowledge both the status of imprecision, due to (finite) random sampling, as well as ignorance, due to incompleteness. Further, both can be combined into uncertainty (Kenward, Molenberghs, and Goetghebeur, 2001).

Another route for sensitivity analysis is to consider pattern-mixture models as a complement to selection models. A third framework consists of so-called shared parameter models, where random effects are employed to describe the relationship between the measurement and dropout processes (Wu and Carroll, 1988; DeGruttola and Tu, 1994).

More detail on some of these procedures can be found in Molenberghs and Verbeke (2005). Let us now turn to the case of sensitivity analysis tools for selection models.

Sensitivity Analysis for Selection Models

Particularly within the selection modeling framework, there has been an increasing literature on MNAR missingness. At the same time, concern has been growing precisely about the fact that models often rest on strong assumptions and relatively little evidence from the data themselves.

A sensible compromise between blindly shifting to MNAR models or ignoring them altogether is to make them a component of a sensitivity analysis. In any case, fitting an MNAR dropout model should be subject to careful scrutiny. The modeler needs to pay attention, not only to the assumed distributional form of the model (Little, 1994b; Kenward, 1998), but also to the impact one or a few influential subjects may have on the dropout and/or measurement model parameters. Because fitting an MNAR dropout model is feasible by virtue of strong assumptions, such models are likely to pick up a wide variety of influences in the parameters describing the nonrandom part of the dropout mechanism. Hence, a good level of caution is in place.

First, an informal sensitivity analysis is applied on the mastitis data. Next, the model of Diggle and Kenward (1994) is adapted to a form useful for sensitivity analysis, whereafter such a sensitivity analysis method, based on local influence (Cook, 1986; Thijs, Molenberghs and Verbeke, 2000), is introduced and applied to the mastitis data.

Informal Sensitivity Analysis of the Mastitis Data

Diggle and Kenward (1994) and Kenward (1998) performed several analyses of the mastitis data described in Section 12.2. In Diggle and Kenward (1994), a separate mean for each group defined by the year of first lactation and a common time effect was considered, together with an unstructured 2×2 covariance matrix. The dropout model included both Y_{i1} and Y_{i2} and was reparameterized in terms of the size variable $(Y_{i1} + Y_{i2})/2$ and the increment $Y_{i2} - Y_{i1}$. Kenward (1998) carried out what we could term a data-driven sensitivity analysis. The right panel of Figure 12.3 reveals that there appear to be two cows, #4 and #5 (black dots), with unusually large increments. Kenward conjectured that this might mean that these animals were ill during the first lactation year, producing an unusually low yield, whereas a normal yield was obtained during the second year. A simple multivariate Gaussian linear model is used to represent the marginal milk yield in the two years (i.e., the yield that would be, or was, observed in the absence of mastitis):

$$\begin{pmatrix} Y_{i1} \\ Y_{i2} \end{pmatrix} = N \left(\begin{pmatrix} \beta_0 \\ \beta_1 \end{pmatrix}, \begin{pmatrix} \sigma_1^2 & \sigma_{12} \\ \sigma_{12} & \sigma_2^2 \end{pmatrix} \right). \tag{12.8.19}$$

Note that β_1 represents the change in average yield between the two years. The probability of mastitis is assumed to follow the logistic regression model:

$$\text{logit}\left[P(\text{dropout})\right] = \psi_0 + \psi_1 y_1 + \psi_2 y_2. \tag{12.8.20}$$

The combined response/dropout model was fitted to the milk yields by using a program analogous to Program 12.7, presented in Section 12.7. In addition, the MAR model ($\psi_2 = 0$) was fitted in the same way. These fits produced the parameter estimates displayed in the All column of Table 12.8.

Using the likelihoods to compare the fit of the two models, we get a difference $G^2 = 3.12$. The corresponding tail probability from the χ_1^2 is 0.07. This test essentially examines the contribution of ψ_2 to the fit of the model. Using the Wald statistic for the same purpose gives a statistic of $(-2.52)^2/0.86 = 7.38$, with corresponding χ_1^2 probability of 0.007. The discrepancy between the results of the two tests suggests that the asymptotic approximations on which these tests are based are not very accurate in this setting, and the standard error probability underestimates the true variability of the estimate of ψ_2.

Table 12.8 Maximum Likelihood Estimates (Standard Errors) of MAR and MNAR Dropout Models under Several Deletion Schemes in the Mastitis Example

	MAR Dropout		
Parameter	All	(7,53,54,66,69,70)	(4,5)
β_0	5.77 (0.09)	5.65 (0.09)	5.81 (0.09)
β_1	0.71 (0.11)	0.67 (0.10)	0.64 (0.09)
σ_1^2	0.87 (0.12)	0.72 (0.10)	0.77 (0.11)
σ_{12}	0.63 (0.13)	0.44 (0.10)	0.72 (0.13)
σ_2^2	1.31 (0.20)	1.00 (0.16)	1.29 (0.20)
ψ_0	-3.33 (1.52)	-4.03 (1.76)	-3.09 (1.57)
ψ_1	0.38 (0.25)	0.50 (0.30)	0.34 (0.26)
ψ_2	0	0	0
-2ℓ	624.13	552.24	574.19
	MNAR Dropout		
Parameter	All	(7,53,54,66,69,70)	(4,5)
β_0	5.77 (0.09)	5.65 (0.09)	5.81 (0.09)
β_1	0.32 (0.14)	0.36 (0.15)	0.64 (0.14)
σ_1^2	0.87 (0.12)	0.72 (0.10)	0.77 (0.11)
σ_{12}	0.55 (0.13)	0.38 (0.11)	0.72 (0.13)
σ_2^2	1.57 (0.28)	1.16 (0.23)	1.29 (0.20)
ψ_0	-0.34 (2.33)	-0.55 (2.57)	-3.10 (1.74)
ψ_1	2.36 (0.79)	2.01 (0.82)	0.32 (0.72)
ψ_2	-2.52 (0.86)	-2.09 (0.98)	0.01 (0.72)
-2ℓ	624.13	551.54	574.19
G^2 for MNAR	3.12	0.70	0.0004

The dropout model estimated from the MNAR setting is as follows:

$$\text{logit}\left[P(\text{mastitis})\right] = -0.34 + 2.36y_1 - 2.52y_2.$$

Some insight into this fitted model can be obtained by rewriting it in terms of the milk yield totals ($Y_1 + Y_2$) and increments ($Y_2 - Y_1$):

$$\text{logit}\left[P(\text{mastitis})\right] = -0.34 - 0.078(y_1 + y_2) - 2.438(y_2 - y_1).$$

The probability of mastitis increases with larger negative increments, i.e., animals that showed (or would have shown) a greater decrease in yield over the two years have a higher probability of getting mastitis. The other differences in parameter estimates between the two models are consistent with this: the MNAR dropout model predicts a smaller average increment in yield (β_1), with larger second year variance and smaller correlation caused by greater negative imputed differences between yields.

12.8.1   Local Influence

The local influence approach, suggested by Cook (1986), can be used to investigate the effect of extending an MAR model for dropout in the direction of MNAR dropout (Verbeke et al., 2001).

Again we consider the Diggle and Kenward (1994) model described in Section 12.7.1 for continuous longitudinal data subject to dropout. Since no data would be observed otherwise, we assume that the first measurement Y_{i1} is obtained for every subject in the study. We denote the probability of dropout at occasion k, given the subject was still in the study up to occasion k by $g(h_{ik}, y_{ik})$. For the dropout process, we now consider an extension of model (12.7.16), which can be written as

$$\text{logit}\,[g(h_{ik}, y_{ik})] = \text{logit}\,[\mathrm{P}(D_i = k | D_i \geq k, y_i)] = h_{ik}\psi + \omega y_{ij}, \qquad (12.8.21)$$

in which h_{ik} is the vector containing the history H_{ik} as well as covariates. When ω equals zero and the model assumptions made are correct, the dropout model is MAR, and all parameters can be estimated using standard software since the measurement and dropout model can then be fitted separately. If $\omega \neq 0$, the dropout process is assumed to be MNAR. Now, a dropout model may be found to be MNAR solely because one or a few influential subjects have driven the analysis. To investigate sensitivity of estimation of quantities of interest, such as treatment effect, growth parameters, or the dropout model parameters, with respect to assumptions about the dropout model, we consider the following perturbed version of (12.8.21):

$$\text{logit}\,[g(h_{ik}, y_{ik})] = \text{logit}\,[\mathrm{P}(D_i = k | D_i \geq k, y_i, W_i)] = h_{ik}\psi + \omega_i y_{ik}, \quad i = 1, \ldots, N. \quad (12.8.22)$$

There is a fundamental difference with model (12.8.21) since the ω_i should not be viewed as parameters: they are local, individual-specific perturbations around a null model. In our case, the null model will be the MAR model, corresponding to setting $\omega = 0$ in (12.8.21). Thus the ω_i are perturbations that will be used only to derive influence measures (Cook, 1986).

This scheme enables studying the effect of how small perturbation in the MNAR direction can have a large impact on key features of the model. Practically, one way of doing this is to construct local influence measures (Cook, 1986). Clearly, not all possible forms of impact resulting from sensitivity to dropout model assumptions will be found in this way, and the method proposed here should be viewed as one component of a sensitivity analysis (e.g. Molenberghs, Kenward, and Goetghebeur, 2001).

When small perturbations in a specific ω_i lead to relatively large differences in the model parameters, it suggests that the subject is likely to drive the conclusions.

Cook (1986) suggests that more confidence can be put in a model which is relatively stable under small modifications. The best known perturbation schemes are based on case deletion (Cook and Weisberg 1982) in which the study of interest is the effect of completely removing cases from the analysis. A quite different paradigm is the local influence approach where the investigation concentrates on how the results of an analysis are changed under small perturbations of the model. In the framework of the linear mixed model Beckman, Nachtsheim, and Cook (1987) used local influence to assess the effect of perturbing the error variances, the random-effects variances, and the response vector. In the same context,

Lesaffre and Verbeke (1998) have shown that the local influence approach is also useful for the detection of influential subjects in a longitudinal data analysis. Moreover, since the resulting influence diagnostics can be expressed analytically, they often can be decomposed in interpretable components, which yield additional insights in the reasons why some subjects are more influential than others.

We are interested in the influence of MNAR dropout on the parameters of interest. This can be done in a meaningful way by considering (12.8.22) as the dropout model. Indeed, $\omega_i = 0$ for all i corresponds to an MAR process, which cannot influence the measurement model parameters. When small perturbations in a specific ω_i lead to relatively large differences in the model parameters, then this suggests that these subjects may have a large impact on the final analysis. However, even though we may be tempted to conclude that such subjects drop out non-randomly, this conclusion is misguided since we are not aiming to detect (groups of) subjects that drop out non-randomly but rather subjects that have a considerable impact on the dropout and measurement model parameters. Indeed, a key observation is that a subject that drives the conclusions towards MNAR may be doing so, not only because its true data-generating mechanism is of an MNAR type, but also for a wide variety of other reasons, such as an unusual mean profile or autocorrelation structure. Earlier analyses have shown that this may indeed be the case. Likewise, it is possible that subjects, deviating from the bulk of the data because they are generated under MNAR, go undetected by this technique. This reinforces the concept that we must reflect carefully upon which anomalous features are typically detected and which ones typically go unnoticed.

Key Concepts

Let us now introduce the key concepts of local influence. We denote the log-likelihood function corresponding to model (12.8.22) by

$$\ell(\gamma|\omega) = \sum_{i=1}^{N} \ell_i(\gamma|\omega_i),$$

in which $\ell_i(\gamma|\omega_i)$ is the contribution of the ith individual to the log-likelihood, and where $\gamma = (\theta, \psi)$ is the s-dimensional vector, grouping the parameters of the measurement model and the dropout model, not including the $N \times 1$ vector $\omega = (\omega_1, \omega_2, \dots, \omega_N)'$ of weights defining the perturbation of the MAR model. It is assumed that ω belongs to an open subset Ω of $I\!R^N$. For ω equal to $\omega_0 = (0, 0, \dots, 0)'$, $\ell(\gamma|\omega_0)$ is the log-likelihood function which corresponds to a MAR dropout model.

Let $\widehat{\gamma}$ be the maximum likelihood estimator for γ, obtained by maximizing $\ell(\gamma|\omega_0)$, and let $\widehat{\gamma}_\omega$ denote the maximum likelihood estimator for γ under $\ell(\gamma|\omega)$. The local influence approach now compares $\widehat{\gamma}_\omega$ with $\widehat{\gamma}$. Similar estimates indicate that the parameter estimates are robust with respect to perturbations of the MAR model in the direction of non-random dropout. Strongly different estimates suggest that the estimation procedure is highly sensitive to such perturbations, which suggests that the choice between an MAR model and a non-random dropout model highly affects the results of the analysis. Cook (1986) proposed to measure the distance between $\widehat{\gamma}_\omega$ and $\widehat{\gamma}$ by the so-called likelihood displacement, defined by

$$LD(\omega) = 2[\ell(\widehat{\gamma}|\omega_0) - \ell(\widehat{\gamma}_\omega|\omega_0)].$$

This takes into account the variability of $\widehat{\gamma}$. Indeed, $LD(\omega)$ will be large if $\ell(\gamma|\omega_0)$ is strongly curved at $\widehat{\gamma}$, which means that γ is estimated with high precision, and small otherwise. Therefore, a graph of $LD(\omega)$ versus ω contains essential information on the influence of perturbations. It is useful to view this graph as the geometric surface formed by the values of the $N + 1$ dimensional vector $\xi(\omega) = (\omega', LD(\omega))'$ as ω varies throughout Ω.

Since this influence graph can be depicted only when $N = 2$, Cook (1986) proposed to look at local influence, i.e., at the normal curvatures C_h of $\xi(\omega)$ in ω_0, in the direction of some N dimensional vector h of unit length. Let Δ_i be the s dimensional vector defined by

$$\Delta_i = \left.\frac{\partial^2 \ell_i(\gamma|\omega_i)}{\partial \omega_i \partial \gamma}\right|_{\gamma=\widehat{\gamma}, \omega_i=0}$$

and define Δ as the $(s \times N)$ matrix with Δ_i as its ith column. Further, let $\ddot{L}$ denote the $(s \times s)$ matrix of second order derivatives of $\ell(\gamma|\omega_0)$ with respect to γ, also evaluated at $\gamma = \widehat{\gamma}$. Cook (1986) has then shown that C_h can be easily calculated by

$$C_h = 2|h'\Delta'\ddot{L}^{-1}\Delta h|.$$

Obviously, C_h can be calculated for any direction h. One evident choice is the vector h_i containing one in the ith position and zero elsewhere, corresponding to the perturbation of the ith weight only. This reflects the influence of allowing the ith subject to drop out non-randomly, while the others can drop out only at random. The corresponding local influence measure, denoted by C_i, then becomes $C_i = 2|\Delta_i'\ddot{L}^{-1}\Delta_i|$. Another important direction is the direction $h_{\max}$ of maximal normal curvature $C_{\max}$. It shows how to perturb the MAR model to obtain the largest local changes in the likelihood displacement. It is readily seen that $C_{\max}$ is the largest eigenvalue of $-2\,\Delta'\ddot{L}^{-1}\Delta$, and that $h_{\max}$ is the corresponding eigenvector.

Local Influence in the Diggle-Kenward Model

As discussed in the previous section, calculation of local influence measures merely reduces to evaluation of Δ and $\ddot{L}$. Expressions for the elements of $\ddot{L}$ in case $\Sigma_i = \sigma^2 I$ are given by Lesaffre and Verbeke (1998), and can easily be extended to the more general case considered here. Further, it can be shown that the components of the columns Δ_i of Δ are given by

$$\left.\frac{\partial^2 \ell_{i\omega}}{\partial \theta \partial \omega_i}\right|_{\omega_i=0} = 0, \quad \left.\frac{\partial^2 \ell_{i\omega}}{\partial \psi \partial \omega_i}\right|_{\omega_i=0} = -\sum_{j=2}^{n_i} h_{ij} y_{ij} g(h_{ij})[1 - g(h_{ij})], \tag{12.8.23}$$

for complete sequences (no dropout) and by

$$\left.\frac{\partial^2 \ell_{i\omega}}{\partial \theta \partial \omega_i}\right|_{\omega_i=0} = [1 - g(h_{id})]\frac{\partial \lambda(y_{id}|h_{id})}{\partial \theta}, \tag{12.8.24}$$

$$\left.\frac{\partial^2 \ell_{i\omega}}{\partial \psi \partial \omega_i}\right|_{\omega_i=0} = -\sum_{j=2}^{d-1} h_{ij} y_{ij} g(h_{ij})[1 - g(h_{ij})]$$
$$- h_{id}\lambda(y_{id}|h_{id})g(h_{id})[1 - g(h_{id})]. \tag{12.8.25}$$

for incomplete sequences. All the above expressions are evaluated at $\widehat{\gamma}$, and where $g(h_{ij}) = g(h_{ij}, y_{ij})|_{\omega_i=0}$, is the MAR version of the dropout model.

Let $V_{i,11}$ be the predicted covariance matrix for the observed vector $(y_{i1}, \ldots, y_{i,d-1})'$, $V_{i,22}$ is the predicted variance for the missing observation y_{id}, and $V_{i,12}$ is the vector of predicted covariances between the elements of the observed vector and the missing observation. It then follows from the linear mixed model (12.7.15) that the conditional expectation for the observation at dropout, given the history, equals

$$\lambda(y_{id}|h_{id}) = \lambda(y_{id}) + V_{i,12}V_{i,11}^{-1}[h_{id} - \lambda(h_{id})], \tag{12.8.26}$$

which is used in (12.8.24).

The derivatives of (12.8.26) with respect to the measurement model parameters are

$$\frac{\partial \lambda(y_{id}|h_{id})}{\partial \beta} = x_{id} - V_{i,12}V_{i,11}^{-1}X_{i,(d-1)}, \tag{12.8.27}$$

$$\frac{\partial \lambda(y_{id}|h_{id})}{\partial \alpha} = \left[\frac{\partial V_{i,12}}{\partial \alpha} - V_{i,12}V_{i,11}^{-1}\frac{\partial V_{i,11}}{\partial \alpha}\right] V_{i,11}^{-1}[h_{id} - \lambda(h_{id})] \tag{12.8.28}$$

where x'_{id} is the dth row of X_i, and where $X_{i,(d-1)}$ indicates the first $(d-1)$ rows X_i. Further, α indicates the subvector of covariance parameters within the vector θ.

In practice, the parameter θ in the measurement model is often of primary interest. Since $\ddot{L}$ is block-diagonal with blocks $\ddot{L}(\theta)$ and $\ddot{L}(\psi)$, we can write for any unit vector h, $C_h = C_h(\theta) + C_h(\psi)$. It now immediately follows from (12.8.24) that influence on θ only arises from those measurement occasions at which dropout occurs. In particular, from expression (12.8.24), it is clear that the corresponding contribution is large only if (1) the dropout probability is small but the subject disappears nevertheless and (2) the conditional mean strongly depends on the parameter of interest. This implies that complete sequences cannot be influential in the strict sense ($C_i(\theta) = 0$) and that incomplete sequences only contribute only at the actual dropout time.

Implementation of Local Influence Analysis in SAS

Program 12.8 relies on SAS/IML to calculate the normal curvature C_h of $\xi(\omega)$ in ω_0, in the direction the unit vector h.

As in Program 12.7, we first need to create the X and Z matrices, Y and INITIAL vectors, and NSUB and NTIME parameters (this can be accomplished in PROC IML using a program analogous to Program 12.6). The initial parameters should be the estimates of the parameters of the MAR model, fitted using a program analogous to Program 12.7. Next, we need the INTEGR and LOGLIK modules introduced in Program 12.7. These are called to calculate the log-likelihood function of the Diggle and Kenward (1994) model under the MNAR assumption. This is needed for the evaluation of Δ as well as for $\ddot{L}$. Program 12.8 also calls the DELTA module to calculate the Δ vector, whereas $\ddot{L}$ is calculated using the NLPFDD module of SAS/IML which was introduced earlier in Section 12.7.2. Finally, Program 12.8 created the C_MATRIX data set that contains the following normal curvatures in the direction of the unit vector h_i containing one in the ith position and zero elsewhere,

$$\text{C} = C_i, \quad \text{C1} = C_i(\beta), \quad \text{C2} = C_i(\alpha), \quad \text{C12} = C_i(\theta), \quad \text{C3} = C_h(\psi),$$

and the normal curvature in the direction of HMAX= $h_{\max}$ of maximal normal curvature CMAX= $C_{\max}$. The C_MATRIX data set can now be used to picture the local influence measures. The complete SAS code of Program 12.8 is provided on the book's companion Web site.

Program 12.8 Local influence sensitivity analysis

```
proc iml;
    use x; read all into x;
    use y; read all into y;
    use nsub; read all into nsub;
    use ntime; read all into ntime;
    use initial; read all into initial;
    g=j(nsub,1,0);
    start integr(yd) global(psi,ecurr,vcurr,lastobs);
    ...
    finish integr;
```

13.2.1   Inter-Rater Reliability

A measure is said to have *inter-rater reliability* if ratings by different individuals at the same time on the same subject are similar. Inter-rater reliability is not only assessed when validating a newly created instrument, but is also used prior to beginning a clinical trial to quantify the reliability of the measure as used by the raters participating in the trial. In the latter case, these evaluation sessions are used to both train and quantify inter-rater reliability, even when a scale with established reliability is being used.

13.2.2   Kappa Statistic

The *Kappa statistic* (κ) is the most commonly used tool for quantifying the agreement between raters when using a dichotomous or nominal scale, while the intraclass correlation coefficient (ICC) or concordance correlation coefficient (CCC) are the recommended statistics for continuous measures. The use of κ will be discussed in this section while examples of the use of ICC and CCC are presented later (see Sections 13.2.5 and 13.2.7).

The κ statistic is the degree of agreement above and beyond chance agreement normalized by the degree of attainable agreement above what would be predicted by chance. This measure is superior to simple percent agreement as it accounts for chance agreement. For instance, if two raters diagnose a set of 100 patients for the presence or absence of a rare disease, they would have a high percentage of agreement even if they each randomly chose one subject to classify as having the disease and classified all others as not having the disease. However, the high agreement would likely not result in a high value for κ.

Table 13.1 Notation Used in the Definition of the κ Statistic

	Rater B		
Rater 1	Yes	No	Total
Yes	n_{11}	n_{12}	$n_{1.}$
No	n_{21}	n_{22}	$n_{2.}$
Total	$n_{.1}$	$n_{.2}$	n

To define κ, consider the notation for a 2×2 table displayed in Table 13.1 and let $p_{ij} = n_{ij}/n$, $p_{i.} = n_{i.}/n$ and $p_{.j} = n_{.j}/n$. The value of κ can be estimated by using the observed cell proportions as follows:

$$\kappa = \frac{p_a - p_c}{1 - p_c},$$

where p_a is given by $p_a = \sum_i p_{ii}$ and represents the observed proportion of agreement in these data. Further, p_c is defined as $p_c = \sum_i p_{i.}p_{.i}$ and represents the proportion of agreement expected by chance. Confidence intervals can be computed using the asymptotic standard deviation of κ provided by Fleiss et al. (1969):

$$s_\kappa = \left[\frac{A + B - C}{(1 - p_c)^2 n} \right],$$

where

$$A = \sum_i p_{ii}[1 - (p_{i.} + p_{.i})(1 - \kappa)]^2,$$

$$B = (1 - \kappa)^2 \sum_{i \neq j} p_{ij}(p_{i.} + p_{.i})^2,$$

$$C = [\kappa - p_c(1 - \kappa)]^2.$$

Fleiss (1981) provided the variance of κ under the null hypothesis ($\kappa = 0$) which can be used to construct a large sample test statistic:

$$s_{\kappa 0} = \frac{1}{\sqrt{n}(1 - p_c)} \left[p_c + p_c^2 - \sum_i p_{i.} p_{.i} (p_{i.} + p_{.i}) \right]^{1/2}.$$

For exact inference calculations, the observed results are compared with the set of all possible tables with the same marginal row and column scores. See Agresti (1992) or Mehta and Patel (1983) for details on computing exact tests for contingency table data.

EXAMPLE: Schizophrenia diagnosis example

The simplest example of inter-rater reliability study is when two raters rate a group of subjects as to whether certain criteria (e.g., diagnosis) are met using a diagnostic instrument. The κ statistic is the traditional approach to assessing interrater reliability in such circumstances. PROC FREQ provides for easy computation of the κ statistic, with inferences based on either large sample assumptions or an exact test.

Bartko (1976) provided an example where two raters classified subjects as to whether they had a diagnosis of schizophrenia. These data, modified for this example, are displayed in Table 13.2 and will be used to demonstrate the computation of the κ statistic in SAS.

Table 13.2 Schizophrenia Diagnosis Example Data Set ($N = 30$)

Rater A	Rater B Yes	No
Yes	5	3
No	1	21

Program 13.1 uses the FREQ procedure to compute the κ statistic in the schizophrenia diagnosis example.

Program 13.1 Computation of the κ statistic

```
data schiz;
   input rater_a $ rater_b $ num;
   cards;
   yes yes   5
   yes no    3
   no  yes   1
   no  no   21
   ;
proc freq data=schiz;
   weight num;
   tables rater_a*rater_b;
   exact agree;
   run;
```

Output from Program 13.1

```
            Simple Kappa Coefficient

      Kappa (K)              0.6296
      ASE                    0.1665
      95% Lower Conf Limit   0.3033
      95% Upper Conf Limit   0.9559
```

```
                    Test of H0: Kappa = 0

        ASE under H0                    0.1794
        Z                              3.5093
        One-sided Pr >  Z              0.0002
        Two-sided Pr > |Z|             0.0004


        Exact Test
        One-sided Pr >=  K             0.0021
        Two-sided Pr >= |K|            0.0021


        Sample Size = 30
```

Note that in this example the observed agreement is $p_{11} + p_{22} = 26/30 = 86.7\%$. However, a 64% agreement can be expected by chance given the observed marginals, i.e.,

$$\frac{1}{N} \sum_i n_i. n_{.i} = 19.2/30 = 64\%.$$

Output 13.1 shows that the estimated κ is 0.6296 with a 95% confidence interval of $(0.3033, 0.9559)$. This level of agreement (in the range of 0.4 to 0.8) is considered moderate, while κ values of 0.80 and higher indicate excellent agreement (Stokes et al., 2000). Output 13.1 also provides for testing the null hypothesis of $\kappa = 0$. By including the EXACT AGREE statement in Program 13.1, we obtained both large sample and exact inferences. Both tests indicate the observed agreement is greater than chance ($p = 0.0004$ using the z-statistic and $p = 0.0021$ using the exact test).

Two extensions of the κ measure of agreement, the weighted κ and stratified κ, are also easily estimated using PROC FREQ. A weighted κ is applied to tables larger than 2 by 2 and allows different weight values to be placed on different levels of disagreement. For example, on a 3-point scale, i.e., $(0, 1, 2)$, a weighted κ would penalize cases where the raters differed by 2 points more than cases where the difference was one. Both Cicchetti and Allison (1971) and Fleiss and Cohen (1973) have published weight values for use in computing a κ statistic. Stokes et al. (2000) provides an example of using SAS to compute a weighted κ statistic.

When reliability is to be assessed across several strata, a weighted average of strata specific κ's can be produced. SAS uses the inverse variance weighting scheme. See the PROC FREQ documentation for details.

The κ statistic has been criticized because it is a function of not only the sensitivity (proportion of cases identified or true positive rate) and specificity (proportion of non-cases correctly identified) of an instrument, but also of the overall prevalence of cases (base rate). For instance, Grove et al. (1981) pointed out that with sensitivity and specificity held constant at 0.95, a change in the prevalence can greatly influence the value of κ. Spitznagel and Helzer (1985) noted that a diagnostic procedure showing a high κ in a clinical trial with a high prevalence will have a lower value of κ in a population based study, even when the sensitivity and specificity are the same. Thus, comparisons of agreement across different studies where the base rate (prevalence) differs are problematic with the κ statistic.

13.2.3   Proportion of Positive Agreement and Proportion of Negative Agreement

One approach to overcoming this drawback with the κ statistic is to use a pair of statistics: the proportion of positive agreement, p_+, and the proportion of negative agreement, p_-. Consider the situation where two raters classify a set of patients as having or not having a specific disease. The first measure, p_+, is the conditional probability that a rater will give a positive diagnosis given that the other rater has given a positive diagnosis. Similarly, p_- is the conditional probability that a rater will give a negative diagnosis given that the other rater has given a negative diagnosis. These conditional probabilities are defined as follows

$$p_+ = 2p_{11}/(2p_{11} + p_{12} + p_{21}), \quad p_- = 2p_{22}/(2p_{22} + p_{12} + p_{21})$$

and are analogous to characterizing a diagnostic test's sensitivity (the proportion of patients with a disease correctly identified) and specificity (the proportion of patients without the disease correctly identified). Note that p_+ can be viewed as a weighted average of the sensitivities computed considering each rater as the gold standard separately, and p_- can be viewed as a weighted average of the specificities computed considering each rater as the gold standard separately. In addition, sensitivity and specificity are not even applicable in situations where there is no gold standard test. However, p_+ and p_- are still useful statistics in such situations, e.g., p_+ is the conditional probability that a rater will give a positive diagnosis given that the other rater has given a positive diagnosis.

Program 13.2 demonstrates how to compute p_+, p_-, and associated confidence intervals in the schizophrenia diagnosis example. The confidence intervals are computed using formulas based on the delta method provided by Graham and Bull (1998).

Program 13.2 Computation of the proportion of positive agreement, proportion of negative agreement, and associated confidence intervals

```
data agreement;
    set schiz nobs=m;
    format p_pos p_neg lower_p lower_n upper_p upper_n 5.3;
    if rater_a='yes' and rater_b='yes' then d=num;
    if rater_a='yes' and rater_b='no'  then c=num;
    if rater_a='no'  and rater_b='yes' then b=num;
    if rater_a='no'  and rater_b='no'  then a=num;
    n=a+b+c+d;
    p1=a/n; p2=b/n; p3=c/n; p4=d/n;
    p_pos=2*d/(2*d+b+c);
    p_neg=2*a/(2*a+b+c);
    phi1=(2/(2*p4+p2+p3))-(4*p4/((2*p4+p2+p3)**2));
    phi2_3=-2*p4/(((2*p4+p2+p3)**2));
    gam2_3=-2*p1/(((2*p1+p2+p3)**2));
    gam4=(2/(2*p1+p2+p3))-(4*p1/((2*p1+p2+p3)**2));
    sum1=(phi1*p4+phi2_3*(p2+p3))**2;
    sum2=(gam4*p1+gam2_3*(p2+p3))**2;
    var_pos=(1/n)*(p4*phi1*phi1+(p2+p3)*phi2_3*phi2_3-4*sum1);
    var_neg=(1/n)*(p1*gam4*gam4+(p2+p3)*gam2_3*gam2_3-4*sum2);
    lower_p=p_pos-1.96*sqrt(var_pos);
    lower_n=p_neg-1.96*sqrt(var_neg);
    upper_p=p_pos+1.96*sqrt(var_pos);
    upper_n=p_neg+1.96*sqrt(var_neg);
    retain a b c d;
    label p_pos='Proportion of positive agreement'
          p_neg='Proportion of negative agreement'
          lower_p='Lower 95% confidence limit'
          lower_n='Lower 95% confidence limit'
          upper_p='Upper 95% confidence limit'
          upper_n='Upper 95% confidence limit';
    if _n_=m;
```

```
proc print data=agreement noobs label;
    var p_pos lower_p upper_p;
proc print data=agreement noobs label;
    var p_neg lower_n upper_n;
    run;
```

Output from Program 13.2

Proportion of positive agreement	Lower 95% confidence limit	Upper 95% confidence limit
0.714	0.446	0.983

Proportion of negative agreement	Lower 95% confidence limit	Upper 95% confidence limit
0.913	0.828	0.998

It follows from Output 13.2 that in the schizophrenia diagnosis example we observed 71.4% agreement on positive ratings and 91.3% agreement on negative ratings. Using the Graham-Bull approach with these data, we compute the following 95% confidence intervals: $(0.446, 0.983)$ for agreement on positive ratings, and $(0.828, 0.998)$ for agreement on negative ratings.

13.2.4   Inter-Rater Reliability: Multiple Raters

Fleiss (1971) proposed a generalized κ statistic for situations with multiple (more than two) raters for a group of subjects. It follows the form of κ above, where percent agreement is now the probability that a randomly chosen pair of raters will give the same diagnosis or rating.

EXAMPLE: Psychiatric diagnosis example

Sandifer et al. (1968) provided data from a study in which six psychiatrists provided diagnoses for each of 30 subjects. Subjects were diagnosed with either depression, personality disorder, schizophrenia, neurosis, or classified as not meeting any of the diagnoses.

Program 13.3 computes the expected agreement (p_c), percent agreement (p_a), kappa statistic (κ), and a large sample test statistic for testing the null hypothesis of no agreement beyond chance in the psychiatric diagnosis example. These statistics were calculated using the formulas provided by Fleiss (1971):

$$p_c = \sum_i p_i^2, \quad p_a = \frac{1}{Nr(r-1)} \left[\sum_{ij} n_{ij}^2 - Nr \right], \quad \kappa = \frac{p_a - p_c}{1 - p_c},$$

$$\text{Var}(\kappa) = \frac{2}{Nr(r-1)} \frac{\sum_i p_i^2 - (2r-3) \left[\sum_i p_i^2\right]^2 + 2(r-2)\sum_i p_i^3}{\left[1 - \sum_i p_i^2\right]^2},$$

where N is the total number of subjects, r is the number of raters, and the subscripts i and j refer to the subject and diagnostic category, respectively.

Program 13.3 Computation of agreement measures

```
data diagnosis;
    input patient depr pers schz neur othr @@;
    datalines;
1  0 0 0 6 0 2  0 3 0 0 3 3  0 1 4 0 1 4  0 0 0 0 6 5  0 3 0 3 0
6  2 0 4 0 0 7  0 0 4 0 2 8  2 0 3 1 0 9  2 0 0 4 0 10 0 0 0 0 6
11 1 0 0 5 0 12 1 1 0 4 0 13 0 3 3 0 0 14 1 0 0 5 0 15 0 2 0 3 1
16 0 0 5 0 1 17 3 0 0 1 2 18 5 1 0 0 0 19 0 2 0 4 0 20 1 0 2 0 3
21 0 0 0 0 6 22 0 1 0 5 0 23 0 2 0 1 3 24 2 0 0 4 0 25 1 0 0 4 1
26 0 5 0 1 0 27 4 0 0 0 2 28 0 2 0 4 0 29 1 0 5 0 0 30 0 0 0 0 6
;
%let nr=6; * Number of raters;
data agreement;
    set diagnosis nobs=m;
    format pa pc kappa kappa_lower kappa_upper 5.3;
    sm=depr+pers+schz+neur+othr;
    sum0+depr; sum1+pers; sum2+schz; sum3+neur; sum4+othr;
    pi=((depr**2)+(pers**2)+(schz**2)+(neur**2)+(othr**2)-sm)/(sm*(sm-1));
    tot+((depr**2)+(pers**2)+(schz**2)+(neur**2)+(othr**2));
    smtot+sm;
    if _n_=m then do;
        avgsum=smtot/m;
        n=sum0+sum1+sum2+sum3+sum4;
        pdepr=sum0/n; ppers=sum1/n; pschz=sum2/n; pneur=sum3/n; pothr=sum4/n;
        * Percent agreement;
        pa=(tot-(m*avgsum))/(m*avgsum*(avgsum-1));
        * Expected agreement;
        pc=(pdepr**2)+(ppers**2)+(pschz**2)+(pneur**2)+(pothr**2);
        pc3=(pdepr**3)+(ppers**3)+(pschz**3)+(pneur**3)+(pothr**3);
        * Kappa statistic;
        kappa=(pa-pc)/(1-pc);
    end;
    * Variance of the kappa statistic;
    kappa_var=(2/(n*&nr*(&nr-1)))*(pc-(2*&nr-3)*(pc**2)+2*(&nr-2)*pc3)/((1-pc)**2);
    * 95% confidence limits for the kappa statistic;
    kappa_lower=kappa-1.96*sqrt(kappa_var);
    kappa_upper=kappa+1.96*sqrt(kappa_var);
    label pa='Percent agreement'
          pc='Expected agreement'
          kappa='Kappa'
          kappa_lower='Lower 95% confidence limit'
          kappa_upper='Upper 95% confidence limit';
    if _n_=m;
proc print data=agreement noobs label;
    var pa pc;
proc print data=agreement noobs label;
    var kappa kappa_lower kappa_upper;
    run;
```

Output from Program 13.3

	Percent agreement	Expected agreement
	0.556	0.220

Kappa	Lower 95% confidence limit	Upper 95% confidence limit
0.430	0.408	0.452

Output 13.3 lists the percent agreement ($p_a = 0.556$), expected agreement ($p_c = 0.22$), and kappa ($\kappa = 0.43$). It also shows the 95% confidence limits for κ. The lower limit is well above 0 and thus there is strong evidence that agreement in the psychiatric diagnosis example is in excess of chance. As there is evidence for agreement beyond chance, as a follow-up analysis one may be interested in assessing agreement for specific diagnoses. Fleiss (1971) describes such analyses—computing the conditional probability that a randomly selected rater assigns a patient with diagnosis j, given that a first randomly selected rater chose diagnosis j. This is analogous to the use of p_+ described in the previous section.

13.2.5   Inter-Rater Reliability: Rating Scale Data

The *inter-rater reliability* of a rating scale is a measure of the consistency of ratings from different raters at the same point in time. The intra-class correlation coefficient (ICC) or the concordance correlation coefficient (CCC) are the recommended statistics for quantifying inter-rater reliability when using continuous measures. Shrout et al. (1979) presented multiple forms of the ICC along with discussions of when to use various forms. Here we present the statistics for estimating ICCs from a two-way ANOVA model:

$$\text{Model outcome} = \text{Subjects} + \text{Raters}.$$

We will consider the case when randomly selected raters rate each subject (ICC1) and also the case when the selected raters are the only ones of interest (ICC2). ICC1 may be of interest when it can be assumed the raters in a study represent a random sample of a population for which the researchers would like to generalize to. ICC2 may be used in rater training sessions prior to a clinical trial with pre-selected raters. The two coefficients are defined as follows:

$$\text{ICC1} = \frac{M_s - M_e}{M_s + (r-1)M_e + r(M_r - M_e)/n},$$

$$\text{ICC2} = \frac{M_s - M_e}{M_s + (r-1)M_e},$$

where r is the number of raters, n is the number of subjects, and M_s, M_r, and M_e are the mean squares for subjects, raters, and error, respectively.

Shrout et al. (1979) also provides formulas for estimating the variances for each form of the ICC and a discussion of a one-way ICC when different sets of raters score each subject. The null hypothesis of a zero ICC can be tested with an F-test given by $F = M_s/M_e$ with $(n-1)$ and $(r-1)(n-1)$ degrees of freedom. As the ICC statistic and the associated tests and confidence intervals are based on summary statistics produced in a typical ANOVA, these can be easily computed using either the GLM or MIXED procedure. The SAS library of sample programs provides a macro that computes all versions of the ICC per Shrout's formulas as well as others:

`http://ftp.sas.com/techsup/download/stat/intracc.html`

Further discussion of the ICC and related statistics is provided later in Section 13.2.7.

EXAMPLE: Rater training session example

We will now describe an example of a rater training session performed at the beginning of a clinical trial. While in the previous examples a small number of raters assessed many subjects, in this example many raters assess only a small number of subjects (often one or two). This situation is more commonplace in clinical trials. That is, a rater training session is a good example of this setting.

Prior to implementing an antidepressant clinical trial, Demitrack et al. (1998) assessed the inter-rater reliability of 85 raters for the Hamilton Rating Scale for Depression (HAMD) (Hamilton, 1960) as part of a rater training session. The HAMD total score is a measure of the severity of a patient's depression with higher scores indicating more severe depression. The main goal of the session was to establish consistency in the administration and interpretation of the scale across multiple investigational sites. From a statistical perspective, the goals were to reduce variability (and thus increase power) by training raters to score using the scale in a similar fashion and to identify potential outlier raters prior to the study.

Raters independently scored the HAMD from a videotaped interview of a depressed patient. Fleiss' generalized κ, as described in the previous section, was used to summarize the inter-rater reliability for this group of raters. As only one subject was assessed, agreement was computed across items of a scale rather than across subjects. The percent agreement now represents the probability that two randomly selected raters provided the same rating score for a randomly selected item. The observed agreement and κ for these data are 0.55 and 0.4.

While an overall summary of agreement provides information about the scale and raters, it is also of importance in a multi-site clinical trial to identify raters who are outliers. These raters have the potential to increase variability in the outcome variable and thus reduce the power to detect treatment differences. Using the mode of the group of raters as the gold standard, Demitrack et al. (1998) computed the percent agreement and inter-class correlation coefficient for each rater (relative to the gold standard). Rater ICCs ranged from 0.40 to 1.0, with only two raters scoring below 0.70. Percentage agreement was also assessed by item in order to identify areas of disagreement for follow-up discussion.

Rater training sessions followed the discussion of the initial videotape results, and the session concluded with raters scoring a videotape of the same patient in an improved condition. Demitrack et al. concluded that the training did not clearly improve the agreement of the group, though they were able to use the summary agreement statistics to identify raters who did not score the HAMD consistent with the remainder of the group prior to implementing the trial. While Demitrack et al. used percent agreement and ICC in their work, the average squared deviation (from the gold standard) is another statistic that has been proposed for use in summarizing inter-rater reliability data (Channon and Butler, 1998).

13.2.6   Internal Consistency

Measures of internal consistency assess the degree to which each item of a rating scale measures the same construct. Cronbach's α, a measure of the average correlation of items within a scale, is a commonly used statistic to assess internal consistency (Cronbach, 1951):

$$\alpha = \frac{k}{k-1}\left[1 - \frac{\sum_i V_i}{V}\right],$$

where k is the number of items, V_i is the variance of the ith item, and V is the variance of the total score.

Cronbach's α ranges from 0 to 1, with higher scores suggesting good internal consistency. It is considered to be a lower bound of the scale's reliability (Shrout, 1998). Perrin suggested 0.7 as a minimal standard for internal consistency when making group comparisons, with a higher standard of at least 0.90 when individual comparisons are to be used (Perrin et al., 1997). Cronbach's α is a function of the average inter-item correlation and the number of items. It increases as the number of related items in the scale increases. If α is very low, the scale is either too short or the items have very little in common (delete items which do not correlate with the others). On the other hand, extremely high scores may indicate some redundancy in the rating scale. Cronbach's α can be computed using the CORR procedure.

EXAMPLE: EESC rating scale example

Kratochvil et al. (2004) reported on the development of a new rating scale with 29 items to quantify expression and emotion in children (EESC). Data from an interim analysis of a validation study involving 99 parents of children with attention-deficit hyperactivity disorder are used here to illustrate the assessment of internal consistency.

Program 13.4 uses PROC CORR for computing internal consistency and item-to-total correlations in the EESC rating scale example. The EESC data set used in the program can be found on the book's companion Web site.

Program 13.4 Computation of Cronbach's α

```
proc corr data=eesc alpha;
    var eesc1 eesc2 eesc3 eesc4 eesc5 eesc6 eesc7 eesc8 eesc9 eesc10
        eesc11 eesc12 eesc13 eesc14 eesc15 eesc16 eesc17 eesc18 eesc19 eesc20
        eesc21 eesc22 eesc23 eesc24 eesc25 eesc26 eesc27 eesc28 eesc29;
    run;
```

Output from Program 13.4

```
                    Cronbach Coefficient Alpha

              Variables                  Alpha

              Raw                      0.900146
              Standardized             0.906284


        Cronbach Coefficient Alpha with Deleted Variable

            Raw Variables                 Standardized Variables

Deleted     Correlation                   Correlation
Variable    with Total      Alpha         with Total      Alpha

EESC1       0.291580      0.899585        0.281775      0.906622
EESC2       0.545988      0.895571        0.564959      0.901673
EESC3       0.591867      0.894167        0.593887      0.901158
EESC4       0.672391      0.892861        0.672410      0.899748
EESC5       0.247457      0.901908        0.240618      0.907326
EESC6       0.458779      0.896995        0.448555      0.903730
EESC7       0.478879      0.896641        0.472999      0.903300
```

EESC8	0.665566	0.893793	0.675876	0.899685
EESC9	0.505510	0.896072	0.498409	0.902853
EESC10	0.651696	0.894064	0.652332	0.900110
EESC11	0.666862	0.893457	0.677098	0.899663
EESC12	0.380065	0.898422	0.395765	0.904652
EESC13	0.368793	0.898586	0.363621	0.905211
EESC14	0.623236	0.893495	0.614713	0.900785
EESC15	0.392515	0.898165	0.407936	0.904440
EESC16	0.389488	0.898200	0.396298	0.904643
EESC17	0.624384	0.894600	0.637290	0.900380
EESC18	0.485442	0.896484	0.480872	0.903162
EESC19	-.015219	0.907669	-.000957	0.911384
EESC20	0.707713	0.891929	0.710875	0.899052
EESC21	0.550139	0.895331	0.540735	0.902104
EESC22	0.639571	0.894357	0.649545	0.900160
EESC23	0.436383	0.897693	0.435663	0.903956
EESC24	0.423187	0.897806	0.424149	0.904157
EESC25	0.294496	0.901074	0.292439	0.906439
EESC26	0.313600	0.899842	0.315250	0.906047
EESC27	0.514513	0.896137	0.512612	0.902602
EESC28	0.362475	0.898664	0.367987	0.905135
EESC29	0.509206	0.896090	0.508525	0.902674

Output 13.4 provides estimates of Cronbach's α using both raw data and items standardized to have a standard deviation of 1.0. The standardized scores may be useful when there are large differences in item variances. The results indicate an acceptable level of internal consistency ($\alpha > 0.90$).

The output also lists the correlation of each item to the remaining total score, as the value of Cronbach's α with each individual item deleted from the scale, as well as the correlation of each item score with the remaining total score. The item-to-total correlations were at least moderate for all items except number 19. The low item-to-total for item 19 led to further investigation. It was found that one item ("My child shows a range of emotions") was being interpreted in both a positive and negative manner by different parents, leading to the poor correlation. Thus, this item was removed in the revised version of the scale.

Despite the acceptable properties, it may still be of interest to consider a more concise scale, one that would be less burdensome on the respondent. From Output 13.3, the removal of any single item would not bring about a substantial change in Cronbach's α. Inter-item correlations (also produced by Program 13.3 but not included in the output to save space) could be examined to make scale improvements in this situation.

The PROC CORR documentation contains another example of the computation of Cronbach's α.

13.2.7 Test-Retest Reliability

Test-retest reliability indicates the stability of ratings on the same individual at two different times while in the same condition. For test-retest assessments, the retest ratings should be taken long enough after the original assessment to avoid recall bias and soon enough to limit the possibility of changes in the condition of the subject. For continuous measures, Pearson's correlation coefficient is often used to quantify the relationship between two measures. However, as this captures only the correlation between measures, and not whether the measures are systematically different, the ICC or CCC are preferred measures for quantifying test-retest reliability. For instance, if the second (retest) measurement is always 2 points higher than the first, Pearson's correlation coefficient will be 1 while the ICC and CCC will be reduced due to the systematic differences. Considering these data as a scatterplot with the x and y axis representing the first and second measurements on the subjects, Pearson's correlation coefficient assesses deviations from the best fitting regression line. The CCC is a measure that assesses departures from the line of identity (a 45 degree line). Thus, systematic departures will reduce the CCC. The closely related ICC statistic is the proportion of the total variability in scores that is due to the variability among subjects. Thus, an ICC close to 1 indicates that little of the variability in scores is due to rater differences, and that thus the test-retest reliability is excellent.

The (2-way) ICC and CCC statistics are given by (Shrout and Fleiss, 1979; Lin et al., 2002)

$$\text{ICC} = \frac{M_s - M_e}{M_s + (m-1)M_e}, \quad \text{CCC} = \frac{2rs_1s_2}{s_1^2 + s_2^2 + (\bar{y}_1 + \bar{y}_2)^2},$$

where M_s and M_e are the mean squares for subjects and error, respectively, r is Pearson's correlation coefficient between observations at the first and second time points, i.e., Y_{i1} and Y_{i2}, and s_1^2 and s_2^2 are the sample variances of the outcomes at the first and second time points. The 95% confidence limits for ICC are given below

$$\left(\frac{F_L - 1}{F_L + k - 1}, \frac{F_U - 1}{F_U + k - 1} \right),$$

where k is the number of occasions,

$$F_L = F_0 / F_{1-\alpha/2}[n-1, (n-1)(k-1)],$$
$$F_U = F_0 F_{1-\alpha/2}[(n-1)(k-1), n-1]$$

and $F_0 = M_s / M_e$.

Lin (1989) demonstrated that

$$Z = \frac{1}{2} \ln \frac{1 - CCC}{1 + CCC}$$

is asymptotically normal with mean

$$\frac{1}{2} \ln \frac{1 - C}{1 + C}$$

and variance

$$\frac{1}{n-2} \left[\frac{(1 - r^2)C^2}{r^2(1 - C^2)} + \frac{4C^3(1 - C)\mu^2}{r(1 - C^2)^2} - \frac{2C^4\mu^4}{r^2(1 - C^2)^2} \right],$$

where C is the true concordance correlation coefficient, r is Pearson's correlation coefficient defined above, and $\mu = (\mu_1 - \mu_2)/\sqrt{\sigma_1\sigma_2}$. Confidence intervals can be formed by replacing the parameters with their estimates.

An alternative testing approach to assessing inter-rater reliability is provided by the Bradley-Blackwood test. This approach simultaneously tests the null hypothesis of equal means and variances at the first and second assessments (Sanchez and Binkowitz, 1999). The Bradley-Blackwood test statistic B, which has an F distribution with 2 and $n-2$ degrees of freedom, is as follows:

$$B = \frac{1}{2M_r} \left[\sum_i D_i^2 - S_r \right],$$

where $D_i = Y_{i1} - Y_{i2}$, and S_r and M_r are the error sum of squares and mean square error of the regression of D_i on $A_i = (Y_{i1} + Y_{i2})/2$.

EXAMPLE: Sickness inventory profile example

Deyo et al. (1991) provided data on the sickness inventory profile (SIP) from a clinical trial for chronic lower back pain. Data were obtained from 34 patients whose symptoms were deemed unchanged in severity, with measures taken two weeks apart.

Program 13.5 computes the ICC and associated large sample 95% confidence limits. As the ICC is computed from mean squares in an ANOVA model, it is easily computable using output from PROC MIXED as demonstrated below.

Program 13.5 Computation of the ICC, CCC and Bradley-Blackwood test

```
options ls=70;
data inventory;
    input patient time sip @@;
    datalines;
1 1  9.5  9 1  7.3 17 1  4.0 25 1 10.1 33 1  4.6
1 2 12.3  9 2  6.1 17 2  3.6 25 2 15.7 33 2  4.3
2 1  3.6 10 1  0.5 18 1  6.6 26 1 10.2 34 1  4.6
2 2  0.4 10 2  1.0 18 2  7.0 26 2 15.2 34 2  4.3
3 1 17.4 11 1  6.0 19 1 13.8 27 1  9.5
3 2  7.4 11 2  3.1 19 2 10.1 27 2  8.2
4 1  3.3 12 1 31.6 20 1  9.8 28 1 19.1
4 2  2.2 12 2 16.6 20 2  8.3 28 2 21.9
5 1 13.4 13 1    0 21 1  4.8 29 1  5.6
5 2  6.0 13 2  2.3 21 2  2.9 29 2 10.6
6 1  4.1 14 1 25.0 22 1  0.9 30 1 10.7
6 2  3.5 14 2 12.2 22 2  0.4 30 2 13.1
7 1  9.9 15 1  3.4 23 1  8.0 31 1  8.6
7 2  9.9 15 2  0.7 23 2  2.8 31 2  6.1
8 1 11.3 16 1  2.8 24 1  2.7 32 1  7.5
8 2 10.6 16 2  1.7 24 2  3.8 32 2  5.2
    ;
* Intraclass correlation coefficient;
proc sort data=inventory;
    by patient time;
proc mixed data=inventory method=type3;
    class patient time;
    model sip=patient time;
    ods output type3=mstat1;
data mstat2;
    set mstat1;
    dumm=1;
    if source='Residual' then do; mserr=ms; dferr=df; end;
    if source='patient' then do; mspat=ms; dfpat=df; end;
    if source='time' then do; mstime=ms; dftime=df; end;
    retain mserr dferr mspat dfpat mstime dftime;
    keep mserr dferr mspat dfpat mstime dftime dumm;
data mstat3;
    set mstat2;
    by dumm;
    format icc lower upper 5.3;
    if last.dumm;
    icc=(mspat-mserr)/(mspat+(dftime*mserr));
    fl=(mspat/mserr)/finv(0.975,dfpat,dfpat*dftime);
    fu=(mspat/mserr)*finv(0.975,dfpat*dftime,dfpat);
    lower=(fl-1)/(fl+dftime);
    upper=(fu-1)/(fu+dftime);
```

```
          label icc='ICC'
                lower='Lower 95% confidence limit'
                upper='Upper 95% confidence limit';
proc print data=mstat3 noobs label;
     var icc lower upper;
     run;
* Concordance correlation coefficient;
data transpose;
     set inventory;
     by patient;
     retain base;
     if first.patient=1 then base=sip;
     if last.patient=1 then post=sip;
     diff=base-post;
     if last.patient=1;
proc means data=transpose noprint;
     var base post diff;
     output out=cccout mean=mn_base mn_post mn_diff var=var_base var_post var_diff;
data cccout;
     set cccout;
     format ccc 5.3;
     ccc=(var_base+var_post-var_diff)/(var_base+var_post+mn_diff**2);
     label ccc='CCC';
proc print data=cccout noobs label;
     var ccc;
     run;
* Bradley-Blackwood test;
data bb;
     set transpose;
     avg=(base+post)/2; difsq=diff**2; sdifsq+difsq;
     dum=1;
     run;
proc sort data=bb;
     by dum;
data bbs;
     set bb;
     by dum;
     if last.dum;
     keep sdifsq;
proc mixed data=bb method=type3;
     model diff=avg;
     ods output type3=bb1;
data bb1;
     set bb1;
     if source='Residual';
     mse=ms; sse=ss; dfe=df;
     keep mse sse dfe;
data test;
     merge bb1 bbs;
     format f p 5.3;
     f=0.5*(sdifsq-sse)/mse;
     p=1-probf(f,2,dfe+2);
     label f='F statistic'
           p='P-value';
proc print data=test noobs label;
     var f p;
     run;
```

Output from Program 13.5

ICC	Lower 95% confidence limit	Upper 95% confidence limit
0.727	0.520	0.854

CCC
0.706

F statistic	P-value
4.194	0.024

Output 13.5 shows that the ICC for the data collected in the sickness inventory profile study is 0.727, with a 95% confidence interval of $(0.520, 0.854)$. Landis and Koch (1977) indicate that ICCs above 0.60 suggest satisfactory stability, and ICCs greater than 0.80 correspond to excellent stability. For reference, the CCC based on these data is 0.706, very similar to the ICC. See Deyo (1991) for reformulations of the formula for computing the CCC which shows the similarity between the CCC and ICC.

In addition, Output 13.5 lists the F statistic and p-value of the Bradley-Blackwood test. The test rejects the null hypothesis of equal means and variances ($F = 4.194$, $p = 0.024$). Thus, the Bradley-Blackwood test would seem to be contradictory to the relatively high ICC value. However, one must consider that the various statistics are differentially sensitive to the strength of the linear relationship, location, and scale shifts between the first and second measurement. Sanchez et al. (1999) discussed these issues in depth and performed a simulation study with multiple reliability measures. They conclude that the CCC comes the closest among all the measures to satisfying all the components of a good test-retest reliability measure. Regardless, they recommend computing estimates of scale and location shifts along with the CCC.

13.3 Validity and Other Topics

A measure is said to be *valid* if it accurately reflects the concept it is intended to measure. Assessing the validity of a measure typically includes quantifying convergent validity, divergent validity, and discriminant validity. These three topics, which comprise the assessment of the construct validity of an instrument, can be addressed in a quantitative fashion and will be discussed below.

Content validity is another topic which is typically addressed in the development stage of a measurement tool. *Content validity* refers to the degree to which a measure covers the range of meanings that are part of the concept to be measured. The most common approaches to assessing content validity are expert reviews of the clarity, comprehensiveness, and redundancy of the measurement tool. Content validity will not be discussed further in this chapter.

However, responsiveness, factor analysis, and minimal clinically relevant differences are three additional topics that are discussed. Responsiveness and factor structure of a measurement tool are two important practical concepts to assess when validating an instrument. Establishing a minimal clinically relevant difference addresses a common need for applied researchers designing studies using and interpreting data from studies with a measurement tool.

13.3.1 Convergent and Divergent Validity

Convergent validity is established by showing a strong relationship between the scale under review and another validated scale thought to measure the same construct. *Divergent validity* is the lack of correlation between the scale under review and scales thought to assess different constructs. Pearson's correlation coefficient is the most commonly used statistic to quantify convergent and divergent validity. Thus, SAS can be easily used to summarize convergent and divergent validity through the CORR procedure. A Pearson's correlation of at least 0.4 has been used (Cappelleri et al., 2004) as evidence for convergent validity. Similarly, correlations less than 0.3 indicate evidence for divergent validity, with correlations between 0.3 and 0.4 taken as no evidence to establish or dismiss convergent or divergent validity. Confidence intervals are useful here to provide information beyond simple point estimates.

Cappelleri et al. (2004) provides an example of the assessment of convergent and divergent validity in assessing a new scale for self-esteem and relationships (SEAR) in patients with erectile dysfunction. As part of the assessment of convergent/divergent validity, they included a global quality of life scale (SF-36) in a validation study. They hypothesized low correlations between the Confidence domain of the SEAR with physical factors of the SF-36 and thus used these comparisons to assess divergent validity. Similarly for convergent validity, the developers hypothesized higher correlations with SF-36 mental health domains. Table 13.3 presents the results.

Table 13.3 Correlation of SF-36 Components with SEAR Confidence Domain

SF-36 component	Correlation
Physical Functioning	0.30
Role-Physical	0.30
Bodily Pain	0.32
Mental Health	0.44
Role-Emotional	0.45
Mental Component Summary	0.45

The correlations greater than 0.4 with the mental health measures were taken as evidence for convergent validity while the correlations with physical domains were borderline evidence (at least no evidence to dismiss) for divergent validity.

13.3.2 Discriminant Validity

Discriminant validity indicates the ability of a scale to distinguish different groups of subjects. An instrument for assessing the severity of a disease should clearly be able to distinguish between subjects with and without the corresponding diagnosis. Often, however, many disease symptoms overlap with other diseases, such as patients diagnosed with depression versus anxiety disorder. Thus in establishing validity of a measure for a particular disease, ability to distinguish between a disease state with overlapping symptoms may be extremely important. Note that there is some inconsistency in the literature regarding the use of the term discriminant validity—with references defining this in the same meaning as divergent validity above (for example, Guyatt and Feeny, 1991). However, in this discussion, discriminant validity is defined as the ability to distinguish between relevant subject groups.

Thurber et al. (2002) assessed the discriminant validity of the Zung self-rating depression scale in 259 individuals referred to a state vocational rehabilitation service. The Zung scale was administered as part of a test battery, including the depression subscale of the Minnesota Multiphasic Personality Inventory-2 (MMPI-2), following a diagnostic interview. Using logistic regression and stepwise logistic regression (such as the LOGISTIC procedure), they demonstrated that the Zung scale was a predictor (the strongest) of depressed versus non-depressed individuals, as well as between individuals diagnosed with depression or substance abuse. The Zung scale was also predictive of a diagnosis of depression even after the forced inclusion of the MMPI-2 depression subscale into the model.

Once the scale has been demonstrated to be predictive of a diagnosis, the ability of the scale to predict the diagnosis for individual subject is often assessed using receiver operator characteristic (ROC) curves. The ROC is a graph of sensitivity versus (1-specificity) for various cutoff scores. For these data, based on a ROC curve, Thurber chose a cutoff score of 60 on the Zung scale for identifying subjects with and without a diagnosis of depression. With this cutoff score, sensitivity (the proportion of depressed subjects correctly identified) was 0.57, while specificity (the proportion of subjects without depression who were correctly identified) was 0.83. See Deyo et al. (1991) for a further discussion of ROC and an example using changes in SIP scores (see Section 13.2.7) to predict improvement status.

13.3.3 Responsiveness

Responsiveness is the ability of an instrument to detect small but clinically important changes. Responsiveness is often referred to as *sensitivity to change* and is often viewed as part of construct validity. Because the main purpose of a clinical trial is to detect a treatment effect, it is important to assess the responsiveness of a scale prior to using it in a clinical trial. A scale that is not responsive may not be able to detect important treatment changes and therefore mislead the experimenter to conclude no treatment effect. The methods discussed above are predominantly *point in time* analyses (except for test-retest reliability which focuses on stability of scores) and do not fully demonstrate that an instrument would be effective for a clinical trial designed to detect differences in change scores. The standardized response mean (the mean change divided by the standard deviation in change scores) is a common unitless statistic for summarizing responsiveness (Stratford et al., 1996). Change scores may be summarized following an intervention expected to produce a clinically relevant change. When a control is available, effect sizes for an effective intervention relative to the control may be compared using multiple measures.

For example, Faries et al. (2001) assessed the responsiveness of the ADHD rating scale when administered by trained clinicians. The scale had been previously validated as a parent scored tool. As no control group was available, the SRM was used to compare the changes observed on the new version of the scale with other validated instruments. Results showed the observed SRM for the clinician scored scale (1.21) was in the range of SRMs observed from other clinician and parent measures (1.13 to 1.40). As the SRM is based on simple summary statistics, PROC UNIVARIATE provided the information necessary for the computation of the SRM.

Responsiveness of an instrument may also be assessed using correlations with a global improvement scale. The global improvement question usually asks the subjects to rate the improvement on their condition on an ordinal scale. An example of such scale would be "very much better", "much better", "a little better", "no change", "a little worse", "much worse" and "very much worse". Mean changes in the instrument scores are calculated for subjects in each global scale response category regardless of the treatment. A monotone increasing (or decreasing, depending on the direction of improvement) function of the mean scores is a desirable property for a scale that is responsive to treatment. An analysis of variance (ANOVA) can be performed to test the mean differences in the mean instrument scores among subjects in different global scale response categories. The model should include the change in the instrument scores as the dependent variables and the global scale response categories as the class variable. A responsive scale should be able to discriminate reasonably well between the response categories of a global rating scale.

13.3.4 Identifying Subscales

It is often desired to identify subscales of a multi-item questionnaire in addition to its overall total score. *Factor analysis* is a technique commonly used to explore the existence of such subscales. It describes the covariance relationships among many items in terms of a few underlying, but unobservable variables called factors. Each factor may correspond to a subscale. If items can be grouped by their correlations and all items within a particular group are highly correlated among themselves but have relatively small correlations with items in a different group, it is feasible that each group of items represents a single underlying construct, or a subscale.

One approach is to come up with an a priori subscale structure proposed by the experts in that field and test the appropriateness of that structure by using the confirmatory factor analysis. In this case, the experts should identify the subscales first and allocate the items to each subscale based on their opinion. Then a confirmatory factor analysis model can be used to test the fit of the model. In a confirmatory factor analysis model, the researcher must have an idea about the number of factors and know which items load on which factors. The model parameters are defined accordingly prior to fitting the model. For example if an item is hypothesized to load on a specific factor, the corresponding factor loading will be estimated and the loadings corresponding to this item on the other factors will be set to zero. After estimating the parameters, the fit of this model is tested to assess the appropriateness of the model. Fitting a confirmatory factor analysis model requires some more detailed knowledge about factor analysis. Details about confirmatory factor analysis and using the CALIS procedure to fit the model can be found in Hatcher (1994).

Another way is to use an exploratory factor analysis to identify the number of subscales and the items that will be allocated to each subscale. As opposed to the confirmatory factor analysis, the researcher is usually unsure about the number of factors and how the items will load on the factors. After identifying the subscales, the experts should approve the face validity of the subscales. Face validity is not validity in technical sense. It is not concerned with what the test actually measures, but what it appears superficially to measure. For example, a psychiatrist could review the items loaded on a "Sleep Problems" subscale of a depression symptom questionnaire to see if any of those items appear to be related to measuring sleep disturbances on a depressed patient.

Exploratory factor analysis usually involves two stages. The first is to identify the number of factors and estimate the model parameters. There are several methods of estimation of the model parameters. The most commonly used are the Principal Component and Maximum Likelihood methods. As initial factors are typically difficult to interpret, a second stage of rotation makes the final result more interpretable. A factor rotation is carried out to look for a pattern of loadings such that each item loads highly on a single factor (subscale) and has small to moderate loadings on the remaining factors. This is called *simple structure*. Orthogonal and oblique factor rotations are two types of transformations may be needed to achieve the simple structure. Orthogonal rotations are appropriate for a factor model in which the common factors are assumed to be uncorrelated and oblique to be correlated. The type of transformation (orthogonal versus oblique) can be decided by using a graphical examination of factor loadings (whether a rigid or nonrigid transformation make the factor loadings close to the axes). For detailed information about factor analysis, see Johnson and Wichern (1982) and Hatcher (1994).

EXAMPLE: Incontinence Quality of Life questionnaire

Incontinence Quality of Life questionnaire (I-QOL) is a validated incontinence-specific quality of life questionnaire that includes 22 items (Wagner et al., 1996; Patrick et al., 1999). The I-QOL yields a total score and three subscale scores: Avoidance and Limiting Behaviors, Psychosocial Impacts, and Social Embarrassment. For simplicity, we selected nine of the items in two subscales to demonstrate how an exploratory factor analysis can be used to define two of the three subscales. The data were obtained from a pre-randomization visit of a pharmacotherapy clinical trial for treatment of women with urinary incontinence.

Program 13.6 uses the FACTOR procedure to perform an exploratory factor analysis of the I-QOL data (the IQOL data set used in the program can be found on the book's companion Web site). PROC FACTOR can be used to fit a linear factor model and estimate the factor loadings. The SCREE option plots the eigenvalues associated with each factor to help identify the number of factors. An eigenvalue represents the amount of variance that is accounted for by a given factor. The scree test looks for a break or separation between the factors with relatively large eigenvalues and those with smaller eigenvalues. The factors that appear before the break are assumed to be meaningful and are retained. Figure 13.1 produced by Program 13.6 indicates that two factors were sufficient to explain most of the variability. Although the number of factors can be specified in PROC FACTOR for the initial model, this number is just an intuitive feeling and it can be changed in subsequent analyses based on empirical results. Even though the NFACT option specifies a certain number of factors, the output will still include the scree plot and the eigenvalues, with the number of factors extracted being equal to the number of variables analyzed. However, the parts of the output related to factor loadings and the rotations will include only the number of factors specified with the NFACT option.

The ROTATE and REORDER options are used to help interpret the obtained factors. In order to achieve the simple structure, the VARIMAX rotation was carried out in this example. The REORDER option reorders the variables according to their largest factor loadings. The SIMPLE option displays means, standard deviations, and the number of observations. Upon examination of the scree plots, eigenvalues, and the related factors, the number of factors in the NFACT option should be changed to the appropriate level, and then the model should be re-run. This examination may include the scree plot, eigenvalues, and the rotated factors together. In some instances, it could be desirable to keep factors with eigenvalues less than 1, if interpretation of these factors makes sense. The maximum likelihood method (METHOD=ML) is useful since it provides a chi-square test for model fit. However, as the test is a function of the sample size, for large studies the test may reject the hypothesis of sufficient number of factors due to differences that are not clinically relevant. Therefore it is recommended to use other goodness-of-fit indices (Hatcher, 1994).

Program 13.6 Subscale identification in the I-QOL example

```
proc factor data=iqol simple method=ml scree heywood reorder rotate=varimax nfact=2;
    var iqol1 iqol4 iqol5 iqol6 iqol7 iqol10 iqol11 iqol17 iqol20;
    ods output Eigenvalues=scree;
data scree;
    set scree;
    if _n_<=9;
    number=_n_;
* Vertical axis;
axis1 minor=none label=(angle=90 "Eigenvalue") order=(-1 to 11 by 2);
* Horizontal axis;
axis2 minor=none label=("Number") order=(1 to 9);
symbol1 i=none value=dot color=black height=3;
proc gplot data=scree;
    plot eigenvalue*number/vaxis=axis1 haxis=axis2 vref=0 lvref=34 frame;
    run;
    quit;
```

Figure 13.1 Scree plot in the I-QOL example

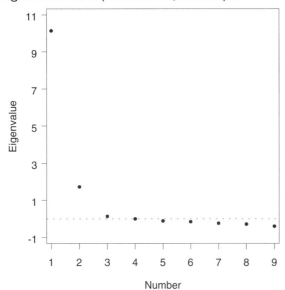

Output from Program 13.6

```
                Initial Factor Method: Maximum Likelihood

                          Factor Pattern

                                       Factor1          Factor2

        IQOL6      IQOL item 6         0.79447          0.10125
        IQOL11     IQOL item 11        0.77077         -0.16251
        IQOL17     IQOL item 17        0.76174          0.39284
        IQOL7      IQOL item 7         0.74458          0.38372
        IQOL4      IQOL item 4         0.69355         -0.46179
        IQOL5      IQOL item 5         0.69353          0.19886
        IQOL20     IQOL item 20        0.65887         -0.20804
        IQOL1      IQOL item 1         0.64754         -0.37380
        IQOL10     IQOL item 10        0.64487         -0.30917

                       Rotated Factor Pattern

                                       Factor1          Factor2

        IQOL4      IQOL item 4         0.80996          0.19553
        IQOL1      IQOL item 1         0.71412          0.22151
        IQOL10     IQOL item 10        0.66487          0.26344
        IQOL11     IQOL item 11        0.64271          0.45545
        IQOL20     IQOL item 20        0.60014          0.34238
        IQOL17     IQOL item 17        0.22891          0.82594
        IQOL7      IQOL item 7         0.22394          0.80715
        IQOL6      IQOL item 6         0.46518          0.65195
        IQOL5      IQOL item 5         0.32498          0.64414
```

Output 13.6 displays selected sections of the output produced by Program 13.6. The top portion of the output (under "Initial Factor Method: Maximum Likelihood") presents the factor loadings after fitting the model with two factors (NFACT=2). All items have high loadings on the first factor and small loadings on the second, suggesting a rotation (each factor retained should represent some of the items). The bottom portion of Output 13.6 (under "Rotated Factor Pattern") displays the factor loadings after the VARIMAX rotation. The transformed factor loading structure suggests that the first factor (subscale) should consist of Items 1, 4, 10, 11 and 20 since those items are heavily loaded on Factor 1. Similarly the second subscale should consist of Items 5, 6, 7, and 17. At this stage, the researcher should decide if the items that load on a given factor share some conceptual meaning and if the items that load on different factors seem to be measuring different constructs. In this example, the first subscale was called as Avoidance and Limiting Behaviors, and the second one was called Psychosocial Impacts in the original version of the I-QOL subscale creation.

13.3.5   Minimal Clinically Important Differences

Another important need is to determine the between- and within-treatment minimum clinically important differences (MCID) for an instrument. The MCID helps clinicians interpret the relevance of changes in the instrument scores. The within-treatment MCID is defined as the improvement in a score with treatment at which a patient recognizes that she or he is improved. The between-treatment MCID is the minimum difference between two treatments that can be considered clinically relevant. One widely accepted way to determine the MCIDs is to anchor the scale to a global rating of change scale such as the one mentioned in the responsiveness discussion. The mean change in the measure of interest for those subjects who rated as "a little better" could be considered as the within-treatment MCID. The difference in the mean changes for subjects who rated as "a little better" and who rated "no change" could be considered as the between-treatment MCID. The between-treatment MCID can be a sound choice for the treatment difference in order to power the clinical studies. These two MCIDs provide guidance to researchers to interpret the change scores for the instrument. They become critical when statistically significant differences needed to be justified as clinically relevant. The choice of a global scale is an important step in determining the MCIDs. Global scales with items that are less sensitive to change may yield larger MCIDs. For example, if the responses to a global scale of improvement are "better," "no change," or "worse," the MCIDs calculated using this scale may be larger than those calculated from the global scale in the previous example. Therefore, the differences could still be clinically important but may not necessarily be minimum using this scale.

13.4   Summary

This chapter summarizes the importance of and methods for establishing the reliability and validity of rating scales. Quality quantitative scientific research must include the use of measurement tools that have been validated for the population under study. The validation of an instrument is not a single test, but a summary of multiple psychometric properties. That is, a scale should have acceptable levels of internal consistency, test-retest reliability, inter-rater reliability, convergent, divergent and discriminant validity, as well as responsiveness.

A rating scale is said to be *reliable* if multiple measurements on a subject agree. Aspects of reliability presented in this chapter include internal consistency, test-retest reliability, and inter-rater reliability. A measure is said to be *valid* if it accurately reflects the concept it is intended to measure. Validity discussions included in this chapter include convergent, divergent, and discriminant validity, as well as responsiveness and assessment of the factor structure of the instrument.

Throughout the chapter, multiple statistical methods for assessing the various aspects of reliability and validity are presented. The statistical methods are presented along with examples from clinical trials, diagnostic studies, rater training sessions, and scale validation studies. Multiple examples include SAS code and output to facilitate understanding and ease of application of these concepts.

References

Agresti, A. (1992). "A Survey of Exact Inference for Contingency Tables." *Statistical Science.* 7, 131–177.

Bartko, J.J., Carpenter, W.T. (1976). "On the Methods and Theory of Reliability." *The Journal of Nervous and Mental Disease.* 163, 307–317.

Bech, P., Allerup, P., Gram, L.F., Reisby, N., Rosenberg, R., Jacobsen, O., Nagy, A. (1981). "The Hamilton Depression Scale: Evaluation of Objectivity Using Logistic Models." *Acta Psychiatrica Scandinavica.* 64, 290–299.

Beretvas, S.N., Pastor, D.A. (2003). "Using Mixed-Effects Models in Reliability Generalization Studies." *Educational and Psychological Measurement.* 63, 75–95.

Cappelleri, J.C., Althof, S.E., Siegel, R.L., Shpilsky, A., Bell, S.S., Duttagupta, S. (2004). "Development and Validation of the Self-Esteem and Relationship (SEAR) Questionnaire in Erectile Dysfunction." *International Journal of Impotence Research.* 16, 30–38.

Channon E., Butler A. (1998). "Comparing investigators' use of rating scales such as PANSS in multi-investigator studies of schizophrenia." *European Neuropsychopharmacology.* 8, S310–311.

Cicchetti, D.V., Allison, T. (1971). "A New Procedure for Assessing Reliability of Scoring EEG Sleep Recordings." *American Journal of EEG Technology.* 11, 101–109.

Cicchetti, D.V., Feinstein, A.R. (1990). "High Agreement But Low Kappa: II. Resolving the Paradoxes." *Journal of Clinical Epidemiology.* 43, 551–558.

Cronbach, L.J. (1951). "Coefficient alpha and the internal structure of tests." *Psychometrika.* 16, 297–334.

Demitrack, M.A., Faries, D., Herrera, J.M., DeBrota, D.J., Potter, W.Z. (1998). "The Problem of Measurement Error in Multisite Clinical Trials." *Psychopharmacology Bulletin.* 34, 19–24.

Deyo, R.A., Diehr, P., Patrick, D.L. (1991). "Reproducibility and Responsiveness of Health Status Measures: Statistics and Strategies for Evaluation." *Controlled Clinical Trials.* 12, S142–158.

Donner, A., Eliasziw, M. (1987). "Sample Size Requirements for Reliability Studies." *Statistics in Medicine.* 6, 441–448.

Evans, K.R., Sills, T., DeBrota, D.J., Gelwicks, S., Engelhardt, N., Santor, D. (2004). "An Item Response Analysis of the Hamilton Depression Rating Scale Using Shared Data From Two Pharmaceutical Companies." *Journal of Psychiatric Research.* 38, 275–284.

Faries, D.E., Yalcin, I., Harder, D., Heiligenstein, J.H. (2001). "Validation of the ADHD Rating Scale as a Clinician Administered and Scored Instrument." *Journal of Attention Disorders.* 5, 107–115.

Fleiss, J.L. (1971). "Measuring Nominal Scale Agreement Among Many Raters." *Psychological Bulletin.* 76, 378–382.

Fleiss, J.L., Cohen, J., Everitt, B.S. (1969). "Large-Sample Standard Errors of Kappa and Weighted Kappa." *Psychological Bulletin.* 72, 323–327.

Fleiss, J.L., Cohen, J. (1973). "The Equivalence of Weighted Kappa and the Intraclass Correlation Coefficient as a Measure of Reliability." *Educational and Psychological Measurement.* 33, 613–619.

Fleiss, J.L. (1981). *Statistical Methods for Rates and Proportions.* Second Edition. New York: Wiley.

Graham, P., Bull, B. (1998). "Approximate Standard Errors and Confidence Intervals for Indices of Positive and Negative Agreement." *Journal of Clinical Epidemiology.* 51, 763–771.

Grove, W.M., Andreasen, N.C., McDonald-Scott, P., Keller, M.B., Shapiro, R.W. (1981). "Reliability Studies of Psychiatric Diagnosis: Theory and Practice." *Archives of General Psychiatry.* 38, 408–413.

Guyatt, G., Patrick, D., Feeny, D. (1991). "Glossary." *Controlled Clinical Trials.* 12, 274S–280S.

Hamilton, M. (1960). "A Rating Scale for Depression." *Journal of Neurology, Neurosurgery and Psychiatry.* 23, 56–62.

Hatcher, L. (1994). *A Step-by-Step Approach to Using the SAS System for Factor Analysis and Structural Equation Modeling.* Cary, NC: SAS Institute Inc.

International Conference on Harmonization (ICH). (1998). "Statistical Principles for Clinical Trials." E-9 Document. Available at `http://www.ich.org/LOB/media/MEDIA485.pdf`.

Johnson, R.A., Wichern, D.W. (1982). *Applied Multivariate Statistical Analysis.* Englewood Cliffs, NJ: Prentice-Hall.

Kraemer, H. (1991). "To Increase Power in Randomized Clinical Trials without Increasing Sample Size." *Psychopharmacology Bulleton.* 27, 217–224.

Kratochvil, C.J., Perwien, A., Vaughan, B., Faries, D., Saylor, K., Busner, J., Buermeyer, C., Hin-Hong, W., Swindle, R. (2004). "Emotional Expression and ADHD Pharmacotherapy in Children: A New Measure." *American Academy of Child and Adolescent Psychiatry.* AACAP poster presentation.

Landis, J.R., Koch, G.G. (1977). "The Measurement of Observer Agreement for Categorical Data." *Biometrics.* 33, 159–174.

Leon, A.C., Marzuk, P.M., Portera, L. (1995). "More Reliable Outcome Measures Can Reduce Sample Size Requirements." *Archives of General Psychiatry.* 52, 867–871.

Lin, L.I. (1989). "A Concordance Correlation Coefficient to Evaluate Reproducibility." *Biometrics.* 45, 255–268.

Lin, L., Hedayat, A.S., Sinha, B., Yang, M. (2002). "Statistical Methods in Assessing Agreement: Models, Issues, and Tools." *Journal of the American Statistical Association.* 97, 257–270.

Mehta, C.R., Patel, N.R. (1983). "A Network Algorithm for Performing Fisher's Exact Test in $r \times c$ Contingency Tables." *Journal of the American Statistical Association.* 78, 427–434.

Patrick, D.L., Martin, M.L., Bushnell, D.M., Yalcin, I., Wagner, T.H., Buesching, D.P. (1999). "Quality of life for women with urinary incontinence: further development of the incontinence quality of life instrument." *Urology.* 53, 71–76.

Perrin, E.S., Aaronson, N.K., Alonso, J., Burnam, A., Lohr, K., Patrick, D.L. (1997). Instrument Review Criteria. Medical Outcomes Trust, March.

Sanchez, M.M., Binkowitz, B.S. (1999). "Guidelines for Measurement Validation in Clinical Trial Design." *Journal of Biopharmaceutical Statistics.* 9, 417–438.

Sandifer, M.G., Hordern, A.M. Timbury, G.C., Green, L.M. (1968). "Psychiatric diagnosis: A comparative study in North Carolina, London, and Glasgow." *British Journal of Psychiatry.* 114, 1–9.

Shrout, P.E. (1998). "Measurement reliability and agreement in psychiatry." *Statistical Methods in Medical Research.* 7, 301–317.

Shrout, P.E., Fleiss, J.L. (1979). "Intraclass correlations: Uses in assessing rater reliability." *Psychological Bulletin.* 86, 420–428.

Spitznagel, E.L., Helzer, J.E. (1985). "A Proposed Solution to the Base Rate Problem in the Kappa Statistic." *Archives of General Psychiatry.* 42, 725–728.

Stokes, M.E., Davis, C.S., Koch, G.G. (2000). *Categorical Data Analysis Using the SAS System, Second Edition.* Cary, NC: SAS Institute Inc.

Stratford, P.W., Binkley, J.M., Riddle, D.L. (1996). "Health Status Measures: Strategies and Analytic Methods for Assessing Change Scores." *Physical Therapy.* 76, 1109–1123.

Thurber, S., Snow M., Honts, C.R. (2002). "The Zung Self-Rating Depression Scale—Convergent Validity and Diagnostic Discrimination." *Assessment.* 9, 401–405.

Wagner, T.H., Patrick, D.L., Bavendam, T.G., Martin, M.L., Buesching, D.P. (1996). "Quality of life of persons with urinary incontinence: development of a new measure." *Urology.* 47, 67–72.

Chapter 14

Decision Analysis in Drug Development

Carl-Fredrik Burman
Andy Grieve
Stephen Senn

Decision analysis is a quantitative methodology for analyzing and optimizing decisions. It has its roots in statistical decision theory as outlined by Raiffa and Schlaifer (1961). The basic idea is to structure the decision problem, quantify uncertainties and preferences, and then solve the resulting optimality problem. In this chapter, we will present some typical decision problems in drug development: go/no go decisions, sample size calculations, dose-finding studies, sequential designs, and project prioritization. The examples will be displayed and analyzed using the dedicated Operations Research module SAS/OR, together with optimization routines in SAS/IML, as well as the statistician's standard SAS arsenal.

14.1 Introduction

We all make thousands of decisions every day. Decision-making is part of most areas of life. Consequently, decision analysis has applications in many areas of society, business, and, not least, medicine. Within the medical fields, decision analysis has, quite naturally, most often been used in order to find the optimal strategy for diagnosing and treating patients

Carl-Fredrik Burman is Statistical Science Director, Technical and Scientific Development, AstraZeneca, Sweden. Andy Grieve is Head, Statistical Research and Consulting Center, Pfizer, United Kingdom. Stephen Senn is Professor, Department of Statistics, University of Glasgow, United Kingdom.

(e.g., Parmigiani, 2002, Chapters 5–6; Bertolli et al., 2003; Liu et al., 2003). The analysis is often made for a population although attempts have been made to also include the preferences of the individual patient (Protheroe et al., 2000; Elwyn et al., 2001). In health economics, the benefits of a treatment are weighed against its costs, and these components are often embedded in a decision analysis model (Cooper et al., 2002; Brown et al., 2003).

Whereas there are plenty of decision analytic applications to medical decision-making and cost-effectiveness analysis in the literature, less has been written about such applications to the core drug development process. One reason is probably the confidentiality of internal company decisions. However, as identified in the FDA's recent Critical Path Initiative (FDA, 2004, 2006), drug development faces large and growing problems, including increased development costs and fewer new drugs that reach the market. One part of the solution to the industry's productivity problem, we believe, is through model-based approaches (FDA, 2004, p. 24) coupled with a decision analysis (Poland and Wada, 2001; Burman et al., 2005). This chapter will give a few examples of decision analysis applications from different parts of drug development, often promoting but sometimes warning against their use. Special attention will be given to the use of decision analysis during clinical development since the use of clinical trials is not only one of the cornerstones of drug development, it is also something that adds a special structure to the problems and thereby differentiates the decision problems in clinical development from those in other parts of science and business.

It is important to recognize that there are many stakeholders in the process of developing a drug and distributing it to the patients in need. It is interesting to compare what is optimal for different stakeholders. Of course, we ought to have a system where industry investments, regulations, evaluations by health care providers, patent laws, guidelines, prescription habits, etc., all serve the good of the patients. The decisions of different agents obviously interact with each other. This chapter will mainly have an industry perspective. However, now and then we will touch upon some other perspectives and the interaction between the interests of different stakeholders.

14.1.1 Outline

We will begin with an introductory example considering a very simplified clinical program (Section 14.2). This example will introduce some of the ideas and notions. In Section 14.3, we will describe the fundamentals of classical statistical decision theory. We will then be ready to describe a number of different applications of decision analysis in drug development:

- Go/no go problems (Section 14.4).
- Sample size calculation from a perspective of maximizing company profits (Section 14.5).
- Sequential design of clinical trials (Section 14.6).
- Finding the best dose in a dose-response study (Section 14.7).
- Project prioritization (Section 14.8).

14.2 Introductory Example: Stop or Go?

Decision trees are very useful for analyzing decision problems. The DTREE procedure can be used to construct, plot, and evaluate such decision trees. In some cases there is an enormous number, or even a continuum, of decision alternatives available. Such problems can often be attacked with optimization routines and/or simulations (see Sections 14.5 and 14.7). In this section, we will introduce a very simple decision tree example, which will be expanded in Sections 14.4 and 14.5.

A classic decision problem is whether to invest or not. In a drug development setting, the question might be if a large and expensive Phase III program should be launched or if the development of the drug should be discontinued. The consequences of a discontinuation are fairly easy to assess. There is, however, considerable uncertainty in the results of a Phase III program, comparing a new drug candidate with the current standard therapy. To simplify things, the result might be that the new drug is regarded to be either superior, non-inferior (but not superior), or inferior to the active control. The new drug cannot be marketed in case of inferiority, as regulatory approval will then not be given. Superior or non-inferior efficacy results will both lead to marketing. However, the regulatory labeling will be different and the market appreciation and sales are likely to be widely different.

Program 14.1 uses PROC DTREE in SAS/OR to create a decision tree for the described go/no go problem. The STAGE1 data set describes the structure of the tree. Each of the five rows in this data set corresponds to a branch, and the four variables define tree characterisics such as node types and branch labels. For example, the _STTYPE_ variable assumes two values, D and C, that identify the decision and chance nodes, respectively. The decision node is a decision to conduct the study or not and the chance node is a (random) outcome of the trial (superiority, non-inferiority, or inferiority). The _OUTCOME_ variable specifies labels for the five branches in this decision tree.

Program 14.1 Simple go/no go problem with an efficacy outcome variable

```
data stage1;
    length _outcome_ $6.;
    input _stname_ $ _sttype_ $ _outcome_ $ _success_ $;
    datalines;
    Decision    D   No_go   .
    .           .   Go      Develop
    Develop     C   Super   .
    .           .   Noninf  .
    .           .   Infer   .
    ;
* Trial's outcome;
symbol1 value=triangle height=10 color=black width=3 line=1;
* Decision point;
symbol2 value=square height=10 color=black width=3 line=1;
* End nodes;
symbol3 value=none height=10 color=black width=3 line=1;
proc dtree stagein=stage1;
    treeplot/graphics norc nolegend
    linka=1 linkb=2 symbold=2 symbolc=1 symbole=3;
    run;
    quit;
```

The decision tree generated by Program 14.1 is displayed in Figure 14.1. This decision tree is almost the simplest one that is meaningful. In general, decision trees that describe the clinical development of a drug involve multiple decisions and have a lot more decision and/or chance nodes. We will consider decision trees arising in clinical trials further in Section 14.4. Here we will refine the tree somewhat on the consequence side.

Efficacy of a drug is not everything. Safety is also important. Assume, for example, that the competitor drug is connected to a specific type of adverse events (AE). Our new drug may or may not have the advantage of a considerably lower rate of these AEs, depending on whether the same biological pathway leading to the safety problem is triggered. We will ignore the possibility of getting a higher rate of the AE, as it is assumed that the safety problem is zero-one. It is, however, straightforward to include a possibility of a safety disadvantage in the analysis.

Figure 14.1 Decision tree in the simple go/no go problem. The square node is the decision to conduct the trial or not and the triangle node is the trial's efficacy outcome (superiority, non-inferiority, or inferiority).

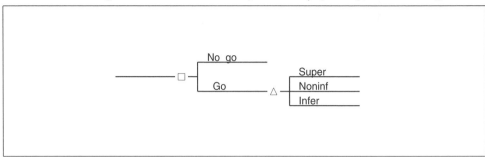

Program 14.2 produces a decision tree (Figure 14.2) which also reflects the outcome of the adverse event comparison in the Phase III trial. The STAGE2 data set defines seven branches and three nodes:

- DECISION (decision node): A decision to conduct the trial or not.
- DEVELOP (chance node): Outcome of the efficacy analysis (superiority, non-inferiority, or inferiority).
- AE (chance node): Outcome of the safety analysis (superiority or equivalence).

The tree does not consider the AE comparison when the experimental drug is inferior to the competitor drug in terms of efficacy because the inferiority outcome is assumed to imply that regulatory approval will not be achieved.

Program 14.2 Simple go/no go problem with multiple outcome variables (efficacy and safety)

```
data stage2;
    length _outcome_ $15.;
    input _stname_ $ _sttype_ $ _outcome_ $ _success_ $;
    datalines;
    Decision D No_go      .
    .        . Go         Develop
    Develop  C Eff_super  AE
    .        . Eff_noninf AE
    .        . Eff_infer  .
    AE       C AE_super   .
    .        . AE_equal   .
    ;
* Trial's outcome;
symbol1 value=triangle height=10 color=black width=3 line=1;
* Decision point;
symbol2 value=square height=10 color=black width=3 line=1;
* End nodes;
symbol3 value=none height=10 color=black width=3 line=1;
proc dtree stagein=stage2;
    treeplot/graphics norc nolegend
    linka=1 linkb=2 symbold=2 symbolc=1 symbole=3;
    run;
    quit;
```

Figure 14.2 Decision tree in the simple go/no go problem with two outcome variables. The square node is the decision to conduct the trial or not and the triangle nodes represent the trial's efficacy outcome (superiority, non-inferiority, or inferiority) and safety outcome (superiority or equivalence).

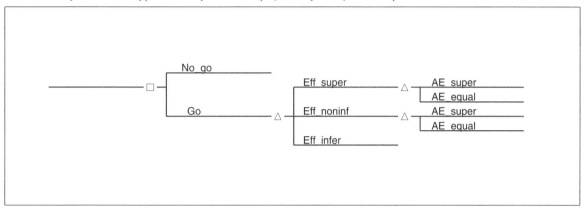

14.2.1   Evaluating a Decision Tree

Structuring a problem is an essential part in solving it. In the above example, this step was rather trivial. For a complex problem, however, it can be quite difficult to find a good description which is as simple as possible and yet captures the essence of the problem. Given the tree structure of a problem, the optimal decision will sometimes be evident. However, it is often necessary to assess the consequences and uncertainties quantitatively.

In the Phase III investment decision, let us say that we can estimate the probabilities of superiority, non-inferiority, and inferiority outcomes in a Phase III trial. This estimation of probabilities is, of course, related to power calculations and depends on a number of factors, such as the precise definition of non-inferiority, the assumed effect size, and variability, etc. We will come back to such topics later and for now take the computation of these probabilities for granted. Uncertainties in a regulatory response and sales can also be included in a decision analysis model but we will ignore these aspects for now. We will also ignore any other potential safety issues apart from the specific kind of AE under consideration.

Let E^+, $E^=$, and E^- denote the events of superiority, non-inferiority, and inferiority, respectively. Similarly, denote the event of proven safety benefit of the new drug with respect to the specified AE by S^+ and the complement event (no proven safety improvement) by $S^=$. For the following example, let

$$P(E^+) = 0.1, \quad P(E^=) = 0.4, \quad P(E^-) = 0.5.$$

We assume that the outcome for the specified AE is independent of the effect outcome and that

$$P(S^+) = 0.7, \quad P(S^=) = 0.3.$$

The independence may not, of course, be the case in practice and the joint probability of efficacy and safety may have to be considered instead. (Compare the conditioning in a similar situation in Program 14.4.)

Note that the probabilities given are for the different conclusions from the Phase III program. It is quite possible that superiority cannot be concluded even if the new drug in fact produces a better effect than the competitor. The probabilities are included in the PROB3 data set in Program 14.3.

In order to compare the alternatives we also need to quantify the value of the possible outcomes. Immediate discontinuation of the project is likely to mean ignorable costs and incomes from the drug (we will not consider the possibilities of out-licensing or limited development). We will therefore regard the discontinuation possibility to have value 0. Previous costs or future unavoidable costs are of no interest in the decision problem. What matters is, of course, only the costs and gains which can be influenced by the decisions made now or in the future.

Expected net profit is taken as the optimality criterion. A more sophisticated analysis should consider some additional factors:

- A small expected gain from an investment may not be worth the risk of losing the investment. The degree of risk aversion is likely to be larger in smaller companies, for which a failure in Phase III may jeopardize the company's survival.

- The resources (e.g., money or personnel) may be limited, forcing the company to choose between several promising projects (see Section 14.8).

- Future cash-flows should be discounted by an appropriate rate. The discount rate should primarily reflect the cost of capital, e.g., the interest rate paid. Some organizations assume a higher discount rate in order to also account for the investment risk and/or limited personnel resources. Although such high discount rates might be used as a simplifying tool, we do not recommend them for important decisions. It is better to explicitly model resource restrictions and risk aversion.

One way of dealing with risk is to define the utility as an appropriate concave function of money (Raiffa, 1968, Chapter 4) and use the expected utility as the optimality criterion. For the sake of simplicity, however, we will assume that cash-flows are discounted and that risk aversion and resource restrictions can be ignored.

Suppose that the cost of running the Phase III program is 250 (the monetary unit in this example can be a million of US dollars, or MUSD). An inferior drug will not be possible to sell and therefore produces no income. Non-inferiority is enough to get some sales but the income will be considerably higher in the case of superiority. Ignoring the development costs, we assume the net profit, as a function of the Phase III program outcome, to be $G(\cdot)$. The net profit here denotes the sales minus promotional and manufacturing costs over the drug's life-cycle. We will let

$$G(E^+, S^+) = 1200, \quad G(E^+, S^=) = 550, \quad G(E^=, S^+) = 450, \quad G(E^=, S^=) = 100$$

and, as stated earlier, $G(E^-, \cdot) = 0$.

Program 14.3 creates a decision tree for the introduced decision analysis problem. The development costs are specified in the STAGE3 data set and the gains are included in the PAYOFF3 data set.

Program 14.3 Evaluated decision tree in the simple go/no go problem with multiple outcome variables (efficacy and safety)

```
data stage3;
    length _outcome_ $10.;
    input _stname_ $ _sttype_ $ _outcome_ $ _reward_ _success_ $;
    datalines;
Phase3      D   No_go       .        .
    .       .   Go          -250     Develop
Develop     C   Eff_super   .        AE
    .       .   Eff_noninf  .        AE
    .       .   Eff_inf     .        .
AE          C   AE_super    .        .
    .       .   AE_equal    .        .
    ;
```

```
data prob3;
    length _event1_ _event2_ _event3_ $10.;
    input _event1_ $ _prob1_ _event2_ $ _prob2_ _event3_ $ _prob3_ ;
    datalines;
    Eff_super 0.2 Eff_noninf 0.5 Eff_inf 0.3
    AE_super  0.3 AE_equal   0.7 .        .
    ;
data payoff3;
    length _state1_ _state2_ $10.;
    input _state1_ $ _state2_ $ _value_;
    datalines;
    Eff_super  AE_super 1200
    Eff_super  AE_equal  550
    Eff_noninf AE_super  450
    Eff_noninf AE_equal  100
    Eff_inf    .           0
    ;
* Trial's outcome;
symbol1 value=triangle height=10 color=black width=3 line=1;
* Decision point;
symbol2 value=square height=10 color=black width=3 line=1;
* End nodes;
symbol3 value=none height=10 color=black width=3 line=1;
proc dtree stagein=stage3 probin=prob3 payoffs=payoff3;
    ods select parameters policy;
    treeplot/graphics norc nolegend compress
    linka=1 linkb=2 symbold=2 symbolc=1 symbole=3;
    evaluate/summary;
    run;
    quit;
```

Output from Program 14.3

```
                        Decision Parameters

            Decision Criterion:    Maximize Expected Value (MAXEV)
        Optimal Decision Yields:    1.5

                        Optimal Decision Policy

                        Up to Stage Phase3

            Alternatives   Cumulative   Evaluating
            or Outcomes      Reward        Value
            ----------------------------------------
            No_go              0           0.0
            Go              -250         251.5*
```

Figure 14.3 displays the decision tree generated by Program 14.3, and cumulative rewards (CR) and expected values (EV) for each branch of the tree. Output 14.3 shows that the optimal value for the go option is +1.5 MUSD. This value is the result of a cumulative cost (negative reward) of 250 MUSD for running the trial and an evaluated expected value at the end node of 251.5 MUSD. The stop option has a zero value.

It should be noted that there are many possible descriptions of the same problem. Instead of using separate chance nodes for efficacy and safety outcomes, one can, of course,

Figure 14.3 Evaluated decision tree in the simple go/no go problem with two outcome variables. The square node is the decision to conduct the trial or not and the triangle nodes represent the trial's efficacy outcome (superiority, non-inferiority, or inferiority) and safety outcome (superiority or equivalence).

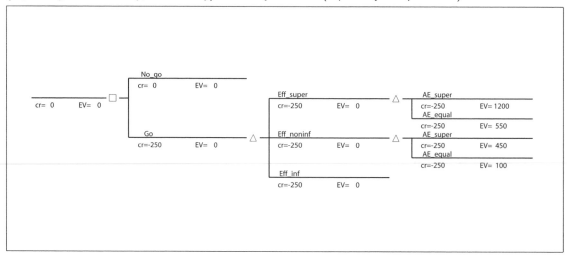

consider only one chance node with an outcome space consisting of five combined events: $\{E^+, S^+\}$, $\{E^+, S^=\}$, $\{E^=, S^+\}$, $\{E^=, S^=\}$ and $\{E^-\}$. Different levels of refinement are possible, and more or less suitable according to the situation. For example, one can average over all possible safety outcomes and just use the probabilities for the efficacy outcomes together with the expected gains given these outcomes, i.e.,

$$G(E^+) = G(E^+, S^+)P(S^+) + G(E^+, S^=)P(S^=)$$

and

$$G(E^=) = G(E^=, S^+)P(S^+) + G(E^=, S^=)P(S^=).$$

This problem formulation would be adequate at the moment. However, the simple decision problem that we have analyzed in this section will be expanded in Section 14.4 and the separation between efficacy and safety outcomes will then be crucial. The previous assumption about independence of efficacy and safety results was made for convenience. Dependence is easily treated by providing conditional probabilities in the data set specified in the PROBIN option of PROC DTREE. This is shown in Program 14.4 (see Section 14.4) in another situation of dependence.

Before expanding the drug development problem, we will give a short general description of decision analysis.

14.3   The Structure of a Decision Analysis

The foundations of statistical decision theory were laid by Wald (1950) and Savage (1954). The theory was mainly concerned with problems within statistical inference, such as estimation and hypothesis testing, rather than with real-life decision problems. A standard formulation (Berger, 1985; Lehmann, 1986; French and Ríos Insua, 2000) is that θ denotes the unknown parameter (or "state of nature"). After observing a random vector X, which distribution is determined by θ, the decision-maker is to make a decision (take an action) denoted by $d(X)$. Statistical decisions may be related to accepting or rejecting a hypothesis or to choosing a point estimate of θ. The consequences of different decisions are modeled by a fully specified loss function $L(\theta, d(X))$. One may proceed by calculating the risk function

$$r(\theta, d) = E_X[L(\theta, d(X))],$$

where the expectation is taken over the distribution of X for the parameter θ.

Since the risk function depends on the unknown parameter, it is not clear which decision rule is optimal. Two possible criteria for optimality are the minimax and Bayes criteria defined below.

Minimax criterion. The minimax rule chooses the decision $d = d(X)$ that minimizes $\max_\theta r(\theta, d)$.

Bayes criterion. The Bayes solution is based on a prior distribution $\pi(\theta)$ for θ and chooses the decision rule that minimizes

$$E_\theta[r(\theta, d)] = \int r(\theta, d)\, \pi(\theta)\, d\theta.$$

Statistical decision theory can also be applied to decision problems which we do not necessarily think of as "statistical", such as deciding whether to invest or not (Raiffa, 1961). In these problems the temporal order is often different from the typical statistical problem: it is common that a decision d (e.g., a go/no go decision in a Phase III trial) has to be taken before observing a random outcome X (e.g., trial results). The decision-maker's preferences are often modeled as a utility $u = u(\theta, d, X)$. To model a utility u instead of a loss L is purely conventional as they are easily interchangeable by the relation

$$u = \text{constant} - L.$$

It is also common to see sequential decision problems. There are often several decisions $d_1, d_2, \ldots$ which are to be taken. Between the decision points, more and more observations $X_1, X_2, \ldots$ can be collected. Each X_k may be a random vector whose distribution is determined by θ. The random outcome X_k can also depend on earlier decisions $d_1, \ldots, d_{k-1}$, as these may relate to which experiments are run. On the other hand, the decisions depend on previously collected information, $d_{k+1} = d_{k+1}(X_1, \ldots, X_k)$. This structure can, at least for simple discrete problems, be illustrated as a decision tree with a number of decision nodes as well as chance nodes.

By writing $d = \{d_1, d_2, \ldots\}$ and $X = \{X_1, X_2, \ldots\}$, we still have a utility function of the form $u = u(\theta, d, X)$. Given a prior distribution for θ, say, $\pi(\theta)$, and using expected utility as the optimality criterion, the problem is to find the decision strategy d which maximizes

$$E[u(d, X)] = \int u(\theta, d, X)\, \pi(\theta)\, d\theta.$$

Backward induction is often useful to solve this problem (Bather, 2000). The minimax optimality criterion, although sometimes fruitful for some statistical problems, is usually of limited value in a real-life decision problem. For example, the minimax solution of the investment problem is almost invariably not to invest. This philosophy could be parodied as one which leads one to spend one's life in bed for fear of being run over by a car if one leaves the house.

A main criticism of fully Bayesian methods that rely on informative priors has been the subjective component. Results that strongly depend on the experimenter's prior beliefs are likely not to be fully accepted by other stakeholders or by the scientific community. This criticism, however, loses its strength when the conclusions are not meant to convince individuals other than those who make the assumptions. Therefore, a semi-Bayesian approach is reasonable: use Bayesian methods internally but classical frequentism externally. For example, a company might use its prior opinion of the effect of a new drug in order to decide on how much resources, if any, to put into its development, how clinical trials should be designed and dimensioned, etc. Still, the company may give traditional frequentist statistical analyses to the regulatory bodies and to the scientific community when publishing the results. As we will see, the utility may even be constructed as a function of the result of the frequentist analysis.

The term "Decision Analysis" was coined in 1965 by Howard (Howard and Matheson, 1984, pages 3 and 97). Decision analysis is partly the application of statistical decision theory but does also pay attention to structuring and modeling a problem. These issues are typically more complicated than the computational optimization, given a specified model. Good accounts of decision analysis are provided by Raiffa (1968) and in a collection of papers edited by Howard and Matheson (1984).

14.4   The Go/No Go Problem Revisited

The go/no go example of Section 14.2, having only two available alternatives, is simplistic. There is usually a larger set of decision alternatives. The company may try to out-license the drug. It can decide to postpone the decision so that more information regarding the drug's likely effect and market potential can be collected. Even if the decision is to start a Phase III program immediately, there are different possible designs for the program. Important design factors are which dose(s) to use and the sample size. All these examples may be modeled by including more decision nodes in the decision tree. In this section we will illustrate the possibility of adding decision nodes with a problem involving the option to purchase information regarding an important AE.

Recall from Section 14.2 that there is good hope that the new drug in the example has an advantage over the existing therapy with respect to a certain adverse event. Assume that a preliminary test (pretest) can be run to investigate this further. If the pretest is positive, then it is highly likely that an AE advantage can be demonstrated in Phase III development. A negative result, on the other hand, predicts no advantage. The question is whether it is worth a relatively modest cost to perform the pretest? We will assume that this investigation may be run in parallel with other necessary activities before making the decision on whether to proceed to a Phase III program (if this were not the case, and running the pretest would delay the project, the payoff data set should reflect that the reward depends on the time of marketing).

One can think of a number of different investigations that could, depending on the situation, possibly serve as pretests, e.g.:

- Use of an AE animal model.
- Studies of the binding between drug and receptor.
- A clinical trial of limited size and duration, perhaps focusing on a surrogate marker in a selected high-risk population.

Often, the results of such investigations are not dichotomous but may be more or less positive or negative. For simplicity, however, we assume that only two outcomes (positive/negative) are possible.

Program 14.4 analyzes the same problem as in Program 14.3 but with the addition of the pretest option. It is assumed that the cost of the pretest is 20 MUSD and that a positive or negative pretest predicts that superiority with respect to the AE can be shown in Phase III with probabilities 0.9 and 0.15, respectively. In order to be consistent with the problem considered in Program 14.3, the probability for a positive pretest must be 0.2, since

$$P(S^+) = P(S^+ \mid \text{Positive pretest}) \, P(\text{Positive pretest})$$

$$+ P(S^+ \mid \text{Negative pretest}) \, P(\text{Negative pretest}).$$

Program 14.4 Evaluated decision tree in the simple go/no go problem with two outcome variables (efficacy and safety) and a pretest

```
data stage4;
   length _stname_ _outcome_ _success_ $10.;
   input _stname_ $ _sttype_ $ _outcome_ $ _reward_ _success_ $;
   datalines;
   Pretest    D   No_test        0    Phase3
   .          .   Test         -20    AEtest
   AEtest     C   AEpos          .    Phase3
   .          .   AEneg          .    Phase3
   Phase3     D   No_go          0    .
   .          .   Go          -250    Develop
   Develop    C   Eff_super      .    AE
   .          .   Eff_noninf     .    AE
   .          .   Eff_inf        .    .
   AE         C   AE_super       .    .
   .          .   AE_equal       .    .
   ;
data prob4;
   length _given1_ _given2_ _event1_ _event2_ _event3_ $10.;
   input _given1_ $ _given2_ $ _event1_ $ _prob1_
         _event2_ $ _prob2_ _event3_ $ _prob3_;
   datalines;
   .        .        AEpos      0.2   AEneg       0.8   .       .
   .        .        Eff_super  0.2   Eff_noninf  0.5   Eff_inf 0.3
   No_test  .        AE_super   0.30  AE_equal    0.70  .       .
   Test     AEpos    AE_super   0.90  AE_equal    0.10  .       .
   Test     AEneg    AE_super   0.15  AE_equal    0.85  .       .
   ;
data payoff4;
   length _state1_ _state2_ $10.;
   input _state1_ $ _state2_ $ _value_;
   datalines;
   Eff_super  AE_super  1200
   Eff_super  AE_equal   550
   Eff_noninf AE_super   450
   Eff_noninf AE_equal   100
   Eff_inf    .            0
   ;
* Trial's outcome;
symbol1 value=triangle height=10 color=black width=3 line=1;
* Decision point;
symbol2 value=square height=10 color=black width=3 line=1;
* End nodes;
symbol3 value=none height=10 color=black width=3 line=1;
proc dtree stagein=stage4 probin=prob4 payoffs=payoff4;
   ods select parameters policy;
   treeplot/graphics norc nolegend compress
   linka=1 linkb=2 symbold=2 symbolc=1 symbole=3 display=(link);
   evaluate/summary;
   run;
   quit;
```

Output 14.4 shows that the optimal decision path is to run the pretest and then go into Phase III development only if the pretest is positive. Comparing the optimal values in Output 14.3 and Output 14.4, we see running the pretest increases the expected value of the project from 1.5 MUSD to 16.9 MUSD. The decision tree produced by Program 14.4 is not shown in order to save space.

Output from Program 14.4

```
                         Decision Parameters

          Decision Criterion:    Maximize Expected Value (MAXEV)
          Optimal Decision Yields:   16.9

                      Optimal Decision Policy

                      Up to Stage Pretest


             Alternatives    Cumulative    Evaluating
             or Outcomes       Reward         Value
             ------------------------------------------
             No_test                 0           1.5
             Test                  -20          36.9*

                      Optimal Decision Policy

                      Up to Stage Phase3

                                        Cumulative    Evaluating
                Alternatives or Outcomes    Reward       Value
             ----------------------------------------------------------
             No_test            No_go             0          0.0
             No_test            Go             -250        251.5*
             Test      AEpos    No_go           -20          0.0
             Test      AEpos    Go             -270        434.5*
             Test      AEneg    No_go           -20          0.0*
             Test      AEneg    Go             -270        205.8
```

It is often crucial to investigate the robustness of the conclusions. The model parameters, especially rewards and probabilities, are typically uncertain. It is good practice to vary the most important parameters and look at how these variations affect the analysis. PROC DTREE is an interactive procedure and allows the user to modify the rewards with the MODIFY statement (see Program 14.17).

However, the most powerful way of investigating robustness properties is via macro programming. A simple SAS macro that evaluates the robustness of the decision analysis model is given in Program 14.5. In this macro, all payoffs are scaled by a common factor f. The output of the program (not displayed) shows that the optimal decision pattern changes considerably when the payoff factor f is varied:

- For sufficiently small payoffs, say, $f = 0.5$, the expected value of a Phase III trial is negative even if the drug has an AE advantage.
- When $f = 0.8$, the AE advantage would motivate a Phase III trial. However, the cost of the pretest is too high in comparison with the information it will provide.
- In the standard scenario, $f = 1.0$, it is optimal to run the pretest and run a Phase III trial if and only if the pretest result is positive.
- For $f = 1.2$, the pretest still gives valuable information but the optimal strategy is to go to Phase III development directly, avoiding the pretest cost.

- Finally, for really high payoffs (e.g., $f = 1.5$), it is worthwhile to run a Phase III trial even when the pretest is negative. Since the Phase III trial will be done irrespective of the result of the pretest, the pretest is obviously redundant in this model and with these parameters. Thus, the analysis shows that the optimal decision is to conduct a Phase III trial without a previous pretest. Note, however, that a pretest may help optimize the design of the Phase III trial. If this is the case, an extended model may still show that the pretest has value.

Program 14.5 Robustness check when varying payoffs

```
/* Refers to data sets from the previous program */
%macro robust(payoff_factor);
    data payoff_changed;
        set payoff4;
        _value_=&payoff_factor*_value_;
    proc dtree stagein=stage4 probin=prob4 payoffs=payoff_changed criterion=maxev;
        evaluate/summary;
        run;
        quit;
%mend robust;


%robust(0.5);
%robust(0.8);
%robust(1.0);
%robust(1.2);
%robust(1.5);
```

14.5   Optimal Sample Size

The Phase III investment example resembles investment problems from other industries, e.g., oil drilling examples described in Raiffa and Schlaifer (1961, Section 1.4.3). One important difference is that Phase III development consists of experiments which we are to design and whose results can be modeled with standard statistical methods. The power, or probability of showing superiority, in one clinical trial is easily determined given the response distributions and test method. The response distributions are typically determined by parameters such as mean effect and standard deviation. Furthermore, these parameters can be modeled using previous clinical (and sometimes preclinical) data (Pallay and Berry, 1999; Burman et al., 2005).

So far, in Sections 14.2 and 14.4, we have taken the Phase III program to be fixed (if it is run at all). The design of these trials is, however, of utmost importance and definitely a field where statisticians have a lot to contribute. In this section we will consider one single design feature: the trial's sample size.

Decision analytic approaches to sample size calculations in clinical trials have been considered in a variety of situations. General Bayesian approaches are outlined by Raiffa and Schlaifer (1961, Section 5.6) and Lindley (1997). Claxton and Posnett (1996) as well as O'Hagan and Stevens (2001) apply such ideas to clinical trials with health economics objectives. Gittins and Pezeshk (2000) and Pallay (2000) discuss the sizing of Phase III trials, considering regulatory requirements. A review of Bayesian approaches to clinical trial sizing is provided by Pezeshk (2003).

Sample size has often been viewed as a relatively simple function of the following parameters:

- "Least clinically relevant effect" δ.
- Standard deviation σ.

- Power $1 - \beta$.
- One-sided significance level $\alpha/2$.

In a two-arm trial of reasonable size, so that a normal approximation is appropriate, the total sample size is given by

$$n = \frac{4\left(\Phi^{-1}(1 - \alpha/2) + \Phi^{-1}(1 - \beta)\right)^2}{(\delta/\sigma)^2},$$

where $\Phi(\cdot)$ is the standard normal cumulative distribution function.

The main difficulty in sample size calculations is the choice of α, β, δ, and σ. The one-sided significance level α is conventionally set at a 2.5% level so that the two-sided size of the test is 5% (FDA, 1998). The standard deviation σ is sometimes hard to estimate at the planning stage and a misspecification may distort the resulting sample size considerably. However, there is often enough data from previous smaller trials with the test drug and from Phase III trials of other drugs with the same response variable for a reasonably robust guess of the value of σ.

More interesting parameters are β and, even more so, the treatment effect δ. We will start by looking at the optimal power when δ is naïvely taken as a fixed known value. After that we will discuss δ.

The power, $1 - \beta$, is often conventionally set at 90%. This is indeed convenient for the statistician performing the sample size calculation as he or she does not have to consider the context around the trial. The importance of the trial, cost of experimentation, recruitment time and other factors are simply ignored. We argue, however, that the choice of the sample size is often a critical business decision. A Phase III program or even a single mortality trial will often cost in the order of 100 MUSD (DiMasi et al., 2003). Changing the power from 80% to 90%, or from 90% to 95%, can result in a considerable increase of the trial's cost, sometimes tens of MUSD. In addition, and often even more important, the time to marketing may be delayed. For such important decisions, a decision analysis framework seems natural.

14.5.1 Optimal Sample Size Given a Fixed Effect

We will start with a very simple decision analysis model and then gradually add some features that can increase the realism of the model. Suppose that a statistically significant trial result in favor of the test drug will lead to regulatory approval and a subsequent monetary gain G for the company. Think of G as the expected discounted net profit during the whole product life-cycle, where sale costs (production, distribution, promotion, etc.) are subtracted from the sales. In the case of a non-significant result, the drug will not be approved and there will be no gain.

Assume that the cost for conducting a clinical trial with $N > 0$ patients is $C(N) = C_0 + cN$, where C_0 is the start-up cost and c is the cost per patient. If no trial is conducted, we have $N = 0$ and $C(0) = 0$. The trial cost and other development costs should not be included in G. The company profit is assumed to be $GI_{\{S\}} - C(N)$, where $I_{\{S\}} = 1$ if the trial is statistically significant and 0 otherwise. Note that this assumption is typically not realistic, as the magnitude of the estimated effect as well as the precision of the estimate are likely to affect the sales. Furthermore, the times of submission, approval, and marketing are likely to depend on N. The model can be made more realistic by incorporating such aspects. This is partly done in Section 14.5.3 below.

Under these assumptions, the expected profit is given by

$$V(N) = E[GI_{\{S\}} - C(N)] = Gp(N) - C(N),$$

where $p(N)$ is the probability of a significant outcome, i.e., power. The values of the parameters vary considerably depending on the drug and therapeutic area. For the sake of illustration, assume that the effect size $\delta/\sigma = 0.2$, value of the trial $G = 1000$, start-up cost $C_0 = 50$ and cost per patient $c = 0.1$. Program 14.6 computes the profit function for the selected parameter values.

Program 14.6 Computation of costs and gains in a Phase III trial

```
data npower;
    alpha=0.05;     /* Two-sided significance level */
    critical=probit(1-alpha/2);
    effsize=0.2;    /* Effect size */
    g=1000;         /* Value of successful (i.e., significant) trial */
    startc=50;      /* Start-up cost */
    c=0.1;          /* Cost per patient */
    do n=0 to 2000 by 10;
        power=probnorm(effsize*sqrt(n/4)-critical);
        egain=g*power;
        cost=startc+c*n;
        profit=egain-cost;
        output;
    end;
run;
axis1 minor=none label=(angle=90 "Utility") order=(0 to 1000 by 200);
axis2 minor=none label=("Sample size") order=(0 to 2000 by 500);
symbol1 i=join width=3 line=1 color=black;
symbol2 i=join width=3 line=20 color=black;
symbol3 i=join width=3 line=34 color=black;
proc gplot data=npower;
    plot (cost egain profit)*n/overlay nolegend haxis=axis2 vaxis=axis1;
    run;
    quit;
```

Figure 14.4, generated by Program 14.6, depicts the resulting profit function. The maximum profit is achieved at $N = 1431$ patients. This corresponds to the probability of a significant outcome (power) of 96.6%.

Figure 14.4 Cost (solid curve), expected gain (dashed curve) and expected profit (dotted curve) as a function of sample size

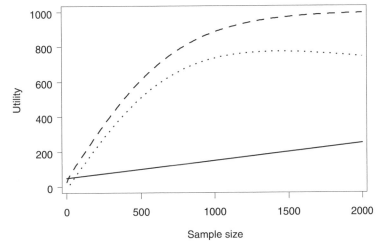

It is clear that Program 14.6 relies on a fairly inefficient algorithm. It simply loops through all values of N between 0 and 2000 to find the sample size that maximizes the

expected net gain. A more efficient approach is to explicitly formulate this problem as an optimization problem and utilize one of many optimization routines available in SAS. In principle, it would have been possible to use a decision tree (i.e., PROC DTREE) for this optimization problem. However, since so many different values of N are possible, it is more convenient to treat it as an approximately continuous variable and use the optimization procedure PROC NLP.

The NOPT macro in Program 14.7 calculates the optimal sample size using the NLP procedure. The procedure maximizes the value of PROFIT (expected net gain) with respect to the N variable (total sample size). The underlying assumptions are identical to the assumptions made in Program 14.6, i.e., the effect size $\delta/\sigma = 0.2$, value of the trial $G = 1000$, start-up cost $C_0 = 50$, and cost per patient $c = 0.1$.

Program 14.7 Optimal sample size

```
%macro nopt(effsize,g,startc,c,alpha);
    proc nlp;
        ods select ParameterEstimates;
        max profit; /* Maximise the expected net gain */
        decvar n;      /* Total sample size */
        critical=probit(1-&alpha/2);
        power=probnorm(&effsize*sqrt(n/4)-critical);
        egain=&g*power;
        cost=&startc+&c*n;
        profit=egain-cost;
    run;
%mend nopt;
%nopt(effsize=0.2,g=1000,startc=50,c=0.1,alpha=0.05);
```

Output from Program 14.7

```
                      PROC NLP: Nonlinear Maximization


                          Optimization Results
                          Parameter Estimates

                                                      Gradient
                                                      Objective
                  N Parameter        Estimate         Function

                  1 n              1431.412256        1.336471E-9
```

Output 14.7 shows that the optimal sample size is 1431 patients. This value is identical to the optimal sample size computed in Program 14.6. Some robustness checks are easily implemented by varying the assumed effect size. For example, it is easy to verify that the optimal sample sizes for $\delta/\sigma = 0.15$ and $\delta/\sigma = 0.25$ are $N = 2164$ and $N = 1018$ patients, respectively.

The model used hitherto in this subsection is generally too simplistic to be of practical value. We will return to the uncertainty in treatment effect in Section 14.5.3 but first, in the next subsection, comment on the number of significant trials needed for regulatory approval.

14.5.2 The Optimal Number of Trials

Often two statistically significant trials are needed for regulatory approval. Assume that two virtually identical trials, each of size N, should be run and that both are required to be significant.

The reasoning used in Section 14.5.1 can then easily be extended to give

$$V_{\text{Two trials}}(N) = G \cdot P(\text{Both trials are significant}) - 2C(N) = Gp(N)^2 - 2(C_0 + cN),$$

which can be optimized using a slightly modified version of Program 14.6.

Provided that the two trials have equal conditions, it is optimal to have the same sample size in both trials. Often the trials are, however, conducted in different countries and in different populations. Costs, recruitment times, and anticipated effects may therefore vary and non-equal sample sizes may be indicated.

The probability of achieving significant results in at least two trials is typically higher if a fixed number of patients is divided into three trials instead of two. In fact, it is easier to get significances in two trials and supportive results in the third (defined as a point estimate in favor of the new drug) when three trials are run than to get significances in both trials when two trials are run. One numerical illustration is provided by Program 14.8. Success of the trial program is defined as all trials giving positive point estimates and at least two trials being statistically significant at the conventional level.

Program 14.8 Optimal number of trials

```
data no_of_trials;
    alpha=0.05;
    effect=20;
    sd=100;
    n_tot=2000;
    critical=probit(1-alpha/2);
    format m n 5.0 p_sign p_supp p_success 5.3;
    do m=1 to 5;
        n=n_tot/m;
        p_sign=probnorm(effect/sd*sqrt(n/4)-critical);
        p_supp=probnorm(effect/sd*sqrt(n/4));
        p_cond=p_sign/p_supp;
        p_success=p_no_neg*(1-probbnml(p_cond,m,1));
        output;
    end;
    label m='Number of trials'
          n='Sample size'
          p_sign='Prob of significance'
          p_supp='Prob of supportive results'
          p_success='Success probability';
    keep m n p_sign p_supp p_success;
proc print data=no_of_trials noobs label;
    run;
```

Output from Program 14.8

Number of trials	Sample size	Prob of significance	Prob of supportive results	Success probability
1	2000	0.994	1.000	0.000
2	1000	0.885	0.999	0.784
3	667	0.733	0.995	0.816
4	500	0.609	0.987	0.798
5	400	0.516	0.977	0.754

Output 14.8 lists several variables, including success probability when a total sample size is divided equally in m trials ($m = 1, \ldots, 5$). With a single trial, the probability of

regulatory success is zero, as two significant trials are needed. Interestingly, the success probability increases when going from two to three trials.

From a regulatory perspective, it is probably undesirable that sponsors divide a fixed sample size into more than two trials without any other reason than optimization of the probability for approval. The regulators should therefore apply such rules that do not promote this behavior. This is a simple example of the interaction between different stakeholders' decision analyses.

In fact, the reason that regulators require two trials to be significant is not entirely clear. Say that two trials are run and each is required to be significant at a one-sided 0.025 level (i.e., $\alpha = 1/40$) and registration follows only when both trials are significant. This corresponds to an overall Type I error rate of $(1/40)^2 = 1/1600 = 0.000625$. However, if centers, patients, and protocols are regarded as exchangeable between the two trials, then an equivalent protection in terms of Type I error rate can be provided to the regulator using fewer patients for equivalent overall power (Senn, 1997; Fisher, 1999; Rosenkranz, 2002; Darken and Ho, 2004). In fact, one could run two such trials and simply pool their results together for analysis. The difference between the resulting pooled trial rule and the conventional two-trials rule in terms of the critical values of the standard normal distribution is illustrated in the accompanying Figure 14.5 taken from Senn (1997).

In this figure, the boundaries for significance are plotted in the $\{Z_1, Z_2\}$ space, where Z_1 is the standardized test statistic for the first trial and Z_2 for the second. For the two-trials rule we require that both $Z_1 > 1.96$ and $\{Z_2 > 1.96\}$. The resulting critical region is that above and to the right of the dashed lines. For the pooled trial rule we require $(Z_1 + Z_2)/\sqrt{2} > 3.227$. The latter rule may be derived by noting that $\text{Var}(Z_1 + Z_2) = 2$ and $1 - \Phi(3.227) \approx 0.000625$. The associated critical region is above and to the right of the solid line.

Figure 14.5 Rejection regions for the pooled-trial and two-trials rules

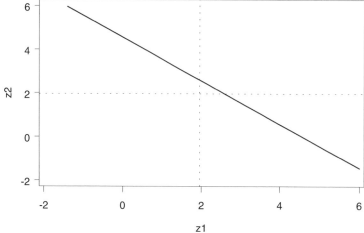

If we calculate power functions for the pooled- and two-trials rules as a function of the standardized non-centrality parameter, $\delta^\star$ for a single trial, then this has the value $2\delta^\star/\sqrt{2} = \sqrt{2}\delta^\star$ for the pooled-trial rule. Bearing in mind that both trials must be significant for the two-trials rule, the power functions are

$$\text{Power}_{\text{Two trials}} = (1 - \Phi(1.96 - \delta^\star))^2$$

$$\text{Power}_{\text{Pooled trial}} = 1 - \Phi(3.227 - \sqrt{2}\,\delta^\star).$$

Figure 14.6 displays a plot of the two power functions. It can be seen from the figure that the power of the pooled-trial rule is always superior to that of the two-trials rule.

Figure 14.6 Power of the pooled-trial (solid curve) and two-trials (dashed curve) rules

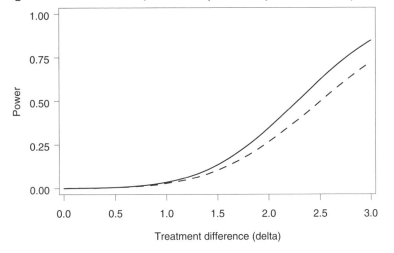

One argument for preferring two trials to such a single trial would be if the trials were not exchangeable and had different design details either in terms of type of centers and patients or in terms of protocol. Two trials that were both significant at a 0.025 level would then be more impressive in terms of robustness than a single trial significant at a 0.000625 level. There is, however, no evidence that in their Phase III programs clinical trial sponsors deliberately choose trials that differ in design nor that regulators require them to. For regulatory thinking on this issue, see Committee for Proprietary Medicinal Products (CPMP) (2001).

14.5.3 A Bayesian Approach to Sample Size Calculations

We have, in Section 14.5.1, looked at the optimal sample size given a fixed treatment effect δ. However, the main problem in deciding on whether to conduct the trial at all and, in that case, how to choose a sample size lies in the uncertainty regarding this parameter. If δ could be objectively established with good precision at this stage, there would be little need to conduct a clinical trial in order to estimate it. Hence, clearly we cannot insert the true value of δ in the power formula, if we regard this as being the effect of the treatment. It may also be misleading to insert a point estimate based on previous data, due to the uncertainty of this estimate. It is often recommended that the sample size calculation should be based on the "least clinically significant effect" but even this is a quantity that cannot be established with absolute precision, since physicians may disagree between themselves and even from time to time, as to what it is.

To decide on whether to invest in a Phase III program a clinical trial sponsor must clearly have some sort of prior opinion about either the size of the effect that would be important or the size that would be likely to obtain. We think that it is reasonable to try to formalize this opinion using a Bayesian prior distribution. However, we warn here that the solution we describe below is not fully Bayesian and may be far from optimal because it does not model utility explicitly as a function of the treatment effect.

Fully Bayesian sample size determinations (with or without utility) have been discussed by several authors, e.g., Lindley (1997) and Pezeshk (2003). We will, however, investigate an approach where Bayesian decision theory is used only for in-house company decisions, such as designing the clinical program, see, for example, Gittins and Pezeshk (2000). Solely frequentist analyses of clinical trial results are communicated externally to regulators and the public. This dichotomy between internal Bayesianism and external frequentism is motivated by the stakeholder situation. The sponsor conducts the trial program in order to provide information to regulators and prescribing physicians on which they can base their decisions. These stakeholders will base their decisions on the data, for convenience often presented as frequentist confidence intervals, and possibly on their own priors. For the sponsor, the optimal program design depends on its own prior and on the anticipated reaction of other stakeholders. The questions to investigate are of the type: What is the probability that the proposed design will lead to trial results that will convince regulators to approve the marketing of our drug and convince physicians to prescribe the drug?

Program 14.9 calculates expected utilities by integrating over the prior distribution for the treatment effect δ (integration is performed using the QUAD function of SAS/IML). This prior distribution is assumed to be normal and its mean and standard deviation are given by the MUE and MUSE variables, respectively (notice that the case when MUSE=0 corresponds to a model with a known treatment effect). SD is the residual standard deviation. The value of MUE, SD, and some other parameters of the program are chosen to facilitate the comparison with the approach described in Section 14.5.1. For example, the effect size is centered around 0.2, the start-up cost is 50, and cost per patient is 0.1.

The program also allows for a possible non-inferior (but non-superior) result. If this option is not applicable, let NIMARGIN=0. The maximal possible value following a non-inferior result is given by the parameter VNIMAX and, similarly, VSUPMAX is the maximal value of a superior result.

In addition to the options of including a prior effect distribution and/or a non-inferiority criterion, a time delay due to increased sample size may also be included. In the model, the maximal profit given marketing (VNIMAX or VSUPMAX) is decreased by a factor TIMEC times the sample size.

In the DO loop in Program 14.9, the expected utility (EUTILITY) is calculated for a number of different sample sizes (given by the vector SIZE). On the negative side, TOT_COST is the total cost for the trial and TIME_FACTOR accounts for the relative loss in value due to the time delay. PSUP and PNI are the probabilities for superiority and non-inferiority, respectively. These are functions of the treatment effect. Given the effect, UTILITY is the expected gain, that is, probability-weighted profit corrected for the time delay via TIME_FACTOR, minus the total cost of running the trial. QUAD is applied to calculate the expected utility over the prior distribution for the effect.

In the model underlying the program, the trial result is important only for the regulatory label (superior, non-inferior, or not approved). Given the label, the profit in this model is independent of the trial result and sample size. If one is trying to apply decision analysis on a practical Phase III sizing problem, we advise that the modeling of the profit be done with greater care. If a better commercial model can be defined in terms of sample size and trial results, it is often relatively straightforward to modify Program 14.9 accordingly.

Program 14.9 Expected utility when the treatment effect is uncertain

```
proc iml;
    muE=20;
    muSE=10;
    sd=100;
    NImargin=10;
    vSUPmax=1000;
    vNImax=400;
    alpha=0.05;
    crit=probit(1-alpha/2);
    timec=0.0001;
    startc=50;
    c=0.1;
```

```
     size=t(do(0,2000,10));
     m=nrow(size);
     Eutility=j(m,1,0);
     do k=1 to m;
         n=size[k,1];
         time_factor=max(1-timec*n,0);
         tot_cost=startc+n*c;
         start h(effect) global(n,muE,muSE,sd,NImargin,crit,
                             vSUPmax,vNImax,time_factor,tot_cost);
             pSup=probnorm(effect/sd*sqrt(n/4)-crit);
             pNI=probnorm((effect+NImargin)/sd*sqrt(n/4)-crit)-pSup;
             utility=time_factor*(vSUPmax*pSup+vNImax*pNI)-tot_cost;
             density=exp(-(effect-muE)**2/(2*muSE**2))/sqrt(2*3.14159*muSE**2);
             h_out=density*utility;
             return(h_out);
         finish;
         call quad(temp,"h",{.M .P}); /* Integrate over the real axis */
         Eutility[k,1]=temp;
     end;
     create plotdata var{size Eutility};
     append;
     quit;
axis1 minor=none label=(angle=90 "Utility") order=(0 to 600 by 200);
axis2 minor=none label=("Sample size") order=(0 to 2000 by 500);
symbol1 i=join width=3 line=1 color=black;
proc gplot data=plotdata;
     plot Eutility*size/haxis=axis2 vaxis=axis1 frame;
     run;
     quit;
```

Figure 14.7 Expected utility as a function of sample size

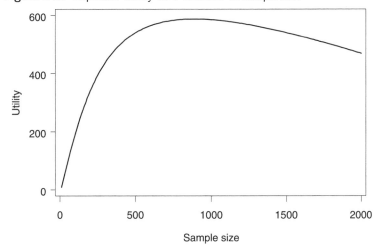

The output of Program 14.9 is displayed in Figure 14.7. The figure depicts the relationship between the expected utility and total sample size. Under the assumptions made in the program, the optimal sample size is 887 patients, which is much smaller than the optimal sample size of 1431 patients for the model considered in Section 14.5.1.

14.5.4   Do Not Sub-Optimize!

A poor decision analysis, which ignores essential components of the problem, is often worse than no formal decision analysis at all. As Einstein said, "Things should be made as simple as possible — but no simpler". We would therefore like to stress that the models presented in Sections 14.5.1 and 14.5.3 must not be applied thoughtlessly to a practical problem. They should be seen just as examples of how a decision model can be built and refined and how to analyze continuous decision problems using SAS. Additional factors which may be important in a practical situation are the demand for safety data, the relation between the amount of Phase III data and sales, alternative investment options in terms of other drugs in the pipeline, etc.

14.5.5   Integrating the Models

Many problems involve a number of decisions, which have to be made at different times. Earlier decisions may lead to information that can be used as input for later decisions. This is obvious in clinical development where the final decisions concerning the design of the Phase III trials will be made after the analysis of a dose-finding trial (Phase IIb), which might follow upon a proof of principle trial (Phase IIa) and so forth. Such problems, involving sequential decisions, are often best attacked by backward induction (Bather, 2000); the analysis starts by solving the decision problems at the latest stage and then works gradually backwards in time to reach an optimal solution of the entire sequential problem.

The drug development decision model of Sections 14.2 and 14.4, focusing on go/no go decisions, is naturally connected with the sample size decisions studied in the present section. Consider for a moment the integrated decision problem with the three questions:

- Should a pretest be run?
- Should a Phase III trial be run?
- What is the optimal sample size in a Phase III trial?

The best way to solve the composite problem is by starting to analyze the latest decision. The optimal sample size is determined for all the different scenarios, e.g.,

- Scenario 1: A positive pretest,
- Scenario 2: A negative pretest,
- Scenario 3: No pretest to be run,

and the corresponding conditional expected utility is calculated. Then, these values can be used as rewards in a decision tree that contains only the pretest and Phase III go/no go decisions. Similarly, PROC DTREE is working by backward induction and analyzes the latest nodes first, taking expectation over chance nodes and maximizing over decision nodes.

Yin (2002) analyses sample sizing in Phase II trials in light of a future go/no go decision.

14.6   Sequential Designs in Clinical Trials

Sequential design and analysis plays a natural role in clinical trials, primarily because of the ability it gives of stopping early either if the treatment under test does not deliver the anticipated effect, or because it is more efficacious than was thought initially. Most sequential approaches are based on frequentist statistics (see, for example, Armitage 1975 and Whitehead 1997) and normally carried out group sequentially, as data are reviewed and monitored on a periodic basis. There is a long tradition of criticizing such frequentist approaches, firstly because by focusing on the consequences of the stopping rules on the Type I error rate they contradict the likelihood principle (Anscombe, 1963; Cornfield, 1969), and secondly because they ignore the fact that stopping is a decision whose losses should be taken into account.

There are a number of approaches that have been looked at to determine an optimal decision rule in a sequential context. The first is traditionally the way that sequential decisions have been tackled and is based on backward induction (Berger, 1985).

The idea in backward induction is extremely simple and begins by considering the last possible decision that can be made at the point that the final data that could be gathered have been gathered. For every possible data set there is an optimal decision to be made. For example, in the case of the comparison of two treatments: Which is the best treatment, or Which can be recommended? Associated with the optimal decision is an expected loss in which the expectation is based on the posterior distribution of the parameters. The penultimate decision is considered next. Now there are three potential decisions that could be taken for each possible data set:

- Stop and recommend the first treatment.
- Stop and recommend the second treatment.
- Continue to the final stage.

If the trial is continued, the loss, which includes the cost of the sampling required to reach the final decision point, can be determined, as can the expected losses of each decision to stop. Consequently, the optimal decision and its expected loss can be determined. In principle, the process can be continued backwards in time to the first decision to be made. Such sequential decision problems are famously difficult because

> ... if one decision leads to another, then to analyse the former, one first needs to analyse the latter, since the outcome of the former depends on the choice of the latter (French, 1989)

A consequence of which is that to completely solve the problem by backward induction involves accounting for exponentially increasing numbers of potential scenarios (Berger, 1985).

The second approach taken to solving sequential decision problems is a forward algorithm based on simulation. Before considering this approach, we will illustrate the idea by using it to determine the optimal sample size as in Section 14.5.

Although the general decision problem formulated in Section 14.3 is seductively simple in its formality, its practical implementation is not always so simple. For complex models the integrations required to determine the optimal Bayes' decision may not be straightforward particularly in multi-parameter, non-linear problems or in sequential decision problems.

Even if the integrations necessary are analytically intractable, the Bayes solution can easily be obtained by simulation. Recall that the Bayesian decision is obtained by minimizing

$$\int_{X,\theta} L(\theta, d(X)) \, p(X \mid \theta) \, \pi(\theta) \, d\theta$$

If the loss function $L(\theta, d(X))$ is available for every combination of θ and $d(X)$, the decision problem can be solved as follows. Assuming that the generation of random variables from both $\pi(X|\theta)$ and $p(\theta)$ is possible, generate samples $\{\theta_j, X_j\}, j = 1, \ldots, M$, and then evaluate for each simulation the loss, $L(\theta_j, d(X_j))$. An estimate of the expected risk can be obtained from

$$\frac{1}{M}\sum_{j=1}^{M} L(\theta_j, d(X_j)).$$

Numerical optimization of the resulting estimated expected risks gives the optimal Bayesian decision.

In the context of finding optimal sample sizes for a clinical trial discussed in Sections 14.5.1, this approach effectively corresponds to replacing the use of the PROBNORM function by simulation, which is hardly worthwhile. However, the same approach allows the simple relaxation of assumptions. For example, suppose that in Program 14.6 we wish

- To use a *t*-test rather than a test based on a known variance.
- To use proper prior information.

Simulation makes this very simple. Program 14.10 uses simulation to determine the optimal sample size based on a *t*-statistic when prior information is available both for the treatment effect (using a normal density) and the residual variance (using an inverse-χ^2 density). Figure 14.8 plots the computed cost, gain, and utility functions.

Program 14.10 Simulating the optimal sample size using *t*-statistic and informative prior

```
data nsample;
    n_sim=1000;        /* Number of simulations*/
    alpha=0.05;        /* 2-sided significance level */
    pr_eff_m=35;       /* Prior mean for treatment effect */
    pr_eff_s=100;      /* Prior SD for treatment effect */
    pr_sd=100;         /* Prior expected value of residual SD */
    pr_v=10;           /* Number of df for prior inverse chi-square */
    g=1000;            /* Value of a successful (i.e., significant) trial */
    startc=50;         /* Start-up cost */
    c=1.0;             /* Cost per patient */
    do n=10 to 1000 by 10;
        t_df=n-2; critical=tinv(1-alpha/2,t_df);
        cost=startc+c*n;
        egain=0;
        do i_sim=1 to n_sim;
            theta=pr_eff_m+pr_eff_s*normal(0);
            sigma_2=pr_v*pr_sd**2/(2*rangam(0,pr_v/2));
            y=theta+sqrt(sigma_2*4/n)*normal(0);
            t_statistic=y/(sqrt(sigma_2*4/n));
            if t_statistic>critical then egain=egain+g;
        end;
        egain=egain/n_sim;
        utility=egain-cost;
        output;
    end;
axis1 minor=none label=(angle=90 "Utility") order=(-1000 to 1000 by 500);
axis2 minor=none label=("Sample size") order=(0 to 1000 by 200);
symbol1 i=spline width=3 line=1 color=black;
symbol2 i=spline width=3 line=20 color=black;
symbol3 i=spline width=3 line=34 color=black;
```

```
proc gplot data=nsample;
    plot (cost egain utility)*n/overlay vaxis=axis1 haxis=axis2 frame;
    run;
    quit;
```

Figure 14.8 Cost (solid curve), expected gain (dashed curve), and utility (dotted curve) as a function of sample size

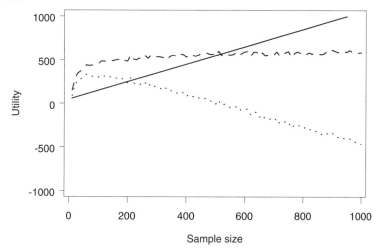

Of course since we are using simulation techniques we need not be restricted to known, simple, convenient parametric forms for the priors. All that is required is an ability to simulate from the priors.

14.6.1   Sequential Decision Problems

To illustrate how a simulation approach can be used to tackle some sequential decision problems we consider the following example based on the approach taken by Carlin et al. (1998). The context is a clinical trial in which the objective is to estimate the treatment effect θ of a new active therapy (A) relative to placebo (P). Large negative values are suggestive of superiority of A over P and conversely large positive values favor P. The protocol allows for a single interim analysis. Along the lines of Freedman and Spiegelhalter (1989), suppose that an indifference region (c_2, c_1) is predetermined such that

- A is preferred if $\theta < c_2$.
- P is preferred if $\theta > c_1$.
- Either treatment is acceptable if $c_2 < \theta < c_1$.

At the final analysis there are two decisions available: d_1 to decide in favor of A and d_2 to decide in favor of P. For each decision, the loss functions given the true value of θ are:

$$l_1(d_1, \theta) = s_1(\theta - c_1), \quad l_2(d_2, \theta) = s_2(c_2 - \theta),$$

where s_1 and s_2 are positive constants. These losses are negatively increasing (gains) for correct decisions. Given the posterior at the final analysis, $p(\theta \mid X_1, X_2)$, the best decision is determined by the smaller of the posterior expected losses:

$$E[l_1(d_1, \theta) \mid X_1, X_2] = s_1(E[\theta \mid X_1, X_2] - c_1),$$
$$E[l_2(d_2, \theta) \mid X_1, X_2] = s_2(c_2 - E[\theta \mid X_1, X_2]).$$

By equating these losses and solving for $E[\theta \mid X_1, X_2]$, it is clear that Decision d_1 will be optimal if

$$E[\theta \mid X_1, X_2] \le \frac{s_1 c_1 + s_2 c_2}{s_1 + s_2}.$$

Otherwise Decision d_2 is optimal.

At the interim analysis, in addition to Decisions d_1 and d_2 to stop in favor of A or P, respectively, based on $p(\theta \mid X_1)$, there is a further decision possible, which is to proceed to the final analysis, to incur the extra cost (c_3) of the patients recruited between the interim and final analyses and then to make the decision.

If σ^2 is known, the forward approach to the decision is algorithmically as follows:

1. Based on the prior distribution and the first stage data X_1, determine the posterior distribution $p(\theta \mid X_1, \sigma^2)$.
2. Determine the expected losses $s_1(E[\theta \mid X_1] - c_1)$ and $s_2(c_2 - E[\theta \mid X_1])$.
3. Let k be the number of simulatons. For $i = 1$ to k,
   (a) Simulate θ_i from $p(\theta \mid X_1, \sigma^2)$.
   (b) Simulate data $X_{2,i}$ from $p(X_2 \mid \theta_i, \sigma^2)$.
   (c) Calculate $s_1(E[\theta \mid X_1, X_{2,i}] - c_1)$ and $s_2(c_2 - E[\theta \mid X_1, X_{2,i}])$.
4. Let l_{final} be the minimum of the following two quantities:

$$\frac{1}{k} \sum_{i=1}^{k} s_1(E[\theta \mid X_1, X_{2,i}] - c_1), \quad \frac{1}{k} \sum_{i=1}^{k} s_2(c_2 - E[\theta \mid X_1, X_{2,i}]).$$

5. Choose the minimum of

$$s_1(E[\theta \mid X_1] - c_1), \quad s_2(c_2 - E[\theta \mid X_1]), \quad \text{and } c_3 + l_{\text{final}}.$$

Obvious modifications need to be made if σ^2 is unknown.

Implementation of this approach is shown in Program 14.11. The output of the program (Figures 14.9 and 14.10) defines regions for the observed treatment difference at the interim for which it is preferable to stop rather than to go on to collect more data and conduct the final analysis.

Program 14.11 Computation of loss functions in the Bayesian sequential design problem

```
data interim_decision;
    n_sim=10000;   /* Number of simulations*/
    s1=1;          /* Cost associated with falsely choosing d1 (active)*/
    s2=1;          /* Cost associated with falsely choosing d2 (placebo)*/
    s3=0.1;        /* Cost associated with sampling after interim*/
    n1=1;          /* Sample size per group at interim */
    n2=1;          /* Sample size per group after interim */
    sigma_2=0.5**2;/* Known sigma**2 */
    c1=0;          /* Upper indifference boundary (gt favours placebo)*/
    c2=-0.288;     /* Lower indifference boundary (lt favours treatment)*/
    pr_m=0.021;    /* Prior mean */
    pr_sd=0.664;   /* Prior standard deviation */


    /* Looping over an observed treatment effect (n1 patients per group) at the interim */
    do x1=-1 to 1 by 0.02;
```

```
        /* Determine posterior mean (interim) and standard deviation*/
        posterior_m_1=(x1*pr_sd**2+2*pr_m*sigma_2/n1)/(pr_sd**2+2*sigma_2/n1) ;
        posterior_sd_1=sqrt(1/(1/pr_sd**2+n1/(2*sigma_2)));
        /* Determine losses at interim (loss_11: decision 1, loss_21: decision 2) */
        loss_11=max(0,s1*(posterior_m_1-c1));
        loss_21=max(0,s2*(c2-posterior_m_1));
        /* Determine minimum of losses at interim */
        loss_interim=min(loss_11,loss_21);

        /* Initialise summation variables for future decisions */
        loss_12_sum=0; loss_22_sum=0;

        /* Simulate future data */
        do i_sim=1 to n_sim;
            /* Simulate treatment parameter from posterior at interim */
            theta=posterior_m_1+posterior_sd_1*normal(0);
            /* Simulate treatment effect from n2 patients on each treatment
            for given theta */
            x2=theta+sqrt(sigma_2*2/n2)*normal(0);
            /* Determine posterior mean (final) and standard deviation*/
            posterior_m_2=(x2*posterior_sd_1**2+2*posterior_m_1*sigma_2/n2)/
                (posterior_sd_1**2+2*sigma_2/n2);
            /* Determine losses at final (loss_12: decision 1, loss_22: decision 2) */
            loss_12=s3+max(0,s1*(posterior_m_2-c1));
            loss_22=s3+max(0,s2*(c2-posterior_m_2));
            /* Accumulate losses */
            loss_12_sum=loss_12_sum+loss_12;
            loss_22_sum=loss_22_sum+loss_22;
        end;

        /* Calculate expected losses */
        loss_12=loss_12_sum/n_sim; loss_22=loss_22_sum/n_sim;
        /* Determine minimum of losses at final */
        loss_final=min(loss_12,loss_22);
        output;
    end;
    keep loss_11 loss_21 loss_12 loss_22 loss_final loss_interim x1;
axis1 minor=none label=(angle=90 "Loss") order=(0 to 0.6 by 0.2);
axis2 minor=none label=("First stage data (X1)") order=(-1 to 1 by 0.5);
symbol1 i=spline width=3 line=20 color=black;
symbol2 i=spline width=3 line=34 color=black;
symbol3 i=spline width=3 line=1 color=black;
/* Plot loss functions for interim analysis */
proc gplot data=interim_decision;
    plot (loss_11 loss_21 loss_interim)*x1/overlay vaxis=axis1 haxis=axis2 frame;
    run;
/* Plot loss functions for final analysis */
proc gplot data=interim_decision;
    plot (loss_12 loss_22 loss_final)*x1/overlay vaxis=axis1 haxis=axis2 frame;
    run;
    quit;
```

In practice, Carlin et al. (1998) use this approach to define a series of critical values for the posterior mean of the treatment effect for each of a number of interims. The advantage of their approach is that the critical values can be determined a priori and there is no need for complex Bayesian simulation during the course of the study. Kadane and Vlachos (2002) develop an alternative approach intended, in their words, to "capitalize on the strengths, and compensate for the weakness of both the backwards and forward strategies".

Figure 14.9 Loss functions for Decision 1 (dashed curve), Decision 2 (dotted curve), and minimum loss function (solid curve) at the interim analysis

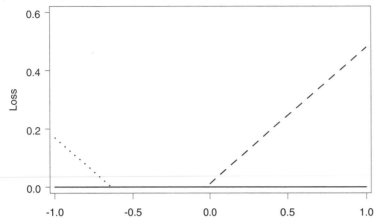

Figure 14.10 Loss functions for Decision 1 (dashed curve), Decision 2 (dotted curve), and minimum loss function (solid curve) at the final analysis

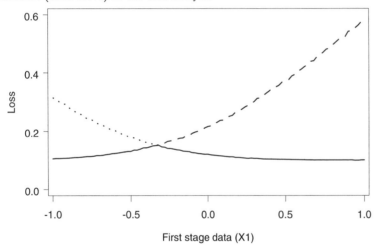

They develop a hybrid algorithm that works backwards as far as is feasible, determining the expected loss of the appropriate optimal continuation as a "callable function". It then proceeds using the forward algorithm from the trial's start to where the backward algorithm becomes applicable. This reduces the size of the space to be covered and hence will increase efficiency. Berry et al. (2000) describe a similar strategy in the context of a dose-response study.

14.7 Selection of an Optimal Dose

One of the most crucial decisions in the development of a new drug is which dose to use. The situation varies between different indications. Sometimes the dose can be titrated for each patient, sometimes the dose is individualized based on gender, body weight, genetic constitution, and other factors. Often one looks for a one-dose-fits-all. The optimal dose is a compromise between beneficial effects and undesired side effects. In general, both effects and, especially, side effects, may be multi-dimensional. As a further complication, which side effects are important may not be clear until after the Phase III program or even after marketing of the drug. We choose to treat only the simple situation with a one-dimensional effect variable and a one-dimensional, well-specified type of adverse event. A practical example of finding the optimal dose, based on decision analysis and modeling of efficacy and safety, is provided by Graham et al. (2002).

Effects and side-effects are typically measured on different scales and defining the optimal dose requires that they be weighted together in some way. It is, however, not obvious how to weigh a decrease in blood pressure against a risk of severe coughing, or how to weigh a decrease in body weight against problems with loose stools. Although such weighting has to be subjective, there is no rational way to avoid it. Given dose-response information for effects and adverse events, to choose a dose more or less implies at least an implicit statement about the relative importance of the two dimensions.

On some occasions, it may occur that effects and side effects can be measured directly on the same scale. One example is when both the effect and the side effect are measured in terms of living and dying. Even though effects and side effects appear to be measured on the same scale, it is not obvious that they should be weighted equally. Different causes of death may give different opportunities for the patient to prepare for death. Even if the outcomes are identical in almost everything, it may subjectively be harder to accept negative events due to the treatment than similar negative events due to the natural propagation of the disease. Furthermore, there may be a legal difference: it is easier to blame the doctor for a treatment's adverse effects than for the negative effects of the untreated disease. Thus, a weighting of adverse effects relative beneficial effects is generally needed.

Suppose that the effect is a continuous variable and the adverse effect is dichotomous. Let $E(D)$ be the expected effect and $p(D)$ the probability that a patient experiences an AE, as functions of the dose D. One reasonable utility function is the simple weighted combination

$$U(D) = E(D) - kp(D).$$

Since it is often hard to choose a value of the weight k, it is useful to investigate the robustness of the conclusions over a range of values of k.

A common model for the effect is the so-called $E_{\max}$ model

$$E(D) = E_0 + E_{\max} \cdot \frac{D^\gamma}{D^\gamma + \mathrm{ED}_{50}^\gamma}. \tag{14.1}$$

The parameters E_0, $E_{\max}$ and ED_{50} have natural interpretations as placebo response, the maximal possible placebo-adjusted effect and the dose for which the adjusted effect is 50% of $E_{\max}$, respectively. The Hill coefficient γ is related to the steepness of the dose-response curve around ED_{50}.

Knowledge of the effect curve alone does not imply what the optimal dose is; safety information is also needed. Assume a logistic model for the probability of AE as a function of the logarithm of the dose,

$$p(D) = \frac{\exp(\alpha + \beta \log(D))}{1 + \exp(\alpha + \beta \log(D))}.$$

14.7.1   Models without Uncertainty

First, assume all parameters to be fixed. Here is an example:

$$E_0 = 0, \quad E_{\max} = 300, \quad \mathrm{ED}_{50} = 0.3, \quad \gamma = 1.0, \quad \alpha = -1.5, \quad \beta = 2.0.$$

Assume also that the weight $k = 1000$, meaning that 0.1% AEs is equalized to one unit of effect. Program 14.12 examines the relationship between the dose of an experimental drug D and the expected effect $E(D)$, adverse event probability $p(D)$, and utility $U(D)$.

Program 14.12 Plot of the dose-effect, dose-Prob(AE), and dose-utility functions

```
/* Define model parameters and macros */
%let k=1000;
%let emax=300;
%let ed50=0.3;
%let gamma=1.0;
%let e0=0.0;
%let alpha=-1.5;
%let beta=2.0;
%macro effmodel(dose,emax,ed50,gamma,e0);
    %global effout;
    %let effout=&e0+&emax*&dose**&gamma/(&dose**&gamma+&ed50**&gamma);
%mend effmodel;
%macro AEmodel(logdose,alpha,beta);
    %global AEout;
    %let AEout=exp(&alpha+&beta*&logdose)/(1+exp(&alpha+&beta*&logdose));
%mend AEmodel;
/* Create data set for plotting */
data window;
    do logd=log(0.01) to log(10.0) by log(10.0/0.01)/100;
        dose=exp(logd);
        %effmodel(dose=dose,emax=&emax,ed50=&ed50,gamma=&gamma,e0=&e0);
        effect=&effout;
        %AEmodel(logdose=logd,alpha=&alpha,beta=&beta);
        AEprob=&AEout;
        AEloss=&k*AEprob;
        utility=effect-AEloss;
        output;
    end;
axis1 minor=none label=(angle=90 "Utility") order=(-300 to 300 by 100);
axis2 minor=none label=("Dose") logbase=10 logstyle=expand;
symbol1 i=join width=3 line=1 color=black;
symbol2 i=join width=3 line=20 color=black;
symbol3 i=join width=3 line=34 color=black;
proc gplot data=window;
    plot (effect AEloss utility)*dose/haxis=axis2 vaxis=axis1
    overlay frame vref=0 lvref=34;
    run;
    quit;
```

Figure 14.11 displays the dose-effect, dose-Prob(AE), and dose-utility functions. It is clear that the first two functions, $E(D)$ and $p(D)$, are monotone functions of the dose. By contrast, the dose-utility function is increasing on $[0, D^*]$ and decreasing on $[D^*, 10]$, where D^* is the dose that maximizes the utility function. The optimal dose, D^*, is easy to find using PROC NLP (see Program 14.13).

Program 14.13 Computation of the optimal dose

```
proc nlp outest=result noprint;
    max utility;
    decvar dose;
    %effmodel(dose=dose,emax=&emax,ed50=&ed50,gamma=&gamma,e0=&e0);
    effect=&effout;
    %AEmodel(logdose=log(dose),alpha=&alpha,beta=&beta);
    AEprob=&AEout;
    utility=effect-&k*AEprob;
```

Figure 14.11 Plot of the dose-effect (solid curve), dose-Prob(AE) (dashed curve), and dose-utility (dotted curve) functions

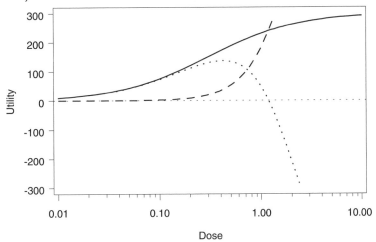

```
data result;
    set result;
    format optdose effect utility 6.2 AEprob 5.3;
    where _type_="PARMS";
    optdose=dose;
    %effmodel(dose=optdose,emax=&emax,ed50=&ed50,gamma=&gamma,e0=&e0);
    effect=&effout;
    %AEmodel(logdose=log(optdose),alpha=&alpha,beta=&beta);
    AEprob=&AEout;
    utility=effect-&k*AEprob;
    label optdose="Optimal dose"
          effect="Effect"
          AEprob="Probability of an AE"
          utility="Utility";
    keep optdose effect AEprob utility;
proc print data=result noobs label;
    run;
```

Output from Program 14.13

Optimal dose	Effect	Utility	Probability of an AE
0.42	175.02	137.13	0.038

Output 14.13 lists the optimal dose, expected response, utility and, finally, probability of observing an AE at this dose. The optimal dose is $D^* = 0.42$, and the associated net utility is given by

$$U(D^*) = E(D^*) - kp(D^*) = 175 - 1000 \times 0.038 = 137.$$

14.7.2   Choosing the Optimal Dose Based on Limited Data

In reality, the efficacy and safety models considered above are uncertain and have to be estimated from the information at hand. We will assume that a relatively small dose-finding trial (124 patients) has been run to study the efficacy and safety of five doses of an experimental drug. The data collected in the study (DOSE, EFF, and AE variables) are contained in the STUDY data set available on the book's companion Web site. The AE variable is 1 for patients who have experienced the AE, and 0 otherwise. The EFF variable is the observed effect. Table 14.1 displays the summary of the efficacy and safety data in the STUDY data set.

Table 14.1 Summary of the Efficacy and Safety Data in the Dose-Finding Trial (STUDY Data Set)

Dose	Number of patients	Number of AEs	Efficacy variable Mean	SD
0.03	26	0	-13.7	291.8
0.1	23	0	135.6	269.3
0.3	25	1	132.6	267.0
1	26	5	293.0	328.3
3	24	10	328.4	339.1

In addition to this limited information on the new drug's effect, there is a lot of literature data for other drugs from the same class. Based on this, we think that we can estimate the E_0, E_{max}, and γ parameters of the E_{max} model, introduced earlier in this section, sufficiently well. (It is assumed that the parameters are shared by all drugs in the class.) Thus, we will take $E_0 = 0$, $E_{max} = 300$, and $\gamma = 1.0$ as fixed. Only the potency, described by ED_{50}, of our new drug remains uncertain. The computations are simplified, as only one of the four parameters in the Emax model must be estimated based on the small in-house trial.

Before estimating the unknown parameters, ED_{50}, α, and β, it is good to consider how reliable the efficacy and safety models are. Even if the models fit the data well, we should not be over-confident. For example, it is often possible to fit logit, probit, and complementary log-log models to the same data set. Still, these models have a large relative difference in the predicted probability of AE at very low doses. In our example, as in many practical situations, there is not enough data to discriminate between different reasonable types of models. A model-independent analysis is therefore a good start.

Program 14.14 calculates estimates (EFF_MEAN, UTIL_EST, P_EST) and their standard errors (EFF_SE, UTIL_SE, P_SE) for mean effect, mean utility, and probability of an AE, respectively, in each of the five dose groups. Note that, depending on the nature of the effect and AE variables, it is plausible that they are positively or negatively correlated. For the sake of simplicity, however, we assume independence throughout this section. This assumption may alter the estimated standard error for the utility but is unlikely to have a major impact on the estimated mean utility.

Program 14.14 Model-free estimates and SEs of $E(D)$, $p(D)$, and $U(D)$

```
%let k=1000;              /* Weighting coefficient */
proc means data=study.study;
    class dose;
    var eff;
    output out=summary_eff(where=(_type_=1)) n=n mean=mean std=std;
proc freq data=study.study;
    table dose*ae/out=summary_ae(where=(ae=1));
data summary;
    merge summary_eff summary_ae;
    by dose;
    if count=. then count=0;
    n_ae=count;
```

```
        eff_mean=mean;
        eff_sd=std;
        keep dose n n_ae eff_mean eff_sd;
data obs_util;
    set summary;
    format eff_mean eff_se util_est util_se 5.1 p_est p_se 5.3;
    eff_se=eff_sd/sqrt(n);
    p_est=n_ae/n;
    p_se=sqrt(p_est*(1-p_est)/n);
    util_est=eff_mean-&k*p_est;
    util_se=sqrt(eff_se**2+(&k*p_se)**2);
    label dose="Dose"
        eff_mean="Effect"
        eff_se="Effect (SE)"
        p_est="Probability of an AE"
        p_se="Probability of an AE (SE)"
        util_est="Utility"
        util_se="Utility (SE)";
    keep dose eff_mean eff_se p_est p_se util_est util_se;
proc print data=obs_util noobs label;
    run;
```

Output from Program 14.14

Dose	Effect	Effect (SE)	Utility	Utility (SE)	Probability of an AE	Probability of an AE (SE)
0.03	-13.7	57.2	-13.7	57.2	0.000	0.000
0.10	135.6	56.2	135.6	56.2	0.000	0.000
0.30	132.6	53.4	92.6	66.2	0.040	0.039
1.00	293.0	64.4	100.7	100.6	0.192	0.077
3.00	328.4	69.2	-88.3	122.1	0.417	0.101

Output 14.14 lists the three estimated quantities (mean effect, mean utility, and probability of an AE) along with the associated standard errors. The highest observed utility is for dose 0.1. For this dose, the utility is significantly better than placebo (as the z-score is $135.6/56.2 \approx 2.42$). Considering the large standard errors, however, it is hard to distinguish between doses in a large range. The only immediate conclusion is that the optimal dose should be higher than 0.03 and probably lower than 3.

After using a model-free approach, we proceed to fit the $E_{\max}$ model to the efficacy data and logistic model to the safety data. Program 14.15 uses the NLIN and LOGISTIC procedures, respectively, to calculate parameter estimates in the two models. The parameter estimates are assigned to macro variables for use later.

Program 14.15 Estimating parameters and posterior distribution

```
data studylog;
    set study;
    logdose=log(dose);
%let e0=0;
%let emax=300;
%let gamma=1.0;
```

```
proc nlin data=study outest=estED50;
    ods select ParameterEstimates;
    parms log_ed50=-1;
    model eff=&e0+&emax*dose**&gamma/(dose**&gamma+exp(log_ed50)**&gamma);
    output out=Statistics Residual=r parms=log_ed50;
data estED50;
    set estED50;
    if _type_="FINAL" then
        call symput('log_est',put(log_ed50, best12.));
    if _type_="COVB" then
        call symput('log_var',put(log_ed50, best12.));
    run;
proc logistic data=studylog outest=est1 covout descending;
    ods select ParameterEstimates CovB;
    model ae=logdose/covb;
data est1;
    set est1;
    if _type_='PARMS' then do;
        call symput('alpha_mean',put(intercept, best12.));
        call symput('beta_mean',put(logdose, best12.));
    end;
    else if _name_='Intercept' then do;
        call symput('alpha_var',put(intercept, best12.));
        call symput('ab_cov',put(logdose, best12.));
    end;
    else if _name_='logdose' then call symput('beta_var',put(logdose, best12.));
    run;
```

Output from Program 14.15

```
                        The NLIN Procedure


                                   Approx      Approximate 95% Confidence
          Parameter     Estimate   Std Error         Limits

          log_ed50       -1.5794    0.5338    -2.6360      -0.5228


                      The LOGISTIC Procedure

              Analysis of Maximum Likelihood Estimates

                                 Standard        Wald
        Parameter   DF   Estimate   Error    Chi-Square    Pr > ChiSq

        Intercept    1    -1.6566   0.3435    23.2639       <.0001
        logdose      1     1.2937   0.3487    13.7647       0.0002


                    Estimated Covariance Matrix

              Variable      Intercept       logdose

              Intercept     0.117968       -0.05377
              logdose      -0.05377         0.1216
```

Output 14.15 lists the estimated model parameters. Note that a direct estimation of ED_{50} (not shown) will result in a 95% confidence interval which contains negative values. Such values are not plausible. As the uncertainty in the parameter estimates will be important later, we prefer to estimate $\log(\text{ED}_{50})$ rather than ED_{50}. The estimate and sample standard error of $\log(\text{ED}_{50})$ are -1.5794 and 0.5338, respectively.

The estimates of α and β in the logistic model for the probability of observing an AE are $\hat{\alpha} = -1.6566$ and $\hat{\beta} = 1.2937$, respectively. The covariance matrix is governed by

$$\Sigma = \left[\begin{array}{cc} 0.118 & -0.054 \\ -0.054 & 0.122 \end{array} \right].$$

One obvious way to consider the parameter uncertainty is to apply a Bayesian viewpoint and consider the posterior distributions for the parameters. The posterior distribution is proportional to the prior multiplied by the likelihood. PROC LOGISTIC and PROC NLIN are likelihood-based procedures and the output from them can be used to get at least rough approximations of the likelihood functions. Assuming that the priors are relatively flat, we may take the two-dimensional normal distribution with mean

$$\left[\begin{array}{c} \hat{\alpha} \\ \hat{\beta} \end{array} \right]$$

and covariance matrix Σ as the approximate posterior for

$$\left[\begin{array}{c} \alpha \\ \beta \end{array} \right].$$

Similarly, the estimate of $\log(\text{ED}_{50})$ and its standard error may, to a reasonable approximation, serve as the mean and standard deviation in a normal prior for $\log(\text{ED}_{50})$.

Program 14.16 uses the previously calculated macro variables to simulate utility curves based on the posteriors for the parameters. Figure 14.12 displays a number of simulated curves together with the simulated expected utility curve.

Program 14.16 Estimating parameters and posterior distribution

```
%let nsim=1000;      /* Number of simulated curves */
%let ndisplay=10;    /* Number of displayed individual curves */
%let doseint=100;    /* Dissolution of dose scale */
%macro out_sim;
    %do __cnt=1 %to &ndisplay.;
        simu&__cnt.=t(util_sim[&__cnt.,]);
    %end;
    create simdata var{dose logdose Eutility
        %do __cnt=1 %to &ndisplay.; simu&__cnt. %str( ) %end;};
    append;
%mend out_sim;
%macro ind_plot;
    %do __cnt=1 %to &ndisplay.;
        simu&__cnt.*dose
    %end;
%mend ind_plot;
/* Simulations */
proc iml;
    seed=257656897;
    z=normal(repeat(seed,&nsim,3));
    alpha_sd=sqrt(&alpha_var);
    beta_sd=sqrt(&beta_var);
    ab_corr=&ab_cov/sqrt(&alpha_var*&beta_var);
    alpha_sim=&alpha_mean+alpha_sd*z[,1];
    beta_sim=&beta_mean+beta_sd*(ab_corr*z[,1]+sqrt(1-ab_corr**2)*z[,2]);
    logED50_sd=sqrt(&log_var);
    logED50_sim=&log_est+logED50_sd*z[,3];
    ED50_sim=exp(logED50_sim);
```

```
        logdose=t(do(log(0.01),log(10),log(10/0.01)/&doseint));
        logdosematrix=j(&nsim,1)*t(logdose);
        dose=exp(logdose);
        dosematrix=exp(logdosematrix);
        onevector=j(1,nrow(logdose));
        alpha_sim_matrix=alpha_sim*onevector;
        beta_sim_matrix=beta_sim*onevector;
        ED50_sim_matrix=ED50_sim*onevector;
        logit=exp(alpha_sim_matrix+beta_sim_matrix#logdosematrix);
        prob_sim=logit/(1+logit);
        Eprob=t(j(1,&nsim)*prob_sim/&nsim);
        effect_sim=&e0+&emax*dosematrix##&gamma/
            (dosematrix##&gamma+ED50_sim_matrix##&gamma);
        Eeffect=t(j(1,&nsim)*effect_sim/&nsim);

        util_sim=effect_sim-&k*prob_sim;
        Eutility=t(j(1,&nsim)*util_sim/&nsim);
        /* Create data set containing simulation results */
        %out_sim;
        quit;
/* Display mean curve and &ndisplay individual simulation curves */
axis1 minor=none label=(angle=90 "Utility") order=(-600 to 200 by 200);
axis2 minor=none label=("Dose") logbase=10 logstyle=expand;
symbol1 i=join width=5 line=1 color=black;
symbol2 i=join width=1 color=black line=34 repeat=&ndisplay.;
proc gplot data=simdata;
    plot Eutility*dose %ind_plot/overlay vaxis=axis1 haxis=axis2 vref=0 lvref=34;
    run;
    quit;
```

Figure 14.12 Plot of simulated utility functions (dashed curves) and simulated expected utility function (solid curve)

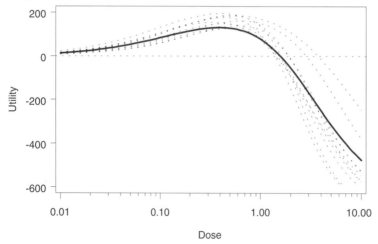

When one single dose has to be chosen, the expected utility is a good criterion. The variability between the different simulated utility curves is, however, of great interest especially when other options are possible. If the variability in optimal dose is small, there is little reason to proceed with more than one dose. However, if the variability in utility between doses that may be optimal is considerable, then there are arguments to bring two or more doses to the next phase of investigations. In our example, the optimal dose is likely to be around 0.4, with limited uncertainty. However, recall that this analysis is dependent on a number of assumptions, such as the logistic and E_{max} models, normal approximations of the posterior, and a flat prior. In practical work, it is important to think about the reliability of the assumptions and to assess the robustness against possible deviations from the assumptions. For example, the optimal dose may be plotted over a range of possible weight coefficients k.

The population model used in Program 14.16 may be refined. Individuals differ in their response to treatment. One reason is pharmacokinetic variability. This can be modeled and sometimes used to improve the benefit-risk relation by individualizing the dose.

14.8   Project Prioritization

An important set of decisions with which a pharmaceutical sponsor is faced is that of the choice of drug development projects and more generally of the prioritization of the projects chosen. This topic has been considered by a number of authors, in particular by Bergman and Gittins (1985) but also by Senn (1996, 1997, 1998), Senn and Rosati (2002), Zipfel (2003), and Burman and Senn (2003).

It is instructive to consider the simpler problem of deciding which drugs to develop, since assessing the value of potential projects prior to inclusion in a portfolio may also suggest how, if at all, those projects that are included might be prioritized. A simple analogy, which is frequently made, is that of packing a number of objects of different values and volumes into a suitcase so that the resulting package is as valuable as possible. The suitcase is a metaphor for the constraints facing drug developers, the volumes of the objects the extent to which the projects contribute individually to reaching or exceeding the constraints and the values of the objects as the expected returns the objects will bring to the sponsor. Put like this, the problem is one of optimization subject to constraints, and it might be thought that linear programming, more specifically integer programming, would be the appropriate tool for solving the problem. In fact, there are various aspects of the problem that suggest that these techniques may be less useful than might at first sight be supposed (this point will be taken up later) and in any case it is useful to examine individual projects in terms of return on investment. We will review this aspect first before returning to optimization considerations.

The simplest useful index that may be used to evaluate projects in development for the purpose of portfolio management is the so-called Pearson Index (Pearson, 1972), which is, effectively, a return on investment index. This calculates for a given project the expected reward as a ratio of the expected cost. It may be defined as follows. First, we assume that the project consists of n stages that have been optimally sequenced. Sunk costs are to be ignored, so that at any stage we can redefine a project, for the purpose of making decisions, as consisting of future stages only. The discounted cost of a given stage i is c_i and the conditional probability of success in that stage given that the project has reached that stage is p_i, with $p_0 = 1$ by convention. The expected current value of the future revenue of a successful project is r. Then the Pearson Index may be written as

$$\mathrm{PI} = \frac{r \prod_{i=1}^{n} p_i - \sum_{i=1}^{n} c_i \prod_{j=0}^{i-1} p_i}{\sum_{i=1}^{n} c_i \prod_{j=0}^{i-1} p_i}$$

Here the denominator is the expected future cost of a project and this reflects the fact that costs will be paid only for a given stage if it is reached and not if the project fails. The first term in the numerator is the unconditional expected future revenue. That is, it is the expected future revenue, r, of a successful project multiplied by the probability of success, $\prod_{i=1}^{n} p_i$. Subtracting the denominator from this term produces the expected return net of expected costs, which is then the numerator of the Pearson Index.

The index has the necessary property that other things being equal it will rank projects with a favorable cost and probability architecture (Senn, 1996, 1998; Zipfel, 2003) more highly than those with an unfavorable architecture. Other things being equal, projects that will fail early (if they fail) are to be preferred to those that will fail late and projects with later costs are to be preferred to those with earlier ones. An example is given by Senn (1996).

Consider four projects, A, B, C, and D in three stages, details for which (including costs and success probabilities) are given in Table 14.2. The value of r (reward) in all four cases is 28. There is nothing to choose among these projects in terms of cost to see to completion $(1 + 2 + 3 = 6)$, nor in terms of return if successful $(28 - 6 = 22)$, nor in terms of probability of success $(0.8 \times 0.6 \times 0.4 = 0.192)$. These figures are the same in all four cases. However, Projects C and D have a favorable cost architecture, and Projects B and D have a favorable probability architecture. In fact, as shown in Table 14.2, the Pearson Index values for the four projects are 0.058 for Project A, 0.331 for Projects B and C, and 1.133 for Project D. The index indicates a ratio of $1.133/0.058 = 19.5$ between Projects D and A and thus the most attractive project (Project D) is nearly 20 times as attractive as the least attractive one (Project A).

Table 14.2 Costs and Success Probabilities for Four Hypothetical Projects

| | | Project | | | |
		A	B	C	D
Cost ($100m)	Stage 1	3	3	1	1
	Stage 2	2	2	2	2
	Stage 3	1	1	3	3
	Total	6	6	6	6
Success probability	Stage 1	0.8	0.4	0.8	0.4
	Stage 2	0.6	0.6	0.6	0.6
	Stage 3	0.4	0.8	0.4	0.8
	Overall	0.192	0.192	0.192	0.192
Reward ($100m)		28	28	28	28
Expected cost		5.1	4.0	4.0	2.5
Expected reward		5.4	5.4	5.4	5.4
Expected profit		0.3	1.3	1.3	2.9
Pearson Index		0.058	0.331	0.331	1.133

The three-stage example of the Pearson Index is simple enough to be solved with a hand calculator. In a more complex situation, PROC DTREE may, of course, be useful. Program 14.17 calculates the expected reward, i.e., the numerator in the Pearson Index, for two of the scenarios. The original PEARSON data set gives the cost architecture for Project A (and B). It shows that the costs of the three successive trials are in order 1, 2, and 3, and that the tree reaches the end as soon as a trial failure is encountered. It also gives the terminal gain 28 in case all the chance events turn out to be OK trial results.

The MODIFY statement in PROC DTREE can be used interactively to modify the costs for the different stages. In the program, the cost of the first trial is changed from 1 to 3 and for the third trial from 3 to 1. The resulting PEARSON data set contains the appropriate costs for trial C (and D). The PROB data set states that the probabilities of failure are 0.2, 0.4, and 0.6 for Trials 1, 2, and 3, respectively. This is the probability architecture for Projects A and C.

Program 14.17 Calculaton of the expecting reward in the Pearson model

```
data pearson;
    input _stname_ $ 1-8 _sttype_ $ 9-12 _outcome_ $ 13-22
        _reward_ 23-26 _success_ $ 27-36;
    datalines;
Start    D   No go        .  .
         .   Go           -1 Stage1
Stage1   C   Failure1     .  .
         .   OK1          -2 Stage2
Stage2   C   Failure2     .  .
         .   OK2          -3 Stage3
Stage3   C   Failure3     .  .
         .   OK3          28 .
;
data prob;
    input _event1_ $10. _prob1_ _event2_ $5. _prob2_;
    datalines;
Failure1  0.2  OK1  0.8
Failure2  0.4  OK2  0.6
Failure3  0.6  OK3  0.4
    ;
proc dtree stagein=pearson probin=prob;
    evaluate/summary;
    save;
    modify Go reward -3;
    modify OK2 reward -1;
    evaluate/summary;
    run;
    quit;
```

Output from Program 14.17

```
                         Decision Parameters

            Decision Criterion:    Maximize Expected Value (MAXEV)
        Optimal Decision Yields:   1.34


                       Optimal Decision Policy

                         Up to Stage Start


                Alternatives    Cumulative   Evaluating
                or Outcomes       Reward        Value
                ----------------------------------------
                No go                  0       0.00
                Go                    -1       2.34*



                         Decision Parameters

            Decision Criterion:    Maximize Expected Value (MAXEV)
        Optimal Decision Yields:   0.296
```

```
            Optimal Decision Policy

             Up to Stage Start

       Alternatives    Cumulative    Evaluating
       or Outcomes         Reward         Value
       ----------------------------------------
       No go                    0         0.000
       Go                      -3         3.296*
```

The way that the Pearson Index could be used in practice would be to rank projects by their ratios and to select them starting with the highest and proceeding as far down the list as constraints permit. This is strongly analogous to the Neyman-Pearson lemma which indicates that selecting points in the sample space to construct a critical region such that a desideratum of power is maximized subject to a constraint of size (Type I error rate) should be done by including points with the highest likelihood ratio[1]. For the project selection problem, the overall research budget is analogous to the tolerated Type I error rate, the expected return is analogous to power, and the Pearson Index to the likelihood ratio. This brings us back to the issue of integer programming discussed previously. As is well known in the context of hypothesis testing, for problems in which the sample space is discrete, improved power can be delivered for a given size under suitable circumstances by going beyond the likelihood ratio criterion. The apparently unending discussion of "superior" alternatives to Fisher's exact test bears witness to this. The analogy for the portfolio selection problem would be that by going beyond the Pearson Index formulation we might, by sacrificing a large project with a high index but replacing it by two smaller projects with less favorable indices, end up with a more valuable portfolio. This is, however, probably not a good idea for similar reasons to those that suggest that "improvements" to Fisher's exact test should not be employed. It is better to look upon the resources available in a drug development project as having some flexibility so that by trading projects or raising capital one could avoid such index-violating decisions.

Nevertheless, for the sake of completeness and in order to illustrate further features of SAS, we will demonstrate the application of integer programming in a simple example below. This should not be taken as an indication that this is necessarily the way we believe that the portfolio management problem should be approached.

Let us look at a simple portfolio planning example using integer programming (Program 14.18). The costs and rewards for ten project candidates are specified in the PROJECTS data set. The budget restriction is given by the _RHS_ variable and set to 100. Finally, all optimization variables are binary. That is, a project is either run (ACTIVE=1) or not run (ACTIVE=0).

Program 14.18 Integer programming to optimize portfolio

```
data projects;
   input _id_ $6. project1-project10 _type_ $ _rhs_;
   datalines;
cost    47 35 22 22 18 16 15 10  9  6 le      100
reward  95 75 40 25 22 20 46 15 30 32 max       .
binary   1  1  1  1  1  1  1  1  1  1 binary    .
;
```

---

[1]It is perhaps worth pointing out that Alan Pearson of the Pearson Index is not to be confused with Egon Pearson of the Neyman-Pearson lemma!

```
proc lp data=projects activeout=results;
proc transpose data=results out=results;
proc sort data=results;
    by _name_;
proc transpose data=projects out=projects;
    id _id_;
proc sort data=projects;
    by _name_;
data together (rename=(_name_=Projects));
    merge projects results;
    by _name_;
    activity=col1;
    index=reward/cost;
    if substr(_name_,1,7)="project";
    keep _name_ cost reward index activity;
proc sort data=together;
    by descending index;
proc print data=together;
    run;
```

Output from Program 14.18

```
                          The LP Procedure

                        Constraint Summary

        Constraint            S/S                    Dual
    Row Name      Type        Col    Rhs  Activity  Activity

      1 cost      LE          11     100    99         0
      2 reward    OBJECTVE    .        0   243         .


    Obs    Projects    cost    reward    activity    index

     1     project10     6       32         1        5.33333
     2     project9      9       30         1        3.33333
     3     project7     15       46         1        3.06667
     4     project2     35       75         0        2.14286
     5     project1     47       95         1        2.02128
     6     project3     22       40         1        1.81818
     7     project8     10       15         0        1.50000
     8     project6     16       20         0        1.25000
     9     project5     18       22         0        1.22222
    10     project4     22       25         0        1.13636
```

Output 14.18 displays some of the output from the LP procedure as well as a list of the projects, ordered by the value of the reward/cost-index, and indicating the optimal set of projects. The output of PROC LP states that the optimal solution gives a reward of 243 at a cost of 99. Note, from the project list, that the fourth most profitable project (project2) according to the index is not chosen while two projects with lower index values are included.

We may try to change the budget restriction and study how the reward is increasing and the optimal set of projects is changing. In our example, the optimal reward is constant, 243, for budget restriction in the interval $[99, 103)$. It is 248 for budgets in $[103, 109)$. Considering risks and cost of capital, it may not be worth an additional investment of $103 - 99 = 4$ to increase expected rewards by $248 - 243 = 5$. However, increasing the budget to 109 increases the reward to 258 and, with a budget of 112, a reward of 278 is possible. The last option corresponds to running the five projects with highest index.

Although the Pearson Index is useful, it is not perfect. Its adequacy depends on the extent to which it captures the options facing a drug developer as regards projects in general. The index captures the obvious option on any project as regards costs: that is to say, to abandon projects that no longer have any chance of being successful. Other options that it does not capture, however, include the option to revise decisions in the light of developing market information and the option to revise investment decisions in the light of changes in the rate of interest (Senn, 1998; Senn and Rosati, 2002). For a general introduction to this field, see Dixit and Pindyck (1994).

14.9 Summary

Decision analysis is a general approach to decisions, based on problem structuring, quantifying, and optimizing. In principle, it can be applied to any decision. We have seen a number of examples from drug development, with emphasis on clinical programs and trials. The examples range from the design of a single trial (sample size, sequential design), over the question about optimal dose, to the design of a program and project prioritization. The list could of course be expanded. In particular, more examples could be taken from discovery and commercial perspectives. An attractive feature of decision analysis is that it often promotes cross-skill cooperation and is a useful tool to facilitate mutual understanding of each others' ideas.

It must be stressed, however, that decision analysis must be handled with care. Within a decision model, it is straightforward to optimize the decision. However, this decision may be quite bad if the decision model is not a good enough model for the real problem. There is a clear risk of getting a suboptimal decision as a result of a model that is too simplified. On the other hand, making models too complicated gives a similar risk. It is then hard to see through the assumptions and to assess their validity. The inherently quantitative nature of a decision analysis may constitute a temptation to overvalue aspects that are easily measured and ignore qualitative aspects. It is important to realize that ethics must get priority over profit, that a project cannot always be analyzed separately from long-term company strategy, that psychology sometimes is more important than cold rationalism. The conclusion is that decision analysis by all means should be applied in pharmaceutical development but that the applications should be made with care.

Acknowledgments

We thank Axel Nilsson and Magnus Kjaer for significant improvements of the SAS code.

References

Anscombe, F.J. (1963). "Sequential medical trials." *Journal of the American Statistical Association.* 58, 365–383.

Armitage, P. (1975). *Sequential Medical Trials.* Second edition. London: Blackwell Scientific Publications.

Bather, J. (2000). *Decision Theory: an Introduction to Dynamic Programming and Sequential Decisions.* Chichester: Wiley.

Berger, J.O. (1985). *Statistical Decision Theory and Bayesian Analysis.* Second edition. New York: Springer-Verlag.

Bergman, S.W., Gittins, J.C. (1985). *Statistical Methods for Pharmaceutical Research Planning.* New York: Marcel Dekker.

Berry, D.A., Muller, P., Grieve, A.P., Smith, M., Parke, T., Balazek, R., Mitchard, N., Krams, M. (2000). "Adaptive Bayesian Designs for Dose-Ranging Drug Trials." In *Case Studies in Bayesian Statistics V.* C. Gatsonis, R. E. Kass, B. Carlin, A. Carriquiry, A. Gelman, I. Verdinelli, and M. West, eds. New York: Springer-Verlag. 99–181.

Bertolli, J., Hu, D.J., Nieburg, P., Macalalad, A., Simonds, R.J. (2003). "Decision analysis to guide choice of interventions to reduce mother-to-child transmission of HIV." *AIDS.* 17, 2089–2098.

Brown, M.M., Brown, G.C., Sharma, S., Landy, J. (2003). "Health care economic analyses and value-based medicine." *Survey of Ophthalmology.* 48, 204–223.

Burman, C.-F., Hamrén, B., Olsson, P. (2005). "Modelling and simulation to improve decision-making in clinical development." *Pharmaceutical Statistics.* 4, 47–58.

Burman, C.-F., Senn, S. (2003). "Examples of option values in drug development." *Pharmaceutical Statistics.* 2, 113–125.

Carlin, B.P., Kadane, J.B., Gelfand, A.E. (1998). "Approaches for optimal sequential decision analysis in clinical trials." *Biometrics.* 54, 964–975.

Claxton, K., Posnett, J. (1996). "An economic approach to clinical trial design and research priority-setting." *Health Economics.* 5, 513–524.

Committee for Proprietary Medicinal Products (CPMP) (2001). "Points to consider on application with 1. Meta-analyses 2. One pivotal study." London: European Evaluation Agency.

Cooper, N.J., Sutton, A.J., Abrams, K.R. (2002). "Decision analytic economic modelling within a Bayesian framework: application to prophylactic antibiotics use for Caesarean section." *Statistical Methods in Medical Research.* 11, 491–512.

Cornfield, J. (1969). "Sequential trials, sequential analysis and the likelihood principle." *The American Statistician.* 20, 18–22.

Darken, P.F., Ho, S.-Y. (2004). "A note on sample size savings with the use of a single well-controlled clinical trial to support the efficacy of a new drug." *Pharmaceutical Statistics.* 3, 61–63.

DiMasi, J.A., Hansen, R.W., Grabowski, H.G. (2003). "The price of innovation: new estimates of drug development costs." *Journal of Health Economics.* 22, 151–185.

Dixit, A.K., Pindyck, R.S. (1994). *Investment under uncertainty.* Princeton: Princeton University Press.

Elwyn, G., Edwards, A., Eccles, M., Rovner, D. (2001). "Decision analysis in patient care." *The Lancet.* 358, 571–574.

FDA (U.S. Department of Health and Human Services. Food and Drug Administration) (1998). "Guidance for industry: Providing clinical evidence of effectiveness for human drug and biological products." http://www.fda.gov/cder/guidance/1397fnl.pdf.

FDA (U.S. Department of Health and Human Services. Food and Drug Administration) (2004). "Challenge and Opportunity on the Critical Path to New Medical Products." http://www.fda.gov/oc/initiatives/criticalpath/whitepaper.html.

FDA (U.S. Department of Health and Human Services. Food and Drug Administration) (2006). "Critical Path Opportunities List." http://www.fda.gov/oc/initiatives/criticalpath/reports/opp_list.pdf.

Fisher, L.D. (1999). "One large, well-designed, multicenter study as an alternative to the usual FDA paradigm." *Drug Information Journal.* 33, 265–271.

Freedman, L.S., Spiegelhalter, D.J. (1989). "Comparison of Bayesian with group sequential methods for monitoring clinical trials." *Controlled Clinical Trials.* 10, 357–367.

French, S (1989). *Readings in Decision Analysis.* London: Chapman and Hall.

French, S., Ríos Insua, D. (2000). *Statistical Decision Theory.* London: Arnold.

Gittins, J. and Pezeshk, H. (2000). "A behavioral Bayes method for determining the size of a clinical trial." *Drug Information Journal.* 34, 355–363.

Graham, G., Gupta, S., Aarons, L. (2002). "Determination of an optimal regimen using a Bayesian decision analysis of efficacy and adverse effect data." *Journal of Pharmacokinetics and Pharmacodynamics.* 29, 67–88.

Howard, R.A., Matheson, J.E. eds. (1984). *The Principles and Applications of Decision Analysis.* Part I–II. Menlo Park, CA: Strategic Decision Group.

Kadane, J.B., Vlachos, P.K. (2002). "Hybrid methods for calculating optimal few-stage sequential strategies: data monitoring for a clinical trial." *Statistics and Computing.* 12, 147–152.

Lehmann, E. L. (1986). *Testing Statistical Hypothesis,* New York: Wiley.

Lindley, D. V. (1997). "The choice of sample size." *The Statistician.* 46, 129–138.

Liu, J.Y., Finlayson, S.R.G., Laycock, W.S., Rothstein, R.L., Trus, T.L., Pohl, H., and Birkmeyer, J.D. (2003). "Determining an appropriate threshold for referral to surgery for gastroesophageal reflux disease." *Surgery.* 133, 5–12.

O'Hagan, A., Stevens, J. (2001). "Bayesian assessment of sample size for clinical trials of cost-effectiveness." *Medical Decision Making.* 21, 219–230.

Pallay, A. (2000). "A decision analytic approach to determining sample sizes in a Phase III program." *Drug Information Journal.* 34, 365–377.

Pallay, A., Berry, S.M. (1999). "A decision analysis for an end of Phase II go/stop decision." *Drug Information Journal.* 33, 821–833.

Parmigiani, G. (2002). *Modeling in Medical Decision Making.* New York: Wiley.

Pearson, A.W. (1972). "The use of ranking formulae in R & D projects." *R & D Management.* 2, 69–73.

Pezeshk, H. (2003). "Bayesian techniques for sample size determination in clinical trials: a short review." *Statistical Methods in Medical Research.* 12, 489–504.

Poland, B., Wada, R. (2001). "Combining drug-disease and economic modelling to inform drug development decisions." *Drug Discovery Today.* 6, 1165–1170.

Protheroe, J., Fahey, T., Montgomery, A.A., Peters, T.J., Smeeth, L. (2000). "The impact of patients' preferences on the treatment of atrial fibrillation: observational study of patient based decision analysis." *British Medical Journal.* 320, 1380–1384.

Raiffa, H. (1968). *Decision Analysis—Introductory Lectures on Choices under Uncertainty,* Reprinted 1997, New York: McGraw Hill.

Raiffa, H., Schlaifer, R. (1961). *Applied Statistical Decision Theory.* Reprinted 2000. New York: Wiley.

Rosenkranz, G. (2002). "Is it possible to claim efficacy if one of two trials is significant while the other just shows a trend?" *Drug Information Journal.* 36, 875–879.

Savage, L.J. (1954). *The Foundations of Statistics.* New York: Wiley.

Senn, S.J. (1996). "Some statistical issues in project prioritization in the pharmaceutical industry." *Statistics in Medicine.* 15, 2669–2702.

Senn, S.J. (1997). *Statistical Issues in Drug Development.* Chichester: Wiley.

Senn, S.J. (1998). "Further statistical issues in project prioritization in the pharmaceutical industry." *Drug Information Journal.* 32, 253–259.

Senn, S.J., Rosati, N. (2002). "Project selection, decision analysis and profitability in the pharmaceutical industry." *Business Briefing: Pharmatech 2002.* 18–21, London. `http://www.wma.net/e/publications/pdf/2001/senn.pdf`

Wald, A. (1950). *Statistical Decision Functions.* New York: Wiley.

Whitehead, J. (1997). *The Design and Analysis of Sequential Medical Trials.* Second edition. New York: Wiley.

Yin, Y. (2002). "Sample size calculation for a proof of concept study." *Journal of Biopharmaceutical Statistics.* 12, 267–276.

Zipfel, A. (2003). "Modeling the probability-cost-profitability architecture of portfolio management in the pharmaceutical industry." *Drug Information Journal.* 37, 185–205.

Index

A

B

Books Available from SAS Press

The Little SAS® Book: A Primer, Third Edition
by **Lora D. Delwiche**
and **Susan J. Slaughter**
(updated to include SAS 9.1 features)

The Little SAS® Book for Enterprise Guide® 3.0
by **Susan J. Slaughter**
and **Lora D. Delwiche**

The Little SAS® Book for Enterprise Guide® 4.1
by **Susan J. Slaughter**
and **Lora D. Delwiche**

Logistic Regression Using the SAS® System:
Theory and Application
by **Paul D. Allison**

Longitudinal Data and SAS®: A Programmer's Guide
by **Ron Cody**

Maps Made Easy Using SAS®
by **Mike Zdeb**

Models for Discrete Data
by **Daniel Zelterman**

Multiple Comparisons and Multiple Tests Using SAS®
Text and Workbook Set
(books in this set also sold separately)
by **Peter H. Westfall, Randall D. Tobias,
Dror Rom, Russell D. Wolfinger,**
and **Yosef Hochberg**

Multiple-Plot Displays: Simplified with Macros
by **Perry Watts**

Multivariate Data Reduction and Discrimination with
SAS® Software
by **Ravindra Khattree**
and **Dayanand N. Naik**

Output Delivery System: The Basics
by **Lauren E. Haworth**

Painless Windows: A Handbook for SAS® Users, Third Edition
by **Jodie Gilmore**
(updated to include SAS 8 and SAS 9.1 features)

Pharmaceutical Statistics Using SAS®: A Practical Guide
Edited by **Alex Dmitrienko, Christy Chuang-Stein,**
and **Ralph D'Agostino**

The Power of PROC FORMAT
by **Jonas V. Bilenas**

PROC SQL: Beyond the Basics Using SAS®
by **Kirk Paul Lafler**

PROC TABULATE by Example
by **Lauren E. Haworth**

Professional SAS® Programmer's Pocket Reference,
Fifth Edition
by **Rick Aster**

Professional SAS® Programming Shortcuts, Second Edition
by **Rick Aster**

Quick Results with SAS/GRAPH® Software
by **Arthur L. Carpenter**
and **Charles E. Shipp**

Quick Results with the Output Delivery System
by **Sunil K. Gupta**

Reading External Data Files Using SAS®: Examples Handbook
by **Michele M. Burlew**

Regression and ANOVA: An Integrated Approach Using
SAS® Software
by **Keith E. Muller**
and **Bethel A. Fetterman**

SAS® for Forecasting Time Series, Second Edition
by **John C. Brocklebank**
and **David A. Dickey**

SAS® for Linear Models, Fourth Edition
by **Ramon C. Littell, Walter W. Stroup,**
and **Rudolf J. Freund**

SAS® for Mixed Models, Second Edition
by **Ramon C. Littell, George A. Milliken, Walter W. Stroup,
Russell D. Wolfinger,** and **Oliver Schabenberger**

SAS® for Monte Carlo Studies: A Guide for Quantitative
Researchers
by **Xitao Fan, Ákos Felsővályi, Stephen A. Sivo,**
and **Sean C. Keenan**

SAS® Functions by Example
by **Ron Cody**

SAS® Guide to Report Writing, Second Edition
by **Michele M. Burlew**

SAS® Macro Programming Made Easy, Second Edition
by **Michele M. Burlew**

SAS® Programming by Example
by **Ron Cody**
and **Ray Pass**

SAS® Programming for Researchers and Social Scientists,
Second Edition
by **Paul E. Spector**

SAS® Programming in the Pharmaceutical Industry
by **Jack Shostak**

SAS® Survival Analysis Techniques for Medical Research,
Second Edition
by **Alan B. Cantor**

SAS® System for Elementary Statistical Analysis,
Second Edition
by **Sandra D. Schlotzhauer**
and **Ramon C. Littell**

SAS® System for Regression, Third Edition
by **Rudolf J. Freund**
and **Ramon C. Littell**

SAS® System for Statistical Graphics, First Edition
by **Michael Friendly**

support.sas.com/pubs

The SAS® Workbook and *Solutions* Set
(books in this set also sold separately)
by **Ron Cody**

Selecting Statistical Techniques for Social Science Data:
A Guide for SAS® Users
by **Frank M. Andrews, Laura Klem, Patrick M. O'Malley,**
Willard L. Rodgers, Kathleen B. Welch,
and **Terrence N. Davidson**

Statistical Quality Control Using the SAS® System
by **Dennis W. King**

A Step-by-Step Approach to Using the SAS® System
for Factor Analysis and Structural Equation Modeling
by **Larry Hatcher**

A Step-by-Step Approach to Using SAS® for Univariate and
Multivariate Statistics, Second Edition
by **Norm O'Rourke, Larry Hatcher,**
and **Edward J. Stepanski**

Step-by-Step Basic Statistics Using SAS®: Student Guide
and *Exercises*
*(*books in this set also sold separately)
by **Larry Hatcher**

Survival Analysis Using SAS®:
A Practical Guide
by **Paul D. Allison**

Tuning SAS® Applications in the OS/390 and z/OS
Environments, Second Edition
by **Michael A. Raithel**

Univariate and Multivariate General Linear Models:
Theory and Applications Using SAS® Software
by **Neil H. Timm**
and **Tammy A. Mieczkowski**

Using SAS® in Financial Research
by **Ekkehart Boehmer, John Paul Broussard,**
and **Juha-Pekka Kallunki**

Using the SAS® Windowing Environment: A Quick Tutorial
by **Larry Hatcher**

Visualizing Categorical Data
by **Michael Friendly**

Web Development with SAS® by Example, Second Edition
by **Frederick E. Pratter**

Your Guide to Survey Research Using the SAS® System
by **Archer Gravely**

JMP® Books

JMP® for Basic Univariate and Multivariate Statistics: A Step-by-
Step Guide
by **Ann Lehman, Norm O'Rourke, Larry Hatcher,**
and **Edward J. Stepanski**

JMP® Start Statistics, Third Edition
by **John Sall, Ann Lehman,**
and **Lee Creighton**

Regression Using JMP®
by **Rudolf J. Freund, Ramon C. LIttell,**
and **Lee Creighton**